Resources Geology & the Environment

GEOS 1024

Eighteenth Edition

Miller | Spoolman

CENGAGE Learning

Australia • Brazil • Japan • Korea • Mexico • Singapore • Spain • United Kingdom • United States

CENGAGE Learning

Resources Geology & the Environment: GEOS 1024, Eighteenth Edition

Living in the Environment, 18th Edition
G. Tyler Miller Jr. | Scott Spoolman

ISBN: 978-1-337-05416-4

Cengage Learning
20 Channel Center Street
Boston, MA 02210
USA

Cengage Learning is a leading provider of customized learning solutions with office locations around the globe, including Singapore, the United Kingdom, Australia, Mexico, Brazil, and Japan. Locate your local office at:
www.international.cengage.com/region.

Cengage Learning products are represented in Canada by Nelson Education, Ltd.

For your lifelong learning solutions, visit **www.cengage.com/custom.**

Visit our corporate website at **www.cengage.com.**

Brief Contents

INDEX

Note: Page numbers in **boldface** refer to boldface terms in the text. Page numbers followed by italicized *f, t, or b* indicate figures, tables, and boxes.

A

N

Q

R

S

Y

Z

6 The Human Population and Its Impact

The problems to be faced are vast and complex, but come down to this: 7 billion people are breeding exponentially. The process of fulfilling their wants and needs is stripping earth of its biotic capacity to support life; a climactic burst of consumption by a single species is overwhelming the skies, earth, waters, and fauna.

PAUL HAWKEN

Key Questions

6-1 How do environmental scientists think about human population growth?

6-2 What factors influence the size of the human population?

6-3 How does a population's age structure affect its growth or decline?

6-4 How can we slow human population growth?

Jaipur, India.

Eastland Photo/Alamy

CORE CASE STUDY Planet Earth: Population 7 Billion

It took about 200,000 years—from the time that the latest version of our species *Homo sapiens sapiens* evolved, to the 1920s—for our population to reach an estimated 2 billion. It took less than 50 years to add the second 2 billion people (by about 1974), and 25 years to add the third 2 billion (by 1999). Twelve years later, in late 2011, we topped 7 billion (Figure 6-1).

So why does it matter that there are now 7.1 billion people on the earth—almost 3 times as many as there were in 1950? Some say it doesn't matter and they contend that we can develop new technologies that could easily support 16 billion people. Many scientists disagree and argue that the current exponential growth of the human population is unsustainable. They point out that as our population grows and as we use more of the earth's natural resources, our ecological footprints also expand (see Figure 1-13, p. 14) and degrade the natural capital that keeps us alive and supports our lifestyles and economies.

In 2009, a team of internationally renowned scientists estimated that we have exceeded the global boundary limits for three of the major components of the earth's life-support system and are close to the boundary limits for five other components (Figure 3-B, p. 73). Some scientists say that because of our strong effect on natural systems, we now live in a new *Anthropocene era* in which our collective actions are moving key parts of our life-support system into danger zones.

Three major factors account for the rapid rise of the human population. *First,* the emergence of early and modern agriculture about 10,000 years ago allowed us to feed more people with each passing century. *Second,* we have developed technologies that have helped us to expand into almost all of the planet's climate zones and habitats (see Figure 1-12, p. 13). *Third,* our death rates have dropped sharply because of improved sanitation and health care and the development of antibiotics and vaccines to help control infectious diseases.

So what is a sustainable level for the human population? Population experts have made low, medium, and high projections for how big the human population will be by the end of this century (Figure 6-1). No one knows whether any of these population sizes are sustainable or for how long.

In this chapter, we examine population growth trends, the environmental impacts of the growing population, proposals for dealing with human population growth, and some stories about countries where population growth has been stabilized.

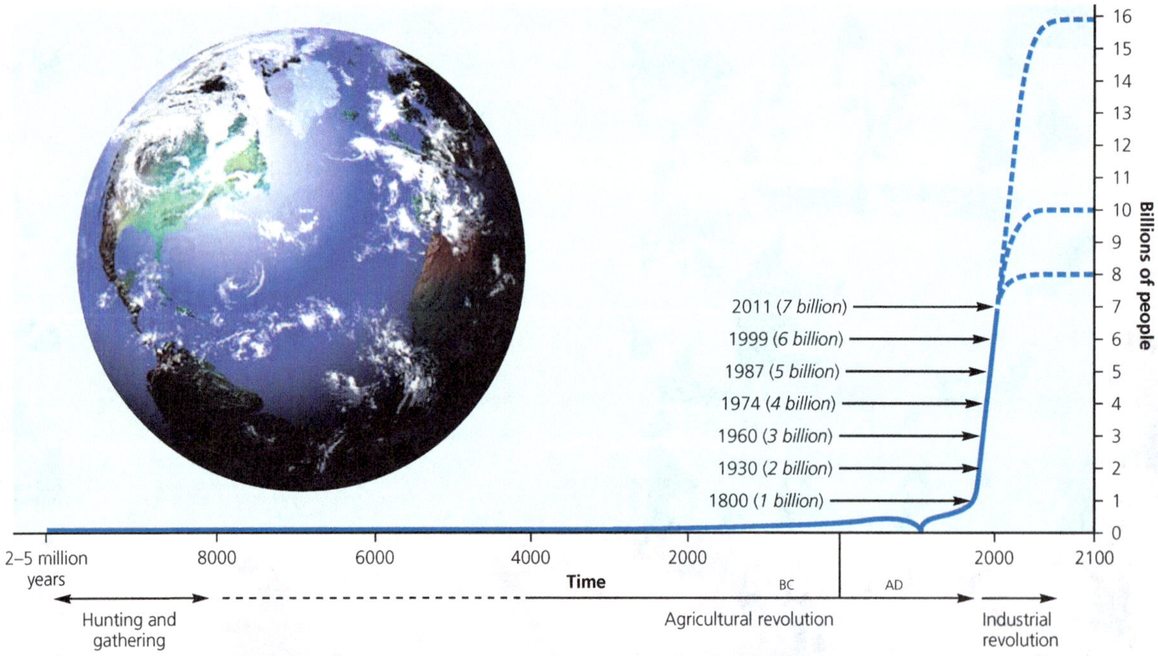

(Compiled by the authors using data from the World Bank and United Nations.)

Figure 6-1 The human population has grown exponentially—showing slow growth throughout most of history and shooting up at a rapid rate within the last 200 years. This graph also shows projections to 2100 that range from 8 billion to 16 billion. (This figure is not to scale.)

Photo: ©NASA

6-1 How Do Environmental Scientists Think about Human Population Growth?

CONCEPT 6-1
The continuing rapid growth of the human population and its impacts on natural capital raise questions about how long the human population can keep growing.

Human Population Growth Shows Certain Trends

Estimates of what the global human population is likely to be in 2050 range from 7.8 billion to 10.8 billion people, with a medium projection of 9.6 billion people. For the year 2100, the projected population size ranges from 8 billion to 16 billion (Figure 6-1). These varying estimates depend on a variety of factors that we consider in more detail later in this chapter. However, *demographers,* or population experts, recognize three important growth trends.

First of all, in recent decades, the rate of population growth has slowed (Figure 6-2), but the world's population is still growing (Figure 6-1) at a rate of about 1.2%. This may not seem like much but in 2012 this growth added about 84 million people to the population—an average of more than 230,000 people each day, or almost 3 more people every second.

Second, demographers recognize that, geographically, human population growth is unevenly distributed and this pattern is expected to continue (Figure 6-3). About 2% of the 84 million new arrivals on the planet in 2012 were added to the world's more-developed coun-

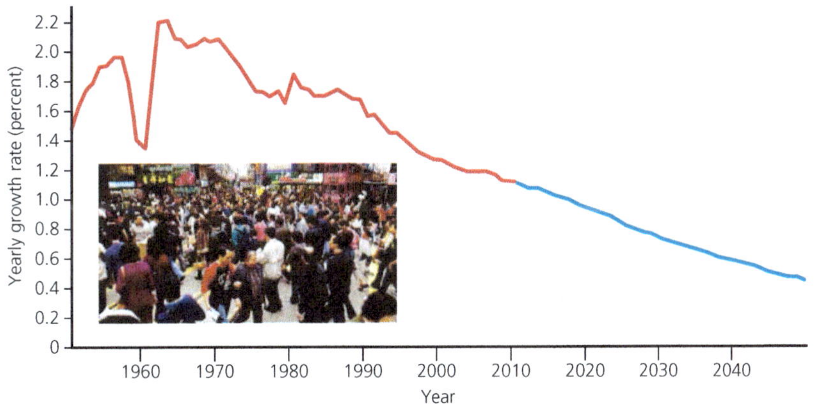

Figure 6-2 The annual growth *rate* of world population has generally dropped since the 1960s, but the population has continued to grow (Figure 6-1).

(Compiled by the authors using data from United Nations Population Division, U.S. Census Bureau, and Population Reference Bureau.)

Photo: JUSTIN GUARIGLIA/National Geographic Creative

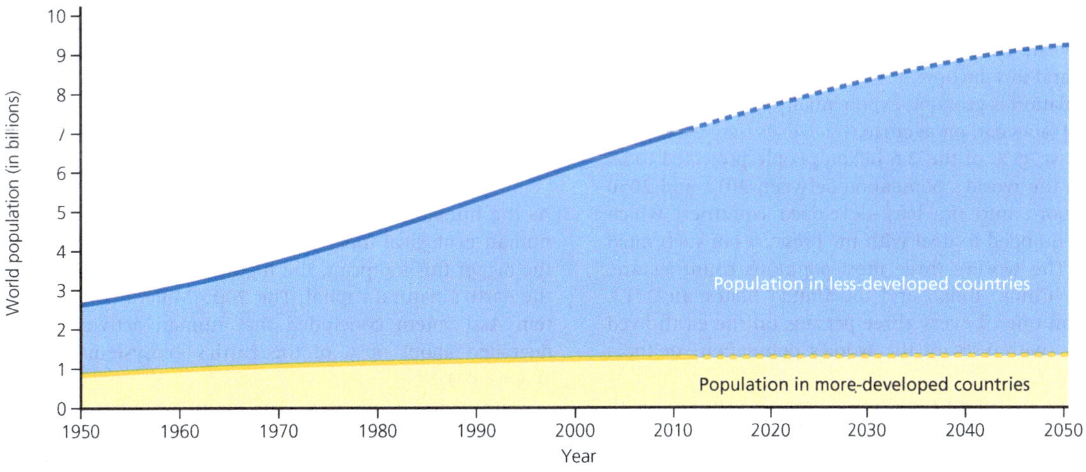

(Compiled by the authors using data from United Nations Population Division and Population Reference Bureau.)

Figure 6-3 Most of the world's population growth between 1950 and 2012 took place in the world's less-developed countries. This gap has been projected to increase between 2012 and 2050.

SCIENCE FOCUS 6.1

HOW LONG CAN THE HUMAN POPULATION KEEP GROWING?

To survive and provide resources for growing numbers of people, humans have modified, cultivated, built on, and degraded a large and increasing portion of the earth's natural systems (Figure 6-A). As a result, our ecological footprints have spread across much of the globe (see Figure 1-12, p. 13) and are projected to expand significantly further by 2030. Some scientists estimate that by then, we would need the equivalent of two planet Earths to sustain the projected population of 9.6 billion people (see Figure 1-13, p. 14).

Scientific studies of the populations of other species tell us that *no population can continue growing indefinitely* (p. 113). How long can we continue to avoid the reality of the earth's carrying capacity for our species by sidestepping many of the factors that sooner or later limit the growth of any population?

The debate over this important question has been going on since 1798 when Thomas Malthus, a British economist, hypothesized that the human population tends to grow exponentially, while food supplies tend to increase more slowly at a linear rate. So far, Malthus has been proven wrong. Food production has grown at an exponential rate instead of at a linear rate because of genetic and other technological advances in industrialized food production.

Environmental scientists are reexamining arguments such as those of Malthus about possible limits to the growth of human populations and economies. One view is that we have exceeded some of those limits, with too many people collectively degrading the earth's life-support system (see Figure 3-B, p. 73). To some scientists and other analysts, the key problem is *overpopulation* because of the sheer number of people in less-developed countries (see Figure 1-14, top, p. 15), which have 82% of the world's population. To others, the key factor is *overconsumption* in affluent, more-developed countries because of their high rates of resource use per person (see Figure 1-14, bottom).

Another view of population growth is that, so far, technological advances have allowed us to overcome the environmental limits that all populations of other species face and that this has had the effect of increasing the earth's carrying capacity for our species. They point out that average life expectancy in most of the world has been steadily rising despite dire warnings from some environmental scientists that we are seriously degrading our life-support system.

Some of these analysts argue that because of our technological ingenuity, there are few, if any, limits to human population growth and resource use per person. They believe that we can continue ever-increasing economic growth and avoid serious damage to our life-support systems by making technological advances in areas such as food production and medicine, and by finding substitutes for resources that we are depleting. As a result, they see no need to slow the world's population growth.

Proponents of slowing and eventually stopping population growth point out that

tries, where population is growing exponentially at about 0.1% a year. The other 98% were added to the world's middle- and low-income, less-developed countries, where the population is growing exponentially, 14 times faster at about 1.4% a year, on average.

At least 95% of the 2.6 billion people projected to be added to the world's population between 2012 and 2050 will be born into the least-developed countries, which are not equipped to deal with the pressures of such rapid growth. The world's three most populous countries are, in order, China, India, and the United States. In 2012, more than one of every three persons on the earth lived in China (with 19% of the world's population) or India (with 18%).

A third important trend in human population growth is the movement of people from rural areas to cities. More than half of the world's people now live in *urban areas,* or cities (see chapter-opening photo) and their surrounding suburbs. The great majority of these urban dwellers live in less-developed countries where resources for dealing with rapidly growing populations are limited. (We cover this and other *urbanization trends* in Chapter 22.)

Human Population Growth Impacts Natural Capital

As the human population grows, so does the global total human ecological footprint (see Figure 1-13, p. 14), and the bigger this footprint, the higher the overall impact on the earth's natural capital. The 2005 Millennium Ecosystem Assessment concluded that human activities have degraded about 60% of the earth's ecosystem services (Figure 6-4).

This raises a question that many scientists are now studying: How many people can the earth support indefinitely (Concept 6-1)? Some analysts contend that we need to define the planet's **cultural carrying capacity**: the maximum number of people who could live in reason-we are not providing the basic necessities for about 1.4 billion people—one of every five on the planet—who struggle to survive on the equivalent of about $1.25 per day. This raises a serious question: How will we meet the basic needs of the additional 2.6 billion people projected to be alive in 2050?

Proponents of slowing growth also warn of two potentially serious consequences that we could face if we do not sharply lower birth rates. First, death rates might increase because of declining health and environmental conditions and increasing social disruption in some areas, as is already happening in parts of Africa. A worst-case scenario for such a trend is a crash of the human population to a more sustainable level, perhaps as low as 2 billion. Second, resource use and degradation of normally renewable resources may intensify as more consumers increase their already large ecological footprints in more-developed countries and in rapidly developing countries such as China, India, and Brazil.

So far, advances in food production and health care have staved off widespread population declines. But there is extensive and growing evidence that we are steadily depleting and degrading much of the earth's irreplaceable natural capital (see Figure 1-3, p. 7) that supports us. We can get away with this for awhile, because the earth's life-support system is very complex and resilient, and because of built-in time delays between disturbances to the system and responses to them from within the system. But like unseen termites eating away the foundation of a house, at some point, such disturbances could reach a *tipping* or *breaking point* (see Figure 3-B, p. 73) beyond which there could be damaging and long-lasting change.

No one knows how close we are to environmental limits that, many scientists say, eventually will control the size of the human population. However, these scientists argue that human population growth is a vital scientific, political, economic, and ethical issue that we must confront.

Figure 6-A Natural capital degradation: Fertile topsoil is an irreplaceable form of natural capital necessary for supplying the world's people with food. According to a joint survey by the UN Environment Programme (UNEP) and the World Resources Institute, topsoil is eroding faster than it forms on about 38% of the world's cropland, including this farmland in Iowa.

Lynn Betts/USDA Natural Resources Conservation Service

Critical Thinking

How close do you think we are to the environmental limits of human population growth? Very close, moderately close, or far away? Explain.

able freedom and comfort indefinitely, without decreasing the ability of the earth to sustain future generations. (See the online Guest Essay by Garrett Hardin on this topic.) This issue has long been a topic of scientific and political debate (Science Focus 6.1).

Animated Figure 6-4 We humans have altered the natural systems that sustain our lives and economies in at least eight major ways to meet the increasing needs and wants of our growing population (Concept 6-1). ***Questions:*** In your daily living, do you think you contribute directly or indirectly to any of these harmful environmental impacts? Which ones? Explain.

Top: ©Dirk Ercken/Shutterstock.com. Center: ©Fulcanelli/Shutterstock.com. Bottom: © Werner Stoffberg/Shutterstock.com.

6-2 What Factors Influence the Size of the Human Population?

CONCEPT 6-2A
Population size increases through births and immigration, and decreases through deaths and emigration.

CONCEPT 6-2B
The average number of children born to the women in a population *(total fertility rate)* is the key factor that determines population size.

The Human Population Can Grow, Decline, or Remain Fairly Stable

The basics of global population change are quite simple. If there are more births than deaths during a given period of time, the earth's population increases, and when the opposite is true, it decreases. When the number of births equals the number of deaths during a particular time period, population size does not change.

Instead of using the total numbers of births and deaths per year, demographers use the **crude birth rate** (the number of live births per 1,000 people in a population in a given year) and the **crude death rate** (the number of deaths per 1,000 people in a population in a given year).

Human populations grow or decline in particular countries, cities, or other areas through the interplay of three factors: *births (fertility), deaths (mortality),* and *migration.* We can calculate the **population change** of an area by subtracting the number of people leaving a population (through death and emigration) from the number entering it (through birth and immigration) during a specified period of time (usually 1 year) (Concept 6-2A):

Population change = (Births + Immigration) − (Deaths + Emigration)

When births plus immigration exceed deaths plus emigration, a population grows; when the reverse is true, a population declines. (see Figure 18, p. S44, in Supplement 6 for a map showing various percentage rates of population growth in the world's countries in 2012.)

Women Are Having Fewer Babies but the World's Population Is Still Growing

Another measurement used in population studies is the **fertility rate**, a measure of how many children are born in a population over a set period of time. Here, we consider two types of fertility rates, the first being the **replacement-level fertility rate**: the average number of children that couples in a population must bear to replace themselves. It is slightly higher than two children per couple (2.1 in more-developed countries and as high as 2.5 in some less-developed countries), mostly because some children die before reaching their reproductive years. Any fertility rate above the replacement level will cause a population to grow.

If we were to reach the replacement-level fertility rate for the world, would it bring an immediate halt to population growth? No, because the number of *future* parents alive has grown dramatically. If all of today's couples were to have an average of 2.1 children, they would not be contributing to population growth. But if all of today's girl children were to grow up to have an average of 2.1 children as well, the world's population would continue to grow for 50 years or more (assuming death rates do not rise) because there are so many girls under age 15 who will be moving into their reproductive years.

The second type of fertility rate, and the key factor affecting human population growth and size, is the **total fertility rate (TFR)**: the average number of children born to the women in a population during their reproductive years (Concept 6-2B). Between 1955 and 2012, the average TFR for more-developed countries dropped from 2.8 to 1.6, and the average TFR for less-developed countries dropped from 6.2 to 2.6. The latter number is projected to continue dropping. If the global TFR, which was 2.4 in 2012, were to drop to 2.1, the world's population would continue to grow for about 50 years and then would level off and gradually decline.

Between 1955 and 2012, the global TFR dropped from 5 to 2.4. Those who support slowing the world's population growth view this as good news. However, to eventually halt population growth, the global TFR will have to drop to 2.1. (See Figure 19, p. S45, in Supplement 6 for a map showing how TFRs vary globally.)

CASE STUDY

The U.S. Population—Third-Largest and Growing

The population of the United States grew from 76 million in 1900 to 314 million by 2012, despite oscillations in the country's TFR and population growth rate (Figure 6-5). It took the country 139 years to reach a population of 100 million people, 52 years to add another 100 million (by 1967), and only 39 years to add the third 100 million (by 2006). During the period of high birth rates between 1946 and 1964, known as the *baby boom,* 79 million people were added to the U.S. population. At the peak of the baby boom in 1957, the average TFR was 3.7 children per woman. In 2012, and in most years since 1972, it has been at or below 2.1 children per woman (1.9 in 2012), compared to a global TFR of 2.4.

The drop in the TFR has slowed the rate of population growth in the United States. But the country's population is still growing faster than those of most other more-developed countries and it is not close to leveling off. According to the U.S. Census Bureau, about 2.3 mil-

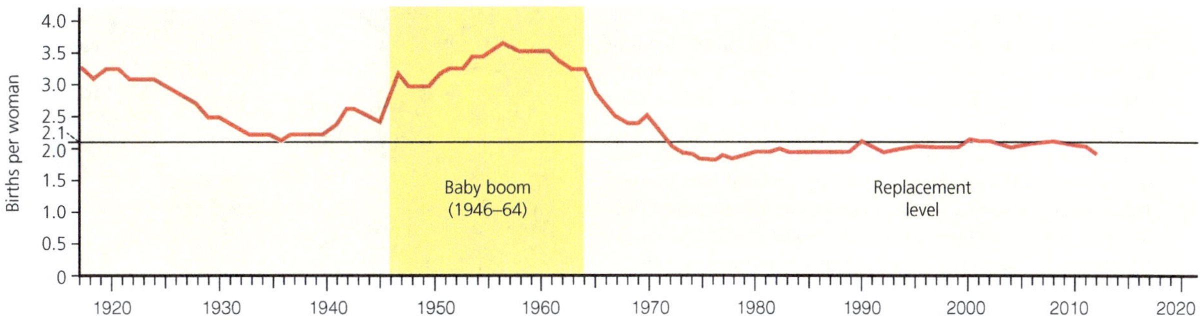

Figure 6-5 The graph shows the total fertility rates for the United States between 1917 and 2012. ***Question:*** The U.S. fertility rate has declined and remained at or below replacement levels since 1972. So why is the population of the United States still increasing?

(Compiled by the authors using data from Population Reference Bureau and U.S. Census Bureau.)

lion people were added to the U.S. population in 2012. About 1.6 million (70% of the total) were added because there were that many more births than deaths, and about 0.7 million (30% of the total) were immigrants.

Immigration has become a political issue in the United States. The country was founded by immigrants and since 1820, it has admitted almost twice as many immigrants and refugees as all other countries combined. The number of legal immigrants (including refugees) has varied during different periods because of changes in immigration laws and rates of economic growth (Figure 6-6).

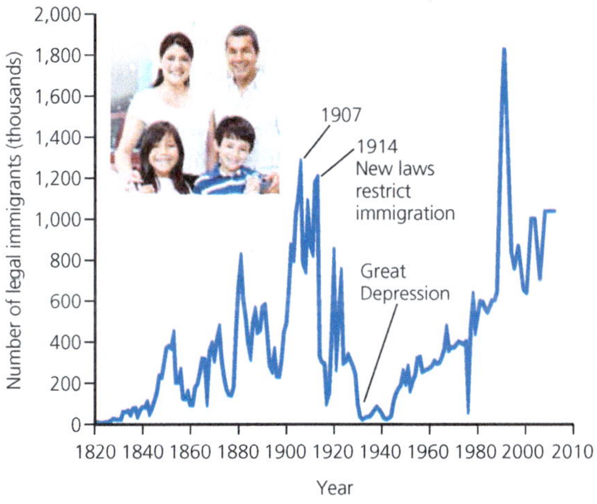

(Compiled by the authors using data from U.S. Immigration and Naturalization Service and the Pew Hispanic Center.)

Figure 6-6 Legal immigration to the United States, 1820–2006 (the last year for which data are available). The large increase in immigration since 1989 resulted mostly from the Immigration Reform and Control Act of 1986, which granted legal status to certain illegal immigrants who could show they had been living in the country before January 1, 1982.

Photo: Andresr/Shutterstock.com

In addition to the fourfold increase in population growth since 1900, some amazing changes in lifestyles took place in the United States during the 20th century (Figure 6-7), which led to Americans living longer. Along with this came dramatic increases in per capita resource use and a much larger total and per capita ecological footprint.

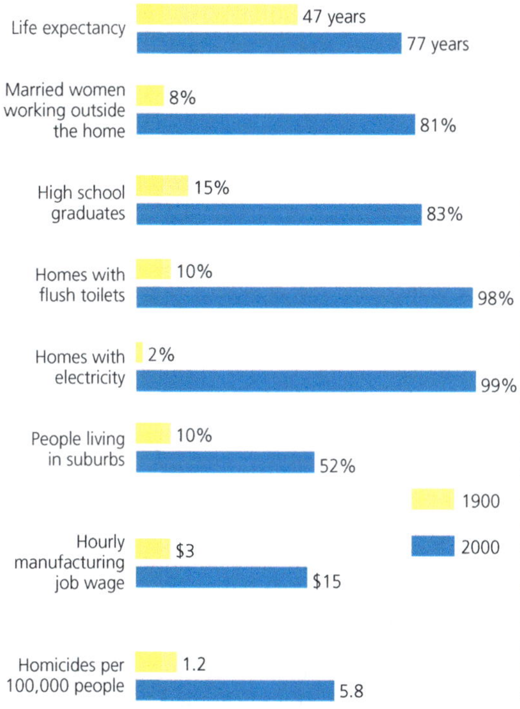

(Compiled by the authors using data from U.S. Census Bureau and Department of Commerce.)

Figure 6-7 Some major changes took place in the United States between 1900 and 2000. ***Question:*** Which two of these changes do you think had the biggest impacts on the U.S. ecological footprint?

Compare the present-day standard of living in the United States to that of 1907 when the three leading causes of death in the United States were pneumonia, tuberculosis, and diarrhea (ailments that are seldom life-threatening now); 90% of U.S. doctors had no college education; one of five adults could not read or write; only 6% of Americans graduated from high school; the average U.S. worker earned a few hundred to a few thousand dollars per year; and there were only 9,000 cars in the country and only 232 kilometers (144 miles) of paved roads.

The U.S. Census Bureau projects that the U.S. population is likely to increase from 314 million in 2012 to 400 million by 2050—an addition of 86 million more Americans over four decades. Because of a high per-person rate of resource use and the resulting waste and pollution, each addition to the U.S. population has an enormous environmental impact (see the map in Figure 8, p. S33, in Supplement 6). Recall that the environmental impact of a population is obtained by multiplying the effects of three factors (see Figure 1-14, p. 15, bottom): population size, affluence (and resulting high rates of resource use per person), and technology. The United States has by far the world's largest total and per capita ecological footprint (see Figure 1-11, p. 13), mostly because of the size of its population multiplied by its very high rate of resource use per person. This explains why some analysts consider the United States to be the world's most overpopulated country.

CONSIDER THIS. . .

THINKING ABOUT The U.S. Population

Considering this information, do you believe that the United States is the world's most overpopulated country? Explain.

Several Factors Affect Birth Rates and Fertility Rate

Many factors affect a country's average birth rate and TFR. One is the *importance of children as a part of the labor force,* especially in less-developed countries. This is a major part of why it makes sense for many poor couples in those countries to have a large number of children. They need help with hauling daily drinking water (Figure 6-8), gathering wood for heating and cooking, and tending crops and livestock.

Another economic factor is the *cost of raising and educating children.* Birth and fertility rates tend to be lower in more-developed countries, where raising children is much more costly because they do not enter the labor force until they are in their late teens or twenties. In the United States, it costs more than $235,000 to raise a middle-class child from birth to age 18. By contrast, many children in poor countries receive little education and instead have to work to help their families survive (Figure 6-9).

The *availability of, or lack of, private and public pension systems* can influence the decisions of some couples on how many children to have, especially the poor in less-developed countries. Pensions reduce a couple's need to have several children to help support them in old age.

Syed Sajjad Ali/Shutterstock.com

Figure 6-8 These young girls in India are carrying water.

Also, there are more *infant deaths* in poorer countries, and over time, this has affected cultural norms related to family size in these countries. The more children a couple has, the more likely it is that at least a few will survive and grow to adulthood.

Urbanization plays a role in birth rate trends. People living in urban areas usually have better access to family planning services and tend to have fewer children than do those living in the rural areas of poorer countries.

Another important factor is the *educational and employment opportunities available for women.* Total fertility rates tend to be low when women have access to education and paid employment outside the home. In less-developed countries, a woman with no education typically has two more children than does a woman with a high school education. In most societies, better-educated women tend to marry later and have fewer children.

Average age at marriage (or, more precisely, the average age at which a woman has her first child) also plays a role. Women normally have fewer children when their average age at marriage is 25 or older.

Birth rates and TFRs are also affected by the *availability of legal abortions.* According to the World Health Organization and the Guttmacher Institute, each year, about 210 million women become pregnant and at least 42 million of them get abortions—about 22 million of them legal and the other 20 million illegal (and often unsafe). The *availability of reliable birth control methods* also allows women to control the number and spacing of the children they have.

Religious beliefs, traditions, and cultural norms also play a role. In some countries, these factors favor large families as many people strongly oppose abortion and some forms of birth control.

PRIIT VESILIND/National Geographic Creative

Figure 6-9 This young girl is breaking granite into gravel in the Kerala State of India.

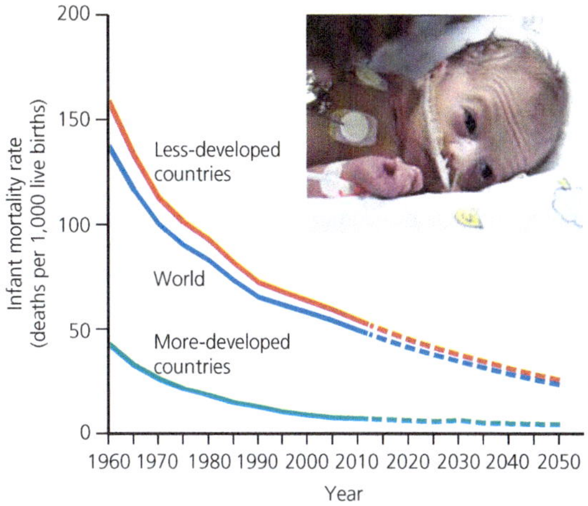

(Compiled by the authors using data from United Nations Population Division.)

Figure 6-10 Infant mortality rates for the world's more-developed countries and less-developed countries, 1950–2012, with projections to 2050 based on medium population projections.

Photo: Gert Vrey/Shutterstock.com

CONSIDER THIS. . .

CONNECTIONS **Preference for Male Children and Social Problems in China**

In China, where families are strongly encouraged to have only one child, there is a strong preference for male children, partly because daughters are likely to marry and leave their parents. Some pregnant Chinese women use ultrasound to determine the gender of their fetuses and get an abortion if the child is female. As a result, baby boys are sometimes kidnapped and sold to families that want boys, and some young girls have been kidnapped and sold as brides for single men. The Chinese government estimates that by 2030, about 30 million Chinese men will not be able to find wives and if the economy slows, they might have poor job prospects. Sociologists say this is a recipe for social upheaval.

Several Factors Affect Death Rates

The rapid growth of the world's population over the past 100 years is not primarily the result of a rise in the birth rate. Instead, it has been caused largely by declining death rates, especially in less-developed countries. More people in some of these countries are living longer and fewer infants are dying because of increased food supplies, improvements in food distribution, better nutrition, medical advances such as immunizations and antibiotics, improved sanitation, and safer water supplies.

A useful indicator of the overall health of people in a country or region is **life expectancy**, which for any given year is the average number of years a person born in that year can be expected to live. Between 1955 and 2012, the average global life expectancy increased from 48 years to 70 years. In 2012, Japan had the world's longest life expectancy of 83 years. In the world's poorest countries, life expectancy is 55 years or less.

GOOD NEWS

Between 1900 and 2012, the average U.S. life expectancy increased from 47 years to 79 years. However, the United States ranks 32nd among nations in life expectancy, down from 5th in 1950. Research indicates that a key factor in this ranking is poor health, even though the United States leads the world in health-care costs per person.

Another important indicator of overall health in a population is its **infant mortality rate**, the number of babies out of every 1,000 born who die before their first birthday. It is viewed as one of the best measures of a society's quality of life because it reflects a country's general level of nutrition and health care. A high infant mortality rate usually indicates insufficient food *(undernutrition)*, poor nutrition *(malnutrition)*, and a high incidence of infectious disease. Infant mortality also affects the TFR. In areas with low infant mortality rates, women tend to have fewer children because fewer of their children die at an early age.

Infant mortality rates in most countries have declined dramatically since 1965 (Figure 6-10) but vary widely among different countries (see Figure 20, p. S45, Supplement 6). Even so, every year, more than 4 million infants (most of them in less developed countries) die of *preventable* causes during their first year of life. This average of nearly 11,000 mostly unnecessary infant deaths per day is equivalent to 55 jet airliners, each loaded with 200 infants younger than age 1, crashing *every day* with no survivors—a tragedy rarely reported in the news.

In 1900 the U.S. infant mortality rate was 165. In 2012 it was 6.0. This sharp decline was a major factor in the marked increase in U.S. average life expectancy during this period. However, in 2012 the United States ranked 44th among all nations in terms of infant mortality rates for two main reasons: the generally inadequate health care for poor women during pregnancy, as well as for their babies after birth, and drug addiction among many pregnant women.

SCIENCE FOCUS 6.2

PROJECTING POPULATION CHANGE

Estimates of the human population size in 2050 range from 7.8 billion to 10.8 billion people—a difference of 3 billion. Why this wide range of estimates? The answer is that there are countless factors that demographers have to consider when making projections for any country or region of the world.

First, they have to determine the reliability of current population estimates. Many of the more-developed countries such as the United States have fairly good estimates. But many countries have poor knowledge of their population size. Some countries even deliberately inflate or deflate the numbers for economic or political purposes.

Second, demographers make assumptions about trends in fertility—that is, whether women in a country will, on average, have fewer babies or more babies in the future. The demographers might assume that fertility is declining by a certain percentage per year. If the estimate of the rate of declining fertility is off by a few percentage points, the resulting percentage increase in population can be magnified over a number of years, and be quite different from the projected population size increase.

For example, UN demographers assumed that Kenya's fertility rate would decline, and based on that, in 2002, they projected that Kenya's total population would be 44 million by 2050. In reality, the fertility rate rose from 4.7 to 4.8 children per woman. The population momentum was therefore more powerful than expected, and in 2010, the UN revised its projection for Kenya's population in 2050 to 70.8 million, which was 60% higher than its earlier projection.

Third, population projections are made by a variety of organizations employing demographers. UN projections tend to be cited most often. But the U.S. Census Bureau also makes world population projections, as does the International Institute for Applied Systems Analysis (IIASA) and the U.S. Population Reference Bureau. The World Bank has also made projections in the past and maintains a database. All of these organizations use differing sets of data and differing methods, and their projections vary (Figure 6-B).

Critical Thinking

If you were in charge of the world and making decisions about resource use based on population projections, which of the projections in Figure 6-B would you rely on? Explain.

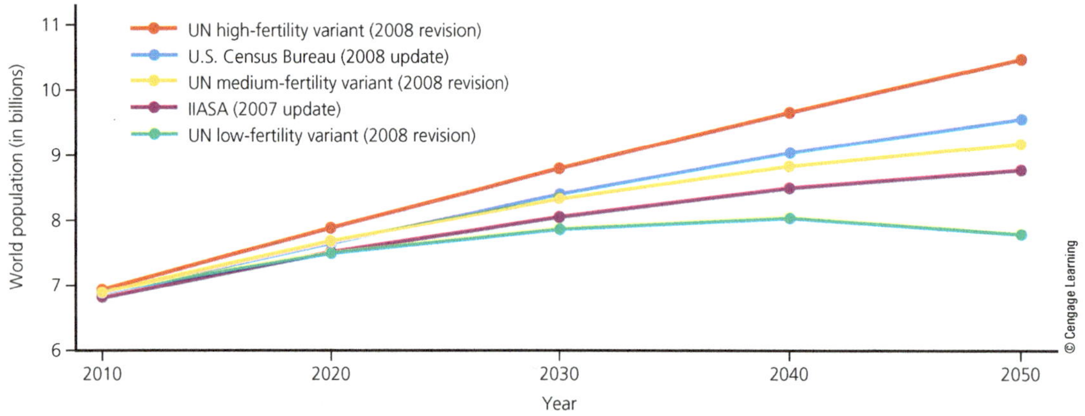

Figure 6-B World population projections to 2050 from three different organizations: the UN, the U.S. Census Bureau, and IIASA. Note that the uppermost, middle, and lowermost curves of these five projections are all from the UN, each assuming a different level of fertility.

Migration Affects an Area's Population Size

The third factor in population change is **migration**: the movement of people into *(immigration)* and out of *(emigration)* specific geographic areas. Most people migrating from one area or country to another seek jobs and economic improvement. But many are driven to migrate by religious persecution, ethnic conflicts, political oppression, or war.

According to a United Nations (UN) study and another study by environmental scientist Norman Myers, in 2008 there were at least 40 million *environmental refugees*—people who had to leave their homes because of water or food shortages, soil erosion, or some other form of environmental degradation or depletion. It is difficult to find more recent, reliable estimates of the number of such refugees, but in 2012, the UN projected that by 2020, there will be 50 million environmental refugees.

Estimates of any population's future numbers can vary considerably. These varying estimates depend mostly on TFR projections. However, demographers also have to make assumptions about death rates, migration, and a number of other variables. If their assumptions are wrong, their population forecasts can be way off the mark (Science Focus 6.2).

6-3 How Does a Population's Age Structure Affect Its Growth or Decline?

CONCEPT 6-3
The numbers of males and females in young, middle, and older age groups determine how fast a population grows or declines.

A Population's Age Structure Helps Us to Make Projections

An important factor determining whether the population of a country increases or decreases is its **age structure**: the numbers or percentages of males and females in young, middle, and older age groups in that population (Concept 6-3).

Population experts construct a population *age-structure diagram* by plotting the percentages or numbers of males and females in the total population in each of three age categories: *prereproductive* (ages 0–14), consisting of individuals normally too young to have children; *reproductive* (ages 15–44), consisting of those normally able to have children; and *postreproductive* (ages 45 and older), with individuals normally too old to have children. Figure 6-11 presents generalized age-structure diagrams for countries with rapid, slow, zero, and negative population growth rates.

A country with a large percentage of its people younger than age 15 (represented by a wide base in Figure 6-11, far left) will experience rapid population growth unless death rates rise sharply. Because of this *demographic momentum*, the number of births in such a country will rise for several decades even if women have an average of only one or two children each, due to the large number of girls entering their prime reproductive years.

In 2012, about 26% of the world's population—29% in the less-developed countries and 16% in more-developed countries—was under age 15. By 2025, the world's current 1.8 billion people under age 15—roughly one of every four persons on the planet—will move into their prime reproductive years. The dramatic differences in population age structure between less-developed and more-developed countries (Figure 6-12) show why most future human population growth will take place in less-developed countries (Figure 6-3).

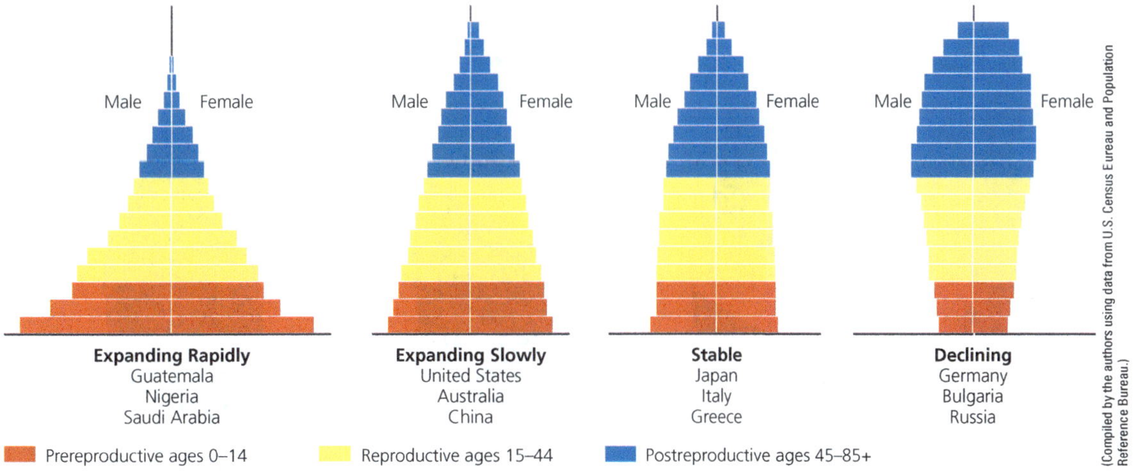

Animated Figure 6-11 Generalized population age-structure diagrams for countries with rapid (1.5-3%), slow (0.3-1.4%), zero (0-0.2%), and negative (declining) population growth rates. ***Question:*** Which of these diagrams best represents the country where you live?

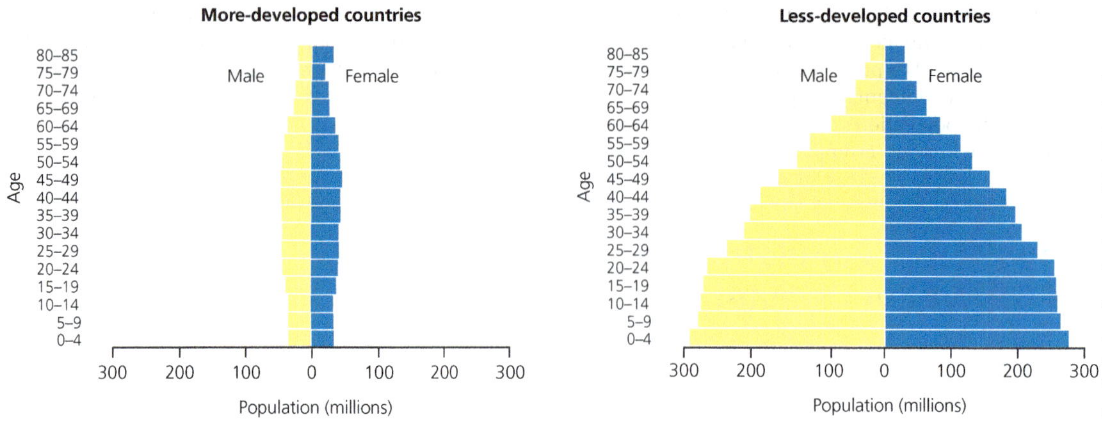

(Compiled by the authors using data from United Nations Population Division and Population Reference Bureau.)

Figure 6-12 Population structure by age and sex in less-developed countries and more-developed countries for 2011. ***Question:*** If all girls under 15 were to have only one child during their lifetimes, how do you think these structures would change over time?

The global population of seniors—people who are 65 and older—is projected to triple by 2050, when one of every six people will be a senior. (See the Case Study that follows.) This graying of the world's population is due largely to declining birth rates and medical advances that have extended life spans. In 2012, the three nations with the largest percentage of their population age 65 or older were, in order, Japan, Germany, and Italy. In such countries, the number of working adults is shrinking in proportion to the number of seniors, which in turn is slowing the growth of tax revenues in these countries. Some analysts worry about how such societies will support their growing populations of seniors.

CASE STUDY

The American Baby Boom

Changes in the distribution of a country's age groups have long-lasting economic and social impacts. For example, consider the American baby boom, which added 79 million people to the U.S. population between 1946 and 1964. Over time, this group looks like a bulge moving up through the country's age structure, as shown in Figure 6-13.

For decades, members of the baby-boom generation have strongly influenced the U.S. economy because they make up about 36% of all adult Americans. Baby boomers created the youth market in their teens and twen-

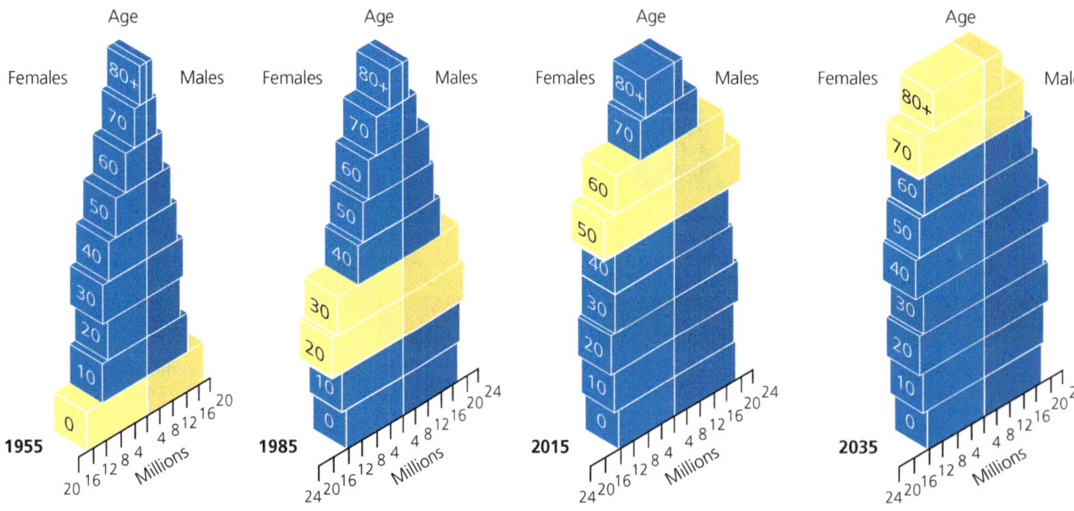

(Compiled by the authors using data from U.S. Census Bureau and Population Reference Bureau.)

Animated Figure 6-13 Age-structure charts tracking the baby-boom generation in the United States, 1955, 1985, 2015 (projected), and 2035 (projected).

ties and are now creating the late middle age and senior markets. In addition to having this economic impact, the baby-boom generation plays an increasingly important role in deciding who gets elected to public office and what laws are passed or weakened.

Since 2011, when the first baby boomers began turning 65, the number of Americans older than age 65 has grown at the rate of about 10,000 a day and will do so through 2030. This process has been called the *graying of America.* As the number of working adults declines in proportion to the number of seniors, there may be political pressure from baby boomers to increase tax revenues to help support the growing senior population. This could lead to economic and political conflicts between younger and older Americans.

CONSIDER THIS. . .

CONNECTIONS Baby Boomers, the U.S. Work Force, and Immigration

According to the U.S. Census Bureau, after 2020, much higher immigration levels will be needed to supply enough workers as baby boomers retire. According to a recent study by the UN Population Division, if the United States wants to maintain its current ratio of workers to retirees, it will need to absorb an average of 10.8 million immigrants every year—more than 10 times the current immigration level—through 2050.

Populations Made Up Mostly of Older People Can Decline Rapidly

As the percentage of people age 65 or older increases, more countries will begin experiencing population declines. If population decline is gradual, its harmful effects usually can be managed. However, some countries are experiencing fairly rapid declines and feeling such effects more severely.

Japan has the world's highest percentage of elderly people (above age 65) and the world's lowest percentage of young people (below age 15). In 2012, Japan's population was 128 million. By 2050, its population is projected to be 95.5 million, a 25% drop. As its population declines, there will be fewer adults working and paying taxes to support an increasing elderly population. Because Japan discourages immigration, it may face a bleak economic future. As a result, some have called for the country to rely more on robots to do its manufacturing jobs and on selling robots in the global economy to help support its aging population.

In China, the growth in numbers of children has slowed because of its one-child policy. As a result, the average age of China's population has been increasing over the past two decades at one of the fastest rates ever recorded. While China's population is not yet declining, the UN estimates that by 2025, China is likely to have too few young workers to support its rapidly aging population. This graying of the Chinese population could lead to a declining work force, higher wages for workers, limited funds for supporting continued economic development, and fewer children and grandchildren to care for the growing number of elderly people. These concerns and other factors may slow economic growth and have led to some relaxation of China's one-child population control policy.

Figure 6-14 lists some of the problems associated with rapid population decline. Countries currently faced with rapidly declining populations include Japan, Russia, Germany, Bulgaria, Hungary, Ukraine, Serbia, Greece, Portugal, and Italy.

Populations Can Decline Due to a Rising Death Rate: The AIDS Tragedy

A large number of deaths from AIDS can disrupt a country's social and economic structure by removing significant numbers of young adults from its population. According to the World Health Organization, between 1981 and 2012, AIDS killed more than 30 million people (617,000 in the United States).

Unlike hunger and malnutrition, which kill mostly infants and children, AIDS kills primarily young adults and leaves many children orphaned, some of whom are also infected with HIV, the virus that can lead to AIDS. Worldwide, AIDS is the leading cause of death for people of ages 15–49.

This pandemic has had a devastating effect in some countries, and has changed their population age structures

Figure 6-14 Rapid population decline can cause several problems. ***Question:*** Which two of these problems do you think are the most important?

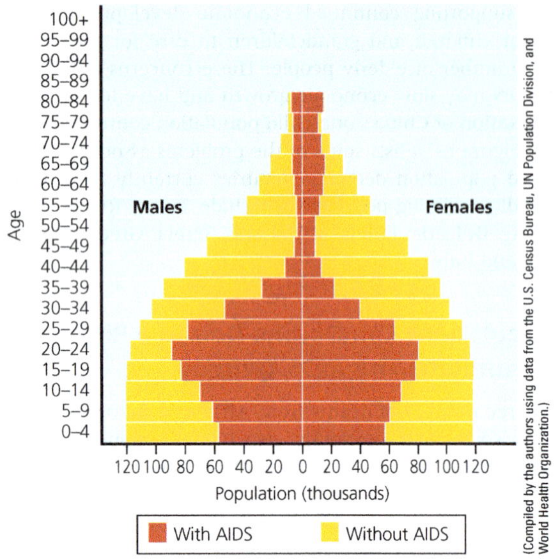

Figure 6-15 In Botswana, more than 25% of people ages 15–49 were infected with HIV in 2011. This figure shows two projected age structures for Botswana's population in 2020—one including the possible effects of the AIDS epidemic (red bars), and the other not including those effects (yellow bars). ***Question:*** How might this affect Botswana's economic development?

(Figure 6-15). This has had a number of harmful effects. One is a sharp drop in average life expectancy, especially in several southern African countries where 15–26% of all people between ages 15 and 49 are infected with HIV. Another is the loss of productive young-adult workers and trained personnel such as scientists, farmers, engineers, and teachers, as well as government, business, and health-care workers. The essential services they could provide are therefore lacking, and there are fewer taxpayers and fewer workers available to support the very young and the elderly. Many experts have called for the creation of a massive international program to help countries ravaged by AIDS.

6-4 How Can We Slow Human Population Growth?

CONCEPT 6-4

We can slow human population growth by reducing poverty, elevating the status of women, and encouraging family planning.

The First Step Is to Promote Economic Development

Scientific studies and experience have shown that the three most effective ways to slow or stop population growth are to reduce poverty, primarily through economic development and universal primary education; to elevate the status of women; and to encourage family planning and reproductive health care (**Concept 6-4**). Let's begin by looking at the role of economic development.

In the world's most desperately poor countries, couples tend to have more children for reasons listed earlier in this chapter (p. 128). Thus, in many less-developed countries, total fertility and population growth rates tend to be high, and large numbers of poor people are increasingly being crowded into unsanitary and difficult living conditions in slums and shantytowns.

Demographers, examining the birth and death rates of western European countries that became industrialized during the 19th century, have developed a hypothesis of population change known as the **demographic transition**: As countries become industrialized and economically developed, their populations tend to grow more slowly. According to the hypothesis, this transition takes place in four stages, as shown in Figure 6-16.

Some analysts believe that most of the world's less-developed countries will make a demographic transition over the next few decades, mostly because newer technologies will help them to develop economically and to raise their per capita incomes. Other analysts fear that rapid population growth, extreme poverty, and increasing environmental degradation and resource depletion in some low-income, less-developed countries could leave these countries stuck in stage 2 of the demographic transition.

Another factor that could hinder the demographic transition in some less-developed countries is that since 1985, economic assistance from more-developed countries has generally dropped. The resulting shortage of funds, coupled with poor economies in some countries, could leave large numbers of people trapped in poverty, which could in turn keep population growth rates high in such countries. Many experts argue that more-developed countries should help less-developed nations to make the demographic transition by aiding them in their economic development. They contend that such an effort could help to stabilize the global population, thereby making for a more sustainable world.

Empowering Women Can Help to Slow Population Growth

A number of studies show that women tend to have fewer children if they are educated, have the ability to control their own fertility, earn an income of their own, and live in societies that do not suppress their rights. Although women make up roughly half of the world's population,

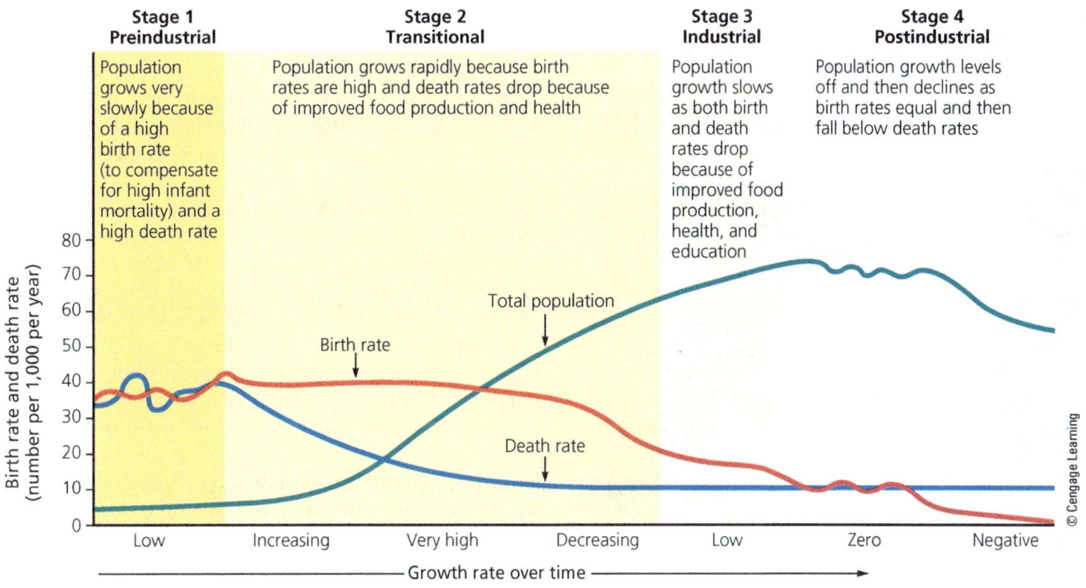

Animated Figure 6-16 The *demographic transition,* which a country can experience as it becomes industrialized and more economically developed, can take place in four stages. ***Question:*** At what stage is the country where you live?

in most societies they have fewer rights and educational and economic opportunities than men have.

Women do almost all of the world's domestic work and child care for little or no pay and provide more unpaid health care (within their families) than do all of the world's organized health-care services combined. In rural areas of Africa, Latin America, and Asia, women do 60–80% of the work associated with growing food, gathering, and hauling wood (Figure 6-17) and animal dung for use as fuel, and hauling water. As one Brazilian woman observed, "For poor women, the only holiday is when you are asleep."

While women account for 66% of all hours worked, they receive only 10% of the world's income and own just 2% of the world's land. They also make up 70% of the world's poor and 66% of its 800 million illiterate adults. Because sons are more valued than daughters in many societies, girls are often kept at home to work instead of being sent to school. Globally, the number of school-age girls who do not attend elementary school is more than 900 million—almost 3 times the entire U.S. population. Poor women who cannot read often have an average of five to seven children, compared to two or fewer children in societies where almost all women can read.

However, an increasing number of women in less-developed countries are taking charge of their lives and reproductive behavior. As it expands, such bottom-up change driven by individual women will play an important role in stabilizing populations, reducing poverty and environmental degradation, and allowing more access to basic human rights.

Family Planning Can Provide Several Benefits

Family planning provides educational and clinical services that help couples choose how many children to have and when to have them. Such programs vary from culture to culture, but most of them provide information on birth spacing, birth control, and health care for pregnant women and infants (Figure 6-18).

Figure 6-17 This woman in Nepal is bringing home firewood. Typically, she spends 2 hours a day, 2 or 3 times a week, on this task.

Helene Rogers/Art Directors & TRIP/Alamy

Figure 6-18 This family planning center is located in Kolkata (Calcutta), India.

Family planning has been a major factor in reducing the number of births throughout most of the world. It has also reduced the number of abortions performed each year and has decreased the numbers of mothers and fetuses dying during pregnancy, according to studies by the UN Population Division and other population agencies.

Such studies indicate that family planning is responsible for a drop of at least 55% in total fertility rates (TFRs) in less-developed countries, from 6.0 in 1960 to 2.6 in 2012. For example, family planning, coupled with economic development, played a major role in the sharp drop in the TFR in Bangladesh from around 6.8 to 2.3 by 2012. Between 1971 and 2012, Thailand also used family planning to cut its annual population growth rate from 3.2% to 0.5%, and to reduce its TFR from 6.4 to 1.6. According to the UN, had there not been the sharp drop in TFRs since the 1970s, with all else being equal, the world's population today would be about 8.5 billion instead of 7 billion.

Family planning also has financial benefits. Studies have shown that each dollar spent on family planning in countries such as Thailand, Egypt, and Bangladesh saves $10–$16 in health, education, and social-service costs by preventing unwanted births.

Despite these successes, certain problems have hindered progress in some countries. There are two major problems. *First,* according to the UN Population Fund, about 42% of all pregnancies in less-developed countries are unplanned and about 26% end with abortion. So ensuring access to voluntary contraception would play a key role in stabilizing the populations and reducing the number of abortions in such countries.

Second, an estimated 215 million couples in less-developed countries want to limit their number of children and determine their spacing, but they lack access to family planning services. According to the UN, providing such services where they are needed would cost about $6.7 billion a year—about what Americans together spend on Halloween each year.

According to the UN Population Fund and the Alan Guttmacher Institute, meeting women's current unmet needs for family planning and contraception could prevent about 53 million unwanted pregnancies, 24 million induced abortions, 1.6 million infant deaths, and 142,000 pregnancy-related deaths of women per year. This could reduce the projected global population size by more than 1 billion people, at an average cost of $20 per couple per year. The Guttmacher Institute estimates that in the United States each year, domestic family planning programs prevent 973,000 unintended pregnancies, of which an estimated 406,000 would end in abortion.

Some analysts call for expanding family planning programs to educate men about the importance of having fewer children and taking more responsibility for raising them. Proponents also call for greatly increased research in order to develop more effective birth control methods for men.

The experiences of countries such as Japan, Thailand, Bangladesh, South Korea, Taiwan, and China show that a country can achieve or come close to replacement-level fertility within a decade or two. Thus, the real story of the past 50 years has been the sharp reduction in the *rate* of population growth (Figure 6-2) resulting from a combination of the reduction of poverty through economic development, empowerment of women, and the promotion of family planning. However, the global population is still growing fast enough to add possibly several billion more people during this century (Figure 6-1).

CASE STUDY

Slowing Population Growth in India

For six decades, India has tried to control its population growth with only modest success. The world's first national family planning program began in India in 1952, when its population was nearly 400 million. In 2012, after 60 years of population control efforts, India had 1.26 billion people—the world's second largest population. Much of this increase occurred because the country's life expectancy rose from 38 years of age in 1952 to 65 in 2012, mostly as a result of declining death rates.

In 1952, India added 5 million people to its population. In 2012, it added 19 million—more than any other country. Also, 31% of India's population is under age 15, which sets the country up for further rapid population growth. The United Nations projects that by 2030, India will be the world's most populous country, and by 2050 it will have a population of 1.69 billion.

Figure 6-19 These people are shopping at Kolkata's (Calcutta's) South City Mall, the largest shopping mall in India.

STEVE RAYMER/National Geographic Creative

India has the world's fourth largest economy and a thriving and rapidly growing middle class of more than 100 million people—a number nearly equal to a third of the U.S. population. This growing class of consumers will enlarge India's ecological footprint, as more Indians use more resources with every passing year (Figure 6-19).

However, the country faces a number of serious poverty, malnutrition, and environmental problems that could worsen as its population continues to grow rapidly. About one-fourth of all people in India's cities live in slums, and prosperity and progress have not touched many of the nearly 650,000 rural villages where more than two-thirds of India's population lives. Nearly half of the country's labor force is unemployed or underemployed and 42% of its population lives in extreme poverty (Figure 6-20).

For decades, the Indian government has provided family planning services throughout the country and has strongly promoted a smaller average family size. Even so, Indian women have an average of 2.5 children.

Two factors help to account for larger families in India. *First,* most poor couples believe they need several children to work and care for them in old age. *Second,* the strong cultural preference in India for male children means that some couples keep having children until they produce one or more boys. The result: even though 90% of Indian couples have access to at least one modern birth control method, only 47% actually use one (compared to 85% in China).

India also faces critical resource and environmental problems. With 18% of the world's people, India has just 2.3% of the world's land resources and 2% of its forests. About half the country's cropland is degraded as a result of soil erosion and overgrazing. In addition, more than two-thirds of its water is seriously polluted, sanitation services often are inadequate, and many of its major cities suffer from serious air pollution.

India is undergoing rapid economic growth, which is expected to accelerate. This not only will help many people in India, but it will also put more pressure on the country's and the earth's natural capital. On the other hand, economic growth may help India to slow its population growth by accelerating its demographic transition.

CASE STUDY

Slowing Population Growth in China: A Success Story

China is the world's most populous country, with 1.35 billion people (Figure 6-21), followed by India with 1.26 billion and the United States with 314 million. In 2011, the U.S. Census Bureau projected that if current trends continue, China's population will increase to about 1.4 billion by 2026 and then will begin a slow decline to as low as 750 million by the end of this century.

In the 1960s, China's large population was growing so rapidly that there was a serious threat of mass starvation. To avoid this, government officials decided to take measures that eventually led to the establishment of the

Samrat35/Dreamstime LLC

Figure 6-20 Homeless people in Kolkata, India in 2011.

Figure 6-21 Thousands of people crowd a street in China, where almost one-fifth of all the people on the planet live.

world's most extensive, intrusive, and strict family planning and birth control program.

China's goal has been to sharply reduce population growth by promoting one-child families. The government provides contraceptives, sterilizations, and abortions for married couples. In addition, married couples pledging to have no more than one child receive a number of benefits, including better housing, more food, free health care, salary bonuses, and preferential job opportunities for their child. Couples who break their pledge lose such benefits.

Since this government-controlled program began, China has made impressive efforts to feed its people and bring its population growth under control. Between 1972 and 2012, the country cut its birth rate in half and reduced the average number of children born to its women from 5.7 to 1.5, compared to 1.9 children per woman in the United States.

Since 1980, China has undergone rapid industrialization and economic growth. According to the Earth Policy Institute, between 1990 and 2010, this reduced the number of people living in extreme poverty by almost 500 million. It also helped at least 300 million Chinese—a number almost equal to the entire U.S. population—to become middle-class consumers. Over time, China's rapidly growing middle class will consume more resources per person, expanding China's ecological footprint (see Figure 1-11, p. 13) within its own borders and in other parts of the world that provide it with resources. This will put a strain on China's and the earth's natural capital unless China steers a course toward more environmentally sustainable economic development.

Some have criticized China for having such a strict population control policy. However, government officials say that the alternative was mass starvation. They estimate that China's one-child policy has reduced its population size by as many as 400 million people.

Big Ideas

- The human population is growing rapidly and may soon bump up against environmental limits.
- Even if population growth is not a problem, the increasing use of resources per person is expanding the overall human ecological footprint and putting a strain on the earth's resources.
- We can slow human population growth by reducing poverty, elevating the status of women, and encouraging family planning.

TYING IT ALL TOGETHER World Population Growth and Sustainability

Jeremy Richards/Shutterstock.com

This chapter began with a discussion of the fact that the world's human population has now passed 7 billion (**Core Case Study**). We noted that this is a result of exponential population growth and that many environmental scientists believe such growth to be unsustainable for the long run. We briefly considered some of the environmental problems brought on by exponential human population growth. We looked at factors that influence the growth of populations, as well as at how some countries have made progress in controlling population growth.

In the first six chapters of this book, you have learned how ecosystems and species have been sustained throughout the earth's history, in keeping with the three scientific **principles of sustainability**, by nature's reliance on solar energy, nutrient cycling, and biodiversity (see Figure 1-2, p. 6 or back cover). These three principles can guide us in dealing with the problems brought on by population growth and decline. That is, by employing solar and other renewable-energy technologies more widely, we can cut pollution and emissions of climate-changing gases that are increasing as the population grows. By reusing and recycling more materials, we could cut our waste and reduce our ecological footprints. And in focusing on preserving biodiversity, we could help to sustain the life-support system on which we and all other species depend.

Chapter Review

Core Case Study

1. Summarize the story of how human population growth has surpassed 7 billion and explain why this is significant to many environmental scientists (Core Case Study). List three factors that account for the rapid increase in the world's human population over the past 200 years.

Section 6-1

2. What is the key concept for this section? What is the range of estimates for the size of the human population in 2050? Summarize the three major population growth trends recognized by demographers. About how many people are added to the world's population each year? What are the world's three most populous countries? List eight major ways in which we have altered natural systems to meet our needs. Define **cultural carrying capacity**. Summarize the debate over whether and how long the human population can keep growing.

Section 6-2

3. What are the two key concepts for this section? Define and distinguish between **crude birth rate** and **crude death rate**. List three variables that affect the growth and decline of human populations. Explain how a given area's **population change** is calculated. Define **fertility rate**, and distinguish between the **replacement-level fertility rate** and the **total fertility rate (TFR)**. How has the global TFR changed since 1955?
4. Summarize the story of population growth in the United States and explain why it is high compared to population growth in most other more-developed countries. About how much of the annual U.S. population growth is due to immigration? List six changes in lifestyles that have taken place in the United States during the 20th century, leading to a rise in per capita resource use. What is the end effect of such changes in terms of the U.S. ecological footprint?
5. List nine factors that affect birth rates and fertility rates. Explain why there are more boys than girls in some countries. Define **life expectancy** and **infant mortality rate** and explain how they affect the population size of a country. Why does the United States have a lower life expectancy and higher infant mortality rate than a number of other more-developed countries?
6. What is **migration**? What are environmental refugees and how quickly are their numbers growing? Describe three major factors that demographers have to consider in making population projections.

Section 6-3

7. What is the key concept for this section? What is the **age structure** of a population? Explain how age structure affects population growth and economic growth. Describe the American baby boom and some of its economic and social effects. What are some problems related to rapid population decline due to an aging population? How has the AIDS epidemic affected the age structure of some countries, especially in Africa?

Section 6-4

8. What is the key concept for this section? What is the **demographic transition** and what are its four stages? What factors could hinder some less-developed countries from making this transition?

9. Explain how the reduction of poverty and empowerment of women can help countries to slow their population growth. What is **family planning** and how can it help to stabilize populations? Describe India's efforts to control its population growth. Describe China's population control program and compare it with that of India.

10. What are this chapter's *three big ideas*? Summarize the story of human population growth and explain how the three scientific **principles of sustainability** (see Figure 1-2, p. 6 or back cover) can guide us in dealing with the problems that stem from population growth and decline.

Note: Key terms are in bold type.

Critical Thinking

1. Do you think that the global population of 7.1 billion (**Core Case Study**) is too large? Explain. If your answer was *yes*, what do you think should be done to slow human population growth? If your answer was *no*, do you believe that there is a population size that would be too big? Explain.

2. If you could say hello to a new person every second without taking a break and working around the clock, how many people could you greet in 1 day? How many in a year? How long would it take you to greet the 84 million people who were added to the world's population this year? How many years would it take you to greet all 7.1 billion people on the planet?

3. Which of the three major environmental worldviews summarized on p. 21 do you believe underlie the two major positions on whether the world is overpopulated (Science Focus 6.1)?

4. Should everyone have the right to have as many children as they want? Explain. Is your belief on this issue consistent with your environmental worldview?

5. Is it rational for a poor couple in a less-developed country such as India to have four or five children? Explain.

6. Identify a major local, national, or global environmental problem, and describe the role that population growth plays in this problem.

7. Some people believe the most important environmental goal is to sharply reduce the rate of population growth in less-developed countries, where at least 92% of the world's population growth is expected to take place between now and 2050. Others argue that the most serious environmental problems stem from high levels of resource consumption per person in more-developed countries, which have much larger ecological footprints per person than do less-developed countries. What is your view on this issue? Explain.

8. Experts have identified population growth as one of the major causes of the environmental problems we face. The population of the United States is growing faster than that of any other more-developed country. However, this fact is rarely discussed, and the U.S. government has no official policy for slowing U.S. population growth. Why do think this is so? Do you think there should be such a policy? If so, explain your thinking and list three steps you would take as a leader to slow U.S. population growth. If not, explain your thinking.

Doing Environmental Science

Prepare an age-structure diagram for your community. You will need to estimate how many people belong in each age category (see p. 131). To do this, interview a randomly drawn sample of the population to find out their ages and then divide your sample into age groups. (Be sure to interview equal numbers of males and females.) Then find out the total population of your community and apply the percentages for each age group from your sample to the whole population in order to make your estimates. Create your diagram and then use it to project future population trends. Write a report in which you discuss some economic, social, and environmental effects that might result from these trends.

Global Environment Watch Exercise

Find three different projections for the size of the global population in 2050 (**Core Case Study**). Explain how the projections were made. To do this, try to find out the assumptions behind each of the projections with regard to total fertility rates, crude death rates, infant mortality rates, life expectancies, and other factors. Based on your reading, choose the projection that you believe to be the closest to reality, and explain why you chose this projection.

Data Analysis

The chart below shows selected population data for two different countries, A and B. Study the chart and answer the questions that follow.

	Country A	Country B
Population (millions)	144	82
Crude birth rate (number of live births per 1,000 people per year)	43	8
Crude death rate (number of deaths per 1,000 people per year)	18	10
Infant mortality rate (number of babies per 1,000 born who die in first year of life)	100	3.8
Total fertility rate (average number of children born to women during their child-bearing years)	5.9	1.3
% of population under 15 years old	45	14
% of population older than 65 years	3.0	19
Average life expectancy at birth	47	79
% urban	44	75

1. Calculate the rates of natural increase (due to births and deaths, not counting immigration) for the populations of country A and country B. Based on these calculations and the data in the table, for each of the countries, suggest whether it is a more-developed country or a less-developed country and explain the reasons for your answers.
2. Describe where each of the two countries may be in the stages of demographic transition (Figure 6-16). Discuss factors that could hinder either country from progressing to later stages in the demographic transition.
3. Explain how the percentages of people under 15 years of age in each country could affect its per capita and total ecological footprints.

13 Water Resources

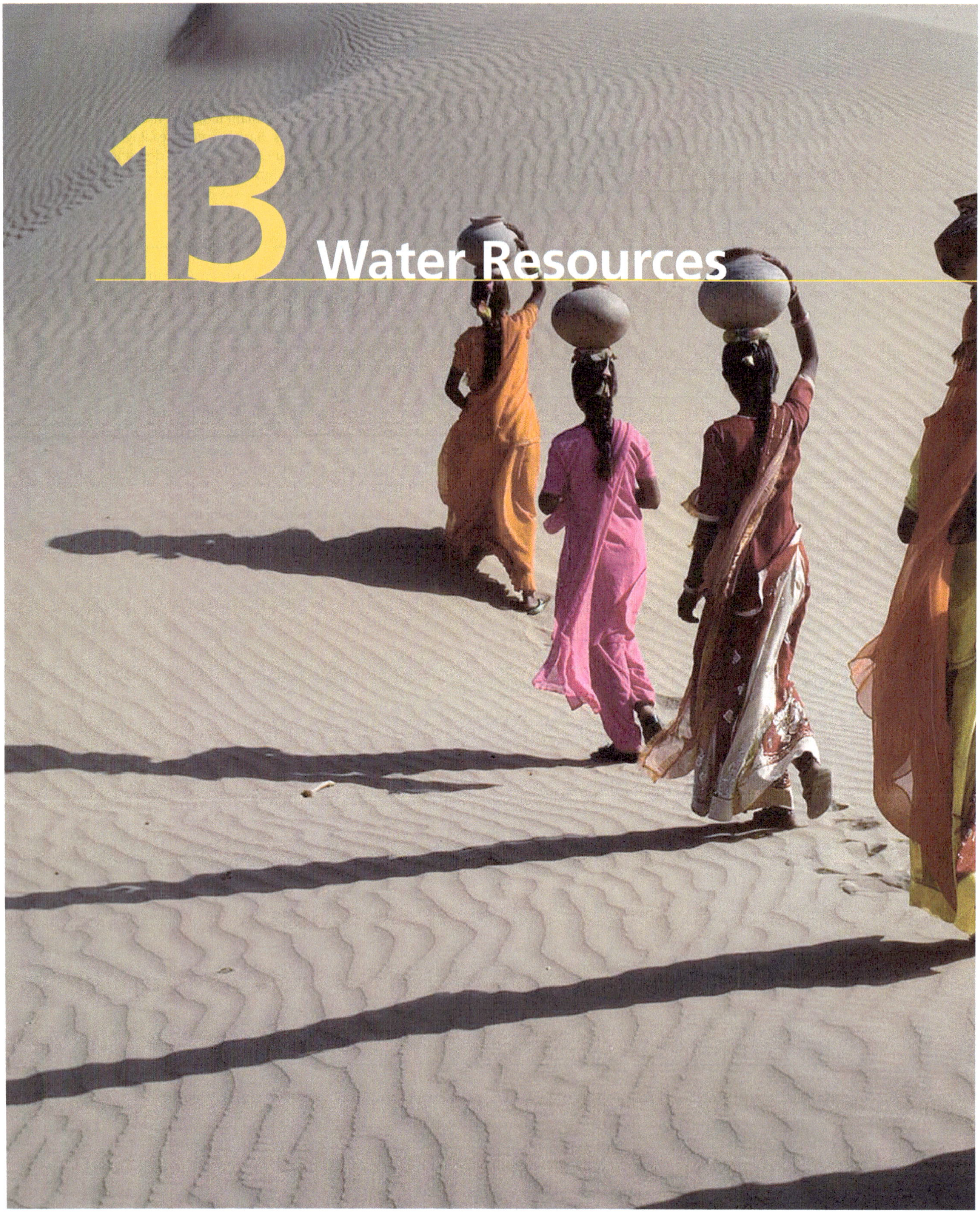

Our liquid planet glows like a soft blue sapphire in the hard-edged darkness of space. There is nothing else like it in the solar system. It is because of water.

JOHN TODD

Key Questions

13-1 Will we have enough usable water?

13-2 Is groundwater a sustainable resource?

13-3 Can surface water resources be expanded?

13-4 Can water transfers be used to expand water supplies?

13-5 Is desalination a useful way to expand water supplies?

13-6 How can we use freshwater more sustainably?

13-7 How can we reduce the threat of flooding?

Each day these women carry water to their village in a dry area of India.

Shiv Ji Joshi/National Geographic Creative

CORE CASE STUDY The Colorado River Story

© Cengage Learning

Figure 13-1 The *Colorado River basin:* The area drained by this river system is more than one-twelfth of the land area of the lower 48 states. This map shows 6 of the river's 14 dams.

The Colorado River, the major river of the arid southwestern United States, flows 2,300 kilometers (1,400 miles) through seven states to the Gulf of California (Figure 13-1). Most of its water comes from snowmelt in the Rocky Mountains. During the past 50 years, this once free-flowing river has been tamed by a gigantic plumbing system consisting of 14 major dams and reservoirs (Figure 13-2), and canals that help control flooding and supply water and electricity to farmers, ranchers, industries, and cities.

This system of dams and reservoirs provides water and electricity from hydroelectric plants at the major dams for roughly 30 million people in seven states—about one of every ten people in the United States. The river's water is used to produce about 15% of the nation's crops and livestock. The system also supplies water to some of the nation's driest and hottest cities. Take away this tamed river, and Las Vegas, Nevada, would be a mostly uninhabited desert area; San Diego and Los Angeles, California, could not support their present populations; and in California's Imperial Valley, vast fields of vegetables would eventually be replaced by cactus and mesquite plants.

So much water is withdrawn from this river to grow crops and support cities in this dry, desertlike climate that very little of it reaches the sea. To make matters worse, since 1999, the system has experienced severe **drought**, a prolonged period in which precipitation is at least 70% lower and evaporation is higher than normal. This overuse of the Colorado River illustrates the challenges faced by governments and people living in arid and semiarid regions with shared river systems, as population growth and economic growth place increasing demands on limited or decreasing supplies of surface water.

To many analysts, emerging shortages of water for drinking and irrigation in several parts of the world represent one of the major environmental challenges that we will face during this century.

PETE MCBRIDE/National Geographic Creative

Figure 13-2 The Glen Canyon Dam was built to create the Lake Powell reservoir on the Colorado River.

13-1 Will We Have Enough Usable Water?

CONCEPT 13-1A

We are using available freshwater unsustainably by wasting it, polluting it, and underpricing this irreplaceable natural resource.

CONCEPT 13-1B

Freshwater supplies are not evenly distributed, and one of every six people on the planet does not have adequate access to clean water.

Freshwater Is an Irreplaceable Resource That We Are Managing Poorly

We live on a planet that is unique to our solar system because of a precious layer of water—most of it saltwater—covering about 71% of its surface (see Figure 8-2, p. 169, and Figure 25, pp. S50–S51, Supplement 6). Look in the mirror. What you see is about 60% water, most of it inside your cells.

Water is an amazing and irreplaceable chemical with unique properties that help to keep us and other species alive (see Science Focus 3.2, p. 64). You could survive for several weeks without food, but for only a few days without **freshwater**, or water that contains very low levels of dissolved salts.

It takes huge amounts of water to supply us with food and with most of the other things that we use to meet our daily needs and wants. Water also plays a key role in determining and moderating the earth's climates, in removing and diluting some of the pollutants and wastes that we produce, and in sculpting the planet's surface to produce rock formations such as the Grand Canyon (Figure 13-3).

Freshwater is one of the earth's most important forms of natural capital, but despite its importance, it is also one of our most poorly managed resources. We use it inefficiently and pollute it, and it is available at too low a cost to billions of consumers. This encourages still greater waste and pollution of this resource, for which we have no substitute (**Concept 13-1A**).

Access to freshwater is a *global health issue*. The World Health Organization (WHO) estimates that an average of 9,300 people die each day from waterborne infectious diseases because they do not have access to safe drinking water.

Access to freshwater is also an *economic issue* because water is vital for producing food and energy and for reducing poverty. According to the United Nations, about 780 million people—almost 2.5 times the entire U.S. population—do not have access to clean water where they live. Because almost half of the world's people do not have water piped to their homes, people in many less-developed countries, usually women and children, have to find and carry daily supplies of freshwater from distant sources (see chapter-opening photo).

Water is also a *national and global security issue* because of increasing tensions within and between some nations over access to limited freshwater resources that they share.

Finally, water is an *environmental issue* because excessive withdrawal of freshwater from rivers and aquifers has resulted in falling water tables, decreasing river flows (**Core Case Study**), shrinking lakes, and disappearing wetlands. This, in combination with water pollution in many areas of the world, has degraded water quality, reduced fish populations, hastened the extinction of some aquatic species, and degraded aquatic ecosystem services (see Figure 8-5, p. 170, and Figure 8-14, p. 178).

Most of the Earth's Freshwater Is Not Available to Us

The earth is often called the blue planet because of its abundance of water in all forms: liquid, solid (ice), and gas (water vapor). However, only a tiny fraction—*about 0.024%* of the planet's enormous water supply—is readily available to us as liquid freshwater, stored in accessible

©kojihirano/Shutterstock.com

Figure 13-3 Water flowing in the Colorado River (**Core Case Study**) over a period of nearly 2 billion years created the spectacular Grand Canyon in the United States.

underground deposits and in lakes, rivers, and streams. The rest is in the salty oceans (about 96.5% of the earth's volume of liquid water), in frozen polar ice caps and glaciers (1.7%), and in underground aquifers (1.7%) (see Figure 25, pp. S50–S51, Supplement 6).

Fortunately, the world's freshwater supply is continually recycled, purified, and distributed in the earth's *hydrologic cycle* (see Figure 3-15, p. 63). This irreplaceable water recycling and purification system works well, unless we alter it, overload it with pollutants, or withdraw freshwater from underground and surface water supplies faster than it can be replenished.

We can also alter the hydrologic cycle through our influence on projected climate change. A 2012 study by oceanographer Paul Durack and his colleagues suggested that atmospheric warming has sped up the water cycle by about 4%, twice as fast as climate models have been projecting, and that by 2100, this acceleration could make the cycle 16–24% faster. As a result, on average, wet places will get wetter with more frequent flooding, dry places will get drier with more intense drought, and many storms will be more violent.

We have paid little attention to our effects on the water cycle mostly because we have thought of the earth's freshwater as essentially a free and infinite resource—an example of the tragedy of the commons (Chapter 1, p. 12). As a result, we have placed little or no economic value on the irreplaceable ecosystem services that water provides (**Concept 13-1A**), and this is a violation of the full-cost pricing **principle of sustainability** (see Figure 1-5, p. 9 or back cover).

On a global basis, we have plenty of freshwater, but it is not distributed evenly (**Concept 13-1B**). Differences in average annual precipitation (Figure 13-4) and economic resources divide the world's countries and people into water *haves* (such as Brazil, Figure 13-5, left) and *have-nots* (such as Saudi Arabia, Figure 13-5, right). For example, Canada, with only 0.5% of the world's population, has 20% of its liquid freshwater, while China, with 19% of the world's people, has only 7% of the supply.

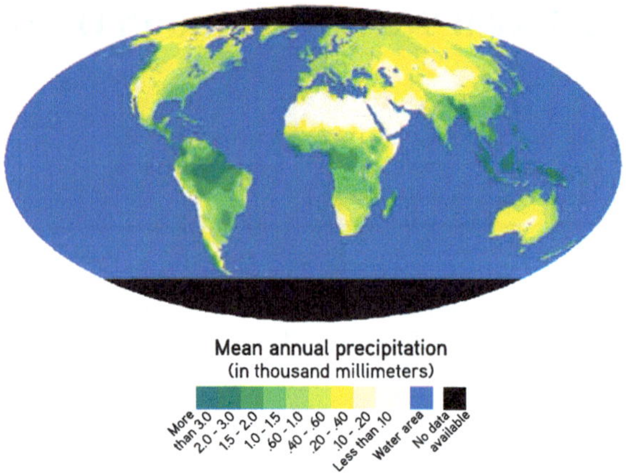

Figure 13-4 Global average annual precipitation.

National Geographic Maps/National Geographic Creative; data from NOAA/NESDIS/NCDC/Satellite Data Service

Groundwater and Surface Water Are Critical Resources

Some precipitation infiltrates the ground and percolates downward through spaces in soil, gravel, and rock until an impenetrable layer of rock or clay stops it. The freshwater in these spaces is called **groundwater**—a key component of the earth's natural capital.

The spaces in soil and rock close to the earth's surface hold little moisture. However, below a certain depth, in the **zone of saturation**, these spaces are completely filled with freshwater. The top of this groundwater zone is the **water table**. It falls in dry weather, or when we remove groundwater faster than nature can replenish it, and it rises in wet weather.

Deeper down are geological layers called **aquifers**, underground caverns and porous layers of sand, gravel, or rock through which groundwater flows. Mostly because of gravity, groundwater normally moves from points of high elevation and pressure to points of lower elevation and pressure. Some caverns have rivers of groundwater flowing through them. However, the porous layers of sand, gravel, or rock in most aquifers are like large, elongated sponges through which groundwater seeps—typically moving only a meter or so (about 3 feet) per year and rarely more than 0.3 meter (1 foot) per day. Watertight layers of rock or clay below such aquifers keep the freshwater from escaping deeper into the earth.

We use pumps to bring this groundwater to the surface for irrigating crops and supplying households and industries. Most aquifers are replenished naturally by precipitation that percolates downward through exposed soil and rock, a process called *natural recharge*. Others are recharged from the side by *lateral recharge* from nearby lakes, rivers, and streams.

According to the U.S. Geological Survey (USGS), there is 100 times more freshwater underground than there is in all of the world's rivers and lakes. However, most aquifers recharge extremely slowly, and because so much of the earth's landscape has been built on or paved over, freshwater can no longer penetrate the ground to recharge aquifers below many urban areas. In addition, in dry areas of the world, there is little precipitation available to recharge aquifers.

Aquifers that get very little, if any, recharge are called *nonrenewable aquifers*. They are found deep underground and were formed tens of thousands of years ago. Withdrawing freshwater from these aquifers amounts to *mining* a nonrenewable resource—an extreme example of the tragedy of the commons (Chapter 1, p. 12).

Another crucial resource is **surface water**, the freshwater from rain and melted snow that flows across the

Figure 13-5 Some countries have ample freshwater. The Iguassu River, marking part of the border between Brazil and Argentina, flows over the world's largest series of falls (left). Other countries such as Saudi Arabia (right) have very little freshwater.

Left: ©jose marques lopes/Shutterstock.com. Right: David Steele/Shutterstock.com.

earth's land surface and into lakes, wetlands, streams, rivers, estuaries, and the oceans. Precipitation that does not infiltrate the ground or return to the atmosphere by evaporation—about 39% of all precipitation—is called **surface runoff**. The land from which surface runoff drains into a particular stream, lake, wetland, or other body of water is called its **watershed**, or **drainage basin**. For example, the drainage basin for the Colorado River is shown in yellow and green on the map in Figure 13-1 (**Core Case Study**).

CONSIDER THIS. . .

CONNECTIONS Groundwater and Surface Water

There is usually a connection between surface water and groundwater because eventually, much groundwater naturally flows into rivers, lakes, estuaries, and wetlands. Thus, if we disrupt the hydrologic cycle by removing groundwater in a particular location faster than it is replenished, some nearby streams, lakes, and wetlands can dry up. This process degrades aquatic biodiversity and other ecosystem services.

We Are Using Increasing Amounts of the World's Reliable Runoff

According to *hydrologists* (scientists who study water and its movements above, on, and below the earth's surface), two-thirds of the annual surface runoff of freshwater into rivers and streams is lost in seasonal floods and is not available for human use. The remaining one-third is **reliable surface runoff**, which we can generally count on as a source of freshwater from year to year.

During the last century, the human population tripled, global water withdrawals increased sevenfold, and per capita withdrawals quadrupled. As a result, we now withdraw about 34% of the world's reliable runoff. This is a global average. In the arid American Southwest, up to 70% of the reliable runoff is withdrawn for human purposes, mostly for irrigation (**Core Case Study** and Case Study that follows). Some water experts project that because of a combination of population growth, rising rates of water use per person, longer dry periods, and failure to reduce unnecessary water losses, we may be withdrawing up to 90% of the reliable freshwater runoff by 2025.

Worldwide, we use 70% of the freshwater we withdraw each year from rivers, lakes, and aquifers to irrigate cropland and raise livestock; that figure rises to 90% in arid regions. Industry uses roughly another 20% of the water withdrawn each year, and cities and residences use the remaining 10%. Our **water footprint** is a rough measure of the volume of freshwater that we use directly and indirectly to stay alive and to support our lifestyles. Agriculture accounts for 92% of humanity's water footprint.

According to the American Water Works Association, the average American (see Case Study that follows) each day directly uses about 260 liters (69 gallons) of freshwater—enough water to fill about 1.7 typical bathtubs of water (a bathtub contains about 151 liters or 40 gallons of water). This water is used mostly for flushing toilets (27%), washing clothes (22%), taking showers (17%), and running faucets (16%), or is lost through leaking pipes, faucets, and other fixtures (14%).

We use even larger amounts of freshwater indirectly. This water is called **virtual water**, the freshwater that is not directly consumed but is used to produce food and other products. It makes up a large part of our water footprints, especially in more-developed nations. Producing and delivering a typical hamburger, for example, takes

These items...	take this many bathtubs full of water to produce
Cup of coffee	1
2-liter soda	3
Loaf of bread	4
1 dozen eggs	14
1 pound of cheese	15
Hamburger	16
T-shirt	17
Blue jeans	72
Small car	2,600
Medium-sized house	16,000

Figure 13-6 Producing and delivering a single one of each of the products shown here requires the equivalent of at least one and usually many bathtubs full of freshwater, called *virtual water. Note:* 1 bathtub = 151 liters (40 gallons).

(Compiled by the authors using data from UN Food and Agriculture Organization, UNESCO-IHE Institute for Water Education, World Water Council, and Water Footprint Network.)

about 2,400 liters (630 gallons) of freshwater—most of which is used to grow grain that is fed to cattle. This water would fill about 16 bathtubs.

Figure 13-6 shows one way to measure the amounts of virtual water used for producing and delivering products. These values can vary depending on how much of the supply chain is included, but they give us a rough estimate of the size of our water footprints.

Because of global trade in food and products, the virtual water used to produce and transport the wheat in a loaf of bread or the coffee beans used to make a cup of coffee (Figure 13-6) is often withdrawn as groundwater or surface water in another part of the world. For some water-short countries, it makes sense to save real freshwater by importing virtual water through food imports, instead of producing food domestically.

Large exporters of virtual water—mostly in the form of wheat, corn, soybeans, and other foods—are the European Union, the United States, Brazil, and Australia. Indeed, Brazil's supply of freshwater per person is more than 8 times the U.S. supply per person, 14 times China's supply, and 29 times India's supply. This explains why Brazil is likely to be one of the world's largest exporters of virtual water in the form of crops such as soybeans (see Figure 1-4, p. 8) to countries such as China.

However, prolonged severe droughts in parts of Australia, the United States, and the European Union are stressing the abilities of these countries to meet the growing global demand for their food exports. If climate change threatens freshwater supplies as is projected in some important crop-growing and food-exporting regions, freshwater scarcity will turn into food scarcity as virtual water supplies drop—another example of connections in nature.

CASE STUDY

Freshwater Resources in the United States

The United States has more than enough renewable freshwater to meet its needs. But it is unevenly distributed and much of it is contaminated by agricultural and industrial practices. The eastern states usually have ample precipitation, whereas many western and southwestern states have little (Figure 13-7).

According to a 2010 study by the USGS, the major uses of groundwater and surface freshwater in the United States are the cooling of electric power plants (41% of total water use), irrigation (37%), public water supplies (13%), industry (5%), and raising livestock (4%). In the eastern United States, most water is used for manufacturing and for cooling power plants (with most of the water heated and returned to its source). In many parts of this area, the most serious water problems are flooding, occasional water shortages as a result of pollution, and drought.

In the arid and semiarid regions of the western half of the United States (**Core Case Study**), irrigation counts for as much as 85% of freshwater use. Much of it is lost to evaporation and a great deal of it is used to grow thirsty crops. The major water problem in much of the western United States is a shortage of freshwater runoff caused by low precipitation (Figure 13-7), high evaporation, and recurring prolonged drought.

Water tables in many water-short areas, especially in the arid and semiarid western states, are dropping quickly

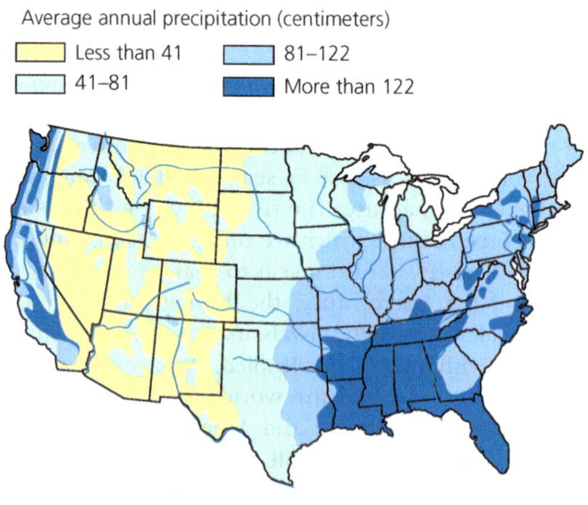

Figure 13-7 This map shows the long-term average annual precipitation and major rivers in the continental United States.

(Compiled by the authors using data from U.S. Water Resources Council and U.S. Geological Survey.)

as farmers and rapidly growing urban areas deplete many aquifers faster than they can be recharged. In 2007, the USGS projected that areas of at least 36 states are likely to face freshwater shortages by 2025 because of a combination of drought, rising temperatures, population growth, urban sprawl, and increased per-person use of water.

The U.S. Department of the Interior has mapped out *water hotspots* in 17 western states (Figure 13-8). In these areas, competition for scarce freshwater to support growing urban areas, irrigation, recreation, and wildlife could trigger intense political and legal conflicts between states and between rural and urban areas within states between now and 2030. Based on 19 global climate models used by the UN Intergovernmental Panel on Climate Change, climate researcher Richard Seager and his colleagues at Columbia University have projected that the southwestern United States and parts of northern Mexico are very likely to have long periods of extreme drought throughout most of the rest of this century.

The Colorado River system (Figure 13-1) will be directly affected by such drought. There are four major problems associated with the use of freshwater from the river (**Core Case Study**). *First,* the Colorado River basin includes some of the driest lands in the United States and Mexico. *Second,* legal pacts signed in 1922 and 1944 between the United States and Mexico allocated more freshwater for human use than the river can supply—even in rare years when there is no drought. In addition, the pacts allocated no water for protecting aquatic and terrestrial wildlife. *Third,* since 1960, the river has rarely flowed all the way to the Gulf of California because of its reduced water flow (due to many dams), increased freshwater withdrawals, and the prolonged drought. And *fourth,* the river receives enormous amounts of pollutants from urban areas, farms, animal feedlots, and industries, as it makes its way toward the sea.

Freshwater Shortages Will Grow

The main factors that cause water scarcity in any particular area are a dry climate, drought, too many people using a freshwater supply more quickly than it can be replenished, and inefficient use of freshwater. Figure 13-9 shows the current degree of *scarcity stress*—a measure based on a comparison of the amount of freshwater available with the amount used by humans—faced by each of the world's major river systems. Like the Colorado River (**Core Case Study**), major rivers including the Nile, Jordan, Yangtze, and Ganges are stressed so highly that their flows regularly dwindle to almost nothing in some locations.

More than 30 countries—most of them in the Middle East and Africa—now face stress from freshwater scarcity. By 2050, some 60 countries, many of them in Asia, with three-fourths of the world's population, are likely to be suffering from such freshwater stress. In 2006, the Chinese government reported that two-thirds of China's 600 major cities faced freshwater shortages.

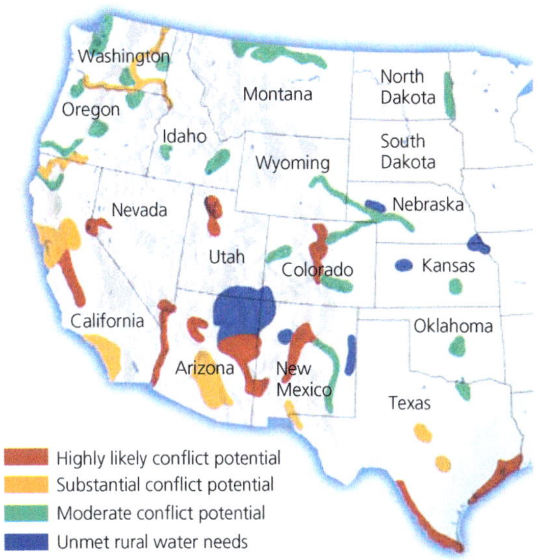

Figure 13-8 This map shows water scarcity hotspots in 17 western states that, by 2025, could face intense conflicts over scarce water needed for urban growth, irrigation, recreation, and wildlife. ***Question:*** Which, if any, of these areas are found in the Colorado River basin (**Core Case Study**)?

(Compiled by the authors using data from U.S. Department of the Interior and U.S. Geological Survey.)

Currently, about 30% of the earth's land area—a total area roughly 5 times the size of the United States—experiences severe drought. By 2059, as much as 45% of the earth's land surface—about 7 times the area of the United States—could experience extreme drought, mostly as a result of projected climate change caused by a warmer atmosphere, according to a 2007 study by climate researcher David Rind and his colleagues.

In each of 263 of the world's water basins, two or more countries share the available freshwater supplies. However, countries in only 158 of those basins have water-sharing agreements. This explains why conflicts among nations over shared freshwater resources are likely to happen more as populations grow, as demand for water increases, and as supplies shrink in many parts of the world.

In 2010, the United Nations reported that about 780 million people—one of every nine in the world—lacked regular access to enough clean water for drinking, cooking, and washing (Figure 13-10), mostly due to poverty (**Concept 1-B**). The report also projected that by 2025, at least 3 billion people—a number roughly equal to the current combined populations of China, India, and the United States—will likely lack such access to clean water. This helps to explain why many analysts view the likelihood of expanding water shortages in many parts of the world as one of the most serious environmental, health, and national security problems.

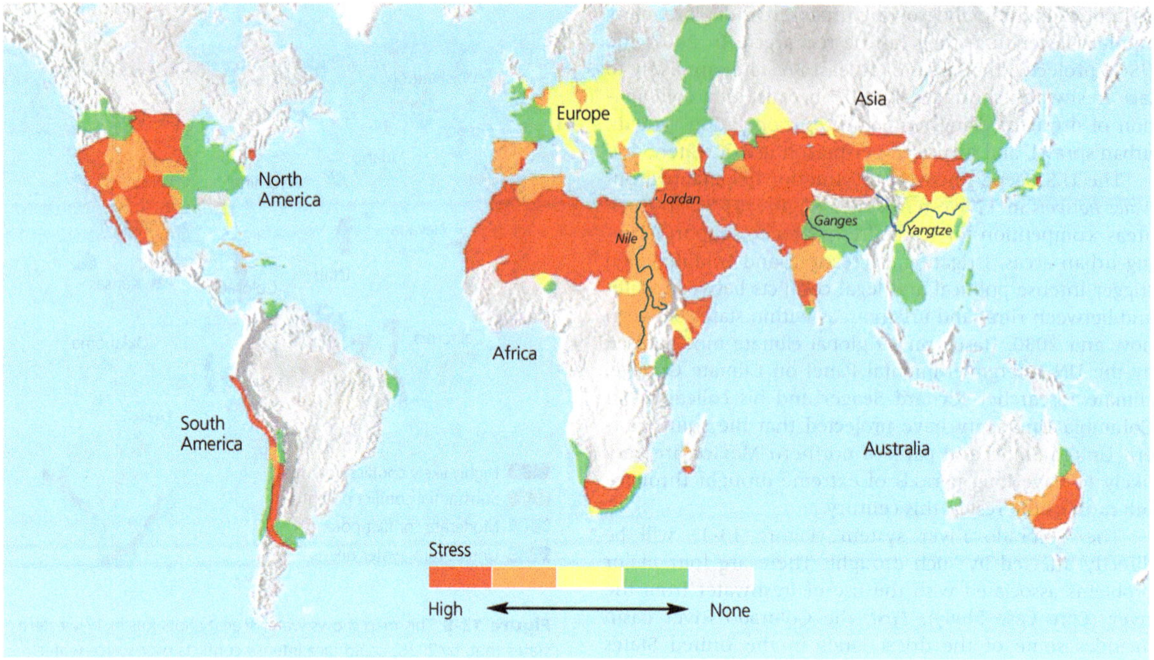

Figure 13-9 Natural capital degradation: The world's major river basins differ in their degree of freshwater *scarcity stress* (Concept 13-1B). ***Questions:*** If you live in a freshwater-stressed area, what signs of stress have you noticed? In what ways, if any, has it affected your life?

(Compiled by the authors using data from World Commission on Water Use for the 21st Century, UN Food and Agriculture Organization, and World Water Council.)

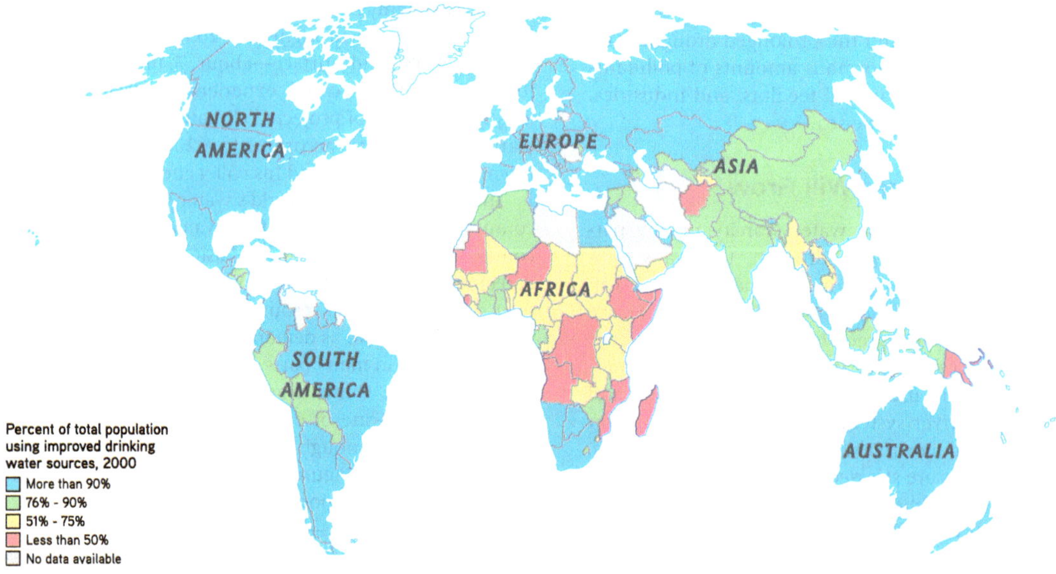

Figure 13-10 Access to freshwater.

National Geographic Maps/National Geographic Creative; World Health Organization World Health Statistics, 2010

SCIENCE FOCUS 13.1

USING SATELLITES TO MONITOR GROUNDWATER SUPPLIES

Figure 13-A These twin satellites measure slight variations in the earth's gravitational field that scientists use to detect and measure changes in large bodies of groundwater.

Scientists use a variety of methods to measure water supplies. For example, the USGS has drilled more than 7,000 observation wells throughout the United States to find and monitor groundwater supplies. Today some scientists are using satellites to measure changes in bodies of groundwater and surface water.

Since 2002, hydrologist Jay S. Famiglietti and his colleagues have been using two twin satellites (Figure 13-A) to measure small variations in the planet's gravitational pull that tell them about changes in ice cover, snow cover, surface water, soil moisture, and groundwater supplies. The two satellites, each the size of a small car, travel in the same orbit, one about 217 kilometers (135 miles) behind the other. They constantly beam microwaves at each other and can detect changes of less than the diameter of a human hair in this distance between the satellites. When the leading satellite speeds up, increasing the distance, it means the mass of the earth beneath the satellite has increased, tugging a bit harder on the satellite. Various forms of water content on and under the surface of the earth can cause such an increase.

By comparing thousands of these satellite measurements to data from ground measurements and computer modeling, Famiglietti and his team have learned what various changes in the data mean. In this way, they have calculated that large aquifers in many areas of the world are being drawn down rapidly. For example, in California's Central Valley, where groundwater is pumped to irrigate crops, Famiglietti estimated that between 2003 and 2010, underlying aquifers were drawn down by nearly 31 cubic kilometers (25 million acre-feet)—more than enough to fill Lake Powell (Figure 13-2), the second-largest reservoir in the United States.

According to the scientists, the satellite data indicate that large aquifers in several other areas, including North Africa, northeastern China, and northern India, are also being overpumped. These are all areas where water shortages are worsening every year.

Critical Thinking

What are some possible problems with relying on satellite data for measuring changes in water supplies?

CONSIDER THIS. . .

CONNECTIONS Virtual Water, Grain, and Hunger

Some water-short countries are reducing their irrigation water needs by importing grain, thereby freeing up more of their own water supplies for industrial and urban development. The result is a competition for the world's grain, which includes indirect competition for water (virtual water used to grow grain). In this global competition for water and grain, financially strong countries hold the advantage. As a result, grain prices are likely to rise, leading to food shortages and more widespread hunger among the poor.

There Are Several Ways to Increase Freshwater Supplies

What can we do to deal with the projected shortages of freshwater in many parts of the world? An important part of any strategy will be getting accurate information about where the water shortages are and where they are getting worse. Scientists have found a number of ways to obtain this information, including using satellites (Science Focus 13.1).

Several approaches to dealing with water shortages have been tried. They include increasing water supplies by withdrawing groundwater; building dams and reservoirs to store runoff in rivers for release as needed (**Core Case Study**); transporting surface water from one area to another; and converting saltwater to freshwater (desalination). The next four sections of this chapter look at these options. Another important solution is to reduce the unnecessary losses of freshwater, an option we consider in Section 13-6.

13-2 Is Groundwater a Sustainable Resource?

CONCEPT 13-2

Groundwater used to supply cities and grow food is being pumped from aquifers in some areas faster than it is renewed by precipitation.

Groundwater Is Being Withdrawn Faster Than It Is Replenished in Some Areas

Aquifers provide drinking water for nearly half of the world's people. Most aquifers are renewable resources unless the groundwater they contain becomes contaminated or is removed faster than it is replenished. In the United States, aquifers supply almost all of the drinking water in rural areas, one-fifth of that in urban areas, and 37% of the country's irrigation water. Relying more on groundwater has advantages and disadvantages (Figure 13-11).

Test wells and satellite data (Science Focus 13.1) indicate that water tables are falling in many areas of the world because the rate at which most of the world's aquifers are being pumped (mostly to irrigate crops) is greater than the rate of natural recharge from rainfall and snowmelt (**Concept 13-2**). The world's three largest grain producers—China, the United States, and India—as well as Mexico, Saudi Arabia, Iran, Yemen, Israel, and Pakistan, are overpumping many of their aquifers.

Figure 13-11 Withdrawing groundwater from aquifers has advantages and disadvantages. ***Questions:*** Which two advantages and which two disadvantages do you think are the most important? Why?

Top: ©Ulrich Mueller/Shutterstock.com

Every day, the world withdraws enough freshwater from aquifers to fill a convoy of large tanker trucks that could stretch 480,000 kilometers (300,000 miles)—well beyond the distance to the moon. Currently, more than 400 million people are being fed by grain produced through this unsustainable use of groundwater, according to the World Bank. This number is expected to grow.

The widespread drilling of inexpensive tube wells by farmers, especially in India and China, has accelerated aquifer overpumping. As water in aquifers is removed faster than it is renewed, water tables fall. Then farmers drill deeper wells, buy larger pumps, and use more energy to pump water to the surface. This process eventually depletes the groundwater in some aquifers or at least removes all the water that can be pumped at an affordable cost.

Saudi Arabia is as water-poor (Figure 13-5, right) as it is oil-rich. Much of its freshwater is pumped from ancient, nonrenewable deep aquifers to irrigate crops such as wheat grown on desert land (Figure 13-12) and to fill large numbers of fountains and swimming pools, which lose a great deal of water through evaporation into the hot, dry desert air. In 2008, Saudi Arabia announced that irrigated wheat production had largely depleted its major deep aquifer and said that it would stop producing wheat by 2016 and import grain (virtual water) to help feed its 30 million people.

In the United States, aquifer depletion is a growing problem, especially in the vast Ogallala aquifer (see Case Study that follows) and in California's arid Central Valley.

CASE STUDY

Overpumping the Ogallala

In the United States, groundwater is being withdrawn from aquifers, on average, 4 times faster than it is replenished, according to the USGS (**Concept 13-2**). Figure 13-13 shows the areas of greatest aquifer depletion. One of the most serious overdrafts of groundwater is in the lower half of the Ogallala, the world's largest known aquifer, which lies under eight Midwestern states from southern South Dakota to Texas (blowup section of Figure 13-13).

The gigantic Ogallala aquifer supplies about one-third of all the groundwater used in the United States and has

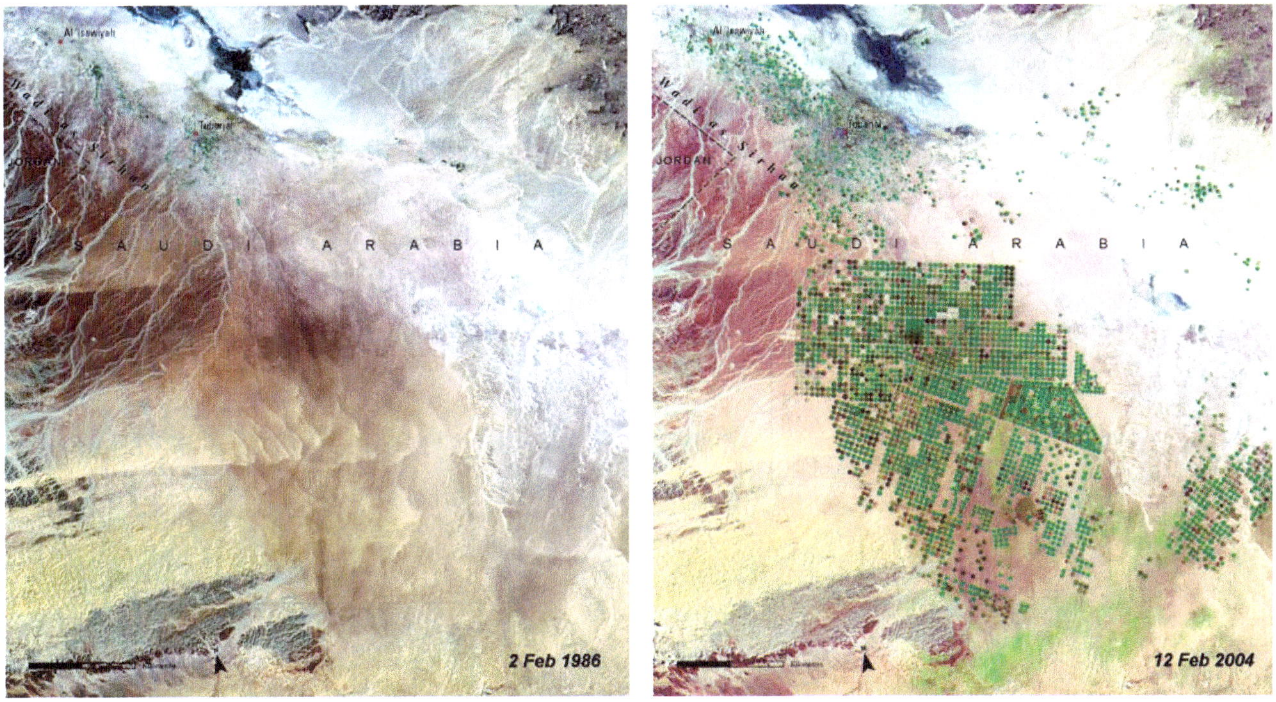

Figure 13-12 Natural capital degradation: Satellite photos of farmland irrigated by groundwater pumped from a deep aquifer in a vast desert region of Saudi Arabia between 1986 (left) and 2004 (right). Irrigated areas appear as green dots (each representing a circular spray system) and brown dots show areas where wells have gone dry and the land has returned to desert.

helped to turn the Great Plains into one of world's most productive irrigated agricultural regions (Figure 13-14). The problem is that the Ogallala is essentially a one-time deposit of liquid natural capital with a very slow rate of recharge.

In parts of the southern half of the Ogallala, groundwater is being pumped out 10–40 times faster than the slow natural recharge rate. This has lowered water tables and raised pumping costs, especially in northern Texas. In parts of Texas and Kansas, farmers cannot always pump enough water to meet their irrigation needs.

Government subsidies designed to increase crop production have encouraged farmers to grow water-thirsty crops in dry areas, which has accelerated depletion of the Ogallala. In particular, corn is a very thirsty crop that has been planted widely on fields watered by the Ogallala. Another resulting problem in addition to aquifer depletion is that nitrates from fertilizers and pesticides commonly sprayed on cornfields are now found in high concentrations in some parts of the aquifer.

The Ogallala also helps to support biodiversity. In various places, groundwater from the aquifer flows out of the ground onto land or onto lake bottoms through exit points called *springs*. In some cases, springs feed wetlands, which are vital habitats for many species, especially birds. When the water tables fall, many of these aquatic oases of biodiversity dry out.

Overpumping of Aquifers Can Have Harmful Effects

Overpumping of aquifers can contribute to limits on food production, rising food prices, and widening gaps between the rich and poor in some areas. As water tables drop, farmers must drill deeper wells, buy larger pumps, and use more electricity to run those pumps. Poor farmers cannot afford to do this and end up losing their land and working for richer farmers, or migrating to cities that are already crowded with poor people struggling to survive.

Withdrawing large amounts of groundwater sometimes causes the sand and rock that is held in place by water pressure in aquifers to collapse. This can cause the land above the aquifer to *subside* or sink, a phenomenon known as *land subsidence*. Extreme, sudden subsidence is sometimes referred to as a *sinkhole*. Once an aquifer becomes compressed by subsidence, recharge is impossible. In addition, land subsidence can damage roadways, water and sewer lines, and building foundations. Since 1925, overpumping of an aquifer to irrigate crops

Animated Figure 13-13 **Natural capital degradation:** This map shows areas of greatest aquifer depletion from groundwater overdraft in the continental United States. The blowup section (right) shows where water levels in the Ogallala aquifer have dropped sharply at its southern end beneath parts of Kansas, Oklahoma, Texas, and New Mexico. ***Questions:*** Do you depend on any of these aquifers for your drinking water? If so, what is the level of severity of overdraft where you live?

(Compiled by the authors using data from U.S. Water Resources Council and U.S. Geological Survey.)

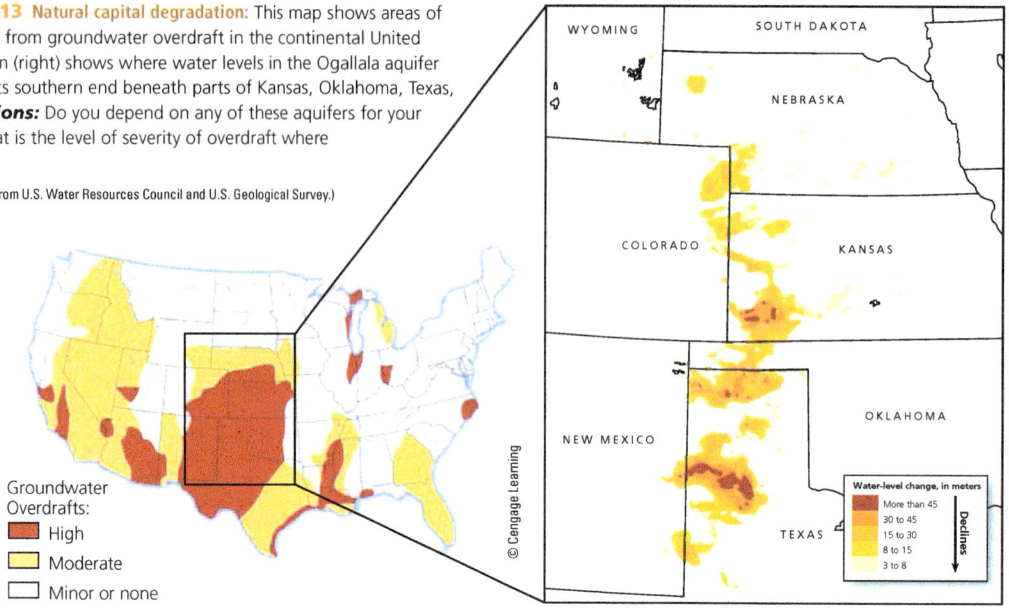

in California's San Joaquin Valley has caused half of the valley's land to subside by more than 0.3 meter (1 foot) and, in one area, by more than 8.5 meters (28 feet) (Figure 13-15).

Mexico City, built on an old lakebed, has one of the world's worst subsidence problems because its groundwater is being increasingly overpumped due to rapid population growth and urbanization. Some parts of the city have sunk as much as 10 meters (33 feet). Parts of Beijing, China, are also sinking as its water table has fallen 61 meters (200 feet) since 1965.

Groundwater overdrafts in coastal areas, where many of the world's largest cities and industrial areas are found, can pull saltwater into freshwater aquifers. The resulting contaminated groundwater is undrinkable and unusable for irrigation. This problem is especially serious in coastal areas of the U.S. states of California, Texas, Florida, Georgia, South Carolina, and New Jersey, as well as in coastal areas of Turkey, Thailand, and the Philippines.

If we do not sharply slow the depletion of aquifers, a growing number of the world's people will have to live on rainwater and suffer from food and drinking water shortages. Figure 13-16 lists ways to prevent or slow the problem of aquifer depletion by using this largely renewable resource more sustainably. **GREEN CAREER:** hydrogeologist

Deep Aquifers Might Be Tapped

With global shortages of freshwater looming, scientists are evaluating *deep aquifers* as future sources of freshwater. Preliminary results suggest that some of these aquifers hold enough freshwater to support billions of people for centuries. In addition, the quality of freshwater in these aquifers may be much higher than the quality of the freshwater in most rivers and lakes.

There are four major concerns about tapping these ancient deposits of freshwater. *First,* they are nonrenewable and cannot be replenished on a human time scale.

Figure 13-14 Center-pivot irrigation systems are moved around in circles to irrigate crops with water that is typically pumped from aquifers such as the Ogallala.

Dick Ireland/U.S. Geological Survey

Figure 13-15 This pole shows subsidence from overpumping of an aquifer for irrigation in California's San Joaquin Central Valley between 1925 and 1977. In 1925, this area's land surface was near the top of the pole. Since 1977, this problem has gotten worse.

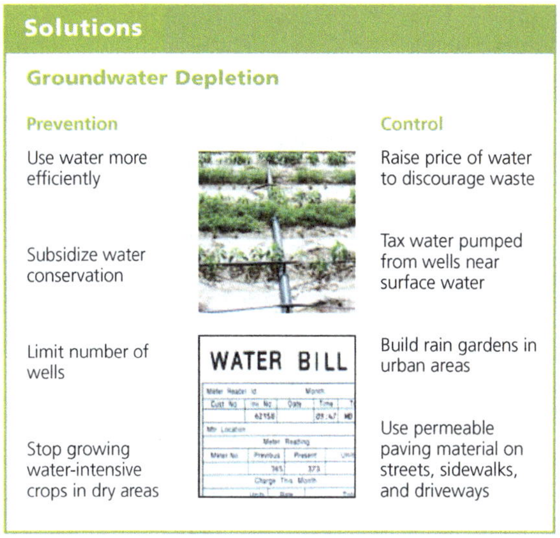

© Cengage Learning

Figure 13-16 There are a number of ways to prevent or slow groundwater depletion by using freshwater more sustainably. ***Questions:*** Which two of these solutions do you think are the most important? Why?

Top: © Anhong | Dreamstime.com. Bottom: Banol2007/Dreamstime.com.

Second, little is known about the geological and ecological impacts of pumping large amounts of freshwater from deep aquifers. *Third,* some deep aquifers flow beneath more than one country and there are no international treaties that govern rights to them. Without such treaties, wars could break out over this resource. *Fourth,* the costs of tapping deep aquifers are unknown and could be high.

13-3 Can Surface Water Resources Be Expanded?

CONCEPT 13-3

Large dam-and-reservoir systems have greatly expanded water supplies in some areas, but have also disrupted ecosystems and displaced people.

Use of Large Dams Provides Benefits and Creates Problems

A **dam** is a structure built across a river to control its flow. Usually, dammed water creates an artificial lake, or **reservoir**, behind the dam (Figure 13-2). The main goals of a dam-and-reservoir system are to capture and store the surface runoff from a river's watershed, and release it as needed to control floods; to generate electricity (hydropower); and to supply freshwater for irrigation and for towns and cities. Reservoirs also provide recreational activities such as swimming, fishing, and boating. Large dams and reservoirs provide benefits but have drawbacks (Figure 13-17).

The world's 45,000 large dams—those that are 15 meters (49 feet) or higher—capture and store about 14% of the world's surface runoff, provide water for almost half of all irrigated cropland, and supply more than half the electricity used in 65 countries. GOOD NEWS The United States has more than 70,000 large and small dams, capable of capturing and storing half of the country's entire river flow.

By using dams, we have increased the annual reliable runoff available for our uses by nearly 33%. As a result, the world's reservoirs now hold 3 to 6 times more freshwater than the total amount flowing at any moment in all of the world's natural rivers. On the down side, this engineering approach to river management has displaced 40–80 million people from their homes, flooded an area of mostly productive land totaling roughly the area of the

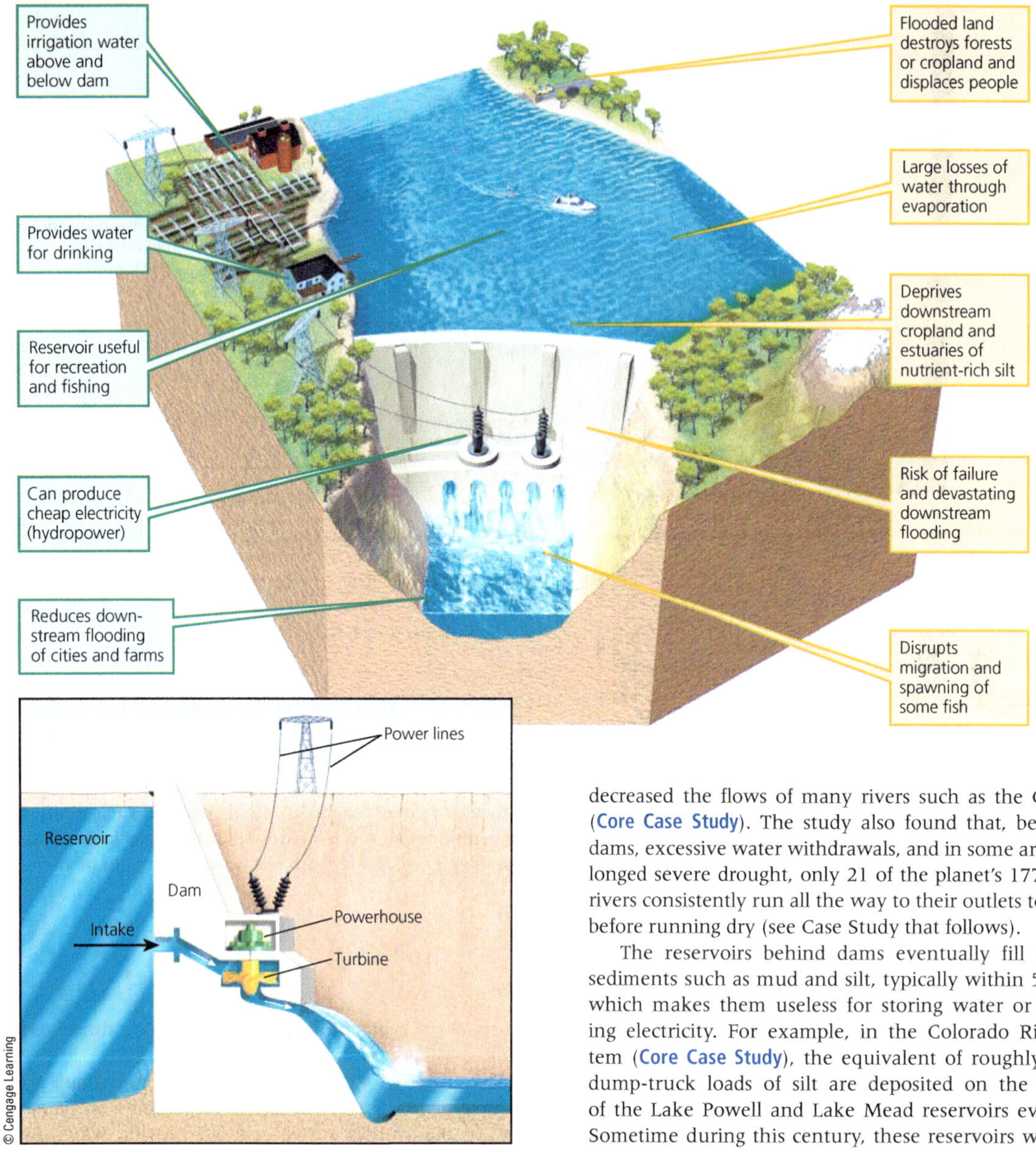

Figure 13-17 Trade-offs: Use of large dams and reservoirs has its advantages (green) and disadvantages (orange) (Concept 13-3). ***Questions:*** Which single advantage and which single disadvantage do you think are the most important? Why?

U.S. state of California, and impaired some of the important ecosystem services that rivers provide (see Figure 8-14, left, p. 178, and Concept 13-3).

A 2007 study by the World Wildlife Fund (WWF) estimated that about one out of five of the world's freshwater fish and plant species are either extinct or endangered, primarily because dams and water withdrawals have sharply decreased the flows of many rivers such as the Colorado (Core Case Study). The study also found that, because of dams, excessive water withdrawals, and in some areas, prolonged severe drought, only 21 of the planet's 177 longest rivers consistently run all the way to their outlets to the sea before running dry (see Case Study that follows).

The reservoirs behind dams eventually fill up with sediments such as mud and silt, typically within 50 years, which makes them useless for storing water or producing electricity. For example, in the Colorado River system (Core Case Study), the equivalent of roughly 20,000 dump-truck loads of silt are deposited on the bottoms of the Lake Powell and Lake Mead reservoirs every day. Sometime during this century, these reservoirs will probably be too full of silt to control floods or to store enough water for generating hydroelectric power. About 85% of all U.S. dam-and-reservoir systems will be 50 years old or more by 2025, and some of those aging dams are being removed because their reservoirs have filled with silt.

If climate change occurs as projected during this century, it will intensify shortages of water in many parts of the world. For example, mountain snows that feed the Colorado River system (Core Case Study) will melt faster and earlier, making less freshwater available to the river system when it is needed during hot summer months.

Nearly 3 billion people in South America, China, India, and other parts of Asia depend on river flows fed by mountain glaciers, which act like aquatic savings accounts. They store precipitation as ice and snow in wet periods and release it slowly during the dry season for use on farms and in cities. In 2010, according to the World Glacier Monitoring Service, many of these mountain glaciers had been shrinking for 19 consecutive years, mostly due to a warming atmosphere. For a while, their melting will likely increase water supplies. However, if these glaciers eventually disappear, most of those 3 billion people will likely be short of water for all purposes, including food production.

CASE STUDY

How Dams Can Kill an Estuary

Since 1905, the amount of water flowing to the mouth of the heavily dammed Colorado River (**Core Case Study**) has dropped dramatically. In most years since 1960, the river has dwindled to a small, sluggish stream by the time it reaches the Gulf of California.

This is the subject of an online video by National Geographic Explorer Alexandra Cousteau (see Individuals Matter 8.1, Chapter 8, p. 183), called *Death of a River.* In that film, Cousteau explains that the river once emptied into a vast jungle *delta,* the wetland area at the mouth of a river containing the river's estuary. The Colorado's delta covered an area of more than 800,000 hectares (2 million acres)—the size of the state of Rhode Island. It hosted forests, lagoons, and marshes rich in plant and animal life and supported a thriving coastal fishery for hundreds of years.

Since the damming of the Colorado—within one human lifetime—this rich ecosystem has collapsed and is now covered by mud flats and desert (Figure 13-18). The dams upstream have cut off the water supply that kept the system alive. All but one-tenth of the river's flow was diverted for use in seven U.S. states. Most of the remaining 10% is assigned to farms and to the growing cities of Mexicali and Tijuana in Mexico. The delta and its wildlife are now mostly gone and its coastal fishery that fed many generations of area residents is disappearing.

Historically, about 80% of the water withdrawn from the Colorado has been used to irrigate crops and raise cattle. That is because the government paid for the dams and reservoirs and has supplied many farmers and ranchers with water at low prices. These government subsidies have led to inefficient use of irrigation water for growing crops such as rice, cotton, and alfalfa that need a lot of water. However, much of this water is lost to the river system before it can be used. For example, there are huge losses of water in the Lake Mead and Lake Powell reservoirs due to evaporation and seepage of water into porous rock beds under the reservoirs.

PeteMcBride/National Geographic Creative

Figure 13-18 The Colorado River delta once contained a rich variety of forests, wetlands, and wildlife. Now it is covered mostly by mud flats and desert.

To deal with the water supply problems in the dry Colorado River basin, water experts call for the seven states using the river to enact and enforce strict water conservation measures and to slow population growth and urban development. They also call for phasing out state and federal government subsidies for agriculture in this region, shifting water-thirsty crops to less arid areas, and banning or severely restricting the use of surface water and groundwater to keep golf courses and lawns green in the desert areas of the Colorado River basin (Figure 13-1). They suggest that the best way to implement such solutions is to sharply raise the historically low price of the river's freshwater over the next decade—another application of the full-cost pricing **principle of sustainability**. According to Cousteau, if just 1% of the river's flow were restored to the delta area, much of it could be partially restored and its fisheries might begin to rebound.

CONSIDER THIS. . .

THINKING ABOUT The Colorado River

What are three steps you would take to deal with the problems of the Colorado River system?

13-4 Can Water Transfers Be Used to Expand Water Supplies?

CONCEPT 13-4

Transferring water from one place to another has greatly increased water supplies in some areas but has also disrupted ecosystems.

Water Transfers Can Be Inefficient and Environmentally Harmful

In some heavily populated dry areas of the world, governments have tried to solve water shortage problems by transferring water to the dry areas from water-rich areas. For example, northern China contains rapidly growing cities, including Beijing with 17.3 million people. These urban populations have helped to deplete underlying aquifers. According to the Chinese Academy of Sciences, two-thirds of China's 669 major cities have water shortages, more than 40% of its rivers are severely polluted, and about 300 million rural residents—a number almost equal to the size of the U.S. population—do not have access to safe drinking water. To help with this serious problem and to boost food production, the Chinese government is implementing its *South–North Water Diversion Project* to transfer 23 trillion liters (6 trillion gallons of water) per year from the Yangtze River in southern China to the thirsty north.

In other cases, water has been transferred to arid areas primarily to irrigate farm fields. For example, when you have lettuce in a salad in the United States, chances are good that it was grown with the use of heavy irrigation in the arid Central Valley of California. In fact, the *California State Water Project* (see map in Figure 13-19) is one of the world's largest freshwater transfer projects. It uses a maze of giant dams, pumps, and lined canals, or *aqueducts* (photo in Figure 13-19), to transport freshwater from the High Sierra Mountains of northeastern California to heavily populated cities and agricultural regions in water-poor central and southern California.

These water transfers have many benefits. Some 440 million Chinese will be served by the South–North Water Diversion Project. California's Central Valley supplies half of the United States' fruits and vegetables, and sprawling southern California cities including San Diego and Los Angeles use water transferred from the north. These areas would be mostly desert without the water transfers.

However, massive water transfers also involve high economic and social costs, large water losses through evaporation and leaks

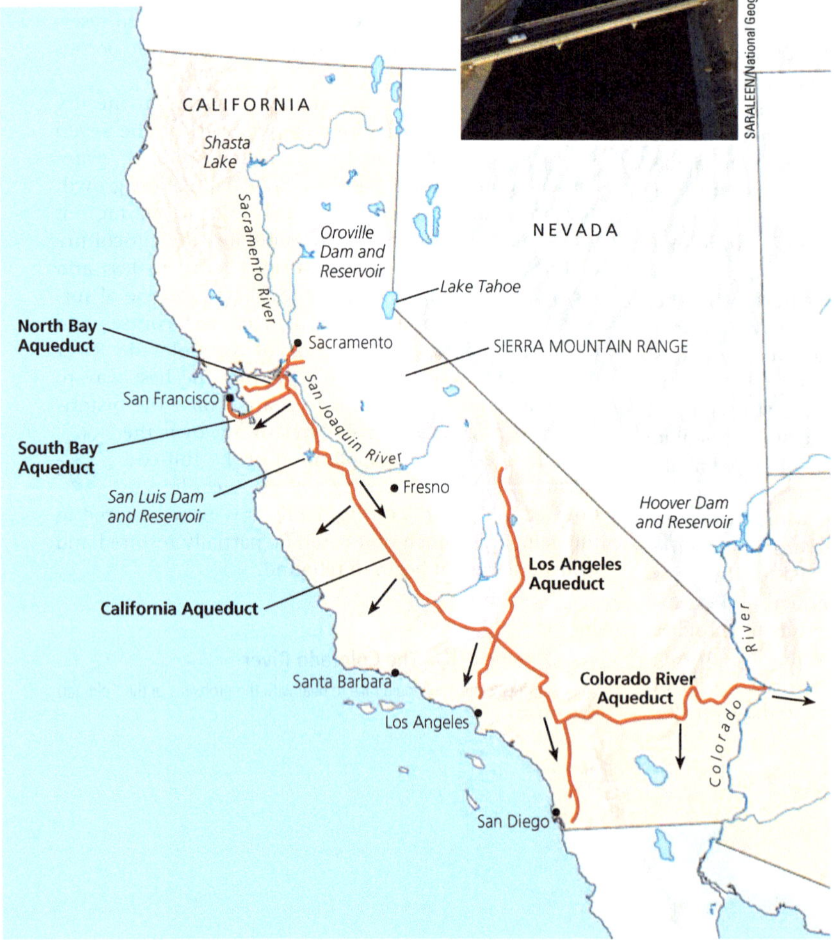

Figure 13-19 The California State Water Project transfers huge volumes of freshwater from one watershed to another. The arrows on the map show the general direction of water flow. The photo shows one of the aqueducts carrying water within the system. ***Questions:*** What effects might this system have on the areas from which the water is taken?

in the water-transfer systems, and degraded ecosystems in areas from which the water was taken (**Concept 13-4**). (See the Case Study that follows.) For example, China's massive water transfer will be quite expensive. To make way for the canals that will carry water, more than 350,000 villagers will have to relocate from lands they have farmed for generations to low-grade farmland or to urban areas. Also, scientists worry that removing huge volumes of water from the Yangtze River could severely damage its ecosystem, which is now suffering from its worst drought in 50 years.

In California, sending water south has degraded the Sacramento River, threatening fisheries and reducing the flushing action that helps to cleanse San Francisco Bay of pollutants. The bay has suffered from pollution, and the flow of freshwater to its coastal marshes and other ecosystems has dropped, putting stress on wildlife species that depend on these ecosystems. Water was also diverted from streams that flow into Mono Lake, an important feeding stop for migratory birds on the eastern slope of the High Sierras. This lake experienced an 11-meter (35-foot) drop in its water level until the diversions were stopped. The water level is now rising again, but Mono Lake's entire ecosystem was in jeopardy. Before that time, Owens Lake, another large lake in the area, was completely drained by the water transfer and its ecosystem was essentially destroyed.

One reason why such damaging water transfers have taken place is that governments have subsidized the costs of such transfers for inefficient uses such as the irrigation of lettuce and other crops that need lots of water in desertlike areas. In central California, agriculture consumes three-fourths of the water that is transferred, and much of it is lost through inefficient irrigation systems. Studies have shown that making irrigation just 10% more efficient would provide all the water necessary for domestic and industrial uses in southern California. Yet, inefficient use of the water continues because taxpayer subsidies make the water inexpensive to farmers and other users. Many scientists have argued for replacing these subsidies with subsidies that would encourage more efficient use of water.

In 2012, the U.S. Bureau of Reclamation was evaluating a proposal to build a 970-kilometer-long (600-mile-long) pipeline to transfer water from the Missouri and Mississippi Rivers to reservoirs in Denver, Colorado, to help supplement the water supply in the Colorado River basin (**Core Case Study**). However, the loss of so much water from these rivers, which require high water flows to sustain large vessel navigation, would face strong political opposition, and it would have harmful effects on these river ecosystems.

According to several studies, projected climate change will make matters worse in many areas where water is being removed for transfers. In southern China, climate change could worsen the drought and create a need for an even larger and more expensive transfer of water. In northern California, many people depend on *snowpacks*, bodies of densely packed, slowly melting snow, in the High Sierras for more than 60% of their freshwater during the hot, dry summer months. Projected atmospheric warming could shrink the snowpack by as much as 40% by 2050 and by as much as 90% by the end of this century. This will sharply reduce the amount of freshwater available for transfer to central and southern California.

CASE STUDY

The Aral Sea Disaster: A Glaring Example of Unintended Consequences

The shrinking of the Aral Sea (Figure 13-20) is the result of a large-scale freshwater transfer project in an area of the former Soviet Union with the driest climate in central Asia. Since 1960, enormous amounts of irrigation water have been diverted from the two rivers that supply water to the Aral Sea. The goal was to create one of the world's largest irrigated areas, mostly for raising cotton and rice. The irrigation canal, the world's longest, stretches more than 1,300 kilometers (800 miles)—roughly the distance

1976

2012

Figure 13-20 Natural capital degradation: The *Aral Sea*, straddling the borders of Kazakhstan and Uzbekistan, was one of the world's largest saline lakes. These satellite photos show the sea in 1976 (left) and in 2012 (right). ***Question:*** What do you think should be done to help prevent further shrinkage of the Aral Sea?

Vladb/Dreamstime.com

Figure 13-21 *Boat cemetery:* These fishing boats, which once plied the Aral Sea, are now stranded in the desert formed by the shrinkage of the sea.

between the two U.S. cities of Boston, Massachusetts, and Chicago, Illinois.

This large-scale freshwater diversion project, coupled with drought and high evaporation rates due to the area's hot and dry climate, has caused a regional ecological and economic disaster. Since 1961, the sea's salinity has risen sevenfold and the average level of its water has dropped by an amount roughly equal to the height of a six-story building. The Southern Aral Sea has lost 90% of its volume of water and is a remnant of the original sea lying in a salt-covered desert (Figure 13-20, right). Water withdrawal for agriculture has reduced the two rivers feeding the sea to mere trickles.

About 85% of the area's wetlands have been eliminated and about half the local bird and mammal species have disappeared. A huge area of the former Southern Aral Sea's lake bottom is now a human-made desert covered with glistening white salt. The sea's greatly increased salt concentration—3 times saltier than ocean water—has caused the presumed local extinction of 26 of the area's 32 native fish species. This has devastated the area's fishing industry, which once provided work for more than 60,000 people. Fishing villages and boats once located on the sea's coastline now sit abandoned in a salty desert (Figure 13-21).

Winds pick up the sand and salty dust and blow it onto fields as far as 500 kilometers (310 miles) away. As the salt spreads, it pollutes water and kills wildlife, crops, and other vegetation. Aral Sea dust settling on glaciers in the Himalayas is causing them to melt at a faster than normal rate—a prime example of unexpected connections and unintended consequences.

The shrinkage of the Aral Sea has also altered the area's climate. The once-huge sea acted as a thermal buffer that moderated the heat of summer and the extreme cold of winter. Now there is less rain, summers are hotter and drier, winters are colder, and the growing season is shorter. The combination of such climate change and severe salinization has reduced crop yields by 20–50% on almost one-third of the area's cropland—the opposite of the project's intended consequences.

Since 1999, the United Nations, the World Bank, and the five countries surrounding the lake have worked to improve irrigation efficiency. They have also partially replaced thirsty crops with other crops that require less irrigation water. Because of a dike that blocks the flow of water from the Northern Aral Sea into the southern sea, the level of the northern sea has risen by 2 meters (7 feet), its salinity has dropped, and dissolved oxygen levels are up. It now supports a healthy fishery that produced 5 times as big a catch (by total weight) in 2011 as it had produced in 2005.

However, the formerly much larger southern sea is still shrinking. By 2012, its eastern lobe was essentially gone, and the European Space Agency projects that the rest of the Southern Aral Sea could dry up completely by 2020.

13-5 Is Desalination a Useful Way to Expand Water Supplies?

CONCEPT 13-5

We can convert salty ocean water to freshwater, but the cost is high, and the resulting salty brine must be disposed of without harming aquatic or terrestrial ecosystems.

Removing Salt from Seawater Is Costly and Has Harmful Effects

Desalination is the process of removing dissolved salts from ocean water or from brackish (slightly salty) water in aquifers or lakes. It is another way to increase supplies of freshwater (**Concept 13-5**).

The two most widely used methods for desalinating water are distillation and reverse osmosis. *Distillation* involves heating saltwater until it evaporates (leaving behind salts in solid form) and condenses as freshwater. *Reverse osmosis* (or *microfiltration*) uses high pressure to force saltwater through a membrane filter with pores small enough to remove the salt and other impurities.

In 2010, according to the International Desalination Association, there were more than 15,000 desalination plants (250 in the United States, half of them in Florida) operating in more than 125 countries, most of them arid nations of the Middle East, North Africa, the Caribbean Sea, and the Mediterranean Sea. Desalination supplies less than 0.3% of the world's demand and only about 0.4% of the U.S. demand for freshwater.

There are three major problems with the widespread use of desalination. *First* is the high cost, because it takes a lot of increasingly expensive energy to remove salt from seawater. A *second* problem is that pumping large volumes of seawater through pipes and using chemicals to sterilize the water and keep down algae growth kills many marine organisms and also requires large inputs of energy (and thus money). A *third* problem is that desalination produces huge quantities of salty wastewater that must go somewhere. Dumping it into nearby coastal ocean waters increases the salinity of those waters, which can threaten food resources and aquatic life, especially if it is dumped near coral reefs, marshes, or mangrove forests. Disposing of it on land could contaminate groundwater and surface water (**Concept 13-5**).

Currently, desalination is practical only for water-short, wealthy countries and cities that can afford its high cost. However, scientists and engineers are working to develop better and more affordable desalination technologies (Science Focus 13.2).

13-6 How Can We Use Freshwater More Sustainably?

CONCEPT 13-6

We can use freshwater more sustainably by cutting water waste, raising water prices, slowing population growth, and protecting aquifers, forests, and other ecosystems that store and release freshwater.

Reducing Freshwater Losses Can Provide Many Benefits

According to water resource expert Mohamed El-Ashry of the World Resources Institute, about 66% of the freshwater used in the world and about 50% of the freshwater used in the United States is lost through evaporation, leaks, and inefficient use. El-Ashry estimates that it is economically and technically feasible to reduce such losses to 15%, thereby meeting most of the world's freshwater needs for the foreseeable future.

So why do we allow for such large losses of freshwater? According to water resource experts, there are two major reasons. First, the *cost of freshwater to users is low.* Such underpricing is mostly the result of government subsidies that provide irrigation water, or the electricity and diesel fuel used by farmers to pump freshwater from rivers and aquifers, at below-market prices. Because these subsidies keep freshwater prices artificially low, due to a failure to apply the full-cost pricing **principle of sustainability** (see Figure 1-5, p. 9 or back cover), users have little or no financial incentive to invest in water-saving technologies. However, farmers, industries, and others who benefit from government freshwater subsidies argue that the subsidies promote the farming of unproductive land, stimulate local economies, and help to keep the prices of food, manufactured goods, and electricity low.

CONSIDER THIS. . .

THINKING ABOUT Government Freshwater Subsidies

Should governments provide subsidies to farmers and cities to help keep the price of freshwater low? Explain.

Higher prices for freshwater encourage water conservation but make it difficult for low-income farmers and city dwellers to buy enough freshwater to meet their needs. When South Africa raised freshwater prices, it dealt with this problem by establishing *lifeline rates,* which give each household a set amount of free or low-priced freshwater to meet basic needs. When users exceed this amount, they pay higher prices as their freshwater use increases. This is a *user-pays* approach.

SCIENCE FOCUS 13.2

THE SEARCH FOR IMPROVED DESALINATION TECHNOLOGY

Reverse osmosis is the favored desalination technology because it requires much less energy than distillation, but it is still energy intensive. In this process, high pressure is applied to seawater in order to squeeze the freshwater out of it. The membrane that filters out the salt must be strong enough to withstand such pressure. Seawater must be pretreated and treated again after desalination to make it pure enough for drinking and for irrigating crops.

Much of the scientific research in this field is aimed at improving the membrane and the pre- and post-treatment processes to make desalination more energy efficient. Scientists are working to develop new, more efficient and affordable membranes that can separate freshwater from saltwater under lower pressure, which would require less energy. According to the World Business Council, since 1995, technological advances have brought the cost of desalination down by 80%, but this is still not affordable enough to make it useful for large-scale irrigation or to meet much of the world's demand for drinking water.

Scientists are considering ways to use solar and wind energy—applying one of the three scientific **principles of sustainability**—in combination with conventional power sources to help bring down the cost of desalinating seawater. In 2012, Saudi Arabia completed the world's largest solar-powered desalination project. It uses concentrated solar energy to power new filtration technology at a plant that will meet the daily water needs of 100,000 people.

Two Australian companies, Energetech and H2AU, have joined forces to build an experimental desalination plant that uses the power generated by ocean waves to drive reverse-osmosis desalination. The plant is located on a barge anchored offshore on the Australian coast. This approach produces no air pollution and conserves energy by using it where it is produced to desalinate water. Thus, costs should be much lower than for other desalination methods.

Some scientists argue for building fleets of such floating desalination plants. These plants would operate out of sight from coastal areas and transfer the water to shore through seabed pipelines or in food-grade shuttle tankers. Because of their distance from shore, the ships could draw water from depths below where most marine organisms are found. The resulting brine could be returned to the ocean and diluted far away from coastal waters, thus helping to solve the brine pollution problem caused by onshore desalination.

Another way to save energy in desalination is to build the plants near coastal electric power plants that use seawater for cooling. The heated water from the power plants could be fed into the desalination process, with some of the waste heat from the power plants used to help power the desalination plants. Scientists estimate that this approach could reduce the cost of today's most efficient desalination process by about 17%.

Despite the still high costs of desalination, analysts expect it to be used more widely in the future as water shortages become worse, especially in arid regions of the world. Given the high costs and other obstacles, however, scientists and engineers still have a lot of work to do before desalination will be a major source of freshwater.
GREEN CAREER: desalination engineer

Critical Thinking

Do you think that improvements in desalination will justify highly inefficient uses of water, such as maintaining swimming pools, fountains, and golf courses in desert areas? Explain.

The second major cause of unnecessary losses of freshwater is a *lack of government subsidies for improving the efficiency of freshwater use.* Withdrawing some of the subsidies that encourage high rates of freshwater use and replacing them with subsidies for more efficient freshwater use would sharply reduce freshwater losses. Understandably, farmers and industries that receive subsidies for freshwater use have vigorously and successfully opposed efforts to eliminate or reduce them.

We Can Improve Efficiency in Irrigation

Since 1980, the amount of food that can be grown per drop of water has roughly doubled. And since the 1970s, the amount of water used per person in the United States has dropped by about 33%, after rising for decades. Most of these water savings have come from improvements to irrigation efficiency in the United States and other more-developed countries. GOOD NEWS

However, there is still a long way to go, especially in less-developed countries. Only about 60% of the world's irrigation water reaches crops, which means that most irrigation systems are highly inefficient. The most inefficient irrigation system, commonly used in less-developed countries, is *flood irrigation,* in which water is pumped from a groundwater or surface water source through unlined ditches where it flows by gravity to the crops being watered (Figure 13-22, left). This method delivers far more water than is needed for crop growth and typically, about 45% of this water is lost through evaporation, seepage, and runoff.

Another inefficient system is the traditional spray irrigation system, a widely used tool of industrialized crop pro-

Figure 13-22 Traditional irrigation methods rely on gravity and flowing water (left). Newer systems such as center-pivot, low-pressure sprinkler irrigation (right) and drip irrigation (center) are far more efficient.

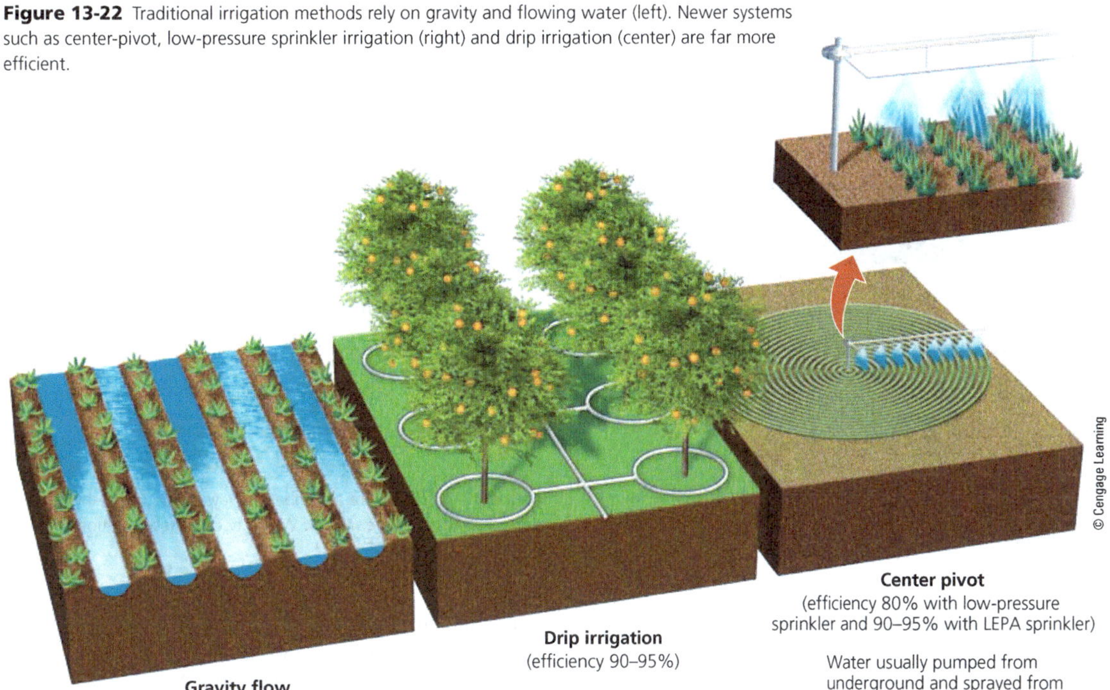

duction (Figure 13-23). It sprays huge volumes of water onto large fields, and as much as 40% of this water is lost to evaporation, especially in dry and windy areas, according to the U.S. Geological Survey. These systems are commonly used in the Midwestern United States and have helped to draw down the Ogallala aquifer (Figure 13-13).

More efficient irrigation technologies greatly reduce water losses by delivering water more precisely to crops—a *more crop per drop* strategy. For example, the *center-pivot, low-pressure sprinkler* (Figure 13-22, right), which uses pumps to spray water on a crop, allows about 80% of the water to reach crops. *Low energy, precision application (LEPA) sprinklers*, another form of center-pivot irrigation, put 90–95% of the water where crops need it.

Drip, or *trickle irrigation*, also called *microirrigation* (Figure 13-22, center) is the most efficient way to deliver small amounts of water precisely to crops. It consists of a network of perforated plastic tubing installed at or below the ground level. Small pinholes in the tubing deliver drops of water at a slow and steady rate, close to the roots of individual plants. These systems drastically reduce water waste because 90–95% of the water input reaches the crops.

Drip irrigation is used on just over 1% of the world's irrigated crop fields and on only about 4% of those in the United States, largely because most drip irrigation systems are costly. This percentage rises to 13% in the U.S. state of California, 66% in Israel, and 90% in Cyprus. However, if freshwater were priced closer to the value of the ecosystem services it provides and if government subsidies for inefficient use of freshwater were reduced or eliminated, drip irrigation could more easily be used to irrigate most of the world's crops.

According to the United Nations, reducing the current global withdrawal of water for irrigation by just 10%

Figure 13-23 Movable spray irrigation systems, widely used to irrigate crops, are inefficient.

Courtesy of International Development Enterprises

Figure 13-24 Solutions: In areas of Bangladesh and India, where water tables are high, many small-scale farmers use treadle pumps to supply irrigation water to their fields.

Solutions

Reducing Irrigation Water Losses

- Avoid growing thirsty crops in dry areas
- Import water-intensive crops and meat
- Encourage organic farming and polyculture to retain soil moisture
- Monitor soil moisture to add water only when necessary
- Expand use of drip irrigation and other efficient methods
- Irrigate at night to reduce evaporation
- Line canals that bring water to irrigation ditches
- Irrigate with treated wastewater

Figure 13-25 There are a number of ways to reduce freshwater losses in irrigation. ***Questions:*** Which two of these solutions do you think are the best ones? Why?

would save enough water to grow crops and meet the estimated additional water demands of the earth's cities and industries through 2025.

Poor Farmers Conserve Water Using Low-Tech Methods

Many of the world's poor farmers use small-scale, low-cost traditional irrigation technologies that are far more sustainable than most large-scale irrigation systems. For example, millions of farmers in Bangladesh and other countries where water tables are high use human-powered treadle pumps to bring groundwater up to the earth's surface and into irrigation ditches (Figure 13-24). These wooden devices are inexpensive (about $25) and easy to build from local materials. One such pump developed by the nonprofit International Development Enterprises (IDE) uses 60–70% less water than a conventional gravity-flow system to irrigate the same amount of cropland.

Other farmers in some less-developed countries use buckets, small tanks with holes, or simple plastic tubing systems for drip irrigation. The cost of a small-scale drip irrigation system developed by IDE is one-tenth as much per unit of land area as that of conventional drip systems (see Individuals Matter 12.1, p. 303).

Rainwater harvesting is another simple and inexpensive way to provide water and is used throughout much of the developing world. It involves using pipes from rooftops, and channels dug in the ground, to direct rainwater that would otherwise run off the land. It can be stored in underground or aboveground storage tanks (cisterns), ponds, and plastic barrels for use during dry seasons. This is especially useful in countries such as India, where much of the rain comes in a short monsoon season.

Farmers in the parched agricultural state of Punjab in India are using a newer tool called a *tensiometer*, a simple tool that measures soil moisture. In a 2011 study, Columbia University scientists provided these devices to 500 Punjabi farmers and requested that they irrigate their crops only when the tensiometer showed soil moisture content to be falling. Those who followed these instructions used 30% less water than they had before using the tool with the same results. The experiment is now being expanded to include 5,000 more farmers.

Other strategies used by poor farmers to increase the amount of crop per drop of rainfall include polyculture farming to create more canopy cover and reduce evaporative water losses; planting deep-rooted perennial crop varieties (see Figure 12-C, p. 310); controlling weeds; and mulching fields to retain more moisture.

Figure 13-25 summarizes several ways to reduce freshwater losses in crop irrigation. Since 1950, Israel has used many of these techniques to slash irrigation water losses by 84% while irrigating 44% more land. Israel now treats and reuses 30% of its municipal sewage water for crop production and plans to increase this to 80% by 2025. The government also gradually eliminated most freshwater subsidies to raise Israel's price of irrigation water, which is now one of the highest in the world.

We Can Cut Freshwater Losses in Industries and Homes

Producers of chemicals, paper, oil, coal, primary metals, and processed foods consume almost 90% of the freshwater used by industries in the United States. Some of these industries recapture, purify, and recycle water to reduce their water use and water treatment costs.

Paul Senyszyn/Escaflowne/iStockphoto.com

Figure 13-26 This yard in Encinitas, a city in a dry area of southern California (USA), uses a mix of plants that are native to the arid environment and require little watering.

For example, more than 95% of the freshwater used to make steel can be recycled. Even so, most industrial processes could be redesigned to use much less freshwater. **GREEN CAREER:** water conservation specialist

Flushing toilets with freshwater (most of it clean enough to drink) is the single largest use of domestic freshwater in the United States and accounts for about one-fourth of home water use. Since 1992, U.S. government standards have required that new toilets use no more than 6.1 liters (1.6 gallons) of freshwater per flush. Even at this rate, just two flushes of such a toilet require more than the daily amount of freshwater available for all uses to many of the world's poor living in arid regions (see chapter-opening photo).

Other water-saving appliances are widely available. Low-flow showerheads can save large amounts of freshwater by cutting the flow of a shower in half. Front-loading clothes washers use 30% less water than top-loading machines use. According to the American Water Works Association, if the typical American household were to stop all water leaks and use these devices, along with low-flow toilets and faucets, it could cut its daily water use by nearly a third.

According to UN studies, 30–60% of the freshwater supplied in nearly all of the world's major cities in less-developed countries is lost, primarily through leakage from water mains, pipes, pumps, and valves. Water experts say that fixing these leaks should be a high priority for water-short countries, because it would cost less than building dams or importing freshwater. Even in advanced industrialized countries such as the United States, these losses to leakage average 10–30%. However, leakage losses have been reduced to about 3% in Copenhagen, Denmark, and to 5% in Fukuoka, Japan.

CONSIDER THIS. . .

CONNECTIONS Water Leaks and Water Bills

Any water leak wastes freshwater and raises water bills. You can detect a silent toilet water leak by adding a few drops of food coloring to the toilet tank and waiting 5 minutes. If the color shows up in the bowl, you have a leak. Also, a faucet leaking one drop per second can drip about 8,200 liters (3,000 gallons) of water down the drain in a year—enough to fill about 75 bathtubs. This also represents money going down the drain.

Many home owners and businesses in water-short areas are using drip irrigation on their properties and copying nature by replacing green lawns with a mix of native plants that need little or no watering (Figure 13-26). Such water-thrifty landscaping saves money by reducing freshwater use by 30–85% and by sharply reducing labor, fertilizer, and fuel requirements. This example of reconciliation ecology (see Chapter 10, p. 242) also helps to preserve biodiversity by providing habitat for threatened honeybee, butterfly, and songbird species—applying the biodiversity **principle of sustainability** (see Figure 1-2, p. 6 or back cover). And it can help landowners to reduce polluted runoff, air pollution, and yard wastes.

GOOD NEWS

SUSTAINABILITY

However, in some of the more-developed countries, people living in arid areas maintain green lawns by watering them heavily. Some communities in water-short areas have even passed ordinances that require green lawns and prevent people from planting native vegetation in their yards. But this might change if people who live in such areas ever have to pay much higher water bills.

Water used in homes can also be reused and recycled. About 50–75% of a typical household's *gray water*—used water from bathtubs, showers, sinks, dishwashers, and clothes washers—could be stored in a holding tank and then reused to irrigate lawns and nonedible plants, to flush toilets, and to wash cars. Such efforts mimic the way nature recycles water, and thus they follow the chemical cycling **principle of sustainability**.

The relatively low cost of water in most communities is one of the major causes of excessive water use in homes and industries. Many water utility and irrigation authorities charge an annual flat fee for water use, and some charge less to the largest users of water. About one-fifth of all U.S. public water systems do not use water meters and charge a single low annual rate for almost unlimited use of high-quality freshwater.

However, simply monitoring water use can help people to reduce their use and losses of freshwater. When the U.S. city of Boulder, Colorado, introduced water meters, water use per person dropped by 40%. In some cities in Brazil, people obtain freshwater on a pay-as-you-go basis. They buy *smart cards*, each of which contains a certain number of water credits that entitle their owners to measured amounts of freshwater. Brazilian officials say this approach saves water and typically reduces household water bills by 40%.

Some colleges and universities are taking the lead in finding ways to use freshwater more sustainably. What students on these campuses are learning can be applied to homes and businesses of all types and sizes.

CONSIDER THIS. . .

CAMPUS SUSTAINABILITY

University of California–Santa Barbara

George Foulsham/University of California, Santa Barbara

At the University of California–Santa Barbara, a team of students and faculty are leading water conservation efforts. They have calculated that student housing is responsible for more than half of the water used on campus, so one effort involves a competition between dorms to reduce water consumption. The school has installed dozens of low-flow water taps and waterless urinals. It is also experimenting with a closed-loop, chilled-water cooling system that has replaced air conditioning units (which lose water to evaporation) in 14 campus buildings. Another important measure has been the use of recycled gray water for irrigating more than 90% of the campus grounds (see photo).

Figure 13-27 lists various ways to use freshwater more efficiently in industries, homes, and businesses (Concept 13-6).

Solutions

Reducing Water Losses

- Redesign manufacturing processes to use less water
- Recycle water in industry
- Fix water leaks
- Landscape yards with plants that require little water
- Use drip irrigation on gardens and lawns
- Use water-saving showerheads, faucets, appliances, and toilets (or waterless composting toilets)
- Collect and reuse gray water in and around houses, apartments, and office buildings
- Raise water prices and use meters, especially in dry urban areas

Figure 13-27 There are a number of ways to reduce freshwater losses in industries, homes, and businesses (Concept 13-6). ***Questions:*** Which three of these solutions do you think are the best ones? Why?

We Can Use Less Water to Remove Wastes

Currently, we use large amounts of freshwater clean enough to drink to flush away industrial, animal, and household wastes. According to the Food and Agriculture Organization of the United Nations (FAO), if current growth trends in population and water use continue, within 40 years, we will need the world's entire reliable flow of river water just to dilute and transport the wastes we produce each year.

We could save much of this freshwater by finding different ways to move the wastes. For example, we could recycle and reuse gray water from homes and businesses and wastewater from sewage treatment plants for purposes such as cleaning equipment and watering lawns and some crops. Israel reuses 70% of its wastewater to irrigate nonfood crops. In Singapore, all sewage water is treated at reclamation plants for reuse by industry. U.S. cities such as Las Vegas, Nevada, and Los Angeles, California, are also beginning to clean up and reuse some of their wastewater. However, only about 7% of the water in the United States is reused, a violation of the chemical cycling **principle of sustainability** that we know how to correct.

Another way to keep freshwater out of the waste stream is to rely more on waterless composting toilets. These devices convert human fecal matter to a small amount of dry and odorless soil-like humus material that can be removed from a composting chamber every year or so and returned to the soil as fertilizer. One of the authors (Miller) used a composting toilet for over a decade with no problems, while living and working deep in the woods in an experimental home and office (see p. 1).

We Can Each Help Out in Using Water More Sustainably

More sustainable water use would include a variety of strategies (Figures 13-16, 13-25, and 13-27) aimed not only at conserving water and using it efficiently, but also at protecting water supplies and the ecosystems that sustain them (Concept 13-6). Such strategies would have to be applied at local and regional levels, as well as national and international levels.

However, in order to be successful, such strategies would also have to be applied at the personal level. Each of us can help to bring about a "blue revolution" in which we could reduce our water footprints by using less freshwater and using water much more efficiently (Figure 13-28).

Finding more sustainable ways to use freshwater is the subject of some major research efforts. One group that is working on this problem is the Global Water Policy Project, founded by the renowned water supply expert, and National Geographic Explorer, Sandra Postel (Individuals Matter 13.1).

What Can You Do?

Water Use and Waste

- Use water-saving toilets, showerheads, and faucets
- Take short showers instead of baths
- Turn off sink faucets while brushing teeth, shaving, or washing
- Wash only full loads of clothes or use the lowest possible water-level setting for smaller loads
- Repair water leaks
- Wash your car from a bucket of soapy water, use gray water, and use the hose for rinsing only
- If you use a commercial car wash, try to find one that recycles its water
- Replace your lawn with native plants that need little if any watering
- Water lawns and gardens only in the early morning or evening and use gray water
- Use drip irrigation and mulch for gardens and flowerbeds

Figure 13-28 Individuals matter: You can reduce your use and waste of freshwater. ***Questions:*** Which of these steps have you taken? Which would you like to take?

13-7 How Can We Reduce the Threat of Flooding?

CONCEPT 13-7

We can lessen the threat of flooding by protecting more wetlands and natural vegetation in watersheds, and by not building in areas subject to frequent flooding.

Some Areas Get Too Much Water from Flooding

Some areas have too little freshwater, but others sometimes have too much because of natural flooding by streams, caused mostly by heavy rain or rapidly melting snow. A flood happens when freshwater in a stream overflows its normal channel and spills into the adjacent area, called the **floodplain**.

People settle on floodplains to take advantage of their many assets, including fertile soil on flat land suitable for crops, ample freshwater for irrigation, and availability of nearby rivers for transportation and recreation. In efforts to reduce the threat of flooding on floodplains, rivers have been narrowed and straightened (or *channelized*), surrounded by protective dikes and *levees* (long mounds of earth along their banks), and dammed to create reservoirs that store and release water as needed (Figure 13-2). However, in the long run, such measures can lead to greatly increased flood damage when heavy snowmelt or prolonged rains overwhelm them.

Floods provide several benefits. They have created some of the world's most productive farmland by depositing nutrient-rich silt on floodplains. They also help recharge groundwater and refill wetlands that are commonly found on floodplains, thereby supporting biodiversity and aquatic ecosystem services.

At the same time, floods kill thousands of people every year and cost tens of billions of dollars in property damage (see the Case Study that follows). Floods are usually considered natural disasters, but since the 1960s, human activities have contributed to a sharp rise in flood deaths and damages, meaning that such disasters are partly human-made.

One such human activity is the *removal of water-absorbing vegetation*, especially on hillsides (Figure 13-29). Once the trees on a hillside have been cut for timber, fuelwood, livestock grazing, or farming, freshwater from precipitation rushes down the denuded slopes, erodes precious topsoil, and can increase flooding and pollution in local streams. Such deforestation can also make landslides and mudflows more likely. A 3,000-year-old Chinese proverb says, "To protect your rivers, protect your mountains."

The second human activity that increases the severity of flooding is the *draining of wetlands* that naturally absorb floodwaters. These areas are then often covered with pavement and buildings that greatly increase runoff. When Hurricane Katrina struck the Gulf Coast of the United States in August 2005 and contributed to the flooding of the city of New Orleans, Louisiana, and

individuals matter 13.1

Sandra Postel: National Geographic Fellow and Freshwater Conservationist

© Mark Thiessen/National Geographic Creative

Sandra Postel is regarded as one of the world's most respected authorities on water issues. In 1994 she founded the Global Water Policy Project, which, through research and education, promotes a deeper knowledge of, and more sustainable use of, the earth's finite freshwater supply. Postel has authored or coauthored four influential books, including *Last Oasis: Facing Water Scarcity* and *Rivers for Life: Managing Water for People and Nature*. She has also written scores of articles for popular and scholarly publications.

In a relentless quest to educate people about water supply issues, Postel has appeared on several television news shows, taken part in numerous environmental documentary films (including the BBC's *Planet Earth*), and addressed the European Parliament. In 2010, she was appointed as Freshwater Fellow of the National Geographic Society, where she serves as the lead water expert for the Society's Freshwater Initiative—a global effort to inspire and empower people to conserve freshwater and to preserve aquatic biodiversity. Postel has also served as a consultant to the Nature Conservancy, the World Bank, and the U.S. National Intelligence Council, and as vice president for research at the Worldwatch Institute.

In 2002, Postel was named one of the "Scientific American 50" for her contributions to science and technology. She has been hailed for her "inspiring, innovative, and practical approach" to promoting the preservation and sustainable use of the earth's freshwater supply.

Background photo: Nikitsin/Dreamstime.com

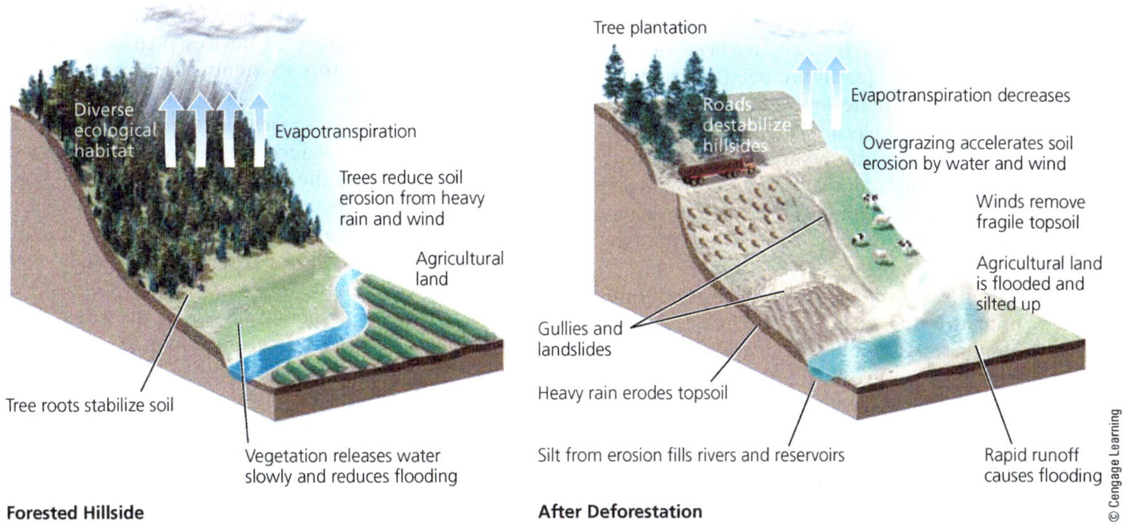

Animated Figure 13-29 **Natural capital degradation:** A hillside before and after deforestation.
Question: How might a drought in this area make these effects even worse?

surrounding areas, the damage was intensified because of the degradation or removal of coastal wetlands. These wetlands had historically helped to absorb water and buffer this low-lying land from storm surges.

Another human-related factor that will likely increase flooding is a rise in sea levels, projected to occur during this century. This is one likely result of climate change related to human activities that have led to the warming of the atmosphere. Climate change models project that, by 2075, as many as 150 million people living in the world's largest coastal cities—a number roughly equal to half of the current U.S. population—could be flooded out by rising sea levels.

CASE STUDY

Living Dangerously on Floodplains in Bangladesh

Bangladesh is one of the world's most densely populated countries. In 2012, its 153 million people were packed into an area roughly the size of the U.S. state of Wisconsin (which has a population of less than 6 million). And the country's population is projected to increase to 226 million by 2050. Bangladesh is a very flat country, only slightly above sea level, and it is one of the world's poorest countries.

The people of Bangladesh depend on moderate annual flooding during the summer monsoon season to grow rice and help maintain soil fertility in their country's delta basin region, which is fed by numerous river systems. The annual floods also deposit eroded Himalayan soil on the country's crop fields. Bangladeshis have adapted to moderate flooding. Most of the houses have flat thatch roofs on which families can take refuge with their belongings in case of rising waters. The roofs can be detached from the walls, if necessary, and floated like rafts. After the waters have subsided, the roof can be reattached to the walls of the house. However, great floods can overwhelm such defenses.

In the past, great floods occurred every 50 years or so. But between 1987 and 2007 there were five severe floods, each of which covered a third or more of the country with water. Bangladesh's flooding problems begin in the Himalayan watershed, where rapid population growth, deforestation, overgrazing, and unsustainable farming on steep and easily erodible slopes have led to increased flows of water during the summer monsoon season. Monsoon rains now run more quickly off the denuded Himalayan foothills, carrying vital topsoil with them (Figure 13-29, right).

This increased runoff of topsoil, combined with heavier-than-normal monsoon rains, has led to more severe flooding along Himalayan rivers as well as downstream in Bangladesh's delta areas. In 1998, a disastrous flood covered two-thirds of Bangladesh's land area, in some places for 2 months, drowning at least 2,000 people and leaving 30 million homeless. It also destroyed more than one-fourth of the country's crops, which caused thousands of people to die of starvation. In 2002, another flood left 5 million people homeless and flooded large areas of rice fields. Yet another major flood occurred in 2004.

Living at sea level on Bangladesh's coastal floodplain also means coping with storm surges, cyclones, and tsunamis. In 1970, a tropical cyclone killed as many as 1 million people. Another cyclone in 2003 killed more than a million people and left tens of millions homeless.

Many of the coastal mangrove forests in Bangladesh (and elsewhere; see Figure 8-9, p. 172) have been cleared for fuelwood, farming, and aquaculture ponds created for raising shrimp. The result: more severe flooding because these coastal wetlands had helped to shelter Bangladesh's low-lying coastal areas from storm surges, cyclones, and tsunamis. In areas of Bangladesh still protected by mangrove forests, damages and death tolls from cyclones have been much lower than they were in areas where the forests have been cleared.

Projected rises in sea level and storm intensity during this century, primarily due to projected climate change caused by atmospheric warming, will likely be a major threat to Bangladeshis who live on the flat delta adjacent to the Bay of Bengal. This would create millions of environmental refugees with no place to go in this already densely populated country.

Bangladesh is one of the few less-developed nations that is preparing and implementing plans to adapt to projected rising sea levels. This includes using varieties of rice and other crops that can better tolerate flooding, saltwater, and drought; shifting to new crops such as corn; developing small vegetable gardens in bare patches between houses to help reduce Bangladeshi dependence on rice; building small ponds that will collect monsoon rainwater to use for irrigating vegetable gardens during dry periods; and creating a network of earthen embankments that might protect against high tides and moderate storm surges when cyclones strike.

Solutions

Reducing Flood Damage

Prevention	Control
Preserve forests in watersheds	Strengthen and deepen streams (channelization)
Preserve and restore wetlands on floodplains	Build levees or floodwalls along streams
Tax development on floodplains	Build dams
Increase use of floodplains for sustainable agriculture and forestry	

Figure 13-30 Methods for reducing the harmful effects of flooding (Concept 13-7). ***Questions:*** Which two of these solutions do you think are the best ones? Why?

Top: allensima/Shutterstock.com. Bottom: ©Zeljko Radojko/Shutterstock.com.

We Can Reduce Flood Risks

Many scientists argue that we could improve flood control by relying less on engineered devices such as dams and levees and more on nature's systems such as wetlands and forests in watersheds.

One engineering approach is the channelizing of streams, which does reduce upstream flooding. However, it also eliminates the aquatic habitats that lie along a meandering stream by taking the water from those systems and sending it in a faster flow straight down a channel. It also reduces groundwater recharge and often leads to downstream flooding.

Similarly, levees or floodwalls along the banks of a river contain and speed up stream flow and can lead to flooding downstream. They also do not protect against unusually high and powerful floodwaters such as those that occurred in 1993 when two-thirds of the levees along the Mississippi River were damaged or destroyed. Similar flooding occurred along the Mississippi in 2011.

Damming, the most common engineering approach, can reduce the threat of flooding by storing water in a reservoir and releasing it gradually. However, dams also have a number of drawbacks (Figure 13-17).

A more ecologically oriented approach to reducing flooding is to *preserve existing wetlands* and *restore degraded wetlands* that lie in floodplains to take advantage of the natural flood control they provide. We would also be wise to sharply reduce emissions of greenhouse gases that contribute to projected atmospheric warming, which will likely raise sea levels and flood many of the world's coastal areas during this century.

On a personal level, we can *think carefully about where we choose to live.* Many poor people live in flood-prone areas because they have nowhere else to go. Most people, however, can choose not to live in areas especially subject to flooding or to water shortages.

Figure 13-30 summarizes these various ways to reduce flooding risks (Concept 13-7).

Big Ideas

- One of the major global environmental problems is the growing shortage of freshwater in many parts of the world.
- We can expand water supplies in water-short areas in a number of ways, but the most important ways are to reduce overall water use and to use water much more efficiently.
- We can use water more sustainably by cutting water losses, raising water prices, and protecting aquifers, forests, and other ecosystems that store and release water.

TYING IT ALL TOGETHER The Colorado River and Sustainability

Atm2003/Shutterstock.com

The **Core Case Study** that opens this chapter discusses the problems and tensions that can occur when a large number of U.S. states share a limited river water resource in a water-short region. Such problems are representative of those faced by many other dry regions of the world, especially areas where the population is growing rapidly and water resources are dwindling for various reasons.

Large dams, river diversions, levees, and other big engineering schemes have helped to provide much of the world with electricity, food from irrigated crops, drinking water, and flood control. However, they have also degraded the aquatic natural capital necessary for long-term economic and ecological sustainability by seriously disrupting rivers, streams, wetlands, aquifers, and other aquatic systems.

The three scientific **principles of sustainability** (see Figure 1-2, p. 6 or back cover) can guide us in using water more sustainably during this century. Scientists hope to use solar energy to desalinate water and expand freshwater supplies. Recycling more water will help us to reduce water losses. Preserving biodiversity by avoiding disruption of aquatic systems and their bordering terrestrial systems is a key factor in maintaining water supplies and water quality.

Chapter Review

Core Case Study

1. Summarize the importance of the Colorado River basin in the United States and how human activities are stressing this system (**Core Case Study**). Define **drought** and explain how it has affected the Colorado River system.

Section 11-1

2. What are the two key concepts for this section? Define **freshwater**. Explain why access to water is a health issue, an economic issue, a national and global security issue, and an environmental issue. What percentage of the earth's freshwater is available to us? Describe how water is recycled by the hydrologic cycle and how human activities can interfere with this cycle. Define **groundwater**, **zone of saturation**, **water table**, and **aquifer**, and explain how aquifers are recharged. Define and distinguish between **surface water** and **surface runoff**. What is a **watershed (drainage basin)**?

3. What is **reliable surface runoff**? What percentage of the world's reliable runoff are we using and what percentage are we likely to be using by 2025? How is most of the world's water used? Define and give an example of a **water footprint**. Define **virtual water** and give three examples of it. Describe the availability and use of freshwater resources in the United States and the water shortages that could occur during this century. What are four major problems resulting from the way people are using water from the Colorado River basin?

4. How many countries face water scarcity today and how many could face water scarcity by 2050? What percentage of the earth's land suffers from severe drought today and how might this change by 2059? How many people in the world lack regular access to clean water today and how high might this number grow by 2025? Why do many analysts view the likelihood of steadily worsening water shortages as one of the world's most serious environmental problems? Explain the connection between water shortages and hunger. How do scientists use satellites to measure changes in water supplies?

Section 11-2

5. What is the key concept for this section? What are the advantages and disadvantages of withdrawing groundwater? Summarize the problem of groundwater depletion in the world and in the United States, especially in the Ogallala aquifer. List two problems that result from the overpumping of aquifers. Describe the potential for using deep aquifers and list the four main concerns about using this resource.

Section 11-3

6. What is the key concept for this section? What is a **dam**? What is a **reservoir**? What are the advantages and disadvantages of using large dams and reservoirs? What has happened to water flows in the Colorado River (**Core Case Study**) since 1960? Explain how the damming of this river has affected its delta. What other problems are likely to further shrink this supply of water? List three possible solutions to the supply problems in the Colorado River basin.

Section 11-4

7. What is the key concept for this section? What is a water transfer? Describe two large water-transfer programs, explain how they came about, and summarize the controversy around each of these programs. Summarize the story of the Aral Sea water-transfer project and its disastrous consequences.

Section 11-5

8. What is the key concept for this section? Define **desalination** and list three major limitations on the widespread use of desalination. What are scientists doing to try to deal with these problems?

Section 11-6

9. What is the key concept for this section? What percentage of available freshwater is lost through inefficient use and other causes, in the world and in the United States? What are two major reasons for those losses? Describe three major irrigation methods and list ways to reduce water losses in irrigation. What are three ways in which people in less-developed countries conserve water? List four ways to reduce water waste in industries and homes. How can we use less water to remove wastes? List four ways in which you can reduce your use and waste of water.

Section 11-7

10. What is the key concept for this section? What is a **floodplain** and why do people like to live on floodplains? What are the benefits and harms of flooding? List two human activities that increase the risk of flooding. Describe the flooding risks that many people in Bangladesh face and what they are doing about it. List and compare two engineering approaches to flood control and two ecologically oriented approaches. What are this chapter's *three big ideas*? Explain how the scientific **principles of sustainability** (see Figure 1-2, p. 6 or back cover) can guide us in using water more sustainably during this century.

Note: Key terms are in bold type.

Critical Thinking

1. What do you believe are the three most important priorities for dealing with the water resource problems of the Colorado River basin, as discussed in the **Core Case Study** that opens this chapter? Explain your choices.
2. List three ways in which human activities are affecting the water cycle. How might these changes to the water cycle affect your lifestyle? How might your lifestyle be contributing to these effects?
3. Explain how our current use of the earth's water resources can be viewed as a good example of the tragedy of the commons (see Chapter 1, p. 12).
4. What role does population growth play in water supply problems? Relate this to water supply problems of the Colorado River basin (**Core Case Study**).
5. Explain why you are for or against **(a)** raising the price of water while providing lower lifeline rates for poor consumers, **(b)** withdrawing government subsidies that provide farmers with water at low cost, and **(c)** providing government subsidies to farmers for improving irrigation efficiency.
6. Calculate how many liters (and gallons) of water are lost in 1 month by a toilet or faucet that leaks 2 drops of water per second. (One liter of water equals about 3,500 drops and 1 liter equals 0.265 gallon.) How many bathtubs (each containing about 151 liters or 40 gallons) could be filled with this lost water?
7. List the three most important ways in which you could use water more efficiently. Which, if any, of these measures do you already take?
8. List three ways in which human activities increase the harmful effects of flooding. What is the best way to prevent each of these human impacts? Do you think they should be prevented? Why or why not?

Doing Environmental Science

Investigate water use at your school. Try to determine all specific sources of any water losses, taking careful notes and measurements for each of them, and estimate how much water is lost per hour, per day, and per year from each source. Sum these estimated amounts to arrive at an estimate of total water losses for a year at your school. Develop a water conservation plan for your school and submit it to school officials.

Global Environment Watch Exercise

Search for *Ogallala* and research the overpumping of this aquifer. Plot the decline of this aquifer on a graph, and write a report explaining the causes and possible ways to slow the decline of the Ogallala aquifer.

Ecological Footprint Analysis

In 2005 (latest available data), the population of the U.S. state of Florida consumed 24.5 billion liters (6.5 billion gallons) of freshwater daily. It is projected that in 2025, the daily consumption will increase to 32.1 billion liters (8.5 billion gallons) per day. Between 2005 and 2025, the population of Florida is projected to increase from 17.5 million to 25.9 million.

1. Based on total freshwater use:
   - **a.** Calculate the per capita consumption of water per day in Florida in 2005 and the projected per capita consumption per day for 2025.
   - **b.** Calculate the per capita consumption of water for the year in Florida in 2005 and the projected per capita consumption per year for 2025.
2. In 2005, how did the Florida *average water footprint* (consumption per person per year), based only on water used within the state, compare with the average U.S. water footprint of approximately 249,000 liters (66,000 gallons) per person per year, and with the global average water footprint of 123,770 liters (32,800 gallons) per person per year?

14 Nonrenewable Mineral Resources

Civilization exists by geological consent, subject to change without notice.

WILL DURANT

Key Questions

14-1 What are the earth's major geological processes and what are mineral resources?

14-2 How long might supplies of nonrenewable mineral resources last?

14-3 What are the environmental effects of using nonrenewable mineral resources?

14-4 How can we use mineral resources more sustainably?

14-5 What are the earth's major geological hazards?

Climax Molybdenum mine site in the Rocky Mountains near Leadville, Colorado.

Pete Mcbride/National Geographic Creative

CORE CASE STUDY The Crucial Importance of Rare-Earth Metals

Figure 14-1 The highly reactive rare-earth metal lanthanum (La) and its compounds are used in a variety of products, including batteries in electric and hybrid-electric cars (Figure 14-2) and lenses in cameras and video cameras.

Mineral resources extracted from the earth's crust are processed into an amazing variety of products that can make our lives easier and provide us with economic benefits and jobs. Most people are aware of our use of mineral resources such as gold, copper, aluminum, sand, and gravel. But fewer people have heard of *rare-earth metals and oxides* that are crucial to the major technologies that support today's lifestyles and economies.

The 17 rare-earth metals, also called *rare earths,* include scandium, yttrium, and 15 lanthanide chemical elements (see periodic table, Figure 1, p. S12, Supplement 4) such as lanthanum (Figure 14-1). Because of their superior magnetic strength and other unique properties, these elements and their compounds are very important for a number of widely used technologies.

Rare earths are used to make liquid crystal display flat screens for computers and television sets, energy-efficient compact fluorescent and LED lightbulbs, solar cells, fiber-optic cables, cell phones, and digital cameras. They are also important in the manufacture of batteries and motors for electric and hybrid-electric cars (Figure 14-2), catalytic converters in car exhaust systems, jet engines, and the powerful magnets in wind turbine generators. Rare earths also go into missile guidance systems, smart bombs, aircraft electronics, and satellites.

Without affordable supplies of these metals, industrialized nations could not develop the current versions of cleaner energy technology and other high-tech products that will be major sources of economic growth and profits during this century. Nations also need these metals to maintain their military strength.

In this chapter we look at the nature and supplies of mineral resources, the rock cycle that produces them, the harmful environmental impacts of their use, and how we can use these resources more sustainably. We also look at the earth's natural geological processes and major geological hazards.

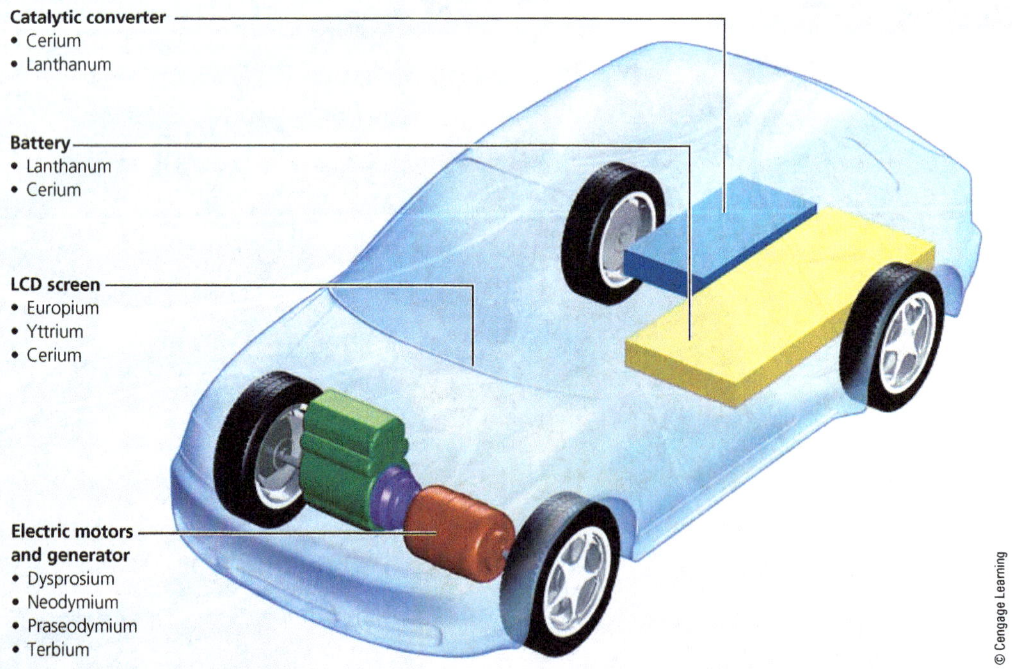

Figure 14-2 Manufacturers of all-electric and hybrid-electric cars use a variety of rare-earth metals.

14-1 What Are the Earth's Major Geological Processes and What Are Mineral Resources?

CONCEPT 14-1A
Dynamic processes within the earth and on its surface produce the mineral resources we depend on.

CONCEPT 14-1B
Mineral resources are nonrenewable because they are produced and renewed over millions of years mostly by the earth's rock cycle.

The Earth Is a Dynamic Planet

Geology is the science devoted to the study of dynamic processes taking place on the earth's surface and in its interior. As the primitive earth cooled over eons, its interior separated into three major concentric zones: the *core*, the *mantle*, and the *crust* (Figure 14-3).

The **core** is the earth's innermost zone. It is extremely hot and has a solid inner part, surrounded by a thick layer of *molten rock*, or hot liquid rock, and semisolid material. Surrounding the core is a thick zone called the **mantle**—a zone made mostly of solid rock that can be soft and pliable at very high temperatures. The outermost part of the mantle is solid rock, and under that part is a zone called the **asthenosphere**—a volume of hot, partly melted rock that flows. Tremendous heat within the core and mantle generates *convection cells* or *currents* that slowly move large volumes of rock and heat in loops within the mantle like gigantic conveyor belts (Figure 14-3). Some of the molten

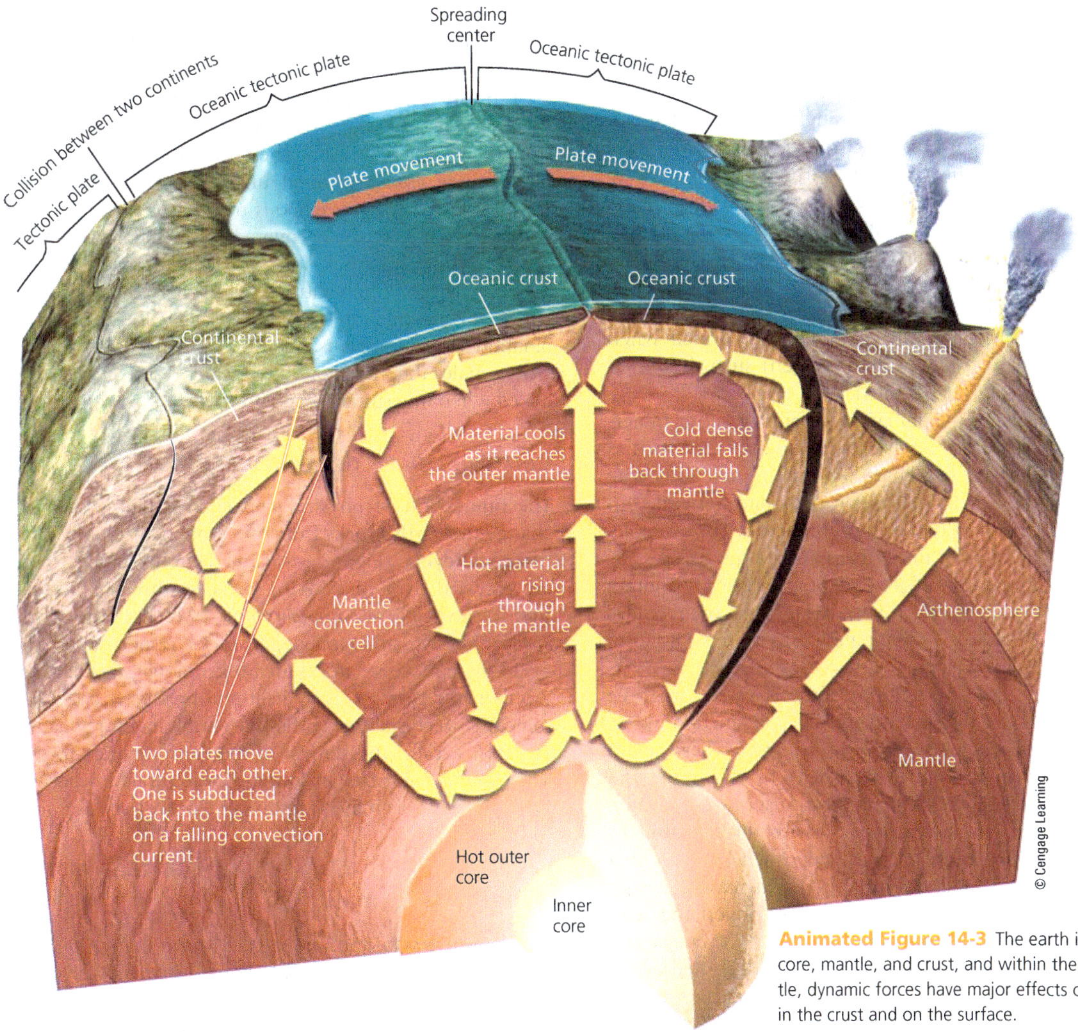

Animated Figure 14-3 The earth is composed of a core, mantle, and crust, and within the core and mantle, dynamic forces have major effects on what happens in the crust and on the surface.

rock in the asthenosphere flows upward into the crust, where it is referred to as *magma*.

The outermost and thinnest zone of solid material is the earth's **crust**. It consists of the *continental crust*, which underlies the continents (including the continental shelves extending into the oceans), and the *oceanic crust*, which underlies the ocean basins and makes up 71% of the earth's crust. The combination of the crust and the rigid, outermost part of the mantle is called the **lithosphere**. It is in this zone that we find the mineral resources on which we depend (Concept 14-1A).

What Are Minerals and Rocks?

The earth's crust beneath our feet consists mostly of minerals and rocks. A **mineral** is a naturally occurring chemical element or inorganic compound that exists as a solid with a regularly repeating internal arrangement of its atoms or ions (a *crystalline solid*). A **mineral resource** is a concentration of one or more minerals in the earth's crust that we can extract and process into raw materials and useful products at an affordable cost. Because minerals take millions of years to form, they are *nonrenewable resources*, and their supplies can be depleted (Concept 14-1B).

A few minerals consist of a single chemical element. They include mercury (see Figure 2-3, left, p. 36), gold (see Figure 2-3, right), and rare-earth metals (Core Case Study) such as lanthanum (Figure 14-1). However, most of the more than 2,000 identified mineral resources that we use occur as inorganic compounds formed by various combinations of elements. Examples include salt (sodium chloride, or NaCl; see Figure 2-6, p. 38), quartz (silicon dioxide, or SiO_2), and rare-earth oxides (formed when rare-earth metals combine with oxygen) such as lanthanum oxide (La_2O_3, Figure 14-4).

Rock is a solid combination of one or more minerals found in the earth's crust. Some kinds of rock such as limestone (calcium carbonate, or $CaCO_3$) and quartzite (silicon dioxide, or SiO_2) contain only one mineral. But most rocks consist of two or more minerals. For example, granite is a mixture of mica, feldspar, and quartz crystals. Deposits of rare earths (Core Case Study) typically contain a mixture of the metals (Figure 14-1) and their oxides (Figure 14-4) that are difficult and costly to separate from one another and to purify to an acceptable level.

Based on the way it forms, rock is placed in three broad classes: sedimentary, igneous, or metamorphic. **Sedimentary rock** is made of *sediments*—dead plant and animal remains and tiny particles of weathered and eroded rocks. These sediments are transported by water, wind, or gravity to downstream, downwind, downhill, or underwater sites. There they are deposited in layers that accumulate over time. Eventually, the increasing weight and pressure on the underlying layers convert the sedimentary layers into rock. Examples include *sandstone* and *shale* (formed from pressure created by deposited layers made primarily of sand), *dolomite* and *limestone* (formed from the compacted shells, skeletons, and other remains of dead aquatic organisms), and *lignite* and *bituminous coal* (derived from compacted plant remains).

Import Export Bulletin Board

Figure 14-4 Oxides of rare-earth metals, such as lanthanum oxide shown here, are often converted to rare-earth metals and used in a variety of products and technologies.

Igneous rock forms below or on the earth's surface under intense heat and pressure when magma wells up from the earth's mantle and then cools and hardens. Examples include *granite* (formed underground) and *lava rock* (formed aboveground). Igneous rock forms the bulk of the earth's crust but is usually covered by sedimentary rock.

Metamorphic rock forms when an existing rock is subjected to high temperatures (which may cause it to melt partially), high pressures, chemically active fluids, or a combination of these agents. Examples include *slate* (formed when shale and mudstone are heated) and *marble* (produced when limestone is exposed to heat and pressure).

Earth's Rocks Are Recycled Very Slowly

The interaction of physical and chemical processes that change the earth's rocks from one type to another is called the **rock cycle** (Figure 14-5 and Concept 14-1B). Rocks are recycled over millions of years by three processes—*erosion, melting,* and *metamorphism*—which produce *sedimentary, igneous,* and *metamorphic* rocks, respectively.

In these processes, rocks are broken down, melted, fused together into new forms by heat and pressure, cooled, and sometimes recrystallized within the earth's interior and crust. The rock cycle is the slowest of the earth's cyclic processes and plays a major role in forming concentrated deposits of the nonrenewable mineral resources that we use.

We Depend on a Variety of Nonrenewable Mineral Resources

We know how to find and extract more than 100 different minerals from the earth's crust. They include *metallic minerals* made of single elements, such as aluminum,

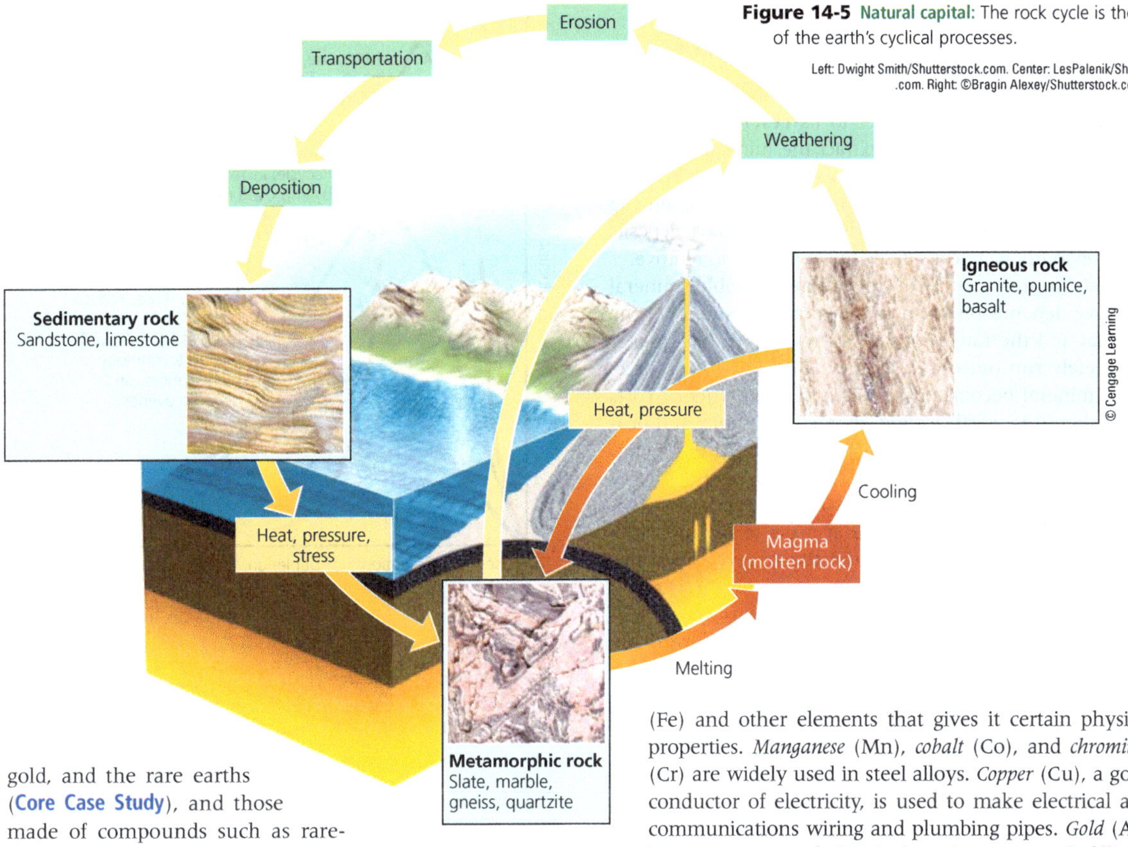

Figure 14-5 Natural capital: The rock cycle is the slowest of the earth's cyclical processes.

Left: Dwight Smith/Shutterstock.com. Center: LesPalenik/Shutterstock.com. Right: ©Bragin Alexey/Shutterstock.com.

gold, and the rare earths (**Core Case Study**), and those made of compounds such as rare-earth oxides (Figure 14-4). Other mineral resources, including sand and limestone, are *nonmetallic minerals* made of nonmetallic elements and compounds.

An **ore** is rock that contains a large enough concentration of a particular mineral—often a metal—to make it profitable for mining and processing. A **high-grade ore** contains a large concentration of the desired mineral, whereas a **low-grade ore** contains a smaller concentration.

Nonrenewable mineral resources are used for many purposes. Today about 60 of the chemical elements in the periodic table (see Figure 1, p. S12, in Supplement 4) are used for making computer chips. *Aluminum* (Al) is used as a structural material in beverage cans, motor vehicles, aircraft, and buildings. *Steel,* an essential material used in buildings, machinery, and motor vehicles, is a mixture (alloy) of iron (Fe) and other elements that gives it certain physical properties. *Manganese* (Mn), *cobalt* (Co), and *chromium* (Cr) are widely used in steel alloys. *Copper* (Cu), a good conductor of electricity, is used to make electrical and communications wiring and plumbing pipes. *Gold* (Au) is a component of electrical equipment, tooth fillings, jewelry, coins, and some medical implants. Molybdenum (Mo) (see chapter-opening photo) is widely used to harden steel and to make it more resistant to corrosion. And many of today's important new technologies make use of various rare-earth elements (**Core Case Study**).

There are several widely used nonmetallic mineral resources. *Sand,* which is mostly silicon dioxide (SiO_2), is used to make glass, bricks, and concrete for the construction of roads and buildings. *Gravel* is used for roadbeds and to make concrete. Another common nonmetallic mineral is *limestone* (mostly calcium carbonate, or $CaCO_3$), which is crushed to make concrete and cement. Still another is *phosphate salts,* used to make inorganic fertilizers and certain detergents.

14-2 How Long Might Supplies of Nonrenewable Mineral Resources Last?

CONCEPT 14-2A

Nonrenewable mineral resources exist in finite amounts and can become economically depleted when it costs more than it is worth to find, extract, and process the remaining deposits.

CONCEPT 14-2B

There are several ways to extend supplies of mineral resources, but each of them is limited by economic and environmental factors.

Supplies of Nonrenewable Mineral Resources Can Be Economically Depleted

Most published estimates of the supply of a given nonrenewable mineral resource refer to its **reserves**: identified deposits from which we can extract the mineral profitably at current prices. Reserves can be expanded when we find new, profitable deposits or when higher prices or improved mining technologies make it profitable to extract deposits that previously were considered too expensive to remove.

The future supply of any nonrenewable mineral resource depends on the actual or potential supply of the mineral and the rate at which we use it. We have never completely run out of a nonrenewable mineral resource, but a mineral becomes *economically depleted* when it costs more than it is worth to find, extract, transport, and process the remaining deposits (**Concept 14-2A**). At that point, there are five choices: *recycle or reuse existing supplies, waste less, use less, find a substitute,* or *do without.*

Depletion time is the time it takes to use up a certain proportion—usually 80%—of the reserves of a mineral at a given rate of use. When experts disagree about depletion times, it is often because they are using different assumptions about supplies and rates of use (Figure 14-6).

The shortest depletion-time estimate assumes no recycling or reuse and no increase in reserves (curve A, Figure 14-6). A longer depletion-time estimate assumes that recycling will stretch existing reserves and that better mining technology, higher prices, or new discoveries will increase reserves (curve B). The longest depletion-time estimate (curve C) makes the same assumptions as A and B, but also includes reuse and reduced consumption to further expand reserves. Finding a substitute for a resource leads to a new set of depletion curves for the new resource.

The earth's crust contains fairly abundant deposits of nonrenewable mineral resources such as iron and aluminum. But concentrated deposits of important mineral resources such as manganese, chromium, cobalt, platinum, and some of the rare earths (**Core Case Study**) are relatively scarce. In addition, deposits of many mineral resources are not distributed evenly among countries.

Five nations—the United States, Canada, Russia, South Africa, and Australia—supply most of the nonrenewable mineral resources used by modern societies. South Africa, for example, is the world's largest producer of chromium and platinum.

Since 1900, and especially since 1950, there has been a sharp rise in the total and per capita use of mineral resources in the United States. According to the U.S. Geological Survey, each person in the United States uses an average of 22 metric tons (24 tons) of mineral resources per year. As a result, the United States has economically depleted some of its once-rich deposits of metals such as lead, aluminum, and iron. Currently, the United States imports all of its supplies of 20 key nonrenewable mineral resources and more than 90% of its supplies of 4 other key minerals.

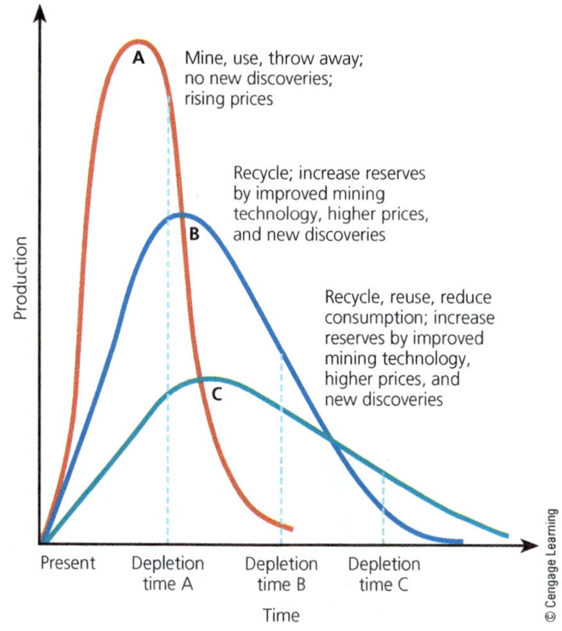

Figure 14-6 Natural capital depletion: Each of these *depletion curves* for a mineral resource (such as aluminum, copper, or a rare earth) is based on a different set of assumptions. Dashed vertical lines represent the times at which 80% depletion occurs.

Most U.S. imports of metallic mineral resources come from reliable and politically stable countries. But there are serious concerns about access to adequate supplies of four *strategic metal resources*—manganese, cobalt, chromium, and platinum—which are essential for the country's economic and military strength. The United States has little or no reserves of these metals.

Global and U.S. Rare-Earth Supplies

Rare-earth elements (**Core Case Study**) are not actually rare, for the most part, but they are hard to find in concentrations high enough to extract and process at an affordable price. There is no global shortage of rare-earth minerals. According to the U.S. Geological Survey, in 2010 China had roughly 50% of the world's known rare-earth reserves, Russia had 15%, and the United States 13%.

Currently, China produces about 97% of the world's rare-earth metals and oxides. This is partly because China does not strictly regulate the environmentally disruptive mining and processing of rare earths. This means that Chinese companies have lower costs than do companies in countries that seek to protect their environments from the disruptions of rare-earth mining and processing. China can thus afford to sell its rare earths at a lower price.

The United States and Japan, which are also heavily dependent on rare earths and their oxides, produce none. Japan has no rare-earth reserves, and U.S. production

Figure 14-7 Improved mining technology has expanded the mining of taconite iron ore in northern Minnesota.

stopped in 2002. A large rare-earth mine in California—the only one in the United States—used to be the world's largest supplier of rare-earth metals. However, it closed down because of the expense of meeting pollution regulations, and because China had driven the prices of rare-earth metals down to a point where the mine was too costly to operate. There are plans to reopen the California mine in the near future, if it can meet government environmental standards at an affordable cost. However, it contains mostly lighter rare earths (with lower atomic numbers), which are more abundant and thus are not as valuable. Meanwhile, the lack of production in the United States makes the country vulnerable to supply interruptions.

Since 2010, China has been reducing its exports of rare-earth metals and their oxides to other countries and has sharply raised its prices on such exports. China's growing use of these resources means that it will use most of the rare earths it produces and may soon need to import some types of rare earths from other countries to meet its industrial needs.

Market Prices Affect Supplies of Nonrenewable Mineral Resources

Geological processes determine the quantity and location of a nonrenewable mineral resource in the earth's crust, but economics determines what part of the known supply is extracted and used. An increase in the price of a scarce mineral resource can often lead to increased supplies and can encourage more efficient use, but there are limits to this effect (**Concept 14-2B**).

According to standard economic theory, in a competitive market system when a resource becomes scarce, its price rises. This can encourage exploration for new deposits, stimulate development of better mining technology, and make it profitable to mine lower-grade ores. It can also encourage a search for substitutes and promote resource conservation. For example, shortly after World War II, high-grade iron ore from rich deposits in northern Minnesota was economically depleted. But by the 1960s, a new process had been developed for mining *taconite* (Figure 14-7), a low-grade and plentiful ore that had been viewed as waste rock in the iron mining process. This improvement in mining technology for iron expanded the reserves in Minnesota and supports a taconite mining industry today.

CONSIDER THIS. . .

CONNECTIONS Metal Prices and Thievery

Resource scarcity can also promote thievery. For example, because of increasing demand, copper prices have risen sharply in recent years. As a result, in several U.S. communities, people have been stealing copper to sell it—stripping abandoned houses of copper pipe and wiring and stealing outdoor central air conditioning units for their copper coils. They have also stolen electrical wiring from beneath city streets and copper piping from farm irrigation systems. In one Oklahoma town, someone cut down several utility poles and stole the copper electrical wiring, causing a blackout.

According to some economists, this price effect may no longer apply very well in most of the more-developed countries. Governments in such countries often use subsidies, tax breaks, and import tariffs to control the supply, demand, and prices of key mineral resources to such an extent that a truly competitive free market does not exist. In the United States, for instance, mining companies get various types of government subsidies, including *depletion allowances*—permission to deduct from their taxable incomes the costs of developing and extracting mineral resources. These allowances amount to 5–22% of their gross income gained from selling the mineral resources.

Mining company representatives insist that they need taxpayer subsidies and low taxes to keep the prices of minerals low for consumers. They also claim that, without the subsidies, their companies might move their operations to other countries where they would not have to pay taxes or comply with strict mining and pollution control regulations.

Can We Expand Reserves by Mining Lower-Grade Ores?

Some analysts contend that we can increase supplies of some minerals by extracting them from lower-grade ores. They point to the development of new earth-moving equipment, improved techniques for removing impurities from ores, and other technological advances in mineral extraction and processing that can make lower-grade ores accessible, sometimes at lower costs. For example, in 1900, the copper ore mined in the United States was typically about 5% copper by weight. Today, it is typically about 0.5%, yet copper costs less (when prices are adjusted for inflation).

However, several factors can limit the mining of lower-grade ores (**Concept 14-4B**). For one, it requires mining and processing larger volumes of ore, which takes more energy and costs more. Another factor is the dwindling supplies of freshwater needed for the mining and processing of some minerals, especially in arid and semiarid areas. A third limiting factor is the growing environmental impacts of land disruption, along with waste material and pollution produced during mining and processing (see chapter-opening photo). We discuss this further in the next section of this chapter.

One way to improve mining technology and reduce its environmental impact is to use a biological approach, sometimes called *biomining*. Miners use natural or genetically engineered bacteria to remove desired metals from ores through wells bored into the deposits. This leaves the surrounding environment undisturbed and reduces the air and water pollution associated with removing the metal from metal ores. On the down side, biomining is slow. It can take decades to remove the same amount of material that conventional methods can remove within months or years. So far, biomining methods are economically feasible only with low-grade ores for which conventional techniques are too expensive.

Can We Get More Minerals from the Oceans?

Most of the chemical elements and compounds found in seawater occur in such low concentrations that recovering these mineral resources takes more energy and money than they are worth. Currently, only magnesium, bromine, and sodium chloride are abundant enough to be extracted profitably from seawater. On the other hand, in sediments along the shallow continental shelf and adjacent shorelines, there are significant deposits of minerals such as sand, gravel, phosphates, copper, iron, tungsten, silver, titanium, platinum, and diamonds.

Another potential ocean source of some minerals is *hydrothermal ore deposits* that form when superheated, mineral-rich water shoots out of vents in volcanic regions of the ocean floor. As the hot water comes into contact with cold seawater, black particles of various metal sulfides precipitate out and accumulate as chimney-like structures, called *black smokers*, near the hot water vents (Figure 14-8). These deposits are especially rich in minerals such as copper, lead, zinc, silver, gold, and some of the rare-earth metals. A variety of exotic forms of life—including giant clams, six-foot tubeworms, and eyeless shrimp—exist in the dark depths around these black smokers. These life-forms are supported by bacteria that produce food by chemosynthesis from sulfur compounds discharged by the vents.

Because of the rapidly rising prices of many of these metals, there is growing interest in deep-sea mining. In 2011, China began using remote-controlled underwater equipment and a manned deep-sea craft to evaluate the mining of mineral deposits around black smokers in the Indian Ocean near Madagascar as well as in other deep-sea areas. Also that year, Japanese explorers found large deposits of rare-earth minerals (**Core Case Study**) on the floor of the Pacific Ocean. The UN International Seabed Authority, established to manage seafloor mining in international waters, began issuing mining permits in 2011.

Some analysts say that seafloor mining is less environmentally harmful than mining on land. However, marine biologists are concerned that the sediment stirred up by such mining could harm or kill organisms that feed by filtering seawater. Proponents of the mining say that the

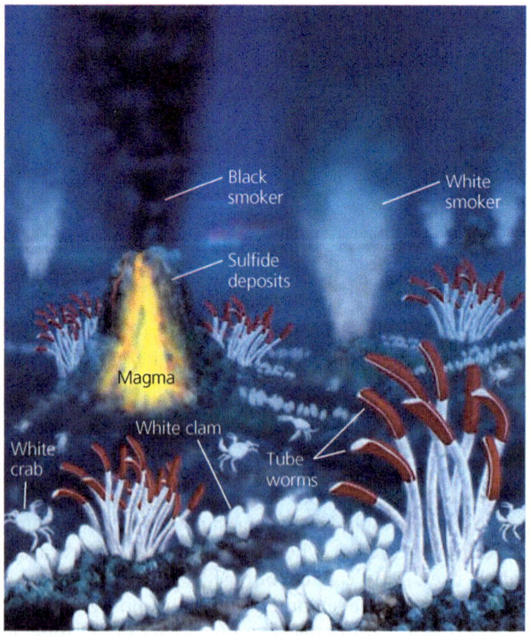

Figure 14-8 **Natural capital:** *Hydrothermal deposits,* or *black smokers,* are rich in various minerals.

number of potential mining sites is quite small and that many of these organisms can live elsewhere.

Another possible source of metals from the ocean is the potato-size *manganese nodules* that cover large areas of the Pacific Ocean floor and smaller areas of the Atlantic and Indian Ocean floors. They also contain some low concentrations of various rare-earth minerals (**Core Case Study**). These modules could be sucked up by giant vacuum pipes or scooped up by underwater mining machines.

So far, obtaining minerals from the ocean bottom has been hindered by the high costs involved, the potential threat to marine ecosystems, and arguments over rights to these minerals in deep-ocean areas that belong to no one country.

14-3 What Are the Environmental Effects of Using Nonrenewable Mineral Resources?

CONCEPT 14-3

Extracting minerals from the earth's crust and converting them into useful products can disturb the land, erode soils, produce large amounts of solid waste, and pollute the air, water, and soil.

Mineral Use Creates Environmental Impacts

Every metal product has a *life cycle* that includes mining the mineral, processing it, manufacturing the product, and disposal or recycling of the product (Figure 14-9). This process makes use of large amounts of energy and water, and results in pollution and waste (**Concept 14-3**).

The environmental impacts of mining a metal ore are determined partly by the ore's percentage of metal content, or *grade*. The more accessible higher-grade ores are usually exploited first. Mining lower-grade ores takes more money, energy, water, and other resources, and leads to more land disruption, mining waste, and pollution.

Removing Mineral Deposits Has Harmful Environmental Effects

Several different mining techniques are used to remove mineral deposits. Shallow mineral deposits are removed by **surface mining**, in which vegetation, soil, and rock overlying a mineral deposit are cleared away. The soil and rock, called **overburden**, are usually deposited in piles of waste material called **spoils** (Figure 14-10). Surface mining is used to extract about 90% of the nonfuel mineral resources and 60% of the coal used in the United States.

The type of surface mining used depends on two factors: the resource being sought and the local topography. In **open-pit mining**, machines are used to dig very large holes and remove metal ores containing copper (Figure 14-11), gold (see Case Study, p. 360), or other metals, or sand, gravel, or stone.

Strip mining is any form of mining involving the extraction of mineral deposits that lie in large horizontal beds close to the earth's surface. In **area strip mining**, used where the terrain is fairly flat, a gigantic earthmover strips away the overburden, and a power shovel—which can be as tall as a 20-story building—removes the mineral deposit or an energy resource such as coal (Figure 14-12). The resulting trench is filled with overburden, and a new cut is made parallel to the previous one. This process is repeated over the entire site.

Contour strip mining (Figure 14-13) is used mostly to mine coal and various mineral resources on hilly or mountainous terrain. Huge power shovels and bulldozers cut a series of terraces into the side of a hill. Then, earthmovers remove the overburden, an excavator or power shovel extracts the coal, and the overburden from each

Figure 14-9 Each metal product that we use has a *life cycle*.

Left: ©kaband/Shutterstock.com. Second to left: ©Andrey N Bannov/Shutterstock.com. Center left: ©Vladimir Melnik/Shutterstock.com. Center: ©mares/Shutterstock.com. Center right: Zhu Difeng/Shutterstock.com. Second to right: ©Michael Shake/Shutterstock.com. Right: Pakhnyushcha/Shutterstock.com.

LianeM/Shutterstock.com

Figure 14-10 Natural capital degradation: This spoils pile in Zielitz, Germany, is made up of waste material from the mining of potassium salts used to make fertilizers.

new terrace is dumped onto the one below. Unless the land is restored, what is left are a series of spoils banks and a highly erodible bank of soil and rock called a *highwall.*

Another surface mining method is **mountaintop removal**, in which the top of a mountain is removed to expose seams of coal, which are then extracted (Figure 14-14). This method is commonly used in the Appalachian Mountains of the United States.

With this method, after a mountaintop is blown apart, enormous machines plow waste rock and dirt into valleys below the mountaintops. This destroys forests, buries mountain streams, and increases the risk of flooding. Wastewater and toxic sludge, produced when the coal is processed, are often stored behind dams in these valleys, which can overflow or collapse and release toxic substances such as arsenic and mercury.

In the United States, more than 500 mountaintops in West Virginia and other Appalachian states have been removed to extract coal. According to the U.S. Environmental Protection Agency (EPA), the resulting spoils have buried more than 1,100 kilometers (700 miles) of stream—a total roughly equal in length to the distance between the two U.S. cities of New York and Chicago.

The U.S. Department of the Interior estimates that at least 500,000 surface-mined sites dot the U.S. landscape, mostly in the West. Such sites can be cleaned up and restored, but it is costly and is rarely done even when required by law.

Deep deposits of minerals are removed by **subsurface mining**, in which underground mineral resources are removed through tunnels and shafts. This method is

Figure 14-11 Natural capital degradation: Kennecott's Bingham Canyon Mine in the U.S. state of Utah has produced more copper than any mine in history. This gigantic open-pit mine is almost 5 kilometers (3 miles) wide and 1,200 meters (4,000 feet) deep, and it is getting deeper.

Lee Prince/Shutterstock.com

Figure 14-12 Natural capital degradation: This area strip mining for coal is taking place in the U.S. state of Wyoming.

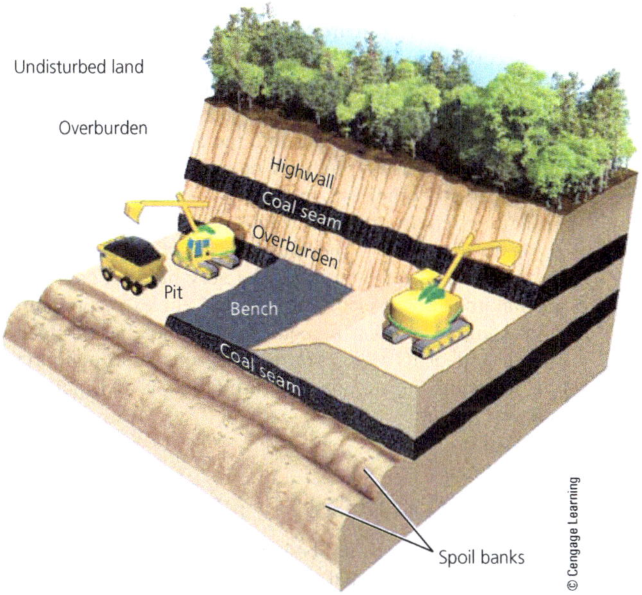

Figure 14-13 Natural capital degradation: Contour strip mining is used in hilly or mountainous terrain.

used to remove metal ores and coal that are too deep to be extracted by surface mining. Miners dig a deep, vertical shaft, blast open subsurface tunnels and chambers to reach the deposit, and use machinery to remove the resource and transport it to the surface.

Subsurface mining disturbs less than one-tenth as much land as surface mining disturbs, and it usually produces less waste material. However, it creates hazards such as cave-ins, explosions, and fires. Miners often get lung diseases caused by prolonged inhalation of mineral or coal dust in subsurface mines. Another problem is *subsidence*—the collapse of land above some underground mines. It can damage houses, crack sewer lines, break gas mains, and disrupt groundwater systems.

Surface and subsurface mining operations also produce large amounts of solid waste—three-fourths of all U.S. solid waste—and cause major water and air pollution. For example, *acid mine drainage* occurs when rainwater that seeps through an underground mine or a spoils pile from a surface mine carries sulfuric acid (H_2SO_4, pro-

Figure 14-14 Natural capital degradation: Mountaintop removal coal mining near Whitesville, West Virginia.

duced when aerobic bacteria act on minerals in the spoils) to nearby streams (Figure 14-15) and groundwater.

According to the EPA, mining has polluted mountain streams in about 40% of the western watersheds in the United States, and it accounts for 50% of all the country's emissions of toxic chemicals into the atmosphere. In fact, the mining industry produces more of such toxic emissions than any other U.S. industry.

Where environmental regulations and enforcement have not been strict, the mining and processing of rare earths (**Core Case Study**) has had a very harmful environmental impact. For example, in China, the extraction and processing of rare-earth metals and oxides has stripped much land of its vegetation and topsoil while polluting the air, acidifying streams, and leaving toxic and radioactive waste piles.

CASE STUDY

The Real Cost of Gold

Many newlyweds would be surprised to know that mining enough gold to make their wedding rings produced roughly enough mining waste to equal the total weight of more than three mid-size cars. This waste is usually left piled near the mine site and can pollute the air and nearby surface water.

In 2012, the world's top five gold-producing countries were, in order, China, Australia, the United States, South Africa, and Russia. In Australia and North America, mining companies dig up massive amounts of rock containing only small concentrations of gold. At about 90% of the world's gold mines, the mineral is extracted with the use of a solution of highly toxic cyanide salts sprayed onto piles of crushed rock. The solution reacts with the gold and then drains off the rocks, pulling some gold with it, into settling ponds (Figure 14-16). After the solution is recirculated a number of times, the gold is removed from the ponds.

Until sunlight breaks down the cyanide, it is extremely toxic to birds and mammals drawn to these ponds in search of water. The ponds can also leak or overflow, which poses threats to underground drinking water supplies and to fish and other forms of life in nearby lakes and streams. Special liners in the settling ponds can help prevent leaks, but some have failed. According to the EPA, all such liners will eventually leak.

After extracting the gold from a mine, some mining companies have declared bankruptcy. This has allowed them to walk away from cleaning up their mining operations, leaving behind large amounts of cyanide-laden water in leaky holding ponds.

Since 1980, millions of poverty-stricken miners have streamed into tropical forests in search of gold and this influx has increased greatly since 2001 when the price of gold began rising sharply. As a result, illegal deforestation to make way for mines has also increased rapidly in parts of the Amazon Basin. These miners use toxic mercury illegally to separate gold from its ore, and they heat the mixture of gold and mercury to vaporize the mercury and leave the gold, causing dangerous air and water pollution. Many of these miners and villagers near their mines then inhale toxic mercury, drink polluted water, and eat fish contaminated with mercury.

There are nontoxic methods for extracting gold, but the cyanide process is strongly entrenched in the industry. For gold and other mineral resource use to become less environmentally harmful, consumers will have to start demanding more sustainably produced mineral products and substitutes, and producers will have to find ways to produce them—a subject of the next section of this chapter.

Figure 14-15 Acid mine drainage from an abandoned open-pit coal mine in Portugal.

Removing Metals from Ores Has Harmful Environmental Effects

Ore extracted by mining typically has two components: the ore mineral, containing the desired metal, and waste material. Removing the waste material from ores produces **tailings** that are left in piles or put into tailings ponds where they settle out. The photo that opens this chapter shows one of two tailings ponds that contain toxic heavy metals at a molybdenum mine in Colorado, located at the headwaters of two mountain streams that flow into the Colorado River (Figure 13-1, p. 318). Particles of toxic metals in tailings piles can be blown by the wind or washed out by rainfall and can contaminate surface water and groundwater. Tailings ponds can also leak and contaminate surface water and groundwater.

Creatas/SuperStock

Figure 14-16 This gold mine in the Black Hills of the U.S. state of South Dakota produces cyanide leach piles and settling ponds.

After the waste material is removed, heat or chemical solvents are used to extract the metals from mineral ores. Heating ores to release metals is called **smelting** (Figure 14-9). Without effective pollution control equipment, smelters emit enormous quantities of air pollutants, including sulfur dioxide and suspended toxic particles, which damage vegetation and acidify soils in the surrounding area. Smelters also cause water pollution and produce liquid and solid hazardous wastes that require safe disposal. A 2012 study by Blacksmith Institute and Green Cross Switzerland found that lead smelting is the world's second most toxic industry after the recycling of lead–acid batteries. Using chemicals to extract metals from their ores can also create numerous problems, as we saw in the case of using cyanide to remove gold (Case Study, p. 360).

14-4 How Can We Use Mineral Resources More Sustainably?

CONCEPT 14-4
We can try to find substitutes for scarce resources, reduce resource waste, and recycle and reuse minerals.

We Can Find Substitutes for Some Scarce Mineral Resources

Some analysts believe that even if supplies of key minerals become too expensive or too scarce due to unsustainable use, human ingenuity will find substitutes (Concept 14-4). They point to the current *materials revolution* in which silicon and other materials are replacing some metals for common uses. They also point out the possibilities of finding substitutes for scarce minerals through nanotechnology (Science Focus 14.1), as well as through other emerging technologies.

For example, fiber-optic glass cables that transmit pulses of light are replacing copper and aluminum wires in telephone cables, and nanowires may eventually replace fiber-optic glass cables. High-strength plastics and materials, strengthened by lightweight carbon, hemp, and glass fibers, are beginning to transform the automobile and aerospace industries. These new materials do not need painting (which reduces pollution and costs), can be molded into any shape, and can increase fuel efficiency by greatly reducing the weights of motor vehicles and airplanes. Such new materials are even being used to build bridges. One such material, called *graphene*, promises to be another new breakthrough material (Science Focus 14.2).

But resource substitution is not a cure-all. For example, platinum is currently unrivaled as a catalyst and is used in industrial processes to speed up chemical reactions, and chromium is an essential ingredient of stainless steel. We can try to find substitutes for such scarce resources, but this may not always be possible.

We Can Use Mineral Resources More Sustainably

Figure 14-17 lists several ways to use mineral resources more sustainably (Concept 14-4). One strategy is to focus on recycling and reuse of nonrenewable mineral resources, especially valuable or scarce metals such as gold, iron, copper, aluminum, and platinum. Recycling, an application of the chemical cycling **principle of sustainability** (see Figure 1-2, p. 6 or back cover), has a much lower environmental impact than that of mining and processing metals from ores. For example,

SCIENCE FOCUS 14.1

THE NANOTECHNOLOGY REVOLUTION

Nanotechnology, or *tiny tech,* uses science and engineering to manipulate and create materials out of atoms and molecules at the ultra-small scale of less than 100 nanometers. The diameter of the period at the end of this sentence is about 1 million nanometers.

At the nanometer level, conventional materials have unconventional and unexpected properties. For example, scientists have learned to link carbon atoms together to form carbon nanotubes that are 60 times stronger than high-grade steel. An incredibly thin and nearly invisible thread of this material is strong enough to suspend a pickup truck. Using carbon nanotubes to build cars would make them stronger and safer and would improve gas mileage by making them up to 80% lighter.

Currently, nanomaterials are used in more than 1,300 consumer products and the number is growing rapidly. Such products include stain-resistant and wrinkle-free clothes, odor-eating socks, self-cleaning glass surfaces and exterior surfaces on buildings, self-cleaning sinks and toilets, sunscreens, waterproof coatings for cell phones and other electrical devices, some cosmetics, some processed foods, and food containers that release nanosilver ions to kill bacteria, molds, and fungi. A new one-atom-thick nanotextile material that can conduct electricity could be incorporated into clothing, and this could allow you to charge your cellphone or start your laptop by plugging it into your jeans or your tee shirt.

Nanotechnologists envision technological innovations such as a supercomputer smaller than a grain of rice, thin and flexible solar-cell films that could be attached to or painted onto almost any surface, motors smaller than a human cell, biocomposite materials that would make our bones and tendons super strong, nanovessels filled with medicines that could be delivered to cells anywhere in the body, and nanomolecules specifically designed to seek out and kill cancer cells. Computer-controlled tabletop nanofactories could produce items from computerized blueprints within minutes.

Nanotechnology has many potential environmental benefits. Designing and building existing and new products from the molecular level up would greatly reduce the need to mine many materials. It would also require less material and energy and reduce waste production. We may be able to use nanoparticles to remove industrial pollutants in contaminated air, soil, and groundwater. Nanofilters might someday be used to desalinate seawater at an affordable cost and to purify water by removing chemicals, bacteria, and even viruses. **GREEN CAREER:** environmental nanotechnology

So what's the catch? The main problem is the serious concerns about the possible harmful health effects of nanotechnology on humans. The tiny size and large combined surface area of the huge numbers of these nanoparticles involved in any particular application make them more chemically reactive and potentially more toxic to humans and other animals than are conventional materials composed of much larger particles. Laboratory studies involving mice and other test animals reveal that nanoparticles can

- be inhaled deeply into the lungs and absorbed into the bloodstream and can penetrate cell membranes, including those in the brain;
- move across the placenta from a mother to her fetus and can move from the nasal passage to the brain;
- result in lung damage similar to that caused by mesothelioma, a deadly cancer resulting from the inhalation of asbestos particles;
- and cause changes in the rate and rhythm of a rodent's heartbeat, similar to such changes caused by heart disease.

We know far too little about these and other and risks at a time when the use of untested, unregulated, and unlabeled nanoparticles is growing rapidly. In 2009, an expert panel of the U.S. National Academy of Sciences said that the federal government was not doing enough to evaluate the potential health and environmental risks of using engineered nanomaterials. For example, the U.S. Food and Drug Administration does not even maintain a list of the food products and cosmetics that contain nanomaterials.

By contrast, the European Union (EU) takes a precautionary approach to the use of nanomaterials and untested chemicals. While U.S. regulators generally assume that nanoparticles and chemicals are innocent until shown to be harmful, EU laws require that manufacturers demonstrate the safety of their products before they can enter the marketplace.

Many analysts say we need to take three steps before unleashing nanotechnology more broadly. *First,* greatly increase research on the potential harmful health effects of nanoparticles. *Second,* develop guidelines and regulations for controlling its growing applications until we know more about the potentially harmful effects of this new technology. *Third,* require labeling of all products containing nanoparticles.

Critical Thinking

Do you think the potential benefits of nanotechnology products outweigh their potentially harmful effects? Explain.

SCIENCE FOCUS 14.2

GRAPHENE: A REVOLUTIONARY MATERIAL

Graphene is made from graphite—the material in the center of a pencil. Both are made of carbon. However, ultrathin graphene consists of a single layer of carbon atoms packed into a two-dimensional hexagonal lattice somewhat like chicken wire and that can be applied as a transparent film to surfaces (Figure 14-A).

Graphene is one of the world's thinnest and strongest materials. A sheet of graphene is 150,000 times thinner than a human hair and 200 times stronger than structural steel. A piece of this amazing material stretched over a coffee mug could support the weight of a car. It is also a good conductor of electricity and conducts heat better than any known material. In 2010, the two scientists who had discovered how to make graphene 6 years earlier won the Nobel Prize in Physics for their discovery.

The use of graphene could revolutionize the electric car industry by leading to the production of batteries that can be recharged 10 times faster and hold 10 times more power than current car batteries. Graphene composites can also be used to make stronger and lighter plastics, lightweight aircraft and motor vehicles, flexible computer tablets, and TV screens as thin as a magazine. Within 5 years, it might also be used to make flexible, more efficient, less costly solar cells that can be attached to almost anything. Graphene could also replace the silicon transistors that are used as chips in computers and other electronic devices, and this could make almost any electronic device run much faster and use less power. Indeed, some scientists contend that use of graphene will change the world more than any technological development since the invention of the silicon chip.

Graphene is made from very high purity and expensive graphite. According to the U.S. Geological Survey, China controls about 73% of the world's high-purity graphite production. As with rare-earth metals (**Core Case Study**), the United States mines very little natural graphite and imports most of its graphite from China, which may restrict U.S. product exports as the use of graphene grows.

Geologists are looking for deposits of graphite in the United States. However, in 2011, a team of Rice University chemists, led by James M. Tour, found ways to make large sheets of high-quality graphene from inexpensive materials found in garbage and from dog feces. If such a process becomes economically feasible, concern over supplies of graphite could vanish, along with any harmful environmental effects of the mining and processing of graphite.

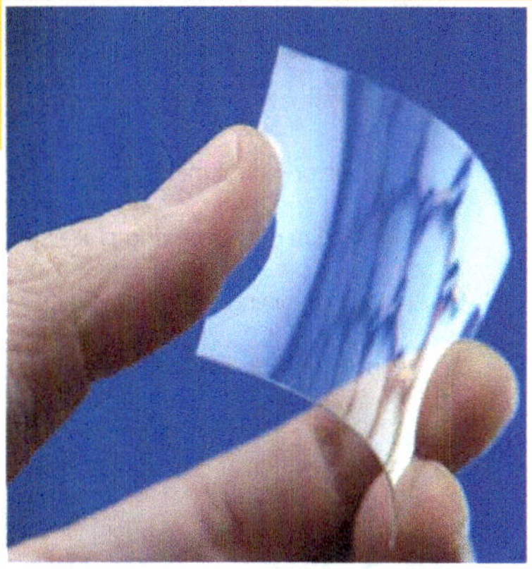

Science News

Figure 14-A Graphene, which consists of a single layer of carbon atoms linked together in a hexagonal lattice, is a revolutionary new material.

Critical Thinking

Would you invest money in a company that mines and purifies graphite and converts it to graphene? Explain.

---

recycling aluminum beverage cans and scrap aluminum produces 95% less air pollution and 97% less water pollution, and uses 95% less energy, than mining and processing aluminum ore. Cleaning up and reusing items instead of recycling them has an even lower environmental impact.

Researchers are working hard to try to ensure adequate supplies of rare earths on a short-term basis, and on a long-term basis, to find alternatives to these materials (**Core Case Study**). One way to increase supplies is to extract and recycle rare-earth metals from the massive amounts of electronic wastes that are being produced.

Companies that build batteries for electric cars are beginning to switch from making nickel–metal-hydride batteries, which require the rare-earth metal lanthanum, to manufacturing lighter-weight lithium-ion batteries, which researchers are now trying to improve (Individuals Matter 14.1). Lithium-ion batteries are also used in cell phones, iPods, iPads, and laptop computers. The problem is that some countries, including the United States, do not have adequate supplies of lithium.

CONSIDER THIS. . .

CONNECTIONS Lithium and U.S. Energy Dependence

Lithium (Li), the world's lightest metal, is a vital component of lithium-ion batteries, which are being used in a variety of devices on a rapidly growing basis. The South American countries of Bolivia, Chile, and Argentina have about 80% of the global reserves of lithium. Bolivia alone has about 50% of these reserves, while the United States holds only about 3%. Japan, China, South Korea, and the United Arab Emirates are buying up access to global lithium reserves to ensure their ability to sell lithium or batteries to the rest of the world. Within a few decades, the United States may be heavily dependent on expensive imports of lithium and lithium batteries.

Scientists are also searching for substitutes for rare-earth metals that could be used to make increasingly important powerful magnets. In Japan and the United States, researchers are developing a variety of devices that use no rare-earth minerals, are light and compact, and can deliver more power with greater efficiency at a reduced cost. These include electric motors for hybrid-electric vehicles and electromagnets for use in wind turbines.

Solutions

Sustainable Use of Nonrenewable Minerals

- Reuse or recycle metal products whenever possible
- Redesign manufacturing processes to use less mineral resources
- Reduce mining subsidies
- Increase subsidies for reuse, recycling, and finding substitutes

Figure 14-17 We can use nonrenewable mineral resources more sustainably (Concept 14-4). ***Questions:*** Which two of these solutions do you think are the most important? Why?

14-5 What Are the Earth's Major Geological Hazards?

CONCEPT 14-5

Dynamic processes move matter within the earth and on its surface and can cause volcanic eruptions, earthquakes, tsunamis, erosion, and landslides.

The Earth Beneath Your Feet Is Moving

We tend to think of the earth's crust, mantle, and core as fairly static. However, according to geologists, the flows of energy and heated material within the earth's convection cells (Figure 14-3) are so powerful that they have caused the lithosphere to break up into a dozen or so huge rigid plates, called **tectonic plates**, which move extremely slowly atop the asthenosphere (Figure 14-18).

These gigantic plates are somewhat like the world's largest and slowest-moving surfboards on which we ride without noticing their movement. Their typical speed is about the rate at which fingernails grow. Throughout the earth's history, continents have split apart and joined as tectonic plates drifted atop the earth's asthenosphere (Figure 4-9, p. 86).

Much of the geological activity at the earth's surface takes place at the boundaries between tectonic plates as they separate, collide, or grind along against each other (Figure 14-19). The tremendous forces produced at these plate boundaries can cause mountains to form, earthquakes to shake parts of the crust, and volcanoes to erupt.

Volcanoes Release Molten Rock from the Earth's Interior

An active **volcano** occurs where magma rising in a plume through the lithosphere reaches the earth's surface through a central vent or a long crack, called a *fissure* (Figure 14-20). Magma that reaches the earth's surface is called *lava* and often builds into a cone.

Many volcanoes form along the boundaries of the earth's tectonic plates when one plate slides under or moves away from another plate. Volcanoes can *erupt*, releasing large chunks of lava rock, glowing hot ash, liquid lava, and gases (including water vapor, carbon dioxide, and sulfur dioxide) into the environment (**Concept 14-5**). Eruptions can be extremely destructive, causing loss of life and obliterating ecosystems and human communities.

While volcanic eruptions can be destructive, they do provide some benefits. They can result in the formation of majestic mountains and lakes such as Crater Lake (see Figure 8-15, p. 179), and the weathering of lava contributes to fertile soils. Hundreds of volcanoes have erupted on the ocean floor, building cones that have reached the ocean's surface, eventually to form islands that have become favorable for human settlement.

We can reduce the loss of human life and some of the property damage caused by volcanic eruptions in several ways. We use historical records and geological measurements to identify high-risk areas, so that people can avoid living in those areas. We also use monitoring devices that warn us when volcanoes are likely to erupt, and in some areas prone to volcanic activity, evacuation plans have been developed.

Earthquakes Are Geological Rock-and-Roll Events

Forces inside the earth's mantle put tremendous stress on rock within the crust. Such stresses can be great enough to cause sudden breakage and shifting of the rock, producing a *fault*, or fracture in the earth's crust (Figure 14-19). When a fault forms, or when there is abrupt movement on an existing fault, energy that has accumulated over time is released in the form of vibrations, called *seismic waves*, that move in all directions through the surrounding rock—an event called an **earthquake** (Figure 14-21 and Concept 14-5). Most earthquakes occur at the boundaries of tectonic plates.

Seismic waves move upward and outward from the earthquake's *focus* (Figure 14-21, left) like ripples in a pool

individuals matter 14.1

Yu-Guo Guo: Designer of Nanotechnology Batteries and National Geographic Emerging Explorer

©Jinsong Hu

Yu-Guo Guo is a professor of chemistry and a nanotechnology researcher at the Chinese Academy of Sciences in Beijing. He has invented nanomaterials that can be used to make lithium-ion battery packs smaller, more powerful, and less costly, which makes them more useful for powering electric cars and electric bicycles. This is an important scientific advance, because the battery pack is the most important and expensive part of any electric vehicle.

Guo's innovative use of nanomaterials has greatly increased the power of lithium-ion batteries by enabling electric current to flow more efficiently through what he calls "3-D conducting nanonetworks." With this promising technology, lithium-ion battery packs in electric vehicles can be fully charged in a few minutes. They also have twice the energy storage capacity of today's batteries, and thus will extend the range of electric vehicles by enabling them to run longer.

Guo predicts that within 5 years, the electric-vehicle market will be well established and, in some cities, will account for 10% of all cars. He is also interested in developing nanomaterials for use in solar cells and fuel cells that could be used to generate electricity and to power vehicles.

Background photo: Pi-Lens/Shutterstock.com

of water. Scientists measure the severity of an earthquake by the *magnitude* of its seismic waves. The magnitude is a measure of ground motion (shaking) caused by the earthquake, as indicated by the *amplitude*, or size of the seismic waves when they reach a recording instrument, called a *seismograph*.

Scientists use the *Richter scale*, on which each unit has an amplitude 10 times greater than the next smaller unit. Seismologists rate earthquakes as *insignificant* (less than 4.0 on the Richter scale), *minor* (4.0–4.9), *damaging* (5.0–5.9), *destructive* (6.0–6.9), *major* (7.0–7.9), and *great* (over 8.0). The largest recorded earthquake occurred in Chile on May 22, 1960, and measured 9.5 on the Richter scale. Each year, scientists record the magnitudes of more than 1 million earthquakes, most of which are too small to feel.

The primary effects of earthquakes include shaking and sometimes a permanent vertical or horizontal displacement of a part of the crust. These effects can have serious consequences for people and for buildings, bridges, freeway overpasses, dams, and pipelines. A major earthquake is a very large rock-and-roll geological event.

One way to reduce the loss of life and property damage from earthquakes is to examine historical records and make geological measurements to locate active fault zones. We can then map high-risk areas and establish building codes that regulate the placement and design of buildings in such areas. Then people can evaluate the

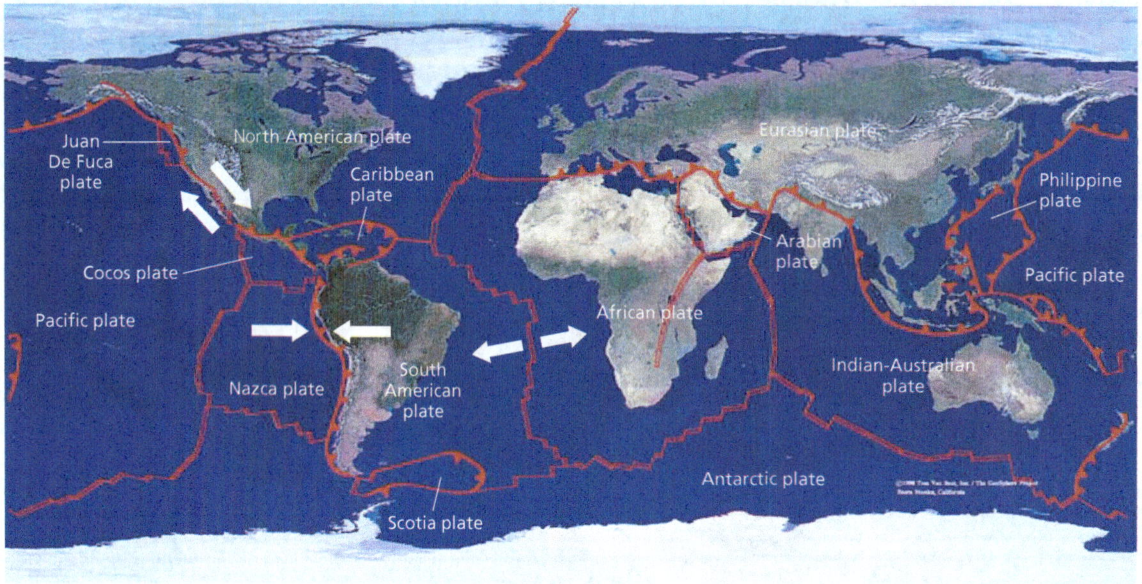

Animated Figure 14-18 The earth's crust has been fractured into several major tectonic plates. White arrows indicate examples of where plates are colliding, separating, or grinding along against each other in opposite directions. ***Question:*** Which plate are you riding on?

Figure 14-19 The San Andreas Fault, created by the North American Plate and the Pacific Plate sliding very slowly past each other, runs almost the full length of California (see map). It is responsible for earthquakes of various magnitudes, which have caused rifts on the land surface in some areas (photo).
Kevin Schafer/Age Fotostock

risk and factor it into their decisions about where to live. Also, engineers know how to make homes, large buildings, bridges, and freeways more earthquake resistant, although this is costly. See Figure 29, p. S55, in Supplement 6, for a map comparing earthquake risks in various areas of the United States, and Figure 30, p. S55, in Supplement 6 for a map of such areas throughout the world.

Earthquakes on the Ocean Floor Can Cause Huge Waves Called Tsunamis

A **tsunami** is a series of large waves generated when part of the ocean floor suddenly rises or drops (Figure 14-22). Most large tsunamis are caused when certain types of faults in the ocean floor move up or down as a result of a large underwater earthquake. Other causes are landslides

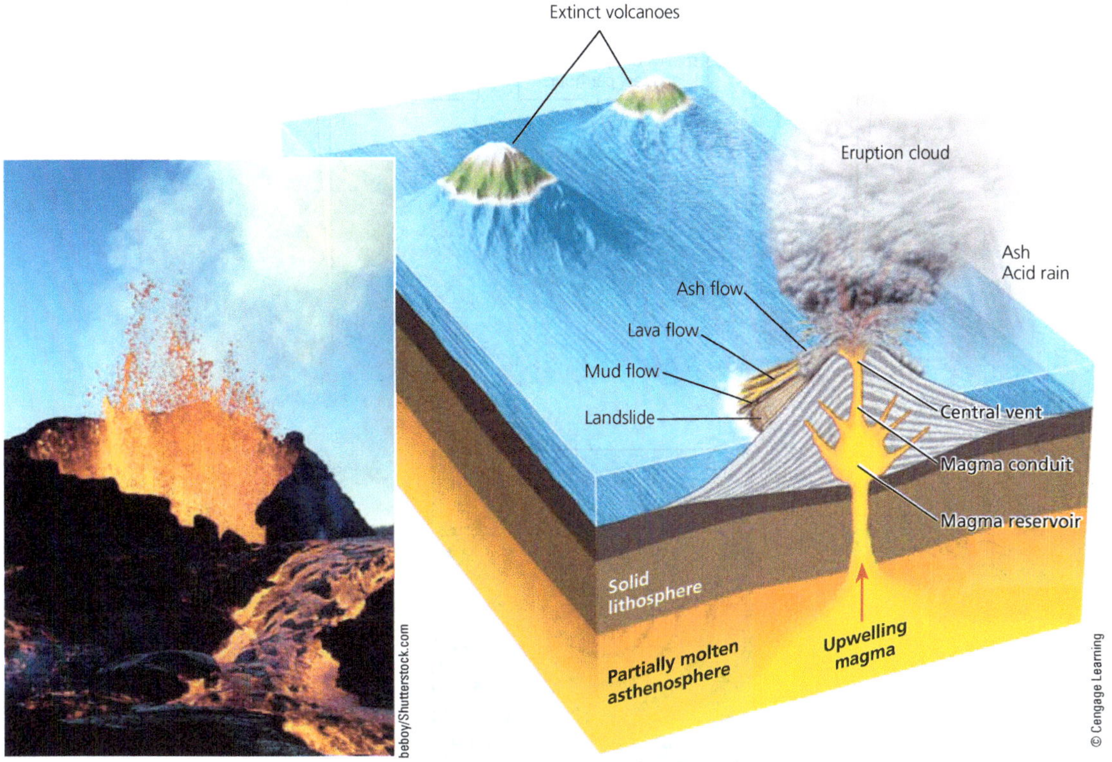

Figure 14-20 Sometimes, the internal pressure in a volcano is high enough to cause lava, ash, and gases to be ejected into the atmosphere (photo inset) or to flow over land, causing considerable damage.

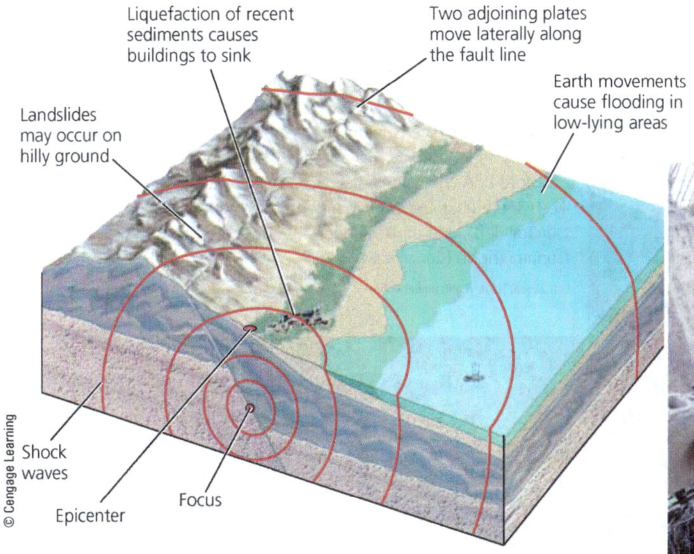

Figure 14-21 An *earthquake* (left) is one of nature's most powerful events.The photo shows damage from a 2010 earthquake in Port-au-Prince, Haiti.

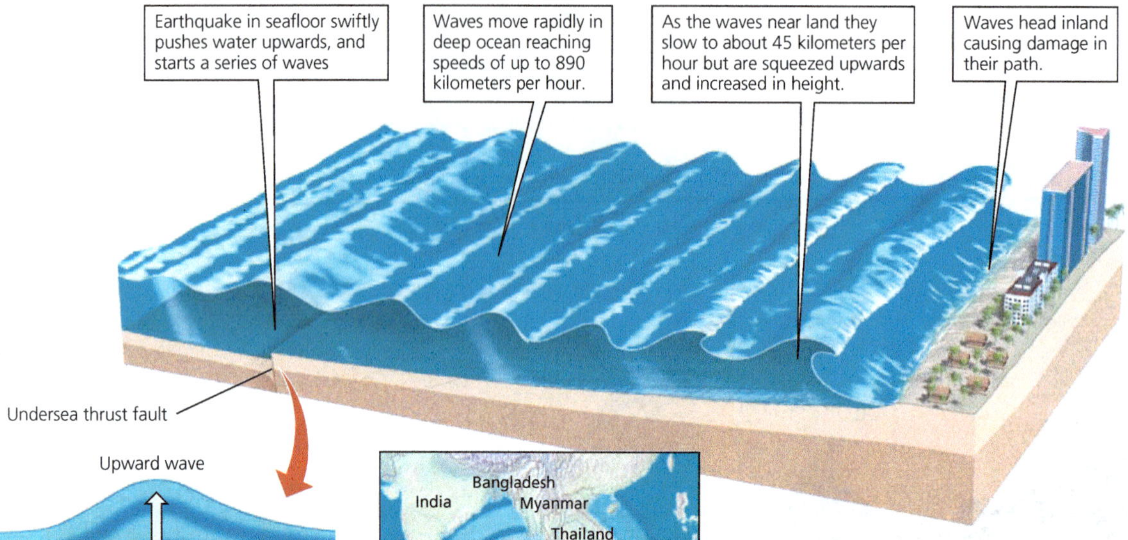

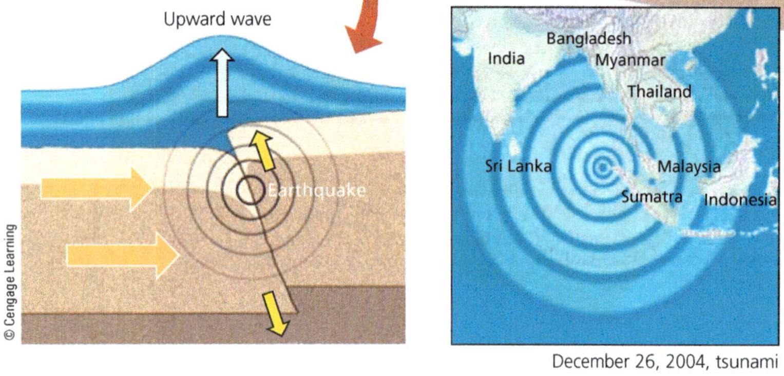

Figure 14-22 This diagram illustrates how a tsunami forms. The map shows the area affected by a large tsunami in December 2004—one of the largest ever recorded.

generated by earthquakes and volcanic eruptions (**Concept 14-5**).

Tsunamis are often called *tidal waves,* although they have nothing to do with tides. They can travel far across the ocean at the speed of a jet plane. In deep water the waves are very far apart—sometimes hundreds of kilometers—and their crests are not very high. As a tsunami approaches a coast with its shallower waters, it slows down, its wave crests squeeze closer together, and their heights grow rapidly. It can hit a coast as a series of towering walls of water that can level buildings.

The largest recorded loss of life from a tsunami occurred in December 2004 when a great underwater earthquake in the Indian Ocean with a magnitude of 9.15

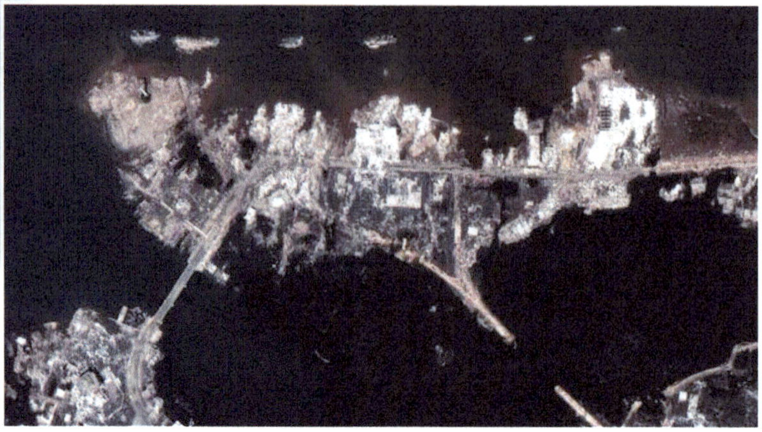

Figure 14-23 The tsunami of December 2004 killed 168,000 people in Indonesia, as well as tens of thousands more in other countries bordering the Indian Ocean. These photos show the Banda Aceh Shore near Gleebruk in Indonesia on June 23, 2004, before the tsunami (left), and on December 28, 2004, after it was stuck by the tsunami (right) (**Concept 14-5**).

Left: DigitalGlobe. Right: DigtialGlobe.

caused a tsunami with towering waves that killed around 279,900 people and devastated many coastal areas of Indonesia (see map in Figure 14-22 and Figure 14-23), Thailand, Sri Lanka, South India, and even eastern Africa. It also displaced about 1.8 million people (1.3 million of them in India and Indonesia), and destroyed or damaged about 470,000 buildings and houses. There were no recording devices in place to provide an early warning of this tsunami.

In 2011, a large tsunami caused by a powerful earthquake off the coast of Japan generated 3-story-high waves that killed almost 19,000 people, displaced more than 300,000 people, and destroyed or damaged 125,000 buildings. It also heavily damaged three nuclear reactors, which then released dangerous radioactivity into the surrounding environment.

In some areas, scientists have built networks of ocean buoys and pressure recorders on the ocean floor to collect data that can be relayed to tsunami emergency warning centers. However, these networks are far from complete.

Big Ideas

- **Dynamic forces that move matter within the earth and on its surface recycle the earth's rocks, form deposits of mineral resources, and cause volcanic eruptions, earthquakes, and tsunamis.**
- **The available supply of a mineral resource depends on how much of it is in the earth's crust, how fast we use it, the mining technology used to obtain it, its market prices, and the harmful environmental effects of removing and using it.**
- **We can use mineral resources more sustainably by trying to find substitutes for scarce resources, reducing resource waste, and reusing and recycling nonrenewable minerals.**

TYING IT ALL TOGETHER Rare-Earth Metals and Sustainability

M. Dykstra/Shutterstock.com

In the **Core Case Study** that opened this chapter, we learned about the importance of various rare-earth metals that are used in a variety of modern technologies, including electric motor vehicles and the energy-efficient compact fluorescent lightbulbs (photo at left) that are replacing energy-inefficient incandescent bulbs.

In this chapter, we looked at technological developments that could help us to expand supplies of mineral resources and to use them more sustainably. For example, if we develop it safely, we could use nanotechnology (Science Focus 14.1) to make new materials that could replace scarce mineral resources and greatly reduce the environmental impacts of mining and processing such resources. We looked at how biomining makes use of microbes to extract mineral resources without disturbing the land or polluting air and water as much as conventional mining operations do. Another emerging technology uses graphene to replace conventional transistors and to produce more efficient and affordable solar cells to generate electricity—an application of the solar energy **principle of sustainability** (see Figure 1-2, p. 6 or back cover).

We can also use mineral resources more sustainably by reusing and recycling them, and by reducing unnecessary resource use and waste—applying the chemical cycling **principle of sustainability**. In addition, industries can mimic nature by using a diversity of ways to reduce the harmful environmental impacts of mining and processing mineral resources, thus applying the biodiversity **principle of sustainability**.

Chapter Review

Core Case Study

1. Explain the importance of the rare-earth metals.

Section 14-1

2. What are the two key concepts for this section? Define **geology**, **core**, **mantle**, **asthenosphere**, **crust**, and **lithosphere**. Define **mineral**, **mineral resource**, and **rock**. Define and distinguish among **sedimentary rock**, **igneous rock**, and **metamorphic rock** and give an example of each. Define and describe the nature and importance of the **rock cycle**. Define **ore** and distinguish between a **high-grade ore** and a **low-grade ore**. List five important nonrenewable mineral resources and their uses.

Section 14-2

3. What are the two key concepts for this section? What are the **reserves** of a mineral resource and how can they be expanded? What two factors determine the future supply of a nonrenewable mineral resource? Explain how the supply of a nonrenewable mineral resource can be economically depleted and list the five choices we have when this occurs. What is **depletion time** and what factors affect it?

4. What five nations supply most of the world's nonrenewable mineral resources? How dependent is the United States on other countries for important nonrenewable mineral resources? Explain the concern over U.S. access to rare-earth mineral resources. Describe the conventional view of the relationship between the supply of a mineral resource and its market price. Explain why some economists believe this relationship no longer applies. Summarize the pros and cons of providing government subsidies and tax breaks for mining companies.

5. Summarize the opportunities and limitations of expanding mineral supplies by mining lower-grade ores. What are the advantages and disadvantages of biomining? Describe the opportunities and possible problems that could result from deep-sea mining.

Section 14-3

6. What is the key concept for this section? Summarize the life cycle of a metal product.

7. What is **surface mining**? Define **overburden** and **spoils**. Define **open-pit mining** and **strip mining**, and distinguish among **area strip mining**, **contour strip mining**, and **mountaintop removal mining**. Describe three harmful environmental effects of surface mining. What is **subsurface mining**? Describe the harmful effects of gold mining. Describe the environmental effects of mining and processing rare earths. Define **tailings** and explain why they can be hazardous. What is **smelting** and what are its major harmful environmental effects?

Section 14-4

8. What is the key concept for this section? Describe the opportunities and limitations of finding substitutes for key mineral resources. What is **nanotechnology** and what are some of its potential environmental and other benefits? What are some problems that could arise from the widespread use of nanotechnology? Describe the potential of using graphene as a new resource. Explain the benefits of recycling and reusing valuable metals. List five ways to use nonrenewable mineral resources more sustainably. What are two examples of research into substitutes for rare-earth metals? What is the relationship between lithium and U.S. energy independence?

Section 14-5

9. What is the key concept for this section? What are **tectonic plates**, and what typically happens when they collide, move apart, or grind against one another? Define **volcano** and describe the nature and major effects of a volcanic eruption. Define **earthquake** and describe its nature and major effects. What is a **tsunami** and what are its major effects?

10. What are the three big ideas of this chapter? Explain how we can apply the three scientific **principles of sustainability** to obtain and use rare-earth metals and other nonrenewable mineral resources in more sustainable ways.

Note: Key terms are in bold type.

Critical Thinking

1. Give three reasons why rare-earth metals (**Core Case Study**) are important to your lifestyle.

2. Would you favor giving the owners of the California rare-earth metals (**Core Case Study**) mine significant government tax breaks and subsidies to put the mine back into production? Explain. Would you favor reducing the environmental regulations for the mining and processing of these metals? Explain.

3. You are an igneous rock. Describe what you experience as you move through the rock cycle. Repeat this exercise, assuming you are a sedimentary rock and again assuming you are a metamorphic rock.
4. What are three ways in which the rock cycle benefits your lifestyle?
5. Use the second law of thermodynamics (see Chapter 2, p. 43) to analyze the scientific and economic feasibility of each of the following processes:
   a. Extracting certain minerals from seawater
   b. Mining increasingly lower-grade deposits of minerals
   c. Continuing to mine, use, and recycle minerals at increasing rates
6. Suppose you were told that mining deep-ocean mineral resources would mean severely degrading ocean-bottom habitats and life-forms such as giant tube worms and giant clams. Do you think that such information should prevent or put an end to ocean-bottom mining? Explain.
7. List three ways in which a nanotechnology revolution could benefit you and three ways in which it could harm you.
8. What are three ways to reduce the harmful environmental impacts from the mining and processing of nonrenewable mineral resources? What are three aspects of your lifestyle that contribute to these harmful impacts?

Doing Environmental Science

Do research to determine which nonrenewable mineral resources go into the manufacture of each of the following items and how much of each of these resources are required to make each item: **(a)** a cell phone, **(b)** a wide-screen TV, and **(c)** a large pickup truck.

Global Environment Watch Exercise

Use the Global Environment Watch site to find and read an article that deals with U.S. rare-earth metal supplies (**Core Case Study**). Is there scientific information cited in the article to support the author's point of view? Give specific examples. Do you think there are any types of supporting scientific data not mentioned in the article that would strengthen the author's point of view? For example, would you add statistical data to support a point, or data in a graph indicating possible cause-and-effect relationships? Be specific and give reasons for your suggestions.

Data Analysis

Rare-earth metals are widely used in a variety of important products (**Core Case Study**). In 2010, China produced about 97% of the world's rare earths and the United States produced none. According to the U.S. Geological Survey, China has 50% of the world's reserves of rare-earth metals and the United States has 13%.

1. In 2010, China had 55 million metric tons of rare-earth metals in its reserves and produced 130,000 metric tons of these metals. At this rate of production how long will China's rare-earth reserves last?
2. In 2010, the global demand for rare-earth metals was about 133,600 metric tons. If China produced all of the world's rare-earth metals, how long would their reserves last?
3. Global demand for rare earths is projected to rise to at least 185,000 metric tons by 2015. If China produced all of the world's rare-earth metals how long would its reserves last?

15 Nonrenewable Energy

Typical citizens of advanced industrialized nations each consume as much energy in 6 months as typical citizens in less developed countries consume during their entire life.

MAURICE STRONG

Key Questions

15-1 What is net energy and why is it important?

15-2 What are the advantages and disadvantages of using oil?

15-3 What are the advantages and disadvantages of using natural gas?

15-4 What are the advantages and disadvantages of using coal?

15-5 What are the advantages and disadvantages of using nuclear power?

Mountaintop removal coal mine site in West Virginia.

Melissa Farlow/National Geographic Creative

CORE CASE STUDY Is the United States Entering a New Oil and Natural Gas Era?

Oil and natural gas are the two most widely used energy resources in the United States (Figure 15-1). There are signs that U.S. oil and natural gas production could increase sharply.

Between 1985 and 2008, oil production in the United States fell while consumption kept rising, and oil imports rose to make up the difference between consumption and production. However, since 2008, U.S. oil production has increased somewhat, largely because high oil prices and improved drilling and extraction technology have made it profitable to extract oil that is dispersed and tightly held in dense formations of shale rock.

According to some oil economists and the International Energy Agency, if oil production increases as projected and oil prices remain at $50 a barrel or higher, the United States could become the world's largest oil producer, probably sometime before 2020. Such a boom in domestic oil production would create large numbers of jobs, stimulate the U.S. economy, and reduce the country's expensive dependence on imported oil.

Since 1999, natural gas drilling and production have also increased dramatically, especially the extraction of natural gas held tightly within shale rock using the same drilling and extraction technology that is used to pull tightly held oil from shale rock. By 2011, this growing supply had led to sharply lower U.S. natural gas prices and made the United States the world's leading natural gas producer. If U.S. natural gas production from shale rock continues to grow as projected, and if natural gas prices do not rise significantly, natural gas could displace environmentally harmful coal as the country's largest source of electricity within three to four decades.

However, there are two major problems with this scenario. One is that the large-scale removal of natural gas and oil held tightly in shale rock requires huge amounts of water and also produces heavily polluted wastewater. This, along with leaks from gas and oil well piping systems, could contaminate shallow aquifers that feed many drinking water wells, as well as deep aquifers, unless the entire drilling, extraction, and wastewater treatment process is strictly monitored and regulated to protect drinking water.

A second problem is that by burning more carbon-containing oil, natural gas, and coal, we will continue to release growing quantities of carbon dioxide (CO_2) and methane (CH_4) into the atmosphere faster than they can be removed by the carbon cycle (see Figure 3-17, p. 66). Computer models project that rising atmospheric levels of these greenhouse gases will play a key role in changing the world's climate in potentially very harmful ways during this century.

In this chapter, we discuss the advantages and disadvantages of using nonrenewable fossil fuels (such as oil, natural gas, and coal) and nuclear power. In the next chapter, we look at the advantages and disadvantages of improving energy efficiency and using a variety of renewable energy resources.

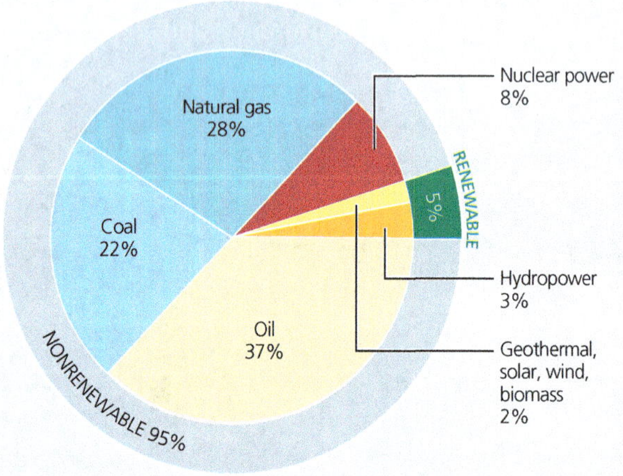

Figure 15-1 Sources of energy used in the United States in 2011. Oil, the most widely used resource, is removed from deposits underground and from deep under the ocean floor in some coastal areas (see photo).

(Compiled by the authors using data from U.S. Energy Information Administration, British Petroleum, and International Energy Agency.)

15-1 What Is Net Energy and Why Is It Important?

CONCEPT 15-1

Energy resources vary greatly in their *net energy yields*—the amount of energy available from a resource minus the amount of energy needed to make it available.

Net Energy Is the Only Energy That Really Counts

It takes energy to produce energy. For example, before oil becomes useful to us, it must be found, pumped up from beneath the ground or ocean floor (see photo in Figure 15-1), transferred to a refinery, converted to gasoline and other fuels and a variety of other widely used chemicals, and delivered to consumers. Each of these steps uses high-quality energy, mostly obtained by burning fossil fuels, especially gasoline and diesel fuel produced from oil. Because of the second law of thermodynamics (see Chapter 2, p. 43), which we cannot violate, some of the high-quality energy used in each step is automatically wasted and degraded to lower-quality energy, mostly heat that ends up in the environment.

The usable amount of high-quality energy available from an energy resource is its **net energy yield.** It is the total amount of high-quality energy available from an energy resource minus the high-quality energy needed to make the energy available (Concept 15-1). It is also related to the *energy return on investment (EROI)*—the energy obtained per unit of energy used to obtain it. Suppose that it takes about 9 units of high-quality energy to produce 10 units of high-quality energy from an energy resource. Then the net energy yield is only 1 unit of energy.

Net energy is like the net profit earned by a business after it deducts its expenses. If a business has $1 million in sales and $900,000 in expenses, its net profit is $100,000.

Figure 15-2 shows generalized net energy yields for energy resources and systems that generate electricity, heat homes and buildings, produce high-temperature heat for industrial processes, and provide transportation. It is based on several sources of scientific data and classifies estimated net energy yields as high, medium, low, or negative (negative being a net energy loss).

Electricity	Net Energy Yield
Energy efficiency	High
Hydropower	High
Wind	High
Coal	High
Natural gas	Medium
Geothermal energy	Medium
Solar cells	Low to medium
Nuclear fuel cycle	Low
Hydrogen	Negative (Energy loss)

High-Temperature Industrial Heat	Net Energy Yield
Energy efficiency (cogeneration)	High
Coal	High
Natural gas	Medium
Oil	Medium
Heavy shale oil	Low
Heavy oil from tar sands	Low
Direct solar (concentrated)	Low
Hydrogen	Negative (Energy loss)

Space Heating	Net Energy Yield
Energy efficiency	High
Passive solar	Medium
Natural gas	Medium
Geothermal energy	Medium
Oil	Medium
Active solar	Low to medium
Heavy shale oil	Low
Heavy oil from tar sands	Low
Electricity	Low
Hydrogen	Negative (Energy loss)

Transportation	Net Energy Yield
Energy efficiency	High
Gasoline	High
Natural gas	Medium
Ethanol (from sugarcane)	Medium
Diesel	Medium
Gasoline from heavy shale oil	Low
Gasoline from heavy tar sand oil	Low
Ethanol (from corn)	Low
Biodiesel (from soy)	Low
Hydrogen	Negative (Energy loss)

Figure 15-2 Generalized *net energy* yields for various energy systems (Concept 15-1). ***Question:*** Based only on these data, which two resources in each category should we be using?

(Compiled by the authors using data from the U.S. Department of Energy; U.S. Department of Agriculture; Colorado Energy Research Institute, *Net Energy Analysis*, 1976; Howard T. Odum and Elisabeth C. Odum, *Energy Basis for Man and Nature*, 3rd ed., New York: McGraw-Hill, 1981, and Charles A.S. Hall and Kent A. Klitgaard, *Energy and the Wealth of Nations*, New York: Springer, 2012.)

Top left: Yegor Korzh/Shutterstock.com. Bottom left: Donald Aitken/National Renewable Energy Laboratory. Top right: Serdar Tibet/Shutterstock.com. Bottom right: Michel Stevelmans/Shutterstock.com.

Some Energy Resources Need Help to Compete in the Marketplace

The following general rule can help us to evaluate the long-term economic usefulness of an energy resource based on its net energy yield: *An energy resource with a low or negative net energy yield can have a hard time competing in the marketplace with other energy alternatives that have medium to high net energy yields unless it receives financial support from the government (taxpayers) or other outside sources.* Such financial support is generally referred to as a *subsidy*, and providing it is called *subsidizing*.

For example, electricity produced by nuclear power has a low net energy yield because large amounts of high-quality energy are needed for each step in the *nuclear power fuel cycle:* to extract and process uranium ore, convert it into nuclear fuel, build and operate nuclear power plants, safely store the resulting highly radioactive wastes for thousands of years, and dismantle each plant after its useful life (typically 40–60 years) and safely store its high-level radioactive parts for thousands of years. The resulting low net energy yield for the whole nuclear fuel cycle is one reason why governments throughout the world heavily subsidize nuclear power to make it available to consumers at an affordable price. Such subsidies help to hide the true cost of the nuclear power fuel cycle and thus violate the full-cost pricing **principle of sustainability** (see Figure 1-5, p. 9 or back cover).

15-2 What Are the Advantages and Disadvantages of Using Oil?

CONCEPT 15-2A
Conventional crude oil is abundant and has a medium net energy yield, but using it causes air and water pollution and releases greenhouse gases to the atmosphere.

CONCEPT 15-2B
Unconventional heavy oil from oil shale rock and tar sands exists in potentially large supplies but has a low net energy yield and a higher environmental impact than conventional oil has.

We Depend Heavily on Oil

Oil is the world's most widely used energy resource (Figure 15-3). We use oil to heat our homes, grow most of our food, transport people and goods, make other energy resources available for use, and manufacture most of the things we use every day, from plastics to cosmetics to asphalt on roads.

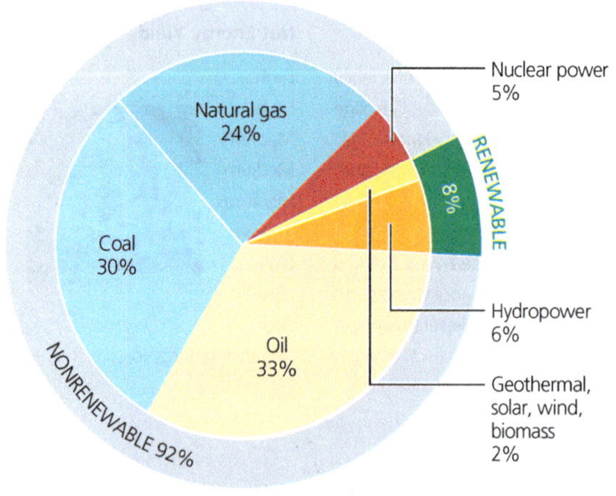

Figure 15-3 Global energy use in 2011.

(Compiled by the authors using data from British Petroleum, U.S. Energy Information Administration, and International Energy Agency.)

Crude oil, or **petroleum**, is a black, gooey liquid consisting mostly of a mix of different combustible hydrocarbons along with small amounts of sulfur, oxygen, and nitrogen impurities. It is also known as conventional or light crude oil. It was formed from the decayed remains of ancient organisms that were crushed beneath layers of rock for millions of years. The resulting liquid and gaseous hydrocarbons migrated upward through porous rock layers to collect as deposits of oil and natural gas, often trapped together beneath layers of impermeable rock.

Scientists identify potential oil deposits by using large machines to pound the earth, sending shock waves deep underground, and they measure how long it takes for the waves to be reflected back. This information is fed into computers and converted into *3-D seismic maps* of the underground that show the locations and sizes of various rock formations. Then oil companies drill holes and remove rock cores from potential oil deposit areas to learn whether there is enough oil to be extracted profitably. If there is, one or more wells are drilled and the light oil is pumped to the surface.

After years of pumping, usually a decade or so, the pressure in a well drops and its rate of crude oil production starts to decline. This point in time is referred to as **peak production** for the well. The same thing can happen to a large oil field when the overall rate of production from its numerous wells begins to drop. Global peak production would occur when the rate of global production of conventional oil begins to decline faster than new oil fields are found and put into production. There is disagreement over whether we have reached global peak production of conventional crude oil and when we might reach it if we have not.

Crude oil from a well cannot be used as it is. It is transported to a refinery by pipeline, truck, rail, or ship (oil tanker) where it is heated to separate it into various

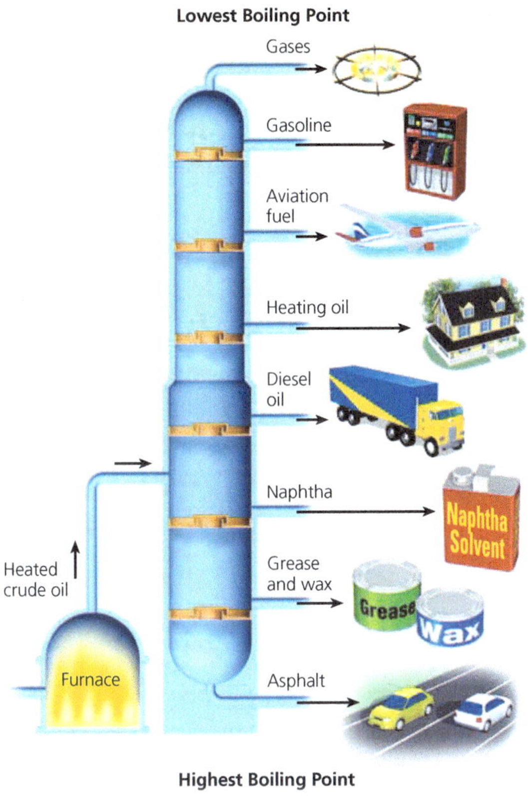

Figure 15-4 When crude oil is refined, many of its components are removed at various levels of a distillation column, depending on their boiling points. The most volatile components with the lowest boiling points are removed at the top of the column, which can be as tall as a nine-story building. The photo shows an oil refinery in Texas.

fuels and other components with different boiling points (Figure 15-4) in a complex process called **refining**. This process, like all other steps in the cycle of oil production and use, requires an input of high-quality energy and decreases the net energy yield of oil. About 2% of the products of refining, called **petrochemicals**, are used as raw materials to make industrial organic chemicals, cleaning fluids, pesticides, plastics, synthetic fibers, paints, medicines, cosmetics, ice cream, and many other products.

Are We Running Out of Conventional Oil?

We use an astonishing amount of oil. Laid end to end, the roughly 32 billion barrels of crude oil used worldwide in 2011 would stretch to about 28 million kilometers (18 million miles)—far enough to reach to the moon and back about 37 times. (One barrel of oil contains 159 liters or 42 gallons of oil.)

According to the 2012 BP Statistical Review of World Energy, in 2011, the world's three largest producers of conventional light oil, in order, were Saudi Arabia (13.2% of world production), Russia (13%), and the United States (8%, see the Case Study that follows). The International Energy Agency projects that by 2017, the United States is likely to be the world's largest oil producer. In 2011, the world's three largest oil consumers were the United States (using 21% of all oil produced), China (11%), and Japan (5%). The Earth Policy Institute projects that, by 2035, China will be using 4 times more oil than the United States.

How much oil is there? No one knows, although geologists have provided us with estimates of amounts existing in identified deposits. However, not all such deposits can be exploited at a profit, and oil that cannot be extracted profitably is not considered to be available. Availability is determined mostly by five factors that can change over time: **(1)** the demand for the oil, **(2)** the technology used to make it available, **(3)** the rate at which we can remove the oil, **(4)** the cost of making it available, and **(5)** its market price.

Available deposits are called **proven oil reserves**—deposits from which the oil can be extracted profitably at current prices with current technology. Proven oil reserves are not fixed. For example, recently improved oil extraction technology (Science Focus 15.1) and higher oil prices have made it profitable to extract light oil that is tightly held in layers of shale rock, and this has increased proven oil reserves.

The world is not about to run out of conventional light oil in the near future, but the easily extracted cheap oil that supports our economies and lifestyles may be running low. Most of the world's oil comes from huge oil fields that were discovered decades ago. Production from many of these fields has begun to decline and new fields are getting harder to find and more expensive to develop.

We can produce more conventional light oil from far offshore in deep ocean seabed deposits and from areas near the Arctic Circle. We can also rely more on unconventional heavy oil—a type of crude oil that does not flow as easily as light oil—from depleted oil wells and other sources. But use of these sources of oil results in lower net energy yields, higher production costs, and higher environmental impacts. Having to rely more on such sources leaves us with three major options: **(1)** learn to live with much higher oil prices and thus higher prices on many other items; **(2)** extend supplies by using oil much more

SCIENCE FOCUS 15.1

REMOVING TIGHTLY HELD OIL AND NATURAL GAS BY DRILLING SIDEWAYS AND FRACKING

Geologists have known for decades about vast deposits of oil and natural gas that are widely dispersed and tightly held in dense layers of shale rock formations found in many areas of the United States, including North Dakota, Texas, and Pennsylvania (see map in Figure 33, p. S58, in Supplement 6).

Until recently, it cost too much to extract such oil and natural gas from shale rock. This situation has changed because of high oil prices along with the use of two newer extraction technologies (Figure 15-A). One is **horizontal drilling**, a method of drilling first vertically to a certain point, then bending the flexible well bore and drilling horizontally. This method is used to gain greater access to oil and gas deposits located within layers of shale or other rock deposits. Usually, wells are drilled vertically for 1.6–2.4 kilometers (1–1.5 miles) or more and then horizontally for up to 1.6 kilometers (1 mile). Two or three horizontally drilled wells can often produce as much oil as 20 vertical wells, which reduces the area of land damaged by drilling operations.

The second technology, called **hydraulic fracturing** or **fracking**, is then used to free the tightly held oil and natural gas. After perforated tubes with explosive charges create fissures in the rock, high-pressure pumps shoot a mixture of water, sand, and chemicals into the well. When the pressure builds enough to fracture the rock, the mixture of water, sand, and chemicals flows into the cracks and creates more cracks and weak spots. The sand allows the cracks to remain open so that the oil or natural gas can flow out of the well to the surface.

Over a period of time this process is repeated 7–10 times per well. Much of the water that is contaminated with fracking chemicals and various other chemicals released from the rock (some of them hazardous) flows back to the surface. This contaminated water can be cleaned up and recycled. Other options are to store it in lined holding ponds or to inject it into deep underground hazardous waste disposal wells. Currently, about 80% of this contaminated water is injected into disposal wells. The rest is either recycled or stored in lined ponds. Using fracking to extract oil or natural gas costs about 5–10 times more than a conventional oil or natural gas well, and the supply is depleted about twice as fast.

The growing use of these two extraction technologies will be a key to the projected new era of oil and natural gas production in the United States (**Core Case Study**), provided the market prices of oil and natural gas remain high enough to make it profitable. However, there are some potentially serious environmental problems related to widespread use of these technologies. We discuss these in Section 15-3.

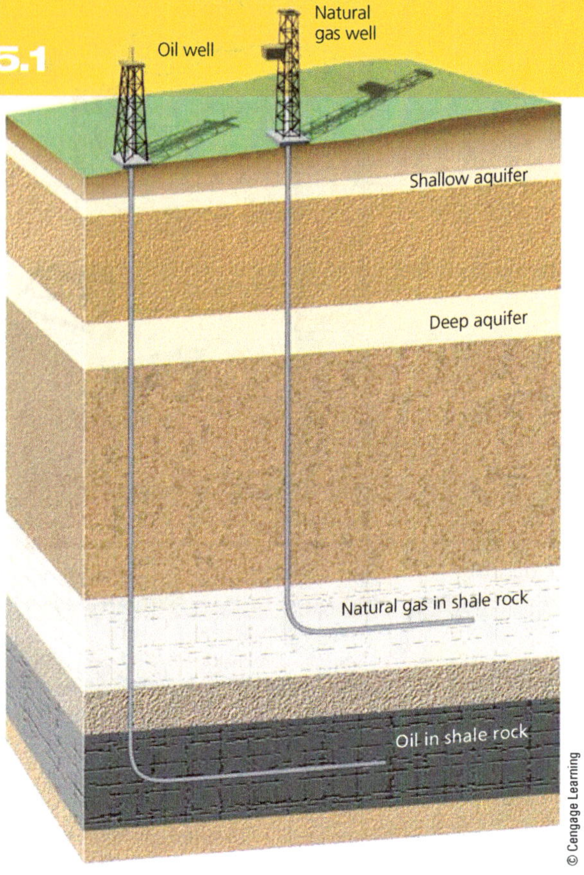

Figure 15-A Horizontal drilling and hydraulic fracturing, or fracking, are being used to release large amounts of oil and natural gas that are tightly held in shale rock formations.

Critical Thinking

Why do you think horizontal drilling allows better access to tightly held oil and natural gas deposits than does drilling vertically into such deposits?

efficiently, for example, by sharply improving vehicle fuel efficiencies; and **(3)** use other energy resources.

The 12 countries that make up the Organization of Petroleum Exporting Countries (OPEC) have about 72% of the world's proven crude oil reserves and thus are likely to control most of the world's oil supplies for many years to come. Today, OPEC's members are Algeria, Angola, Ecuador, Iran, Iraq, Kuwait, Libya, Nigeria, Qatar, Saudi Arabia, United Arab Emirates, and Venezuela. Other countries, including Canada and Russia, also have large oil reserves. According to BP, in 2011, Venezuela had the largest portion (18%) of the world's proven light oil reserves, followed by

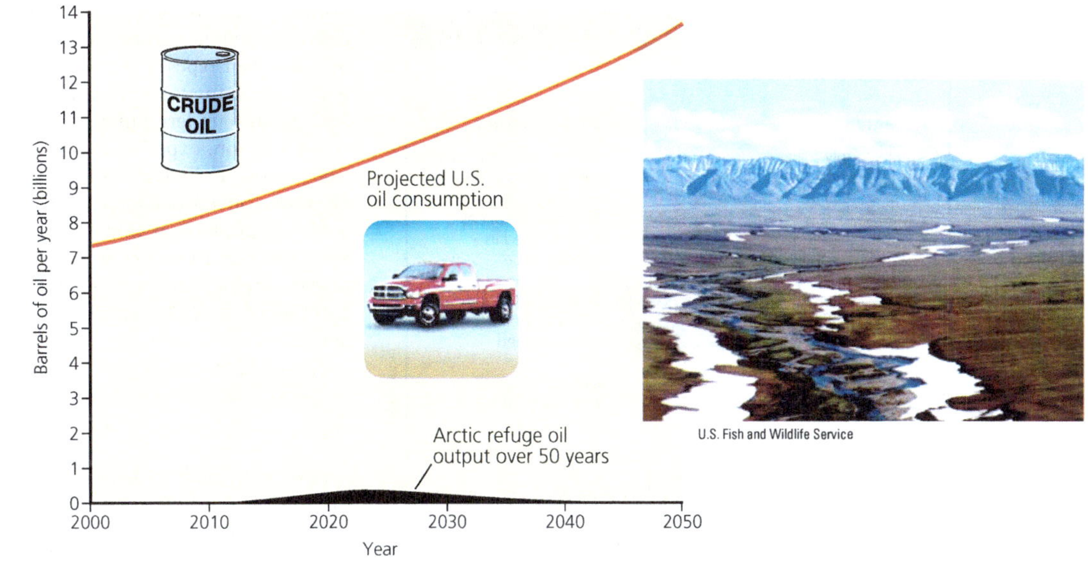

Figure 15-5 The amount of conventional light crude oil that might be found in the Arctic National Wildlife Refuge, if developed and extracted over 50 years, is a tiny fraction of the projected U.S. demand for oil.

(Compiled by the authors using data from U.S. Department of Energy and U.S. Geological Survey.)

Saudi Arabia (16%), Canada (11%), Iran (9%), Iraq (9%), Kuwait (6%), the United Arab Emirates (6%), and Russia (5%). The world's three largest users of light oil—the United States, China, and Japan—have, in order, only about 2%, 1%, and 0.003% of the world's proven crude oil reserves.

Based on data from the U.S. Department of Energy and the U.S. Geological Survey, if global consumption of conventional light oil continues to grow at about 2% per year, then:

- Saudi Arabia, with the world's second largest crude oil reserves, could supply the world's demand for oil for about 7 years.
- Estimated unproven crude oil reserves under Alaska's Arctic National Wildlife Refuge (ANWR) (see Figure 31, p. S56, in Supplement 6) would meet the world's demand for 1–5 months and U.S. demand for 7–24 months (Figure 15-5).
- The Arctic Circle holds enough technically recoverable crude oil to meet the global demand for about 3 years at high production costs.

Bottom line: to keep using conventional light oil at the projected rate of increase, we must expand global proven crude oil reserves by an amount equal to Saudi Arabia's current reserves every 7 years. Most oil geologists say this is highly unlikely.

Use of Conventional Oil Has Environmental Costs

The extraction, processing, and burning of conventional crude oil has severe environmental impacts, including land disruption, greenhouse gas emissions, and other forms of air pollution, water pollution, and loss of biodiversity.

A critical and growing problem is that burning oil or any carbon-containing fossil fuel releases the greenhouse gas CO_2 into the atmosphere. According to most of the world's top climate scientists, this has been warming the atmosphere and will contribute to projected climate change during this century. Currently, burning oil, mostly as gasoline and diesel fuel for transportation, accounts for 43% of global CO_2 emissions, which have been increasing rapidly (see Figure 14, p. S70, in Supplement 7).

Another problem is that, as easily accessible deposits are becoming depleted, oil producers are turning to oil that is buried deep underground in sensitive areas and under the ocean floor in certain coastal areas. As was revealed in the catastrophic oil spill in the Gulf of Mexico in 2010, going to these harder-to-reach deposits greatly increases the risk of severe environmental degradation.

Figure 15-6 lists the advantages and disadvantages of using conventional oil as an energy resource.

CASE STUDY

Oil Production and Consumption in the United States

The United States gets about 87% of its commercial energy from fossil fuels, with 37% coming from oil (Figure 15-1). Currently, oil production in the United States, especially from shale rock, is increasing rapidly and could make the country the world's largest oil producer by 2020. However, oil production from shale rock is likely to decline over the next two decades as the richest deposits are depleted. The long-term problem for the United States is that it uses about 21% of the oil produced globally but produces only 9% of the world's oil

Trade-Offs

Conventional Oil

Advantages	Disadvantages
Ample supply for several decades	Water pollution from oil spills and leaks
Net energy yield is medium but decreasing	Environmental costs not included in market price
Low land disruption	Releases CO_2 and other air pollutants when burned
Efficient distribution system	Vulnerable to international supply interruptions

Figure 15-6 Using conventional light oil as an energy resource has advantages and disadvantages. ***Questions:*** Which single advantage and which single disadvantage do you think are the most important? Why? Do the advantages of relying on conventional light oil outweigh its disadvantages? Explain.

Photo: Richard Goldberg/Shutterstock.com

and has only about 2% of the world's proven conventional oil reserves. Also, much of its supply lies under environmentally sensitive land and coastal areas that are expensive to develop.

Since 1982, oil consumption in the United States has greatly exceeded domestic production. This helps to explain why, in 2012, the United States imported 39% of its crude oil (compared to 24% in 1970 and 60% in 2005). The decrease in dependence on oil imports between 2005 and 2012 resulted from a combination of a weak economy and reduced oil consumption due to improvements in fuel efficiency. Also, there was a slight increase in U.S. oil production between 2008 and 2012, mostly because of increased production of oil from shale rock deposits.

In 2012, the five largest suppliers of oil imported into the United States were, in order, Canada, Saudi Arabia, Venezuela, Mexico, and Iraq. By importing oil, the United States transfers a massive amount of its wealth to these oil-producing countries.

Can the United States significantly reduce its dependence on oil imports by producing its own oil faster than its current oil supply is being depleted? Some say "yes" and project that domestic oil production will increase dramatically over the next few decades—especially from oil found in shale rock. They argue that this will lead to a new era of oil production in the United States (**Core Case Study**). The U.S. Energy Information Agency (EIA) estimates that increased production of oil from shale rock could continue to reduce U.S. dependence on imported oil.

However, we do not know how much of the oil held tightly in shale rock deposits can be extracted profitably and at an acceptable environmental cost. For example, horizontal drilling and fracking produce massive amounts of contaminated wastewater. And there is no guarantee that the well pipes and casing in any fracking operation will not leak the toxic chemicals used in and produced by fracking into underground drinking water supplies.

In addition, experience indicates that production of oil from shale rock beds drops off about twice as fast as it does in most conventional oil fields. Thus, long-term, profitable oil production from these resources may be overestimated. And producing oil from shale rock using current technology will be profitable only as long as the price of oil is at least $50 a barrel. In addition, this oil would be developed with high production costs, lower net energy yields, and potentially high environmental impacts.

Heavy Oil from Oil Shale Rock

A potential supply of heavy oil is *shale oil.* It is produced by mining, crushing, and heating oil shale rock (Figure 15-7, left) to extract a mixture of hydrocarbons called *kerogen* that can be distilled to produce shale oil (Figure 15-7, right). Before the thick shale oil is sent by pipeline to a refinery, it must be heated to increase its flow rate and processed to remove sulfur, nitrogen, and other impurities, which decreases its net energy yield. (These kerogen-containing deposits found in oil shale rock differ from the oil dispersed and tightly held in shale rock, discussed in the **Core Case Study**.)

About 72% of the world's estimated oil shale rock reserves are buried deep in rock formations located primarily under government-owned land in the U.S. states of Colorado, Wyoming, and Utah in an area known as the Green River formation (see map in Figure 31, p. S56, in Supplement 6). The U.S. Bureau of Land Management estimates that these deposits contain an amount of potentially recoverable heavy oil equal to almost 4 times the amount in Saudi Arabia's proven reserves of conventional oil. Estimated potential global supplies of unconventional heavy shale oil are about 240 times larger than estimated global supplies of conventional crude oil.

The problem is that it takes considerable energy, money, and water to extract kerogen from shale rock and convert it to shale oil. Thus, its net energy yield is low. Also, the process pollutes large amounts of water and releases 27–52% more CO_2 into the atmosphere per unit of energy produced than does producing conventional crude oil (**Concept 15-2B**). In 2011, the U.S. Bureau of Land Management stated that unless oil prices rise sharply: "There are no economically viable ways yet known to extract and process oil shale for commercial purposes."

Heavy Oil from Tar Sands

A growing source of heavy oil is **tar sands**, or **oil sands**, which are a mixture of clay, sand, water, and a combustible organic material called *bitumen*—a thick, sticky, tarlike heavy oil (Figure 15-8) with a high sulfur content.

Figure 15-7 Heavy shale oil (right) can be extracted from oil shale rock (left).

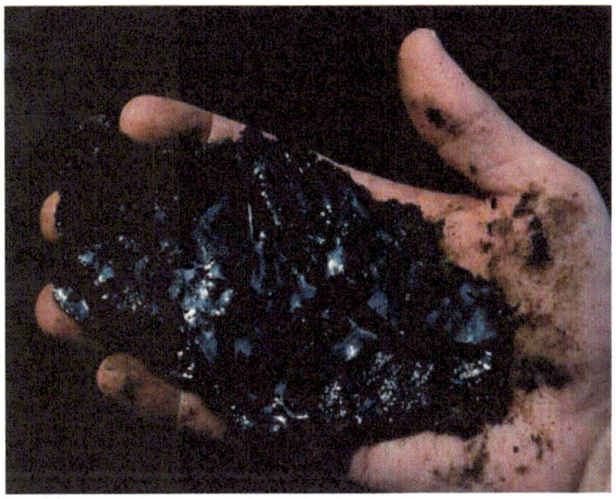

Figure 15-8 This gooey tar, or bitumen, with the consistency of peanut butter, was extracted from tar sands in Alberta, Canada, to be converted to heavy synthetic oil.

Northeastern Alberta in Canada has three-fourths of the world's tar sands resource in sandy soil under a vast area of remote boreal forest (see Figure 7-13, bottom photo, p. 156). If we include its conventional light oil and heavy oil from tar sands, Canada has the world's third largest proven oil reserves.

The big drawback to using tar sands is that developing this resource has major harmful impacts on the land (Figure 15-9), air, water, wildlife, and climate, compared to developing conventional light oil and tightly held oil from shale rock (**Concept 15-2B**). Before the mining takes place, the overlying boreal forest is clear-cut, wetlands are drained, and rivers and streams are sometimes diverted. Next the overburden of sandy soil, rocks, peat, and clay is stripped away to expose the tar sands deposits. Then five-story-high electric power shovels dig up the sand and load it into three-story-high trucks, which carry it to an upgrading plant. There, the tar sands are mixed with hot water and steam to extract the bitumen. Next, the bitumen is heated by natural gas in huge cookers and converted into a low-sulfur, synthetic, heavy crude oil that has to be processed further to allow it to flow through pipelines to a refinery.

According to a 2009 study by CERA, an energy consulting group, the process of extracting, processing, and refining bitumen from tar sands into heavy oil releases 3 to 5 times more greenhouse gases per barrel of oil produced than does extracting and producing conventional light oil. Part of this process is the removal of the boreal forest and peat deposits lying above the tar sands deposits. Left undisturbed, these forests and peatlands help to reduce the threat of projected climate change by storing great amounts of carbon. Much of this carbon is released to the atmosphere as CO_2 when the forests and peatlands are removed, which adds to the threat of global climate disruption.

Although tar sand sites can be planted with vegetation after the strip mining, such restoration is expensive and rare and cannot match the capacity of ancient peatlands and boreal forests for absorbing carbon and helping to offset projected climate dis-

Figure 15-9 Tar sands mining operation in Alberta, Canada.

ruption. Also, some studies have estimated that the conversion of tar sands to heavy oil produces only 5–20% more greenhouse gases than does conventional oil production. However, these studies did not take into account the entire production process, including the loss of carbon-absorbing peatlands and boreal forest.

In addition, this process uses huge amounts of water and creates lake-size tailings ponds containing toxic sludge and wastewater. Many migrating birds die trying to get water and food from these ponds. Also, the dikes of compacted sand surrounding the tailings ponds have the potential to leak and release large volumes of toxic sludge onto nearby land and into streams and rivers.

Finally, it takes a great deal of energy to produce oil from tar sands, which involves burning natural gas to provide heat for the bitumen cookers and using diesel fuel to run the massive vehicles and machinery. This too adds to the pollution of air, water, and land and to the emissions of greenhouse gases.

Figure 15-10 lists the major advantages and disadvantages of using heavy oil from tar sands and from oil shale rock as an energy resource.

Trade-Offs

Heavy Oils from Oil Shale and Tar Sand

Advantages	Disadvantages
Large potential supplies	Low net energy yield
Easily transported within and between countries	Releases CO_2 and other air pollutants when produced and burned
Efficient distribution system in place	Severe land disruption and high water use

Figure 15-10 Using heavy oil from tar sands and from oil shale rock as an energy resource has advantages and disadvantages (Concept 15-2). ***Questions:*** Which single advantage and which single disadvantage do you think are the most important? Why? Do the advantages of relying on heavy oil from these sources outweigh the disadvantages? Explain.

Photo: Christopher Kolaczan/Shutterstock.com

15-3 What Are the Advantages and Disadvantages of Using Natural Gas?

CONCEPT 15-3

Conventional natural gas is more plentiful than oil, has a medium net energy yield and a fairly low production cost, and is a clean-burning fuel, but producing it has created environmental problems.

Natural Gas Is a Useful, Clean-Burning, but Not Problem-Free Fossil Fuel

Natural gas is a mixture of gases of which 50–90% is methane (CH_4). It also contains smaller amounts of heavier gaseous hydrocarbons such as propane (C_3H_8) and butane (C_4H_{10}), and small amounts of highly toxic hydrogen sulfide (H_2S). This versatile fuel has a medium net energy yield (Figure 15-2) and is widely used for cooking, heating space and water, and industrial purposes, including production of most of the world's nitrogen fertilizer. It can also be used as a fuel for cars and trucks and for natural gas turbines used to produce electricity in power plants.

This versatility helps to explain why natural gas provides about 28% of the energy consumed in the United States. It burns cleaner than oil and much cleaner than coal, and when burned completely, it emits about 30% less CO_2 than oil and about 50% less than coal.

Conventional natural gas is often found in deposits lying above deposits of conventional oil. It also exists in tightly held deposits in shale rock and can be extracted through horizontal drilling and fracking (see Science Focus 15.1). See Figure 33, p. S58, in Supplement 6 for a map of major U.S. natural gas shale rock deposits. In the United States and many other countries, natural gas is distributed to users by a large network of underground pipelines.

When a natural gas deposit is tapped, propane and butane gases can be liquefied under high pressure and removed as **liquefied petroleum gas (LPG)**. LPG is stored in pressurized tanks for use mostly in rural areas not served by natural gas pipelines. Natural gas can also be transported across oceans, by converting it to **liquefied natural gas (LNG)** at a high pressure and at a very low temperature. This highly flammable liquid is transported in refrigerated tanker ships. At its destination port, it is heated and converted back to the gaseous state and then distributed by pipeline. LNG has a low net energy yield, because more than a third of its energy content is used to liquefy it, process it, deliver it to users by ship, and convert it back to natural gas.

In 2011, the International Energy Agency estimated that recoverable conventional supplies of natural gas could meet the current global demand for about 120 years, and that potentially recoverable unconventional supplies of natural gas could sustain current global production for 250 years. That year, according to BP's 2012 Annual Review of World Energy, Russia had about 21% of the world's proven conventional natural gas reserves, followed by Iran (16%), Qatar (12%), and Turkmenistan (12%). China and India, with their rapidly growing economies, have only 1.5% and 0.6%, respectively, of the

GOOD NEWS

world's proven natural gas reserves and the United States has only 4%. Japan has no significant natural gas reserves and depends on imports of expensive LNG.

In 2011, the world's three largest producers of natural gas were the United States (with 20% of global total production), Russia (19%), and Canada (5%). In that year, the United States used about 22% of the world's production. (See Figure 5, p. S66, in Supplement 7 for a graph of U.S. natural gas consumption between 1980 and 2012, with projections to 2040.) U.S. natural gas production has been increasing rapidly, mostly because of development of the technology used to extract tightly held natural gas from shale rock (Figure 15-A). This source accounted for about 60% of U.S. natural gas production in 2011 (up from 2% in 2000). Thus, the United States does not have to rely on natural gas imports.

The demand for natural gas in the United States is projected to more than double between 2010 and 2050. If much of this demand is met by increased production of natural gas from shale rock, the United States could continue meeting its needs for natural gas from domestic resources. If natural gas prices remain affordable, such a trend would reduce the use of coal-burning power plants and make new nuclear power plants even more uneconomical than they are now. This could also slow the shift to greater use of renewable solar and wind energy resources.

However, U.S. natural gas producers would like to export natural gas as LPG to countries where natural gas prices are much higher than in the United States. Chemical industries and utilities that use natural gas to provide heat and produce electricity oppose this because it could decrease the supply and raise domestic natural gas prices.

There are some potential problems that could temper this rosy outlook for natural gas. In 2011, the U.S. Geological Survey cut its nationwide estimate of recoverable shale gas by 50% and pointed out that natural gas production from shale rock tends to peak and drop off much faster than does production from conventional natural gas wells. One question is whether the rate of increased production of natural gas from shale rock can exceed the rate of decline in conventional natural gas production from aging fields. Another major potential drawback is the environmental problems related to greatly increasing U.S. production of natural gas from shale rock, as discussed in the Case Study that follows.

Figure 15-11 lists the advantages and disadvantages of using conventional natural gas as an energy resource.

Trade-Offs

Conventional Natural Gas

Advantages	Disadvantages
Ample supplies	Low net energy yield for LNG
Versatile fuel	Production and delivery may emit more CO_2 and CH_4 per unit of energy produced than coal
Medium net energy yield	Fracking uses and pollutes large volumes of water
Emits less CO_2 and other air pollutants than other fossil fuels when burned	Potential groundwater pollution from fracking

Figure 15-11 Using conventional natural gas as an energy resource has advantages and disadvantages. ***Questions:*** Which single advantage and which single disadvantage do you think are the most important? Why? Do you think that the advantages of using conventional natural gas outweigh the disadvantages? Explain.

Photo: Werner Muenzker/Shutterstock.com

CASE STUDY

Natural Gas Production and Fracking in the United States: Environmental Problems and Solutions

The production of natural gas from shale rock deposits involves drilling wells; using huge amounts of water, sand, and chemicals to frack the gas; bringing up the natural gas along with the resulting toxic wastewater; dealing with this wastewater; and transporting the natural gas to users through underground pipelines. At several points in this production process, natural gas can leak into the atmosphere and into underground sources of drinking water for nearby homes and communities. Despite the industry's attempts to solve this problem, natural gas has leaked from loose pipe fittings and faulty cement seals in natural gas well bore holes, as well as from pipelines and cutoff valves used to deliver natural gas.

Drinking water contaminated by natural gas can catch fire (Figure 15-12), and some home owners have had to install expensive systems to remove the natural gas to prevent explosions. According to a 2012 study by the National Academy of Sciences, people living within 900 meters (3,000 feet) of a natural gas well are likely to have up to 17 times more methane in their groundwater than those who live farther away.

Within a decade or two, there may be at least 100,000 more natural gas wells using fracking technology, according to the U.S. Energy Information Administration. Without increased monitoring and regulation of the entire natural gas production process, including fracking, the greatly increased production of natural gas (and oil) from shale rock could have several harmful environmental effects:

- Fracking requires enormous volumes of water. In water-short areas this could help to deplete aquifers, degrade aquatic habitats, and diminish the availability of water for other purposes.

MARK THIESSEN/National Geographic Creative

Figure 15-12 Natural gas fizzing from this faucet in a Pennsylvania home can be lit like a natural gas stove burner. This began happening after an energy company drilled a fracking well in the area, but the company denies responsibility. The home owners have to keep their windows open year-round to keep the lethal and explosive gas from building up in the house.

- Fracking fluids can contain several potentially hazardous chemicals that are used to reduce friction, inhibit corrosion, and stop bacterial growth. Each fracked well produces millions of gallons of wastewater that are brought to the surface along with the released natural gas. This slurry contains a mix of naturally occurring salts, toxic heavy metals, and radioactive materials leached from the rock. It also contains chemicals used in the fracking process, which natural gas companies do not have to reveal to the public. After fracking, the slurry is stored in various ways, some more secure than others, and there are several points in this complex process at which some or all of this toxic slurry could be released to contaminate nearby groundwater, surface waters, or land.
- Because fracking requires a special type of sand, some areas of the country, especially western Wisconsin, are seeing a boom in the mining of this sand. This new form of strip mining has expanded rapidly with little regulation and is destroying large areas of wildlife habitat while creating air and noise pollution in the mining areas.
- According to a 2012 study by the National Academy of Sciences and another study by the U.S. Geological Survey, in recent years, one of the major causes of hundreds of small earthquakes in 13 states has been the shifting of bedrock resulting from the high-pressure injection of large amounts of wastewater from fracking and other industrial activities into deep underground storage wells. The pressure involved is typically 9 times higher than the pressure required to crush a submarine on its deepest dive. Such earthquakes could release hazardous wastewater into aquifers and cause breaks in the steel lining and cement seals of oil and gas well pipes. In 2012, the U.S. state of Ohio shut down a deep well used for the disposal of fracking wastewater after several small earthquakes occurred in the vicinity of the well.
- In 2011, ecologist Robert Howarth and geoscientist David Hughes, in separate studies, both estimated that emissions of climate-changing CH_4 and CO_2 from the entire process for supplying and burning natural gas from shale rock are higher than those from supplying and burning conventional natural gas and coal. If these preliminary estimates are verified, the positive environmental image of natural gas—based mainly on the fact that it is relatively clean-burning and not on its entire cycle of production and use—will be tarnished.

Producers maintain that fracking is necessary for exploiting natural gas from shale deposits at an affordable cost. They point out that increased natural gas production from fracking has lowered U.S. natural gas prices and benefitted the 55% of U.S. consumers who burn natural gas. In addition, the natural gas fracking boom has created thousands of jobs and boosted local economies in some areas. And the resulting shift from coal to natural gas for producing electricity has reduced U.S. air pollution.

Producers also argue that no groundwater contamination directly due to fracking has ever been recorded, mostly because the fracking takes place far below drinking water aquifers. However, critics report that the EPA has found at least one example of drinking water contamination that resulted from fracking. They also contend that natural gas producers have squelched numerous reports of drinking water contamination from fracking by offering financial settlements to people who make such claims with the stipulation that they cannot reveal any information about the alleged contamination. Much of the contamination may have come from leaks and faulty cement seals in well pipes, which indicates inadequate inspection and regulation of the entire natural gas production process by states and the federal government.

Currently, people who rely on aquifers and streams for their drinking water in areas affected by the boom in shale gas production have little protection against pollution of their water supplies resulting from natural gas production. This is because, under political pressure from natural gas suppliers, the 2005 Energy Policy Act excluded the fracking process from certain regulations under the federal Safe Drinking Water Act. Other loopholes have also exempted natural gas production from parts of several other federal environmental laws, including the Clean Water Act, the Clean Air Act, and the National Environmental Policy Act.

In addition, people who live near fracking operations must put up with around-the-clock noise and air pollution from drilling equipment, diesel engines, trucks hauling sand, and explosions set off each time a well is fracked. Without stricter regulation and monitoring, the drilling of another 100,000 natural gas wells during the next 10–20 years will increase the risk of harmful environmental effects from the production of natural gas, which could cause a public backlash against this technology.

According to a 2010 MIT study, the harmful environmental impacts of producing natural gas from shale rock are "manageable but challenging." Some energy analysts, along with the U.S. Energy Information Agency, have suggested several ways to reduce the environmental threats arising from shale gas production (Figure 15-13). Making such changes would help us in implementing the full-cost pricing **principle of sustainability** (see Figure 1-5, p. 9 or back cover).

Solutions

- Step up research on the environmental impact of natural gas production
- Greatly increase monitoring and legal regulation of natural gas production, including regular inspections of the metal casings and concrete seals in well pipes
- Develop federal regulations on disposal, storage, treatment, and reuse of fracking wastewater
- Require complete disclosure of all chemicals used in fracking
- Require use of the least harmful chemicals available in fracking fluids
- Require testing of aquifers and drinking water wells for any chemical contamination from fracking operations before drilling begins and as long as gas extraction continues
- Overturn all exemptions for oil and natural gas production from any and all federal pollution regulations

Figure 15-13 Solutions: Reforms such as these, recommended by several energy analysts, could reduce the harmful impact of shale gas production. ***Questions:*** Which three of these steps do you think are the most important ones to take? Explain.

Unconventional Natural Gas

There are two major sources of unconventional natural gas that are both difficult and costly to exploit without high environmental impacts. One source is *coal bed methane gas* found in coal beds near the earth's surface across parts of the United States and Canada (see the map in Figure 32, p. S57, in Supplement 6). The environmental impacts of using this resource would include scarring of land, depletion of some water sources, and possible pollution of aquifers. So far it has not been economical to exploit this resource.

The other source of unconventional natural gas is *methane hydrate*—methane trapped in icy, cage-like structures of water molecules buried under arctic permafrost in tundra areas of North America, northern Europe, and Siberia. Methane hydrate is also found lying on the ocean floor in several areas of the world. So far, it costs too much to get natural gas from methane hydrates. Also, scientists warn that the projected large-scale release of methane (a potent greenhouse gas) to the atmosphere during removal and processing of this resource would likely speed up atmospheric warming and the resulting projected climate disruption.

15-4 What Are the Advantages and Disadvantages of Using Coal?

CONCEPT 15-4A

Conventional coal is plentiful and has a high net energy yield at low costs, but using it results in a very high environmental impact.

CONCEPT 15-4B

We can produce gaseous and liquid fuels from coal, but they have lower net energy yields and using them would result in higher environmental impacts than those of conventional coal.

Coal Is a Plentiful but Dirty Fuel

Coal is a solid fossil fuel formed from the remains of land plants that were buried 300–400 million years ago and exposed to intense heat and pressure over millions of years (Figure 15-14).

Coal is burned in power plants (Figure 15-15) to generate about 45% of the world's electricity, according to the International Energy Agency. This includes 93% of the electricity used in South Africa and 73% of that used in China. In the United States, mostly because of cheaper natural gas, the percentage of all electricity used that is produced by burning coal dropped from 53% in 1997 to 37% in 2012 and could drop to 30% by 2020.

Coal is also burned in industrial plants to make steel, cement, and other products. In order, the world's five largest users of coal are China, the United States, India, Russia, and Japan. In 2010, China burned 3 times more coal than the United States burned.

Coal is an abundant fossil fuel. Five countries have three-fourths of the world's proven coal reserves. They are the United States with 28% of global coal reserves (see

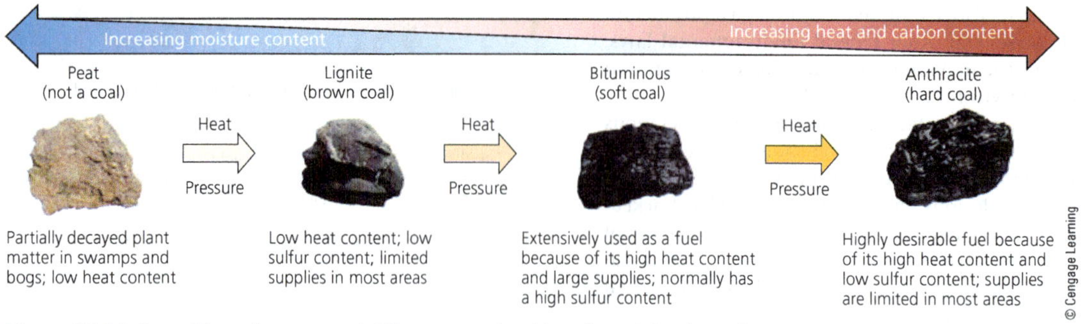

Figure 15-14 Over millions of years, several different types of coal have formed. Peat is a soil material made of moist, partially decomposed organic matter, similar to coal; it is not classified as a coal, although it is used as a fuel. These different major types of coal vary in the amounts of heat, carbon dioxide, and sulfur dioxide released per unit of mass when they are burned.

Figure 31, p. S56, in Supplement 6 for a map of major U.S. coal deposits), followed by Russia (with 18%), China (13%), Australia (9%), and India (7%). The U.S. Geological Survey estimates that identified U.S. coal reserves could last about 250 years at the current consumption rate and that identified and potential global supplies of coal could last for 200–1,100 years, depending on how rapidly they are used.

The problem is that coal is by far the dirtiest of all fossil fuels. Even when costly air-pollution-control technologies are used, burning coal pollutes the air and creates a toxic ash that is difficult to deal with. And the processes of making coal available severely degrade land and pollute water and air (see chapter-opening photo and Figures 14-12 through 14-14, p. 359).

Coal is mostly carbon but contains small amounts of sulfur, which is converted to the air pollutant sulfur dioxide (SO_2) when the coal burns. Burning coal also releases large amounts of black carbon particulates, or *soot* (Figure 15-16), and much smaller, fine particles of air pollutants such as mercury. The fine particles can get past our bodies' natural defenses that help to keep our lungs clean. According to a 2010 study by the Clean Air Task Force, fine-particle pollution in the United States, mostly from the older U.S. coal-burning power plants without the latest air-pollution-control technology, prematurely kills at least 13,000 people a year—an average of nearly 36 people every day. According to a World Bank report, burning coal in China, where air pollution control is far from adequate, causes at least 650,000 deaths a year.

Coal-burning power and industrial plants are among the largest emitters of the greenhouse gas CO_2 (Figure 15-17). China leads the world in such emissions, followed by the United States. Another problem with burning coal is that it emits trace amounts of radioactive materials as well as toxic and indestructible mercury into the atmosphere.

Finally, burning coal and removing some of the pollutants it releases from smokestack emissions produce a highly toxic ash. In the United States, about 57% of the ash is buried in landfills or in active or abandoned mines or is made into a wet slurry that is stored in holding ponds (Figure 15-18). The ash stored underground can slowly leach into groundwater, and the wet slurry can break through a pond's earthen walls, as it did at a coal ash storage pond near Knoxville, Tennessee in 2008. Such a spill can severely pollute nearby surface waters, groundwater, and land.

Coal ash is a major problem in China. According to a 2010 Greenpeace study, coal ash dumped into open landfills (Figure 15-19) is China's largest category of solid industrial waste. From these landfills, toxic chemicals are easily dispersed into the environment by wind and rain.

The Clean Coal Campaign

For decades, economically and politically powerful U.S. coal mining companies, coal-hauling railroad companies, and coal-burning power companies and industries have fought to preserve their profits by opposing measures such as stricter air pollution standards for coal-burning plants and classification of coal ash as a hazardous waste. For more than 30 years, these companies have also led the fight against efforts to classify climate-changing CO_2 as a pollutant that could be regulated by the EPA. Such regulation would likely raise their cost of doing business and make coal less competitive with cheaper sources of electricity such as natural gas and wind.

Since 2008, U.S. coal and electric utility industries have mounted a highly effective, well-financed publicity campaign built around the notion of clean coal. We can burn coal more cleanly by adding costly air-pollution-control devices to power plants. But critics argue that there could never be such a thing as clean coal. Even with stricter air pollution controls, burning coal will always involve some emissions of health-damaging air pollutants and climate-changing CO_2. It will always create indestructible and hazardous coal ash, which will actually increase with better air pollution controls, because such controls

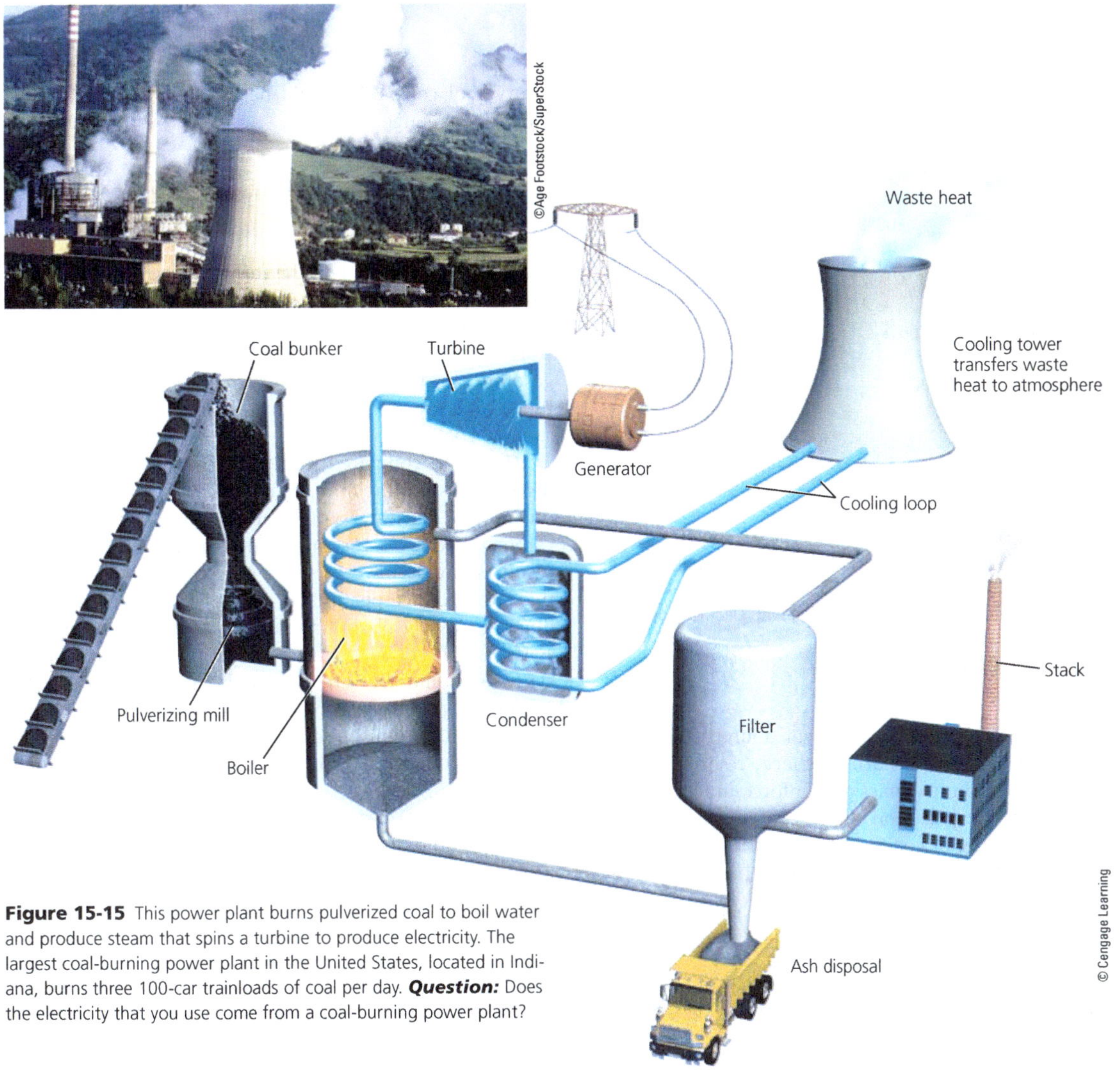

Figure 15-15 This power plant burns pulverized coal to boil water and produce steam that spins a turbine to produce electricity. The largest coal-burning power plant in the United States, located in Indiana, burns three 100-car trainloads of coal per day. ***Question:*** Does the electricity that you use come from a coal-burning power plant?

involve the creation of more coal ash. Also, mining coal will always involve disrupting land—in many cases, vast areas of land—and polluting water and air.

Coal companies and utilities can get away with talking about clean coal, and can continue to produce electricity cheaply by burning coal, primarily because the harmful environmental and health costs of producing and using coal are not included in the market prices of coal and coal-fired electricity. This violates the full-cost pricing **principle of sustainability** (see Figure 1-5, p. 9 or back cover). According to a 2010 study by Harvard Medical School's Center for Health and the Global Environment and a similar 2009 study by the U.S. National Academy of Sciences, including all such costs would double or triple the price of electricity from coal-fired power plants.

Figure 15-20 lists the advantages and disadvantages of using coal as an energy resource (**Concept 15-4A**). An important and difficult question for humanity is whether

Figure 15-16 These smokestacks on a coal-burning industrial plant emit large amounts of air pollution because the plant has inadequate air pollution controls.

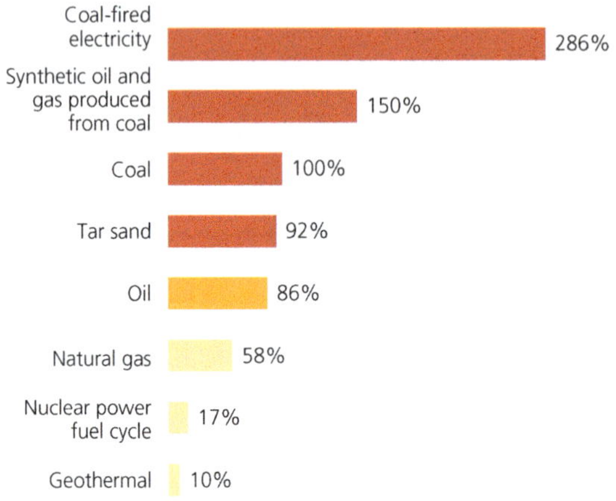

Figure 15-17 CO_2 emissions, expressed as percentages of emissions released by burning coal directly, vary with different energy resources. ***Question:*** Which of these produces more CO_2 emissions per kilogram: burning coal to heat a house, or heating with electricity generated by coal?

(Compiled by the authors using data from U.S. Department of Energy.)

we should begin shifting from use of abundant coal to using less environmentally harmful energy resources. Part of this shift would be to enact and enforce much stricter regulations on air pollution from coal-burning plants and on the handling of coal ash. In countries such as the United States, Russia, China, and India that have large reserves of coal, this would be a difficult economic and political challenge.

We Can Convert Coal into Gaseous and Liquid Fuels

We can convert solid coal into **synthetic natural gas (SNG)** by a process called *coal gasification,* which removes sulfur and most other impurities from coal. We can also convert it into liquid fuels such as methanol and synthetic gasoline through a process called *coal liquefaction.* These fuels, called *synfuels,* are often referred to as cleaner versions of coal.

However, compared to burning coal directly, producing synfuels requires the mining of 50% more coal. Producing and burning synfuels could also add 50% more carbon dioxide to the atmosphere (Figure 15-17). As a result, synfuels have a lower net energy yield and cost more to produce per unit of energy than does coal production. Also, it takes large amounts of water to produce synfuels. Therefore, greatly increasing the use of these synfuels would worsen two of the world's major environmental problems: projected climate disruption caused mostly by CO_2 emissions and increasing water shortages in many parts of the world (see Figure 13-9, p. 324, and Figure 13-10, p. 324).

Figure 15-21 lists the advantages and disadvantages of using liquid and gaseous synfuels produced from coal (**Concept 15-4B**).

Figure 15-18 Coal sludge impoundment in West Virginia above an elementary school.

Jim West/Alamy

GREG GIRARD/National Geographic Creative

Figure 15-19 Coal ash from a power plant in China is dumped into an open landfill where it can be blown by wind and washed by precipitation onto surrounding lands.

Trade-Offs

Coal

Advantages	Disadvantages
Ample supplies in many countries	Severe land disturbance and water pollution
Medium to high net energy yield	Fine particle and toxic mercury emissions threaten human health
Low cost when environmental costs are not included	Emits large amounts of CO_2 and other air pollutants when produced and burned

© Cengage Learning

Figure 15-20 Using coal as an energy resource has advantages and disadvantages. ***Questions:*** Which single advantage and which single disadvantage do you think are the most important? Why? Do you think that the advantages of using coal as an energy resource outweigh its disadvantages? Explain.

Photo: El Greco/Shutterstock.com

Trade-Offs

Synthetic Fuels

Advantages	Disadvantages
Large potential supply in many countries	Low to medium net energy yield
Vehicle fuel	Requires mining 50% more coal with increased land disturbance, water pollution, and water use
Lower air pollution than coal	Higher CO_2 emissions than coal

© Cengage Learning

Figure 15-21 The use of synthetic natural gas (SNG) and liquid synfuels produced from coal as energy resources has advantages and disadvantages (Concept 15-2). ***Questions:*** Which single advantage and which single disadvantage do you think are the most important? Why? Do you think that the advantages of using synfuels produced from coal as an energy source outweigh the disadvantages? Explain.

Photo: ©mironov/Shutterstock.com

15-5 What Are the Advantages and Disadvantages of Using Nuclear Power?

CONCEPT 15-5

Nuclear power has a low environmental impact and a very low accident risk, but its use has been limited by a low net energy yield, high costs, fear of accidents, long-lived radioactive wastes, and its role in spreading nuclear weapons technology.

How Does a Nuclear Fission Reactor Work?

To evaluate the advantages and disadvantages of nuclear power, we must know how a nuclear power plant and its accompanying nuclear fuel cycle work. A nuclear power plant is a highly complex and costly system designed to perform a relatively simple task: to boil water and produce steam that spins a turbine and generates electricity.

Figure 15-22 This water-cooled nuclear power plant, with a pressurized water reactor, produces intense heat that is used to convert water to steam, which spins a turbine that generates electricity. ***Question:*** How does this plant differ from the coal-burning plant in Figure 15-15?

Robert Llewellyn/SuperStock

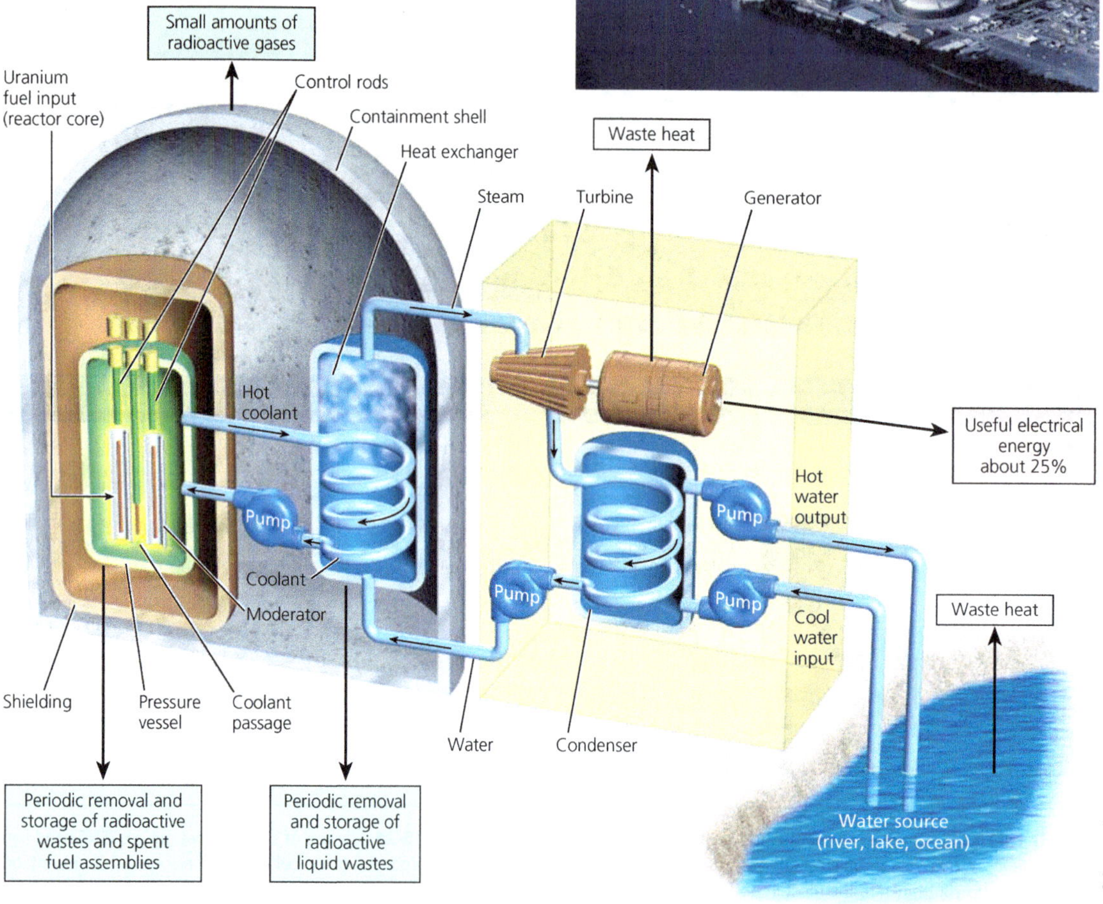

© Cengage Learning

What makes a nuclear power plant complex and costly is the use of a controlled nuclear fission reaction (see Figure 2-9, center, p. 40) to provide the heat. The fission reaction takes place in a *reactor.* The most common reactors, called light-water reactors (LWRs; see Figure 15-22), produce 85% of the world's nuclear-generated electricity (100% in the United States).

The fuel for a reactor is made from uranium ore mined from the earth's crust. After it is mined, the ore must be enriched to increase the concentration of its fissionable uranium-235 by 1–5%. The enriched uranium-235 is processed into small pellets of uranium dioxide. Each pellet, about the size of an eraser on a pencil, contains the energy equivalent of about a ton of coal. Large numbers of the pellets are packed into closed pipes, called *fuel rods,* which are then grouped together in *fuel assemblies,* to be placed in the core of a reactor.

Control rods are moved in and out of the reactor core to absorb neutrons generated in the fission reaction, thereby regulating the rate of fission and the amount of power produced. A *coolant,* usually water, circulates through the reactor's core to remove heat to keep the fuel rods and other reactor components from melting and releasing massive amounts of radioactivity into the environment. An LWR includes an emergency core cooling system as a backup to help prevent such meltdowns. A nuclear reactor cannot explode as an atomic bomb does and cause massive damage. The danger is from smaller explosions that can release radioactivity into the environment or cause a core meltdown.

Some nuclear plants withdraw the large quantities of cooling water they need from a nearby source such as a river or lake and return the heated water to that source. Other nuclear plants transfer the waste heat from the intensely hot water to the atmosphere by using one or more gigantic cooling towers, such as those at the Three Mile Island nuclear power plant near Harrisburg, Pennsylvania (USA), shown in the inset photo of Figure 15-22. There, a serious accident in 1979 caused a partial meltdown of one of the plant's reactors, but no lives were lost.

A *containment shell* with thick, steel-reinforced concrete walls surrounds the reactor core. It is designed to help keep radioactive materials from escaping into the environment, in case there is an internal explosion or a melting of the reactor's core. It also helps protect the core against some external threats such as tornadoes and plane crashes. These essential safety features help to explain why a new nuclear power plant costs as much as $10 billion and why that cost continues to rise.

What Is the Nuclear Fuel Cycle?

Building and running a nuclear power plant is only one part of the **nuclear fuel cycle** (Figure 15-23), which also includes the mining of uranium, processing and enriching the uranium to make fuel, using it in a reactor, safely storing the resulting highly radioactive wastes for thousands of years until their radioactivity falls to safe levels, and retiring the highly radioactive worn-out plant by taking it apart and storing its high- and moderate-level radioactive parts safely for thousands of years.

The final step in the cycle occurs when, after 20–60 years, a reactor comes to the end of its useful life, mostly because of corrosion and radiation damage to its metal parts, and it must be *decommissioned*, or retired. It cannot simply be shut down and abandoned, because its structure contains large quantities of high- and intermediate-level radioactive materials that must be kept out of the environment for thousands of years.

As long as a reactor is operating safely, the power plant itself has a fairly low environmental impact and a very low risk of an accident. However, considering the entire nuclear fuel cycle, the potential environmental impact increases. High-level radioactive wastes must be stored safely for thousands of years, and several points in the cycle are vulnerable to terrorist attack. Also, the uranium-enrichment and other technologies used in the cycle can be used to produce nuclear weapons–grade uranium (Concept 15-5).

Each step in the nuclear fuel cycle adds to the cost of nuclear power and reduces its net energy yield (Concept 15-1). If we add the enormous amount of energy needed to dismantle a plant at the end of its life and transport and safely store its highly radioactive materials, some scientists estimate that using nuclear power will eventually have a negative net energy yield, requiring more energy than it will ever produce.

Proponents of nuclear power tend to focus on the low CO_2 emissions and multiple safety features of the reactors. But in evaluating the safety, economic feasibility, net energy yield, and overall environmental impact of nuclear power, energy experts and economists caution us to look at the entire nuclear fuel cycle, not just the power plant itself. Figure 15-24 lists the major advantages and disadvantages of producing electricity by using the nuclear power fuel cycle (Concept 15-3).

Let's look more closely at some of the challenges involved in using nuclear power.

Storing Radioactive Spent-Fuel Rods Presents Risks

The high-grade uranium fuel in a nuclear reactor lasts for 3–4 years, after which it becomes *spent*, or useless, and must be replaced. On a regular basis, reactors are shut down for refueling, which usually involves replacing about a third of the reactor's fuel rods that contain the spent fuel.

The amount of nuclear waste from nuclear reactors is not huge, but the spent-fuel rods are so intensely hot and highly radioactive that they cannot be simply thrown away. Researchers have found that 10 years after being removed from a reactor, a single spent-fuel rod assembly can still emit enough radiation to kill a person standing 1 meter (39 inches) away in less than 3 minutes.

Thus, after spent-fuel rod assemblies are removed from reactors, they are stored in *water-filled pools* (Figure 15-25, left). After several years of cooling, they can be transferred to *dry casks* made of heat-resistant metal alloys and concrete and filled with inert helium gas (Figure 15-25, right). No one knows how long these casks can be used before they break down. They are licensed for 20 years and could last for 100 or more years—still a tiny fraction of the thousands of years that the waste must be safely stored.

A 2005 study by the U.S. National Academy of Sciences warned that the intensely radioactive waste storage pools and dry casks at 68 nuclear power plants in 31 U.S. states are especially vulnerable to sabotage or terrorist attack because they lie outside of the heavily protected reactor containment buildings. At each of the country's nuclear power plants, a government team of mock terrorists runs a test "attack" about every 3 years to test their security. Government records reveal that, between 2005 and 2010, eight of the roughly 100 attempts to breach security at U.S. nuclear plants were successful.

A 2002 study by the Institute for Resource and Security Studies and the Federation of American Scientists pointed out that in the United States, many millions of people live near aboveground spent-fuel storage sites. For some time, critics have been calling for the construction of much more secure structures to protect spent-fuel storage pools and dry casks and for moving more of the wastes from pools to casks. They charge that this has not been done because it would add billions of dollars to the already high cost of electricity produced by the nuclear power fuel cycle.

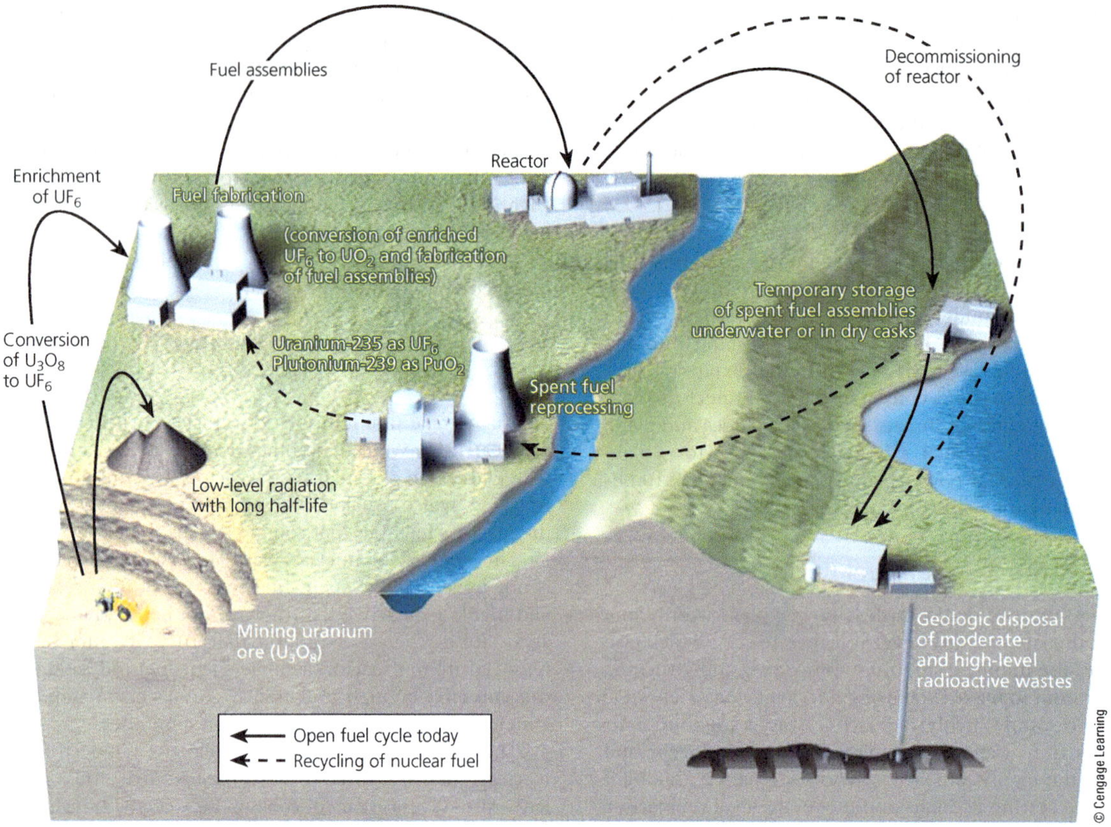

Figure 15-23 Using nuclear power to produce electricity involves a sequence of steps and technologies that together are called the *nuclear fuel cycle*. ***Question:*** Do you think the market price of nuclear-generated electricity should include all the costs of the nuclear fuel cycle, in keeping with the full-cost pricing **principle of sustainability**? (See Figure 1-5, p. 9 or back cover.) Explain.

SUSTAINABILITY

Trade-Offs

Conventional Nuclear Fuel Cycle

Advantages	**Disadvantages**
Low environmental impact (without accidents)	Low net energy yield
	High overall cost
Emits 1/6 as much CO_2 as coal	Produces long-lived, harmful radioactive wastes
Low risk of accidents in modern plants	Promotes spread of nuclear weapons

© Cengage Learning

Figure 15-24 Using the nuclear power fuel cycle (Figure 15-23) to produce electricity has advantages and disadvantages. ***Questions:*** Which single advantage and which single disadvantage do you think are the most important? Why? Do you think that the advantages of using nuclear power outweigh its disadvantages? Explain.

Photo: ©Kletr/Shutterstock.com

CONSIDER THIS. . .

THINKING ABOUT Nuclear Waste Security

Do you favor measures to provide better protection for spent-fuel rods, even if they would raise the cost of electricity? Explain.

Dealing with Radioactive Nuclear Wastes Is a Difficult Scientific and Political Problem

The nuclear waste problem begins with spent-fuel rods. They can be processed to remove radioactive plutonium, which can then be used as nuclear fuel, thus closing the nuclear fuel cycle (Figure 15-23). This reprocessing reduces the storage time for the remaining wastes from up to 240,000 years (longer than the current version of the human species has been around) to about 10,000 years.

However, reprocessing is very costly, and the resulting plutonium could also be used by terrorists or nations to make nuclear weapons, as India did in 1974. This is mainly why the United States, after spending billions of dollars, abandoned this fuel recycling approach in 1977.

Figure 15-25 After 3 or 4 years in a reactor, spent-fuel rods are removed and stored in a deep pool of water contained in a steel-lined concrete basin (left) for cooling. After about 5 years of cooling, the fuel rods can be stored upright on concrete pads (right) in sealed dry-storage casks made of heat-resistant metal alloys and thick concrete. ***Questions:*** Would you be willing to live within a block or two of these casks or have them transported through the area where you live in the event that they were transferred to a long-term storage site? Explain. What are the alternatives?

Photos: U.S. Department of Energy/Nuclear Regulatory Commission

Also, a 2007 study by the nonprofit Institute for Energy and Environmental Research found that nuclear reprocessing increases the volume of nuclear waste sixfold and costs more than using mined uranium, further adding to the high cost of the nuclear fuel cycle. Currently, France, Russia, Japan, India, the United Kingdom, and China reprocess some of their nuclear fuel.

Some analysts have suggested that we could shoot our intensely radioactive wastes into space or into the sun. But the costs of such an effort would be extremely high and a launch accident—such as the 1986 explosion of the Space Shuttle Challenger—could disperse high-level radioactive wastes over large areas of the earth's surface.

Most scientists and engineers agree in principle that deep burial in an underground repository is the safest and cheapest way to store high-level radioactive wastes for thousands of years. Such repositories are in use on a limited basis in the U.S. state of New Mexico, for long-term storage of nuclear waste from the U.S. nuclear weapons program, and in Finland. However, some scientists contend that it is not possible to demonstrate that this or any method will work for thousands of years.

Between 1987 and 2009 the U.S. Department of Energy spent $12 billion on research and testing of a repository for long-term underground storage of high-level radioactive wastes from commercial nuclear reactors on federal land in the Yucca Mountain desert region northwest of Las Vegas, Nevada. In 2010, this project was abandoned for scientific and political reasons and because it was too small to store even the existing radioactive wastes. A government panel is looking for alternative solutions and sites, including two sites for temporary storage of dry casks. Meanwhile these deadly wastes are building up.

Some are calling for reviving the Yucca Mountain site process and for finding another underground storage site to handle future wastes. This presents a political problem, because most states do not want to host a nuclear waste repository, and most people do not want to have highly radioactive wastes transported through their communities on a regular basis from the country's various nuclear reactors (see the map in Figure 34, p. S59 in Supplement 6) to a central nuclear waste repository.

Another radioactive waste problem arises when a nuclear power plant reaches the end of its useful life after about 40 to 60 years and must be closed. Around the world, 285 of the 432 commercial nuclear reactors now operating will need to be decommissioned by 2025.

Eventually all nuclear plants will have to be dismantled and their high-level radioactive materials will have to be stored safely. Scientists have proposed three ways to do this. For any particular plant, one strategy is to store the highly radioactive parts in a permanent, secure repository. A second approach is to install a physical barrier around the plant and set up full-time security for 30–100 years, until the plant can be dismantled after its radioactivity has reached safer levels. These levels would still be high enough to require long-term safe storage of leftover materials.

A third option is to enclose the entire plant in a concrete and steel–reinforced tomb, called a containment structure. This is what was done with a reactor at Chernobyl, Ukraine, that exploded and nearly melted down in 1986, due to a combination of poor reactor design and human operator error. The explosion and the radiation released over a large area killed a number of people and contaminated a vast area of land with long-lasting radioactive fallout in what is viewed as the world's worst nuclear power plant accident. However, within a few years, the containment structure began to crumble, due to the corrosive nature of the radiation inside the damaged reactor, and to leak radioactive wastes. The structure is being rebuilt at great cost and is unlikely to last even several hundred years.

Regardless of the method chosen, the high costs of retiring nuclear plants add to the enormous costs of the nuclear power fuel cycle and reduce its already low net energy yield. Even if all the nuclear power plants in the world were shut down tomorrow, we would still have to find a way to protect ourselves from their high-level radioactive components for thousands of years.

Can Nuclear Power Lessen Dependence on Imported Oil and Help Reduce Projected Climate Change?

Some proponents of nuclear power in the United States claim it will help the country to reduce its dependence on imported crude oil. Other analysts argue that because oil-burning power plants provide only about 1% of the electricity produced in the United States and similarly small amounts in most other countries that use nuclear power, replacing these plants with very costly nuclear power plants would not save much oil.

Nuclear power advocates also contend that increased use of nuclear power will greatly reduce the CO_2 emissions that contribute to projected climate change. Critics argue that the nuclear power industry has mounted a misleading but effective public relations campaign to convince the public that nuclear power does not emit CO_2 and other greenhouse gases.

As scientists point out, this argument is only partially correct. While nuclear plants are operating, they do not emit CO_2. However, during the 10 years that it typically takes to build a plant, especially in the manufacturing of many tons of construction cement, large amounts of CO_2 are emitted. Every other step in the nuclear power fuel cycle (Figure 15-23) also involves CO_2 emissions. Such emissions are much lower than those from coal-burning power plants (Figure 15-17), but they still contribute to atmospheric warming and projected climate disruption.

In 2009, Michael Mariotte, Executive Director of the Nuclear Information and Resource Service, estimated that in order for nuclear power to play an effective role in slowing projected climate disruption over the next 50 years, the world would need to build some 2,000 nuclear reactors—an average of one about every 2 weeks. If power plants could be brought online at such a pace, the amount of high-level radioactive wastes generated would grow dramatically. This would require that new and very costly nuclear waste repositories be built at a similar fast pace.

Experts Disagree about the Future of Nuclear Power

In the 1950s, researchers predicted that by the year 2000, at least 1,800 nuclear power plants would supply 21% of the world's commercial energy (25% of that in the United States) and most of the world's electricity. After almost 60 years of development, a huge financial investment, and enormous government subsidies, some 432 commercial nuclear reactors in 31 countries produced only 5% of the world's commercial energy and 15% of its electricity. In the United States, 104 licensed commercial nuclear power reactors generate about 8% of the country's overall energy and 19% of its electricity.

In 2013, 60 new nuclear reactors were under construction in 13 countries, far from the number needed just to replace the reactors that will have to be decommissioned in coming years. Another 156 reactors are planned, but even if they are completed after a decade or two, they will not replace the 285 aging reactors that must be retired around the world. This helps to explain why nuclear power is now the world's slowest-growing form of commercial energy (see Figure 8, p. S67, in Supplement 7).

The future of nuclear power is a subject of debate. Critics argue that the most serious problem with the nuclear power fuel cycle is that it is uneconomical. They contend that the nuclear power industry could not exist without high levels of financial support from governments and taxpayers, because of the extraordinarily high cost of ensuring safety and the low net energy yield of the nuclear power fuel cycle.

For example, the U.S. government has provided huge research and development subsidies, tax breaks, and loan guarantees to the industry (with taxpayers accepting the risk of any debt defaults) for more than 50 years. It also assumes most of the financial burden of finding ways to store radioactive wastes. In addition, the government provides accident insurance guarantees, because insurance companies have refused to fully insure any nuclear reactor from the consequences of a catastrophic accident. Nuclear power is the only energy resource that receives this government subsidy.

According to the nonpartisan Congressional Research Service, since 1948, the U.S. government has spent more than $95 billion (in 2011 dollars) on nuclear energy research and development (R & D)—more than 4 times the amount spent on R & D for solar, wind, geothermal, biomass, biofuels, and hydropower combined. Critics contend that, without these large and little-known subsidies and tax breaks, the nuclear industry would not exist in the United States. Some question the need for continuing such taxpayer support.

CONSIDER THIS. . .

THINKING ABOUT Government Subsidies for Nuclear Power

Do you think the benefits of nuclear power justify high government (taxpayer) subsidies and tax breaks for the nuclear industry? Explain.

Another obstacle to the growth of nuclear power has been public concerns about the safety of nuclear reactors. Because of the multiple built-in safety features, the risk of exposure to radioactivity from nuclear power plants in the United States and most other more-developed countries is extremely low. However, several explosions and partial or complete meltdowns have occurred (see the Case Study that follows).

CONSIDER THIS. . .

CONNECTIONS Nuclear Power Plants and the Spread of Nuclear Weapons

In the international marketplace, the United States and 14 other countries have been selling commercial and experimental nuclear reactors and uranium fuel-enrichment and purification technology for decades. Much of this information and equipment can be used to produce nuclear weapons. Energy expert John Holdren pointed out that 60 countries that have nuclear weapons or the knowledge to develop them (not including the United States, Great Britain, and the former Soviet Union) have gained most of such information by using civilian nuclear power technology. Some critics see this as the single most important reason for not building more nuclear power plants anywhere in the world.

Proponents of nuclear power argue that governments should continue funding research, development, and pilot-plant testing of potentially safer and less costly new types of reactors. The nuclear industry claims that hundreds of new *advanced light-water reactors (ALWRs)* could be built in just a few years. ALWRs have built-in safety features designed to make meltdowns and releases of radioactive emissions almost impossible. Another proposal is to mix beryllium and uranium to create fuel rods for conventional reactors that would help to prevent meltdowns.

In 2012, the U.S. Nuclear Regulatory Commission approved construction of two new nuclear reactors at an existing power plant in Georgia. Each will have a passive cooling system that can slowly cool the core and prevent a meltdown without the use of pumps when the system has to be shut down. These reactors are being built with the help of loan guarantees from the federal government. Also, the state of Georgia gave permission to the electric utilities to pre-charge their customers for the costs of building these reactors even if they are never completed. Also in 2012, the Department of Energy provided research and development funds for the development and evaluation of small modular reactors that would be built at a factory, hauled to a plant site by train, truck, or barge, placed underground, and wired up.

Some scientists call for replacing today's uranium-based reactors with new ones to be fueled by thorium. They argue that such reactors would be much less costly and safer because they cannot melt down. Also, the nuclear waste they produce cannot be used to make nuclear weapons. To help reduce its dependence on coal, China, which gets less than 2% of its electricity from nuclear power, is building 26 nuclear power plants, some of them fueled with thorium. China is also building a repository for high-level nuclear waste in the country's arid west.

To be environmentally and economically acceptable, some analysts believe that any new-generation nuclear technology should meet the five criteria listed in Figure 15-26. So far, no existing or proposed reactors even come close to doing so. However, even with considerable government financial support and loan guarantees, most U.S. utility companies and money lenders are unlikely to take on the financial risk of building new nuclear plants of any design, as long as electricity can be produced more cheaply with the use of natural gas (Core Case Study) and wind power.

CASE STUDY

The 2011 Nuclear Power Plant Accident in Japan

A major accident occurred on March 11, 2011, at the Fukushima Daiichi Nuclear Power Plant on the northeast coast of Japan. The accident was triggered by a major offshore earthquake that caused a severe tsunami (see Figure 14-22, p. 368). A huge wave of seawater washed over the nuclear plant's protective seawalls and knocked out the circuits and backup diesel generators of the emergency core cooling systems for three of the reactors. Then, explosions (presumably from the buildup of hydrogen gas) blew the roofs off three of the reactor buildings (Figure 15-27) and released radioactivity into the atmosphere and nearby coastal waters. Evidence indicates that the cores of these three reactors suffered full meltdowns.

After some initial confusion and conflicting statements about the severity of the accident, the Japanese government evacuated all residents within a 20-kilometer (12-mile) radius of the plant. Two weeks later, as the severity of the accident became more apparent, people within 30 kilometers (19 miles) were urged to evacuate. More than 110,000 people left their homes, and some areas that now still contain high radiation levels will likely remain unsafe to occupy for up to 20 years.

Preliminary studies indicate that four key human-related factors contributed to this accident: **(1)** failure of the utility company to develop worst-case scenarios that

Solutions

- Reactors must be built so that a runaway chain reaction is impossible.
- The reactor fuel and methods of fuel enrichment and fuel reprocessing must be such that they cannot be used to make nuclear weapons.
- Spent fuel and dismantled structures must be easy to dispose of without burdening future generations with harmful radioactive waste.
- Taking its entire fuel cycle into account, nuclear power must generate a net energy yield high enough so that it does not need government subsidies, tax breaks, or loan guarantees to compete in the open marketplace.
- Its entire fuel cycle must generate fewer greenhouse gas emissions than other energy alternatives.

Figure 15-26 Some critics of nuclear power say that any new generation of nuclear power plants should meet all of these five criteria. ***Question:*** Do you agree or disagree with these critics? Explain.

would have helped speed up their reaction to the crisis, **(2)** the fact that the plant's protective seawalls were not built high enough to withstand huge tsunami waves in this well-known earthquake zone, **(3)** design flaws that exposed the emergency core cooling system controls and backup generators to flooding and that failed to protect the spent-fuel rod storage pools from the damages they suffered, and **(4)** a too-cozy relationship between nuclear plant owners and the government's nuclear regulatory officials.

This accident and the flawed response to it on the parts of plant officials and the government greatly damaged the public confidence of Japanese citizens in the safety of nuclear power. It was a serious accident that contaminated a large area with low to moderate levels of radioactivity. Over the long term, this radiation is projected to kill from 100 to 1,500 people by causing cancers, especially thyroid cancer.

This accident could lead to reduced reliance on nuclear power in Japan. It has prompted Germany, Switzerland, and Belgium to announce plans for phasing out nuclear power. Nuclear power proponents see this as an overreaction, arguing that the annual death toll resulting from the burning of coal is much greater than that of nuclear power accidents.

Is Nuclear Fusion the Answer?

Other proponents of nuclear power hope to develop **nuclear fusion**—a nuclear change at the atomic level in which the nuclei of two isotopes of a light element such as hydrogen are forced together at extremely high temperatures until they fuse to form a heavier nucleus, releasing energy in the process (see Figure 2-9, bottom, p. 40). Some scientists hope that controlled nuclear fusion will provide an almost limitless source of energy.

With nuclear fusion, there would be no risk of a meltdown or of a release of large amounts of radioactive materials, and little risk of the additional spread of nuclear weapons. Fusion power might also be used to destroy toxic wastes and to supply electricity for desalinating water and for decomposing water to produce hydrogen fuel as a very clean-burning energy source.

However, in the United States, after more than 50 years of research and a $25 billion investment, controlled nuclear fusion is still in the laboratory stage. None of the approaches tested so far has produced more energy than they use. In 2006, the United States, China, Russia, Japan, South Korea, India, and the European Union agreed to spend at least $12.8 billion in a joint effort to build a large-scale experimental nuclear fusion reactor by 2026 to determine if it can produce a net energy yield. By 2012, the estimated cost of this project had doubled and it was behind schedule.

If everything goes well, the experimental fusion reactor is supposed to produce enough electricity to run the air conditioners in a small city for a few minutes. Some critics view it as a very costly pie-in-the-sky project that diverts money from more promising energy alternatives like wind and solar energy. Unless there is an unexpected scientific breakthrough, some skeptics will continue to quip that "nuclear fusion is the power of the future and always will be."

AP Photo/AIR PHOTO SERVICE, File

Figure 15-27 An explosion in one of the reactors in the Fukushima Daiichi nuclear power plant severely damaged the reactor.

Big Ideas

- A key factor to consider in evaluating the long-term usefulness of any energy resource is its net energy yield.
- Conventional oil, natural gas, and coal are plentiful and have moderate to high net energy yields, but use of these fossil fuels, especially coal, has a high environmental impact.
- The nuclear power fuel cycle has a low environmental impact and a very low accident risk, but high costs, a low net energy yield, long-lived radioactive wastes, and its role in spreading nuclear weapons technology have limited its use.

TYING IT ALL TOGETHER A New U.S. Oil and Natural Gas Era and Sustainability

Pepik/Shutterstock.com

We began this chapter with a look at the possibility of a new era of oil and natural gas production in the United States. We also learned a guiding scientific principle underlying all energy use—that the long-term usefulness of any energy resource depends on its *net energy yield*. Conventional oil, natural gas, and coal have medium to high net energy yields that will decrease as we use up their easily accessible supplies. Using these energy resources and others derived from them involves high environmental impacts, although these impacts vary. For example, using coal has a very high environmental impact (see photo at left and chapter-opening photo), compared to use of oil and natural gas. Implementing regulations to reduce these harmful environmental impacts will further reduce the net energy yields of these fuels.

We also looked at how we use nuclear power to produce electricity. As we evaluate the net energy yield and environmental impacts of nuclear power, it is important to consider the whole cradle-to-grave nuclear power fuel cycle—from mining the uranium fuel to dismantling and storing the worn-out reactor parts for thousands of years. Considering this whole fuel cycle, nuclear power's environmental impacts are fairly high, and its net energy yield is so low that it is economically unsustainable and must be propped up by various subsidies.

We cannot recycle energy because of the second law of thermodynamics (see Chapter 2, p. 43). However, by reusing and recycling more of the materials we use in our daily lives and in industry and transportation, we could cut our need for energy, thereby raising net energy yields—an application of the chemical cycling **principle of sustainability** (see Figure 1-2, p. 6 or back cover). Also, by using a diversity of energy resources, just as nature relies on the biodiversity **principle of sustainability**, we can further reduce the environmental impacts of our use of energy. Applying the full-cost pricing **principle of sustainability** (see Figure 1-5, p. 9 or back cover) to all energy resources would give us a more realistic understanding of the true economic and environmental costs of using nonrenewable fossils fuels and nuclear power, as well as a variety of renewable energy alternatives that we examine in the next chapter.

Chapter Review

Core Case Study

1. Summarize the potential for greatly increased production of oil and natural gas in the United States and describe two major problems to be overcome.

Section 15-1

2. What is the key concept for this section? What is **net energy yield** and why is it important in evaluating energy resources? Explain why some energy resources need help in the form of subsidies to compete in the marketplace, and give an example.

Section 15-2

3. What are the two key concepts for this section? What is **crude oil (petroleum)**, and how are oil deposits detected and removed? What percentages of the commercial energy used in the world and in the United States are provided by conventional crude oil? What is the **peak production** for an oil well and for the world oil deposits? What is **refining**? What are **petrochemicals** and why are such chemicals important? What countries are the world's three largest producers of oil and what countries are the three largest consumers?

4. What are **proven oil reserves** and what five factors determine such reserves? Define **horizontal drilling** and **hydraulic fracturing** or **fracking** and explain how these two technologies are being used to extract tightly held oil and natural gas from shale rock. What three countries have the largest percentages of the world's proven oil reserves? What percentages are found in the United States and China? Based on data from the U.S. Department of Energy and the U.S. Geological Survey, what are three conclusions that have been drawn concerning consumption of conven-

tional light oil? What are the major environmental costs of using oil? What are the major advantages and disadvantages of using conventional light oil as an energy resource? Describe U.S. dependence on oil and on imported oil. How can U.S. oil production increase dramatically in coming years, according to some analysts? What are two factors that could limit such an increase in domestic oil production?

5. Explain how we can get heavy oil from oil shale rock and from **tar sands (oil sands)**. What are the major advantages and disadvantages of using heavy oils produced from tar sands and from oil shale rock?

Section 5-3

6. What is the key concept for this section? Define **natural gas**, **liquefied petroleum gas (LPG)**, and **liquefied natural gas (LNG)**. What countries have the three largest portions of the world's proven natural gas reserves? What percentages of the world's reserves are held by the United States and China? What are the major advantages and disadvantages of using conventional natural gas as an energy resource? Why has natural gas production risen sharply in the United States and what two factors could hinder this rise? Describe five major problems resulting from increased use of fracking to produce natural gas in the United States and six ways to deal with these problems. What are two other sources of unconventional natural gas and what major problems are related to the use of these resources?

Section 15-4

7. What are the two key concepts for this section? What is **coal** and how is it formed? How does a coal-burning power plant work? What percentage of the electricity used in the world and in the United States comes from burning coal in power plants? What three countries have the largest proven reserves of coal? What are three major problems resulting from the use of coal? Explain why there is no such thing as clean coal. What are the major advantages and disadvantages of using coal as an energy resource? What are the major advantages and disadvantages of using **synthetic natural gas (SNG)** produced from coal?

Section 15-5

8. What is the key concept for this section? How does a nuclear fission reactor work and what are its major safety features? Describe the **nuclear fuel cycle**. What are three ways to decommission a nuclear power plant at the end of its useful life? What are the major advantages and disadvantages of relying on the nuclear fuel cycle as a way to produce electricity? How do nuclear plant operators store highly radioactive spent-fuel rods? Why is dealing with the highly radioactive wastes produced by the nuclear fuel cycle such a difficult problem? Explain why nuclear power is not likely to reduce U.S. dependence on imported oil and why it is not likely to reduce projected climate disruption.

9. Compare the projections for growth of the nuclear industry in the 1950s with its actual role now in generating electricity. What is the role of government subsidies in nuclear power? List and explain three major factors that have kept nuclear power from growing. What is the connection between commercial nuclear power plants and the spread of nuclear weapons? Describe the 2011 nuclear power plant accident in Japan. Define **nuclear fusion**, and summarize the story of attempts to develop it as an energy resource.

10. What are this chapter's *three big ideas*? Explain how the chemical cycling, biodiversity, and full-cost pricing **principles of sustainability** (see Figure 1-2, p. 6, and Figure 1-5, p. 9, or back cover) could be applied to our future energy resource choices.

Note: Key terms are in bold type.

Critical Thinking

1. How might greatly increased production of domestic oil and natural gas in the United States over the next two decades (**Core Case Study**) affect the country's future use of coal, nuclear power, and energy from the sun and wind? How might such an increase affect your life during the next 20 years?

2. Should governments give a high priority to net energy yields when deciding what energy resources to support? What are other factors that should be considered? Explain your thinking.

3. To continue using conventional oil at the current rate, we must discover and add to global oil reserves the equivalent of one new Saudi Arabian reserve every 7 years. Do you think this is possible? If not, what effects might the failure to find such supplies have on your life and on the lives of any child and grandchild that you might have?

4. List three steps you could take to reduce your dependence on oil and gasoline. Which of these things do you already do or plan to do?

5. Some people see imports of Canadian oil produced from tar sands as a way to reduce U.S. dependence on oil imports from potentially unstable Middle Eastern countries and from Venezuela. Others call for developing unconventional oil from shale rock (Science Focus 15.1) as a domestic resource in order to reduce U.S. oil imports. Still others oppose both of these options because they involve environmentally

harmful processes. What is your view on this issue? Explain.

6. Explain why you agree or disagree with the following proposals made by various energy analysts as ways to solve U.S. energy problems: **(a)** find and develop more domestic supplies of conventional crude oil; **(b)** place a heavy federal tax on gasoline and imported oil to help reduce the consumption of crude oil resources and to encourage use of other alternatives; **(c)** increase dependence on coal; **(d)** phase out the use of coal by 2050; **(e)** increase dependence on nuclear power; **(f)** phase out all nuclear power plants by 2040.

7. Explain why you agree or disagree with each of the following proposals made by the U.S. nuclear power industry: **(a)** provide a new round of large government subsidies and loan guarantees to build a large number of better-designed nuclear fission power plants, in order to reduce dependence on imported oil and slow projected climate change; **(b)** prevent the public from participating in hearings on the licensing of new nuclear power plants and on safety issues at the nation's nuclear reactors; **(c)** allow electric utility companies to begin charging consumers for some of the costs of proposed new nuclear power plants before they are built; **(d)** greatly increase federal subsidies for developing nuclear fusion.

8. In 2011, the top five U.S. oil companies generated $135 billion in profits while receiving $4 billion in government subsidies and $24 billion in tax cuts. In 2011, the U.S. Senate failed to pass legislation to eliminate these subsidies and tax breaks. Do you believe that U.S. oil companies should continue to receive these subsidies and tax breaks? Explain.

Doing Environmental Science

Do a study of energy use at your school. Try to determine how the electricity used by the school is generated. If it comes from more than one source (such as coal-fired, gas-fired, or nuclear power plants, and wind farms), determine or estimate the percentages of the total for the various sources. Similarly, determine how buildings are heated, how water is heated, and how vehicles are powered. Write a report summarizing your findings.

Global Environment Watch Exercise

Search *fracking* and use the results to find information on the latest developments in public opposition to, and regulation of, fracking in the United States. Write a report on your findings.

Data Analysis

Use the graph below, comparing U.S. oil consumption, production, and imports, to answer the questions that follow.

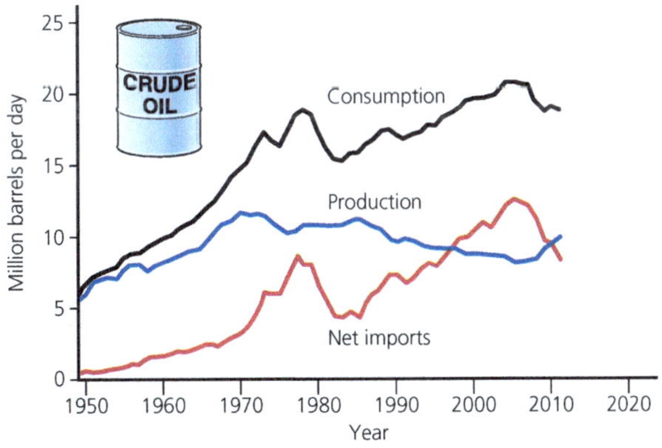

(Compiled by the authors using data from U.S. Energy Information Administration, *Monthly Energy Review*, April 2012, and *Annual Energy Review*, October 2011.)

1. By what percentage did U.S. oil consumption increase between 1982 and 2005?
2. By what percentage did U.S. oil consumption decrease between 2005 and 2011?
3. By what percentage did U.S. oil production decrease between 1985 and 2004?
4. By what percentage did U.S. oil production increase between 2004 and 2011?
5. By what percentage did U.S. net oil imports increase between 1982 and 2005?
6. By what percentage did U.S. net oil imports decrease between 2005 and 2011?

What general conclusions can you draw from these data? Compare notes and discuss your conclusions with your classmates.

16 Energy Efficiency and Renewable Energy

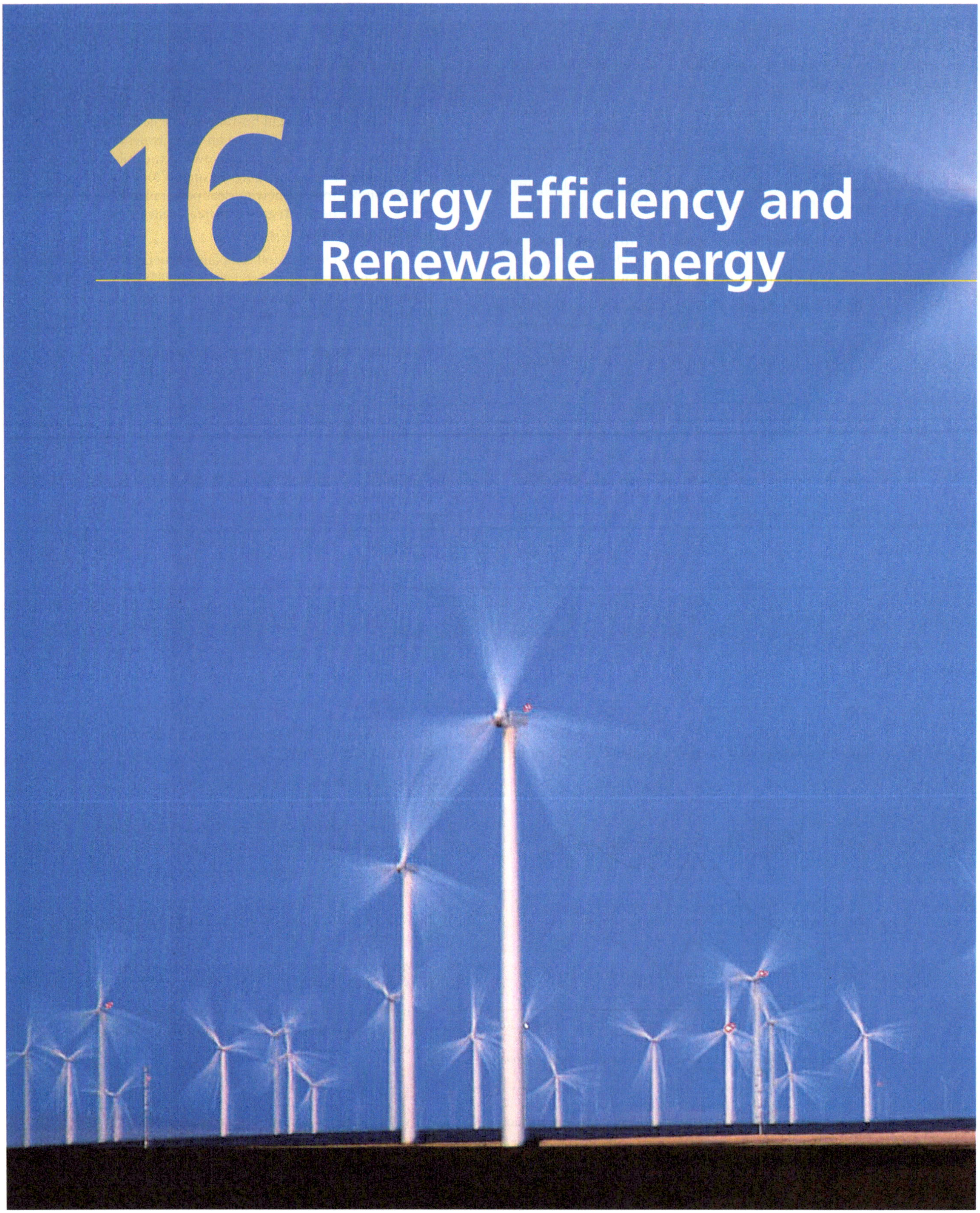

Just as the 19th century belonged to coal and the 20th century to oil, the 21st century will belong to the sun, the wind, and energy from within the earth.

LESTER R. BROWN

Key Questions

16-1 Why is energy efficiency an important energy resource?

16-2 What are the advantages and disadvantages of using solar energy?

16-3 What are the advantages and disadvantages of using hydropower?

16-4 What are the advantages and disadvantages of using wind power?

16-5 What are the advantages and disadvantages of using biomass as an energy source?

16-6 What are the advantages and disadvantages of using geothermal energy?

16-7 What are the advantages and disadvantages of using hydrogen as an energy source?

16-8 How can we make the transition to a more sustainable energy future?

Wind farm in Oregon (USA).

Jim Oleachea/National Geographic Creative

CORE CASE STUDY The Astounding Potential for Wind Power in the United States

Between the earth's equator and its polar regions, the sun's rays strike the earth at different angles. This results in more solar energy at the equator and less at the poles. Together with the earth's rotation, these differences create flows of air called *wind* (see Figure 7-3, p. 146). Because wind power is an indirect form of solar energy, relying more on it is a way to apply the solar energy **principle of sustainability** (see Figure 1-2, p. 6 or back cover).

We can capture this energy with groups of wind turbines called *wind farms* that convert it into electrical energy (Figure 16-1, left) that can be fed into electrical grids. Wind farms can be located on land (see chapter-opening photo) or at sea (Figure 16-1, right).

Unlike nonrenewable fossil fuels, wind power is widely distributed and inexhaustible. The cost of producing electricity from wind is much lower than that of producing it by using the nuclear fuel cycle (Figure 15-23, p. 392) and roughly equal to the cost of producing electricity by burning coal. When we include the harmful environmental and health costs, producing electricity from wind is cheaper than producing it from coal. Using wind power also greatly reduces our emissions of climate-changing carbon dioxide (CO_2) into the atmosphere.

Because today's wind turbine towers can be as tall as a 30-story building and typically have blades as long as 60 meters (200 feet), they can tap into the stronger and more reliable winds found at higher altitudes on land and at sea. The U.S. Department of Energy (DOE) estimates that wind farms at favorable sites in three states—North Dakota, Kansas, and Texas—could more than meet the electricity needs of the lower 48 states. In addition, the National Renewable Energy Laboratory (NREL) estimates that offshore winds—off the Atlantic and Pacific coasts, and off the shores of the Great Lakes—could generate 4 times the electricity currently used in the lower 48 states.

Texas, the nation's leading oil-producing state, is also the nation's leading producer of electricity from wind power, followed by California. These two states produce more electricity from wind than all but five of the world's countries. To some analysts, wind power has more advantages and fewer serious disadvantages than all other energy resources except for energy efficiency, another topic covered in this chapter.

Wind power proponents call for a crash program to develop land-based and offshore wind farms in the United States. Over time, this would help reduce dependence on coal, which, when burned, produces air pollutants that kill at least 24,000 Americans a year and adds huge quantities of CO_2 to the atmosphere. Greatly expanding the U.S. wind turbine industry would also create much-needed new jobs and boost the American economy.

In this chapter, we explore and compare the benefits and drawbacks of improving energy efficiency, relying on wind power, and using a variety of other renewable energy resources.

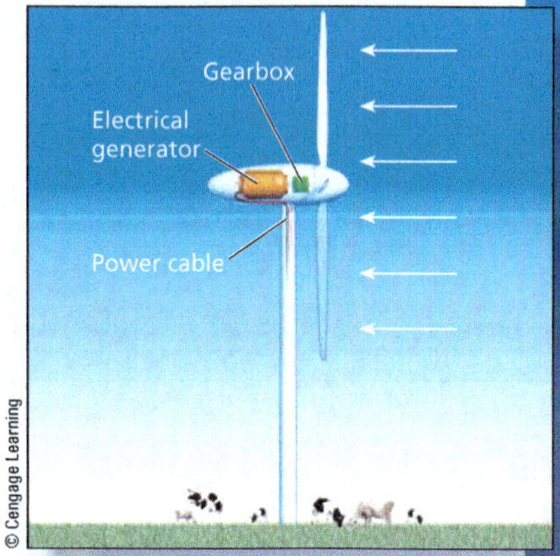

Wind turbine

Figure 16-1 Wind turbines convert the kinetic energy in wind to electricity, another form of kinetic energy (moving electrons). Wind power is an indirect form of solar energy.

Photo: ssuaphotos/Shutterstock.com

16-1 Why Is Energy Efficiency an Important Energy Resource?

CONCEPT 16-1A

Improvements in energy efficiency could save at least a third of the energy used in the world and up to 43% of the energy used in the United States.

CONCEPT 16-1B

We have a variety of technologies for sharply increasing the energy efficiency of industrial operations, motor vehicles, appliances, and buildings.

We Use Energy Inefficiently

Energy efficiency is a measure of how much useful work we can get from each unit of energy we use. Improving energy efficiency means using less energy to provide the same amount of work in the forms of light, heat, transportation, and other benefits. This amounts to a largely untapped source of energy that is abundant, clean, cheap, and readily available.

You may be surprised to learn that roughly 84% of all commercial energy used in the United States does not produce useful work (Figure 16-2). About 41% of this energy unavoidably ends up as low-quality waste heat in the environment because of the degradation of energy quality imposed by the second law of thermodynamics (see Chapter 2, p. 43). The other 43% is wasted unnecessarily, mostly due to the inefficiency of lightbulbs, industrial motors, motor vehicles, and power plants (**Concept 16-1A**).

Another reason for our inefficient and thus wasteful use of energy is that many people live and work in poorly insulated, badly designed buildings that require excessive heating during cold weather and excessive cooling during hot weather. Also, many Americans live in ever-expanding suburban areas around large cities where they must depend on their cars for getting around. Roughly three of every four Americans commute to work, mostly in energy-inefficient vehicles, and only 5% rely on more energy-efficient mass transit. In addition, many people leave lights, computers, and other electrical devices on for long periods when they are not using these devices.

Energy inefficiency costs the United States an average of about $570,000 per minute, according to energy analyst Amory Lovins (see his online Guest Essay on this topic). Much of this cost is due to heavy reliance on four energy-inefficient devices:

- Huge *data centers* (electronic clouds), filled with racks of electronic servers (Figure 16-3) that process information flowing on the Internet, use only about 10% of the electrical energy they pull off of the grid. The other 90% ends up as low-quality heat that flows into the environment. Most of these centers run 24 hours a day at their maximum capacities regardless of the demand. They also require large amounts of energy for cooling and for providing hundreds to thousands of backup batteries.
- The *internal combustion engine* that propels most motor vehicles wastes about 80% of the high-quality energy in its fuel. In other words, only about 20% of the money that people spend on gasoline gets put to good use.
- A *nuclear power plant* (see Figure 15-22, p. 390), which produces electricity, wastes about 75% of the high-quality energy in its nuclear fuel and probably closer to 82% when we include the additional energy used in the nuclear fuel cycle (see Figure 15-23, p. 392).
- A *coal-fired power plant* (see Figure 15-15, p. 387) wastes about 65% of the energy that is released by burning coal to produce electricity, and probably 75–80% if we include the energy used to dig up the coal and transport it to the plant, and to transport and store the ash by-product.

By improving energy efficiency, we can gain numerous economic and environmental benefits (Figure 16-4). To most energy analysts, *it is the quickest, cleanest, and usually the cheapest way to provide more energy, reduce pollution and environmental degradation, and slow projected climate change.* GOOD NEWS

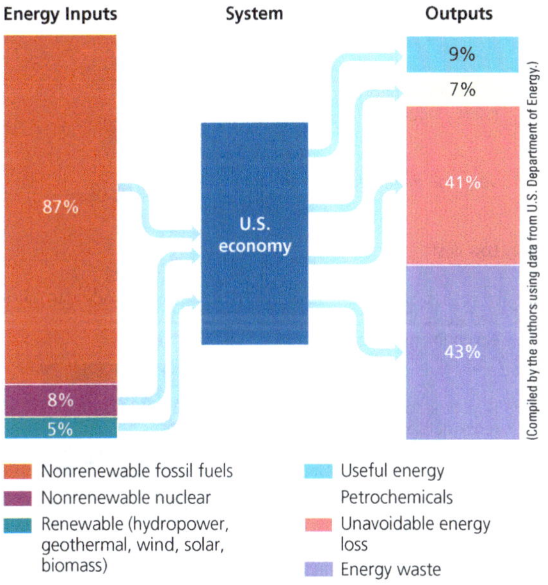

Figure 16-2 Flow of commercial energy through the U.S. economy. Only 16% of the country's high-quality energy ends up performing useful tasks. ***Question:*** What are two examples of unnecessary energy waste?

poetic_disorder/Photos.com

Figure 16-3 Electronic data processing centers that support the web-based global economy typically waste about 90% of the large amount of energy they use.

Solutions

Improving Energy Efficiency

- Prolongs fossil fuel supplies
- Reduces oil imports and improves energy security
- Very high net energy yield
- Low cost
- Reduces pollution and environmental degradation
- Buys time to phase in renewable energy
- Creates local jobs

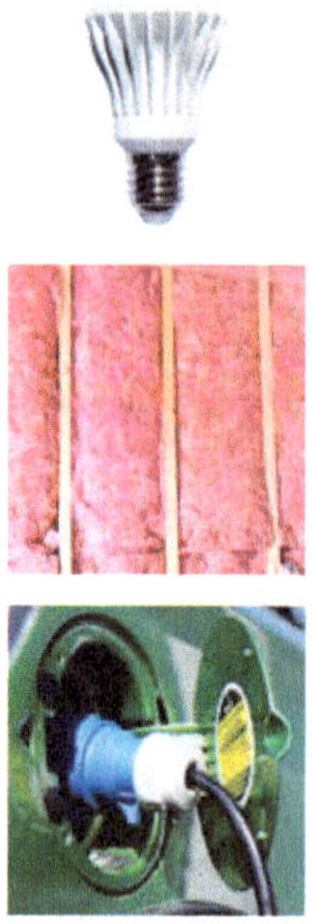

© Cengage Learning

Figure 16-4 Improving energy efficiency has several benefits. ***Questions:*** Which two of these benefits do you think are the most important? Why?

Top: ©Dmitriy Raykin/Shutterstock.com. Center: ©V. J. Matthew/Shutterstock.com. Bottom: ©andrea lehmkuhl/Shutterstock.com.

We Can Improve Energy Efficiency in Industries and Utilities

Industry accounts for about 30% of the world's energy consumption and 33% of U.S. energy consumption, mostly for producing metals, chemicals, petrochemicals, cement, and paper. Here are several ways in which industries can improve their energy efficiency and save money (**Concept 16-1B**).

One way is to use **cogeneration**, which involves using a *combined heat and power (CHP)* system. For example, the steam used for generating electricity in a CHP system can be captured and used again to heat the power plant or other nearby buildings, rather than released into the environment as waste heat. The energy efficiency of these systems is 75–90% (compared to 30–40% for coal-fired, natural gas, and nuclear power plants). Denmark leads the world by getting 52% of its electricity from CHP systems, compared to just 8% in the United States. GOOD NEWS

Industries could also use more *energy-efficient electric motors*. Typical inefficient motors use one-fourth of the electricity produced in the United States and 65% of the electricity used in U.S. industry. Most of these motors run only at full speed with their output throttled to match the task—somewhat like keeping one foot on the gas pedal of a car and the other on the brake pedal to control its speed. They can be replaced with more energy-efficient variable-speed motors, which run at the minimum rate needed for each job.

Recycling materials such as steel and other metals is a third way for industry to save energy and money. For example, producing steel from recycled scrap iron uses 75% less high-quality energy than does producing steel from virgin iron ore and emits 40% less CO_2. A fourth way is to use more *energy-efficient lighting* (see Figure 2-14, p. 44).

Yet another way to save energy would be to redesign the software and cooling systems in government and private electronic data processing systems (Figure 16-3). This could be done with available technology. And the servers could be powered down when they are not in use. We could also improve the efficiency of transmitting electricity to industries and communities (see the following Case Study).

A growing number of major corporations are saving money by improving energy efficiency. For example, the CEO of Dow Chemical Company, which operates 165 manufacturing plants in 37 countries, estimates that between 1996 and 2009, energy efficiency improvements cost Dow about $1.1 billion, but resulted in savings of about $9.4 billion.

CASE STUDY

Building a Smarter and More Energy-Efficient Electrical Grid

Grid systems of high-voltage transmission lines carry electricity from power plants, wind farms (**Core Case Study**), and other electricity producers to users. In the United States, many energy experts place top priority on converting and expanding the outdated regional electrical grid system (Figure 16-5) into what they call a regional and national *smart grid*. This would be an energy-efficient, digitally controlled, ultra-high-voltage (UHV) system with superefficient transmission lines.

A smart grid could adapt to a disaster that causes a power loss in one area by quickly bringing in electricity from other parts of the country. It would also make it easier for power companies and consumers to buy electricity produced from wind, solar, and other renewable forms of energy in different parts of the country. According to the U.S. Department of Energy (DOE), building such a grid would cost the United States up to $800 billion over the next 20 years but would save the U.S. economy up to $2 trillion during that period.

A smart grid system would use smart meters to monitor the amount of electricity used and the patterns of use for each customer. This information would help consumers to reduce their power consumption and power bills. Utilities have learned that when a monthly electric bill lists the average consumption in the neighborhood, this nudges many home owners to significantly reduce their electricity use and waste.

We Can Improve Energy Efficiency and Save Money in Transportation

There is a lot of room for improving energy efficiency in transportation, which accounts for about 28% of the energy consumption and two-thirds of the oil consumption in the United States. In 2011, the U.S. government agreed to raise its fuel economy standards for new cars and light trucks from 12.6 kilometers per liter (kpl) (29.6 miles per gallon or mpg) to 23.3 kpl (54.5 mpg) by 2025. GOOD NEWS Energy experts such as Joseph Romm call for the government to require all new cars and light trucks sold in the United States to get more than 43 kpl (100 mpg) by 2040.

Jimsphotos/Dreamstime.com

Figure 16-5 The outdated high-voltage grid system in the United States could someday be replaced with an ultra-high-voltage smart grid system.

One reason why many consumers buy large and energy-inefficient motor vehicles is that they do not realize that gasoline costs them much more than the price they pay at the pump. The *hidden costs* not included in the price include government subsidies and tax breaks for oil companies, car manufacturers, and road builders; costs of pollution control and cleanup; time wasted idling in traffic jams; and higher medical bills and health insurance premiums resulting from illnesses caused by air and water pollution.

Consumers pay these hidden costs, but not at the gas pump. According to a 2005 study by the International Center for Technology Assessment, the hidden costs of gasoline for U.S. consumers were about $3.18 per liter ($12 per gallon). Thus, when gas costs $4 a gallon, U.S. consumers are really paying about $16 a gallon.

One way to include more of these hidden costs in the market price is through gasoline taxes, which are widely used in Europe but are politically unpopular in the United States. Some analysts call for increasing U.S. gasoline taxes and reducing payroll and income taxes to balance such increases, thereby relieving consumers of any additional financial burden.

Another way for governments to encourage higher energy efficiency in transportation is to give consumers significant tax breaks or other economic incentives to encourage them to buy more fuel-efficient vehicles. Energy expert Amory Lovins has proposed a *fee-bate* program in which buyers of fuel-inefficient vehicles would pay a high fee, and the resulting revenues would be given to buyers of fuel-efficient vehicles as rebates. For example, the fee on a gas-guzzling vehicle averaging about

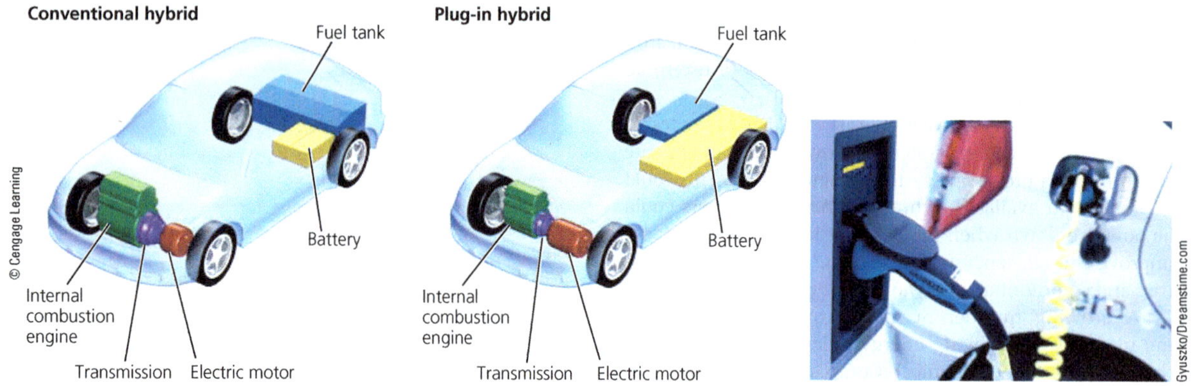

Figure 16-6 Solutions: A *conventional gasoline–electric hybrid* vehicle (left) is powered mostly by a small internal combustion engine with an assist from a strong battery. A *plug-in hybrid electric* vehicle (right) typically has a smaller internal combustion engine with a larger and more powerful battery that can be plugged into a 110-volt or 220-volt outlet and recharged (see photo). An all-electric vehicle (not shown) runs completely on a rechargeable battery. ***Question:*** Would you buy one of these vehicles? Explain.

5 kpl (12 mpg) or less might be $10,000. The government would then give that amount as a rebate to a buyer of a fuel-efficient vehicle that averages 17 kpl (40 mpg) or more. Within a short time, such a program—endorsed by the U.S. National Academy of Sciences—would greatly increase sales of gas-sipping vehicles. It would cost the government (taxpayers) very little because the fees are collected from one set of buyers and given to another set of buyers.

Other ways to save energy and money in transportation include building or expanding mass transit systems within cities, constructing high-speed rail lines between cities (as is done in Japan, much of Europe, and China), and carrying more freight by rail instead of in heavy trucks. Another approach is to encourage bicycle use by building bike lanes along highways and on city streets. The University of New England in Maine offers free, high-quality bikes to new students who agree to leave their cars at home. Also, many companies are saving money by using Internet conferencing as an alternative to flying their employees to meetings.

More Energy-Efficient Vehicles Are on the Way

There is growing interest in developing modern, superefficient, ultralight, and ultrastrong cars that could get up to 130 kpl (300 mpg) using existing technology—a concept that Amory Lovins developed in 1991.

One of these vehicles is the energy-efficient, gasoline–electric *hybrid car* (Figure 16-6, left). These cars have a small, traditional gasoline-powered engine and a battery-powered electric motor used to provide the energy needed for acceleration and hill climbing. The most efficient current models of these cars get a combined city/highway mileage of up to 21 kpl (50 mpg) and emit about 65% less CO_2 per kilometer driven than do comparable conventional cars.

A newer option is the *plug-in hybrid electric vehicle* (Figure 16-6, right). Such vehicles can travel 48–97 kilometers (30–60 miles) on electricity alone. Then the small gasoline motor kicks in, recharges the battery, and extends the driving range to 600 kilometers (370 miles) or more. The battery can be plugged into a conventional outlet and fully charged in 6 to 8 hours. Another option is an all-electric vehicle that runs on a battery only. The problem for the average consumer is that the prices on hybrid, plug-in hybrid, and all-electric cars are high because of the high cost of their batteries.

According to a DOE study, replacing most of the current U.S. vehicle fleet with plug-in hybrid vehicles over two decades would cut U.S. oil consumption by 70–90%, eliminate the need for costly oil imports, save consumers money, and reduce CO_2 emissions by 27%. If the batteries in these cars were recharged mostly by electricity generated by renewable resources such as wind (**Core Case Study**), solar cells, or hydroelectric power, U.S. emissions of CO_2 would drop by 80–90%. This would help to slow projected climate change and save thousands of lives by reducing air pollution from motor vehicles and from coal-burning power plants.

GOOD NEWS

Fully electric cars are promoted as not contributing to climate change because, while operating, they don't emit CO_2. This is correct but misleading, because the process of manufacturing these vehicles does add CO_2 to the atmosphere, requiring about twice as much energy as it takes to build a similar-sized gasoline powered car. Also, when their batteries are recharged with electricity produced by coal-burning, natural gas-burning, or nuclear-power plants, more CO_2 is emitted. Such emis-

SCIENCE FOCUS 16.1

THE SEARCH FOR BETTER BATTERIES

The major obstacle standing in the way of wider use of plug-in hybrid-electric and all-electric vehicles is the lack of an affordable, small, lightweight, and easily rechargeable car battery that can store enough energy to power a vehicle over long distances.

One promising type of battery is the *lithium-ion battery,* commonly used in laptop computers and cell phones. These batteries are light (because lithium is the lightest solid chemical element) and can pack a lot of energy into a small space. By hooking many of them together, we can use them to power hybrid, plug-in hybrid, and all-electric motor vehicles.

Much of the current research on batteries involves nanotechnology (see Science Focus 14.1, p. 362). For example, researchers at the Massachusetts Institute of Technology (MIT) developed a new type of lithium-ion battery using such technology. It is less expensive and can be charged 40 times faster than the nickel–metal hydride batteries that are used to power many of today's hybrid vehicles. GOOD NEWS

In 2010, MIT researcher Michael Strano and his colleagues developed a *thermal wave battery* that produces electricity when a pulse of heat travels through carbon nanotubes. The tubes are coated with a nanosized film of reactive fuel that burns when ignited by a spark or a laser. This creates a fast-moving thermal wave that races along the length of each carbon nanotube 10,000 times faster than the heat produced by the normal burning of the fuel. Prototypes of such batteries put out about 100 times more energy than do lithium-ion batteries of the same weight, and they can be stored indefinitely without leaking any of their charge.

MIT researcher Angela Belcher is working on another new type of battery. She has genetically engineered a virus that can coat itself with electricity-conducting materials to form a miniscule nanowire. Belcher is trying to find ways to link these tiny wires up to form the components of a battery far more compact and powerful than any yet developed. Such viral batteries would essentially grow themselves without producing the toxic wastes often associated with battery production.

Scientists have also developed *ultracapacitors,* which are small mechanical batteries that quickly store and release large amounts of electrical energy, thus providing the power needed for quick acceleration. They can be recharged in minutes and can hold a charge much longer than conventional chemical batteries, which recharge more slowly and have to be replaced more frequently than ultracapacitors. Graphene (see Science Focus 14.2, p. 363) can be used as a component of ultracapacitors. However, ultracapacitors cannot store nearly as much energy as comparable chemical batteries and so far are more expensive per unit of energy produced.

If any one or a combination of these new battery technologies can be mass-produced at an affordable cost, plug-in hybrid and all-electric vehicles could take over the car and truck market. **GREEN CAREER:** battery engineer

Critical Thinking

How might your life change if one or more of the new battery technologies discussed above become a reality?

sions drop sharply when the batteries are recharged with electricity produced by wind turbines, solar cells, or hydropower.

The key to greatly increasing the use of hybrid, plug-in hybrid, and all electric motor vehicles is to ramp up research and development of improved, more affordable batteries (Science Focus 16.1). Another important factor will be to build a network of recharging stations in many convenient locations within and between communities. **GREEN CAREERS:** plug-in hybrid and all-electric car technology

A future stage in the development of superefficient cars might be an electric vehicle that uses a *fuel cell*—a device that uses hydrogen gas (H_2) as a fuel to produce electricity. Fuel cells are at least twice as efficient as internal combustion engines, have no moving parts, and require little maintenance. Their H_2 fuel is usually produced with the use of electricity, and if such electricity could be generated by wind and solar power, the widespread use of fuel cells would drastically reduce our emissions of CO_2 and other pollutants that cause air pollution and climate change—two of the world's most serious environmental, economic, and health problems. However, such cars are unlikely to be widely available until 2020 or later and will probably be very expensive, in part because H_2 has a negative net energy yield, as discussed later in this chapter. **GREEN CAREER:** fuel-cell technology

We can also improve a vehicle's fuel efficiency by reducing its weight. One way to do this is to make car bodies out of *ultralight* and *ultrastrong* composite materials such as fiberglass, carbon fiber, hemp fiber, and graphene (see Science Focus 14.2, p. 363). In addition to being lighter and stronger than some metals, these materials also would not require painting, a process that is costly and environmentally hazardous. The current cost of making such car bodies is high, but mass production would likely help to bring these costs down.

DIANE COOK & LEN JENSHEL/National Geographic Creative

Figure 16-7 This green roof is one of hundreds in New York City. One of the windows shows a reflection of the Empire State Building, which was retrofitted in 2011. Its energy costs were cut by 38% after the renovation.

We Can Design Buildings That Save Energy and Money

Globally, the construction and operation of buildings, most of which have life spans of 50–100 years, accounts for about one-third of the world's resource consumption, 25–40% of its use of energy, 30–40% of all CO_2 emissions, and 30–40% of the solid wastes produced.

We know how to make a transition to using more sustainable buildings over the next few decades. We can apply the solar energy and chemical cycling **principles of sustainability** (see Figure 1-2, p. 6 or back cover) by using energy more efficiently and using more renewable energy resources, reusing and recycling materials, reducing pollution and waste, and in the process, creating jobs and saving money. GOOD NEWS

For example, if a building is oriented to face the sun, it can get more of its heat from solar energy, which can cut heating costs by up to 20%. This is a simple application of the solar energy **principle of sustainability** that people have been using for centuries. Heating costs can be cut by as much as 75% when the building is well-insulated and airtight.

Green architecture, based on energy-efficient, resource-efficient, and money-saving designs, employs several old and new technologies. Along with natural lighting, direct solar heating, insulated windows, and energy-efficient appliances and lighting, it makes use of solar hot water heaters, electricity from solar cells, windows that darken automatically to deflect heat from the sun, thin sheets of aerogel insulation, and recycling of wastewater. **GREEN CAREERS:** sustainable environmental design and architecture

Some green buildings also have *living roofs*, or *green roofs*, covered with specially designed soil and vegetation (Figure 16-7). Such a roof can reduce the costs of cooling and heating a building by absorbing heat from the summer sun and helping to insulate the structure and retain heat in the winter. A living roof also absorbs precipitation, which would normally become part of a city's storm water runoff, adding to the pollution of its waterways.

Several European countries have used green roofs for more than 30 years. The cities of Stuttgart, Germany, Copenhagen, Denmark, and Toronto, Canada, now require green roofs on all new flat-roofed buildings. Green roofs are also sprouting in many American cities, including New York City (Figure 16-7) and Chicago, Illinois, which has 600 green roofs, more than any other U.S. city. White or light-colored roofs are also used to help reduce cooling costs by reflecting incoming solar radiation, especially in hotter climates—another strategy for working with nature that people have used for centuries.

Photos: ©Alison Gannett

Figure 16-8 Solutions: This energy-efficient, Victorian-style straw-bale house was designed and built by Alison Gannett in Crested Butte, Colorado. The left photo was taken during construction; the right photo shows the completed house.

Superinsulation is very important in energy-efficient design. A house can be so heavily insulated and airtight that heat from direct sunlight, appliances, and human bodies can warm it with little or no need for a backup heating system, even in extremely cold climates. Superinsulated houses in Sweden use 90% less energy for heating and cooling than do typical American homes of the same size. One example of superinsulation is straw-bale construction, in which a house's walls are built of straw bales that are covered on the inside and outside with adobe (Figure 16-8). Such walls can have insulating values of 2 to 6 times those of conventional walls. (See the online Guest Essay on this topic by Nancy Wicks.)

Green building certification standards have now been adopted in 21 countries, thanks to the efforts of the World Green Building Council. Between 1999 and 2012, the U.S. Green Building Council's Leadership in Energy and Environmental Design (LEED) program has awarded silver, gold, and platinum standard certificates to more than 14,200 U.S. buildings that meet certain efficiency and environmental standards. In 2009, the Sustainable Building Coalition proposed that we go further and evaluate the sustainability of buildings over their lifetimes according to measurements of resource use, energy use, water use, CO_2 emissions, outputs of solid and hazardous wastes, thermal comfort, and indoor air quality.

CONSIDER THIS. . .

THINKING ABOUT Efficient Building Design

What are three ways in which the building in which you live or work could have been designed to be more energy efficient?

We Can Save Money and Energy in Existing Buildings

There are many ways to save energy and money in existing buildings. The Department of Energy estimates that U.S. home owners can reduce their energy bills by as much as 25% through simple measures such as adding insulation, sealing air leaks, and using energy-efficient appliances. A good first step toward getting such savings is to have an expert make an *energy audit* of a house or other building to suggest ways to improve energy efficiency. Some utility companies provide energy audits at little or no cost. Such an audit might result in some or all of the following recommendations:

GOOD NEWS

- *Insulate the building and plug leaks.* About one-third of the heated air in typical U.S. homes and other buildings escapes through holes, cracks, and closed, single-pane windows (Figure 16-9). During hot weather, these windows and cracks let heat in, which increases the need for air conditioning. Adding insulation and plugging air leaks are two of the quickest and least expensive ways to improve energy efficiency and save money.
- *Use energy-efficient windows.* Energy-efficient windows can pay for themselves within several years, cut expensive heat losses from a house or other building by two-thirds, lower cooling costs in the summer, and reduce heating system CO_2 emissions. Also, *smart windows* that automatically adjust the tint of glass to reflect or absorb light are now available.
- *Stop other heating and cooling losses.* Leaky heating and cooling ducts in attics and unheated basements allow 20–30% of a home's heating and cooling energy to escape, and those in basements can draw unwanted moisture into homes. Careful sealing of duct joints can reduce this loss and save money. Also, use of white or light-colored roofs and living roofs can cut electricity use for air conditioning and reduce CO_2 emissions. Engineers are working on roofs that can change colors—being white and reflecting solar energy in hot weather and turning dark in order to absorb solar energy during cold weather.
- *Heat houses and other buildings more efficiently.* In order, the most energy-efficient ways to heat indoor space or to conserve indoor heat are: superinsulation (which would include plugging leaks); a geo-

Courtesy of HeatSeekers

Courtesy of HeatSeekers

Figure 16-9 A *thermogram,* or infrared photo, of a house in Great Britain before (left) and after it was well insulated (right). Many homes in the United States and other countries are so full of leaks that their heat loss in cold weather and heat gain in hot weather are equivalent to what would be lost through a large, window-sized hole in a wall of the house. ***Question:*** How do you think the place where you live would compare to the house on the left in terms of heat loss and the resulting waste of money on high heating and cooling bills?

thermal heat pump that transfers heat stored from underground into a home; passive solar heating; a high-efficiency, conventional heat pump (in warm climates only); and a high-efficiency natural gas furnace.

- *Heat water more efficiently.* One approach is to use a roof-mounted solar hot water heater. Another option is a *tankless instant water heater* fired by natural gas or LPG. (Using electricity for heating water is not efficient.) These devices, widely used in many parts of Europe and the United States, heat water instantly, providing hot water only when and for as long as it is needed.
- *Use energy-efficient appliances.* A refrigerator with its freezer on the bottom uses about half as much energy as one with the freezer on top or on the side, because on such older models, the dense cold air flows out whenever the door is opened. Microwave ovens use 25–50% less electricity than electric stoves do for cooking and 20% less than convection ovens use. Clothes dryers with moisture sensors cut energy use by 15%. Front-loading clothes washers use 55% less energy and 30% less water than top-loading models use and cut operating costs in half. LED televisions use about 70% less energy than plasma sets of the same size use.
- *Stop using the standby mode.* According to the DOE, keeping TVs, satellite and cable boxes, computers, and other electronic devices on standby when they are not being used consumes about 10% of the electricity used by a typical household. Consumers can reduce their energy use and their monthly power bills by plugging their standby electronic devices into a smart power strip that cuts off power to a device when it detects that the device has been turned off.
- *Use energy-efficient computers.* According to the EPA, if all computers sold in the United States met its Energy Star requirements, consumers would save $1.8 billion a year in energy costs and reduce greenhouse gas emissions by an amount equal to that of taking about 2 million cars off the road.
- *Use energy-efficient lighting.* As energy-wasting incandescent bulbs are being phased out, home owners, colleges and universities, and businesses are switching to more energy-efficient and longer-lasting compact fluorescent light (CFL) bulbs and LED bulbs (see Figure 2-14, p. 44). Halogen lightbulbs are also available and are less expensive than most LED bulbs, but they are only slightly more efficient than incandescent bulbs and run so hot that they can start a fire when they come into contact with flammable materials. As prices come down, highly efficient LED bulbs that use 85% less energy and last about 17 times as long as incandescent bulbs and 4 times as long as CFLS are expected to take over more of the lighting market. In 2011, the Earth Policy Institute estimated that by shifting to energy-efficient fluorescent lights in homes, office buildings, and factories and using LEDs for all traffic lights, we could save enough electricity to close more than 700 of the world's 2,800 coal-burning power plants.
- *Use automatic motion sensors to turn lights on and off as needed.* In 2011, the Earth Policy Institute estimated that by shifting from incandescent bulbs to LEDs and installing automatic motion sensors and dimmers, we could cut the energy used for lighting in buildings by more than 90%.

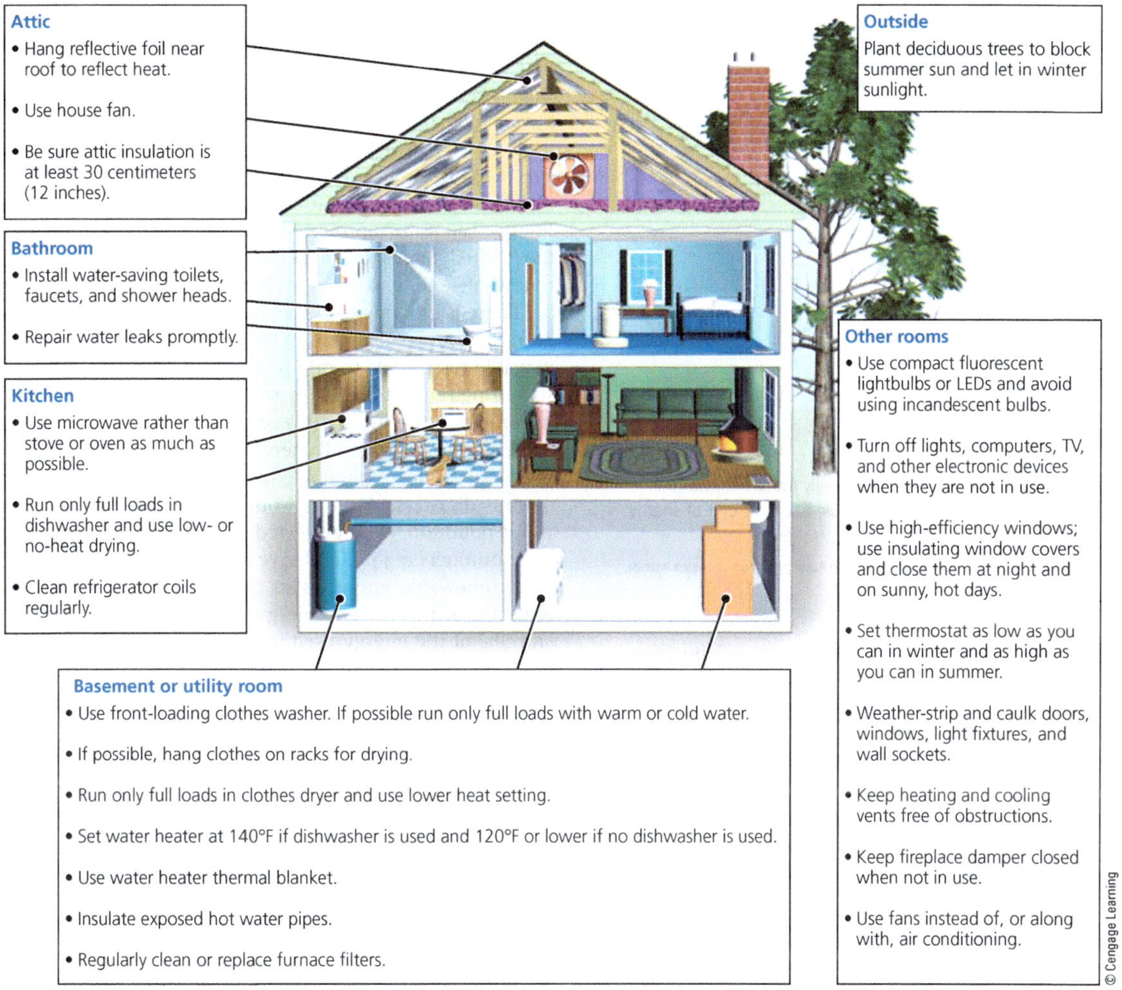

Figure 16-10 Individuals matter: You can save energy where you live. ***Question:*** Which of these things do you do?

Figure 16-10 summarizes ways in which you can cut your energy use in the place where you live. By saving energy, you can reduce your ecological footprint and save more of your hard-earned money.

CONSIDER THIS. . .

CAMPUS SUSTAINABILITY Competing to Save Energy at UNC

In 2010, the Morrison Residence Hall at the University of North Carolina at Chapel Hill won the first prize in the EPA's first Energy Star National Building Competition. Students living there reduced their energy use by 36% in 1 year and saved more than $250,000 in energy costs by keeping track of their energy consumption on a touch-screen computer in the dorm's lobby. Competitions among the dorm floors encouraged students to turn off lights and computers and take shorter showers. The university helped out by improving the building's lighting, ventilation, heating, and cooling systems.

Why Are We Still Wasting So Much Energy and Money?

Considering its impressive array of benefits (Figure 16-4), why is there so little emphasis on improving energy efficiency? One reason is that in the United States and other countries, energy resources such as fossil fuels and nuclear power are artificially cheap, primarily because of the government subsidies and tax breaks they receive and because their market prices do not include the harmful environmental and health costs of their production and use. This is a violation of the full-cost pricing **principle of sustainability** (see Figure 1-5, p. 9 or back cover).

Another reason is that there are too few government tax breaks, rebates, low-interest and long-term loans, and

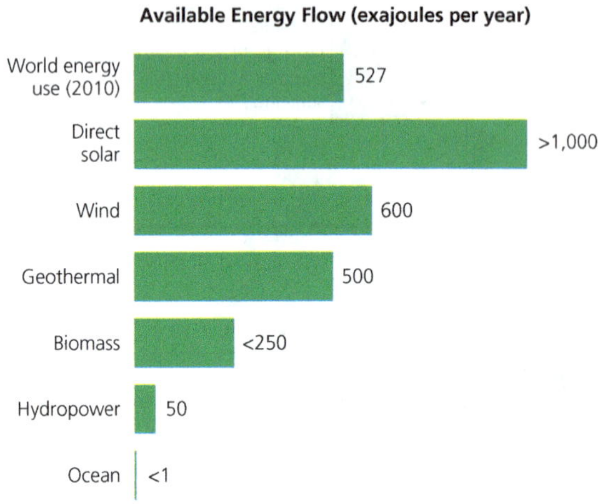

Figure 16-11 Estimated energy potentially available from various renewable energy resources, 2011. (A joule is a basic unit of energy and an exajoule is 1×10^{18} joules.)

(Compiled by the authors using data from Energy Information Agency and the Worldwatch Institute, and T.B. Johansson.)

other economic incentives for consumers and businesses that invest in improving energy efficiency. A third reason is that some governments have not put a high priority on educating the public about the environmental and economic advantages of cutting energy waste.

We Can Use Renewable Energy to Provide Heat and Electricity

The lesson from one of nature's three scientific **principles of sustainability** (see Figure 1-2, p. 6 or back cover) is to *rely mostly on solar energy.* We can get renewable solar energy directly from the sun or indirectly from wind (**Core Case Study**), flowing water, and biomass, none of which would exist without direct solar energy. Another form of renewable energy is geothermal energy from the earth's interior. Figure 16-11 shows the energy potentially available from some major renewable energy resources.

Studies show that with increased and consistent government backing in the form of research and development funds and subsidies and tax breaks, renewable energy could provide 20% of the world's electricity by 2025 and 50% by 2050. In 2012, 13 countries got more than 30% of their electricity from renewable energy. Costa Rica gets more than 95% of its electric power from renewable hydroelectric, wind, and geothermal energy. Germany is phasing out nuclear power and gets 22% of its electricity from wind, solar energy, hydropower, and biomass. Germany has plans to get 80% of its electricity from renewable resources by 2050. Canada gets 63% of its electricity from renewable energy, most of it from hydropower. In 2012, the National Renewable Energy Laboratory projected that, with a crash program, the United States could get 50% of its electricity from renewable energy sources by 2050.

There is growing evidence that a gradual shift from an economy powered primarily by climate-disrupting nonrenewable fossil fuels to one powered by a mix of climate-stabilizing renewable energy sources is underway and will gain momentum during this century. This would result in more decentralized and energy-efficient national economies that are less vulnerable to energy supply cutoffs and natural disasters. Also, it would greatly reduce air and water pollution, slow projected climate disruption, create large numbers of jobs, and save consumers money.

If renewable energy is so great, why does it provide only 8% of the world's energy (see Figure 15-3, p. 376) and 5% of the energy used in the United States (see Figure 15-1, p. 374)? There are three major reasons. *First,* since 1950, government tax breaks, subsidies, and funding for research and development of renewable energy resources have been much lower than those for fossil fuels (especially oil) and nuclear power, although subsidies and tax breaks for renewables have increased in recent years.

Second, although subsidies and tax breaks for fossil fuels and nuclear power have essentially been guaranteed for many decades, those for renewable energy in the United States have to be renewed by Congress every few years. The resulting financial uncertainty makes it risky for companies to invest in renewable energy.

Third, the prices we pay for nonrenewable fossil fuels and nuclear power do not include the harmful environmental and human health costs of producing and using them. This helps to shield them from free-market competition with renewable sources of energy.

Energy analysts such as Amory Lovins (Individuals Matter 16.1) say that if these economic handicaps—*inaccurate pricing* and *unbalanced and intermittent subsidies*—were eliminated, many forms of renewable energy would be cheaper than fossil fuels and nuclear energy, and would quickly take over the energy marketplace.

Throughout the rest of this chapter, we evaluate these renewable energy options.

individuals matter 16.1

Amory B. Lovins: Sustainable Energy Leader

©Rocky Mountain Institute

For over three decades, world-renowned energy expert Amory B. Lovins has been working to demonstrate that we can greatly improve energy efficiency and rely on a variety of renewable energy resources in order to reduce pollution, help preserve biodiversity, and slow projected climate change. He has estimated that in doing so, the world could save at least $1 trillion a year.

In 1976, Lovins published a paper in *Foreign Affairs* titled "Energy Strategy: The Road Not Taken," in which he made a distinction between two possible energy futures. One is a *hard energy path,* built on dependence on fossil fuels and nuclear power. The other is a *soft energy path,* based on using energy more efficiently and depending on a variety of renewable energy sources such as solar, wind, and geothermal energy and biofuels. At the time, critics viewed Lovins as an unrealistic dreamer. But the ground-breaking ideas in his soft-path energy strategy now form the core of many national and corporate energy strategies.

Lovins is co-founder, chairman, and chief scientist at the Rocky Mountain Institute (RMI) in Snowmass, Colorado. RMI is a nonprofit think tank where more than 90 scientists, designers, and engineers conduct cutting-edge research and consultation on energy resources, energy efficiency, and energy policy.

Lovins and the RMI staff are regularly consulted by hundreds of utility companies and major corporations and 60 of the world's governments. *Newsweek* magazine has described Lovins as "one of the Western world's most influential energy thinkers." He has published more than 450 papers, written 31 books, taught at nine major universities, and won essentially all of the world's major environmental awards. In the mid-1990s, RMI, under Lovins' direction, developed cutting-edge designs for hybrid and plug-in hybrid motor vehicles (Figure 16-6).

In 1984, Lovins applied his research and knowledge in designing and building his home and office in Colorado, where winters are extremely cold. This superinsulated, solar-heated, solar-powered, and partially earth-sheltered building incorporated energy-efficiency ideas that were 20 years ahead of the times. Since it was built, solar energy has provided 99% of its heat and hot water and 90% of its household electricity. The building's heating bill is less than $50 a year. These efficiencies were accomplished through the use of energy-efficient lights, appliances, computers, and other electrical devices, and solar cells that generate electricity, along with the building's passive solar design.

Background photo: Chris Gunn/National Renewable Energy Laboratory

16-2 What Are the Advantages and Disadvantages of Using Solar Energy?

CONCEPT 16-2
Passive and active solar heating systems can heat water and buildings effectively, and the costs of using direct sunlight to produce high-temperature heat and electricity are coming down.

We Can Heat Buildings and Water with Solar Energy

Homes and buildings can get all or most of their heat through a **passive solar heating system** (Figures 16-12, top, and 16-13 and Individuals Matter 16.1). Such a system absorbs and stores heat from the sun directly within a well-insulated structure. Water tanks and walls and floors made of concrete, adobe, brick, or stone can store much of the collected solar energy as heat and release it slowly throughout the day and night. One new form of heat storage technology involves a wallboard that contains microscopic capsules of paraffin wax. In a passively heated building, a small backup heating system such as a vented natural gas or propane heater can be used, but it is not usually necessary in a well-designed building (see the online Guest Essay on this topic by Nancy Wicks).

Use of passive solar energy is not new. For thousands of years, people have intuitively followed this principle of sustainability, by orienting their dwellings to take advantage of the sun's heat and light. They have also built thick walls of adobe and stone that collect and store heat, and in hot climates, people have used light colors on their roofs and walls to reflect incoming solar energy and keep their

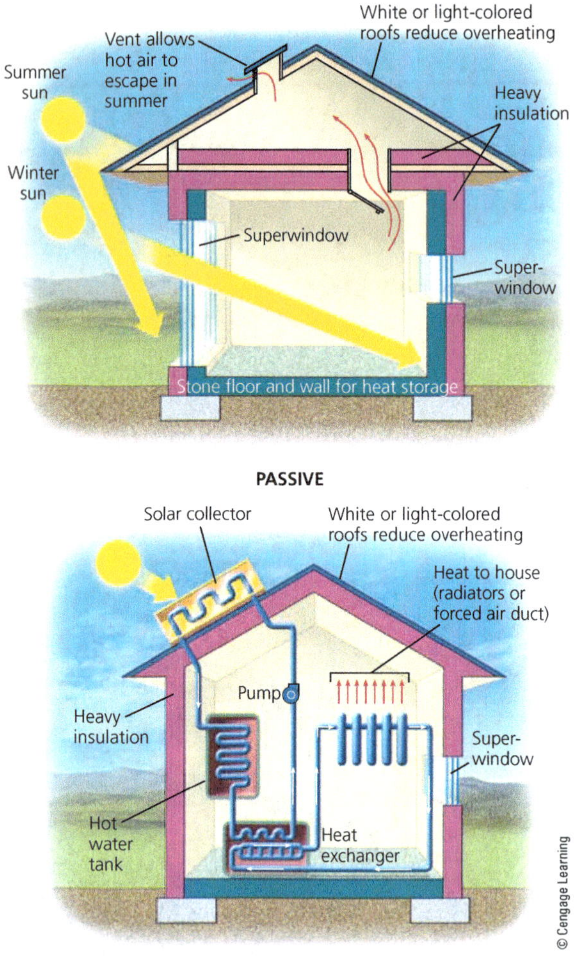

Figure 16-12 Solutions: A home can be heated with a passive or an active solar system.

Figure 16-13 This passive solar home (right) in Golden, Colorado, collects and stores incoming solar energy to provide much of its heat in a climate with cold winters. Notice the solar hot water heating panels in the yard. Some passive solar houses have sunrooms (see inset photo) to help collect incoming solar energy.

Photos: Alan Ford/National Renewable Energy Laboratory

houses cool. Now we are rediscovering this ancient solar wisdom. According to the U.S. Department of Energy and the Passive House Institute, a typical passive U.S. house uses about one-fifth as much energy per year as the average U.S. house uses.

An **active solar heating system** (Figure 16-12, bottom) captures energy from the sun by pumping a heat-absorbing fluid (such as water or an antifreeze solution) through special collectors, usually mounted on a roof or on special racks to face the sun. Some of the collected heat can be used directly. The rest can be stored in a large insulated container filled with gravel, water, clay, or a heat-absorbing chemical and used as needed.

Rooftop active solar collectors are also used to heat water in many homes and apartment buildings. With systems that cost as little as $200, about one in ten houses and apartment buildings in China currently use the sun to provide hot water. By 2030, half of all households in China may get their hot water in this way—an excellent example of applying the solar energy **principle of sustainability** (see Figure 1-2, p. 6 or back cover). Once the fairly low initial cost is paid, the hot water is essentially free. In Spain and Israel, all new buildings must have rooftop systems for heating water and interior space. According to the U.N. Development Programme, solar water heaters could be used to provide half of the world's hot water.

Figure 16-14 lists the major advantages and disadvantages of using passive or active solar heating systems for heating buildings. They can be used to heat new homes in areas with adequate sunlight (see the maps in Figure 35, p. S60, and Figure 36, p. S61, in Supplement 6). But solar energy cannot be used to heat existing homes and other buildings that are not oriented to receive sunlight or that are blocked from sunlight by other buildings or trees.

We Can Cool Buildings Naturally

Direct solar energy actually works against us when we want to keep a building cool, but we can use indirect solar energy (mainly wind) to help cool buildings. For example, we can open windows to take advantage of breezes and use fans to keep the air moving. When there is no breeze, superinsulation and high-efficiency windows help to keep hot air outside. Here are some other ways to keep cool:

- Block the high summer sun with shade trees, broad overhanging eaves, window awnings, or shades (Figure 16-12, top).
- In warm climates, use a light-colored roof to reflect as much as 90% of the sun's heat (compared to only 10–15% for a dark-colored roof), or use a green roof (Figure 16-7).
- Use geothermal heat pumps for cooling (and for heating in winter).

Trade-Offs

Passive or Active Solar Heating

Advantages	Disadvantages
Medium net energy yield	Need access to sun 60% of time during daylight
Very low emissions of CO_2 and other air pollutants	Sun can be blocked by trees and other structures
Very low land disturbance	High installation and maintenance costs for active systems
Moderate cost (passive)	Need backup system for cloudy days

Figure 16-14 Heating a house with passive or active solar energy system has advantages and disadvantages (Concept 16-2). ***Questions:*** Which single advantage and which single disadvantage do you think are the most important? Why? Do you think that the advantages of using these technologies outweigh their disadvantages?

We Can Concentrate Sunlight to Produce High-Temperature Heat and Electricity

One of the problems with direct solar energy is that it is dispersed. **Solar thermal systems**, also known as *concentrated solar power (CSP)*, use different methods to collect and concentrate solar energy in order to boil water and produce steam for generating electricity (Figure 16-15). These systems are used mostly in desert areas with ample sunlight (see the map in Figure 35, p. S60, and Figure 36, p. S61, in Supplement 6).

One such system uses troughs of curved collectors that concentrate solar energy and use it to heat synthetic oil in a pipe that runs through the center of each trough (Figure 16-15, left). This concentrated heat—as high as 400°C (750°F)—is used to boil water and produce steam that powers a turbine that drives a generator to produce electricity. Another system (Figure 16-15, right) uses an array of computer-controlled mirrors to track the sun and focus its energy on a central power tower to provide enough heat to boil water that is used to produce electricity. The heat produced by either of these systems can also be used to melt a certain kind of salt stored in a large insulated container. The heat stored in this *molten salt system* can then be released as needed to produce electricity at night or on cloudy days.

Scientists estimate that we could meet all of the world's electricity needs by using solar thermal systems in combination with high-voltage smart electrical grids (see Case Study, p. 405). Together, these systems would

Figure 16-15 *Solar thermal power:* This solar power plant (left) in a California desert uses curved (parabolic) solar collectors to concentrate solar energy to provide enough heat to boil water and produce steam for generating electricity. In another type of system (right), an array of mirrors tracks the sun and focuses reflected sunlight on a central receiver to boil the water for producing electricity.

Left: ©National Renewable Energy Laboratory; Right: Sandia National Laboratories/National Renewable Energy Laboratory

Trade-Offs

Solar Thermal Systems

Advantages	Disadvantages
High potential for growth	Low net energy yield and high costs
No direct emissions of CO_2 and other air pollutants	Needs backup or storage system on cloudy days
Lower costs with natural gas turbine backup	Requires high water use
Source of new jobs	Can disrupt desert ecosystems

Figure 16-16 Using solar energy to generate high-temperature heat and electricity has advantages and disadvantages (Concept 16-2). ***Questions:*** Which single advantage and which single disadvantage do you think are the most important? Why? Do you think that the advantages of using these technologies outweigh their disadvantages?

Top: Sandia National Laboratories/National Renewable Energy Laboratory. Bottom: ©National Renewable Energy Laboratory.

cover less than 1% of the world's deserts—an area roughly the size of Austria or Wisconsin—according to these estimates. GOOD NEWS

One problem with solar thermal systems is that their net energy yield is low, which means that they need large government subsidies or tax breaks in order to compete in the marketplace with alternatives that have higher net energy yields. Researchers are working on ways to raise the net energy yield for such systems to make them more affordable.

Water shortages in some areas may also limit the production of electricity from solar thermal power plants. These systems require large volumes of cooling water for condensing the steam back to water and for washing the surfaces of the mirrors and parabolic troughs. And they are all built in sunny, arid desert areas where water is scarce.

Figure 16-16 summarizes the major advantages and disadvantages of concentrating solar energy to produce high-temperature heat and electricity.

We can use concentrated solar energy on a smaller scale, as well. In some sunny areas, people use inexpensive *solar cookers* to focus and concentrate sunlight for boiling and sterilizing water (Figure 16-17, left) and for cooking food (Figure 16-17, right). In 2009, inventor Jon Boehner received a $75,000 prize for develop-

Figure 16-17 Solutions: This woman in Tibet is using a solar cooker to boil water (left). The simple solar oven (right) is being used to cook food.

Left: Patstock/Age Fotostock

M. Cornelius/Shutterstock.com

ing a $6 solar cooker made from a cardboard box. Solar cookers can replace wood and charcoal fires, which help to reduce indoor air pollution (a major killer for many of the world's poor people) and deforestation by decreasing the need for firewood and charcoal made from firewood. Also, researchers at Rice University have modified a solar stove so that it can be used as an autoclave to sterilize surgical instruments in areas of less-developed countries not connected to an electrical grid.

We Can Use Solar Cells to Produce Electricity

The dream of using solar energy to produce electricity goes way back. In 1931, Thomas Edison (inventor of the electric lightbulb) told Henry Ford, "I'd put my money on the sun and solar energy." This dream is now a reality.

We can convert solar energy directly into electrical energy using **photovoltaic (PV) cells**, commonly called **solar cells**. Most solar cells are thin wafers of purified silicon (Si) or polycrystalline silicon with trace amounts of metals that allow them to conduct electricity. A typical solar cell has a thickness ranging from less than that of a human hair to that of a sheet of paper. When sunlight strikes these transparent cells, they produce electricity (a flow of electrons), and many cells wired together in a panel can produce electrical power. Such systems can be connected to existing electrical grid systems or to batteries that store the electrical energy until it is needed.

We can mount arrays of solar cells on rooftops and incorporate them into almost any type of roofing material, including tiles of different colors and shapes. The material used to make them can be made into paper-thin, rigid or flexible sheets that can be printed like newspapers and attached to a variety of surfaces such as walls, windows, and clothing. Solar cells have no moving parts and they operate safely and quietly with no emissions of pollutants or greenhouse gases. In 2012, Stanford University scientists developed a "peel and stick" thin solar cell that can be attached to a variety of surfaces. A growing number of businesses and some colleges are covering their parking lots with solar-cell arrays (Figure 16-18). Producing electricity from a network of solar cells on rooftops connected by a smart grid is like connecting millions of laptop computers to the Internet. **GREEN CAREER:** solar-cell technology

Nearly 1.6 billion people, or one of every four in the world, live in less-developed countries in rural villages that are not connected to an electrical grid, but many of them are now getting electrical service with the use of solar cells (Figure 16-19). As these small, off-grid systems reach more rural villages, they will help hundreds of millions of people to lift themselves out of poverty. Eventually, new thin-film solar cells are expected to drastically lower the cost of providing electricity to these areas, eliminating the need for expensive electrical grids of any size.

Large solar-cell power plants are in operation in Portugal, southern Spain, Germany, South Korea, the southwestern United States, and China (Figure 16-20). Energy from some of these plants is stored for use at night or on cloudy days by using the excess electricity they generate to run pumps that compress air into underground caverns. This stored pressure is released as needed to drive turbines and produce electricity.

Another way to store energy is to use electricity from solar-cell power plants and wind farms to recharge the batteries of plug-in hybrid electric cars (Figure 16-6, right)

Figure 16-18 Covering parking lots such as this one in California with solar-cell arrays provides electricity and shade for cars.

and all-electric cars. Such electricity could also be used to decompose water and produce hydrogen gas, which can be used as a fuel, as we discuss later in this chapter. Thus, we are applying the solar energy **principle of sustainability** by indirectly using the sun's energy to power motor vehicles.

Solar cells emit no greenhouse gases, although they are not carbon-free, because fossil fuels are used to produce and transport the panels. However, these emissions are small compared to those resulting from the use of fossil fuels and the nuclear power fuel cycle. Conventional solar cells also contain toxic materials that must be recovered when the cells wear out after 20–25 years of use, or when they are replaced with new systems.

One problem with solar cells is that they are not highly energy efficient and thus have a low to medium net energy yield, depending on the design and the production process. Scientists are rapidly improving the efficiency of solar cells, which will increase their net energy yield. Figure 16-21 lists the major advantages and disadvantages of using solar cells (Concept 16-2).

The main problem with using solar cells to produce electricity has been their high cost, although prices have been dropping sharply. Since 1980 the price of a watt of electricity from solar cells has dropped by a factor of 20. As a result, solar cells have become the world's fastest-growing way to produce electricity (see the graph in Figure 9, p. S67, in Supplement 7). Indeed, such use of solar cells grew by 74% in 2011. This is because of their advantages (Figure 16-21, left) and because of increased government subsidies and tax breaks for solar developers. Production is likely to grow much more, because within the next decade or two, new thin-film nanotechnology and graphene (Science Focus 14.2, p. 363) solar cells are projected to become inexpensive enough to compete with fossil fuels.

In addition, some businesses and home owners are installing solar-cell systems and spreading the costs out over decades by including them in their mortgages. Others are leasing solar-cell systems from companies that install and maintain them. Some communities and neighborhoods are installing shared solar-cell arrays, known as "community solar" or "solar gardens." They can provide electricity for individuals who rent, live in condos, or have their roofs shaded by buildings or trees. And it is often

Figure 16-19 Solutions: A solar-cell panel provides electricity for lighting this hut in rural West Bengal, India. ***Question:*** Do you think your government should provide aid to poor countries for obtaining solar-cell systems? Explain.

Figure 16-20 This solar-cell power plant is located near the city of Shizuishan, China.

cheaper to use solar cells to provide electricity for individual homes in the villages of less-developed countries (Figure 16-19) than to build a central power plant and grid.

Jim Lyons, chief engineer for General Electric, projects that solar cells will be the world's number-one source of electricity by the end of this century. If that happens, it will represent a huge global application of the solar energy **principle of sustainability**. **GREEN CAREER:** solar-cell technology

Trade-Offs

Solar Cells

Advantages	Disadvantages
Medium net energy yield	Need access to sun
Little or no direct emissions of CO_2 and other air pollutants	Some designs have low net energy yield
Easy to install, move around, and expand as needed	Need electricity storage system or backup
Competitive cost for newer cells	Costs high for older systems but dropping rapidly
	Solar-cell power plants could disrupt desert ecosystems

Figure 16-21 Using solar cells to produce electricity has advantages and disadvantages (Concept 16-2). ***Questions:*** Which single advantage and which single disadvantage do you think are the most important? Why? Do you think that the advantages of using this technology outweigh its disadvantages? Why?

16-3 What Are the Advantages and Disadvantages of Using Hydropower?

CONCEPT 16-3

We can use water flowing over dams, tidal flows, and ocean waves to generate electricity, but environmental concerns and limited availability of suitable sites may limit the use of these energy resources.

We Can Produce Electricity from Falling and Flowing Water

Hydropower is any technology that uses the kinetic energy of flowing and falling water to produce electricity. It is an indirect form of solar energy because it depends on heat from the sun evaporating water, which is deposited as rain or snow at higher elevations where it can flow to lower elevations in rivers as part of the earth's solar-powered water cycle (see Figure 3-15, p. 63).

The most common approach to harnessing hydropower is to build a high dam across a large river to create a reservoir. Some of the water stored in the reservoir is allowed to flow through large pipes at controlled rates to spin turbines that produce electricity (see Figure 13-17, p. 330).

Hydropower, the leading renewable energy source, produces about 16% of the world's electricity in 150 countries. In order, the world's top four producers of hydropower are China, Canada, Brazil, and the United States. In 2010, hydropower supplied about 6% of the electricity used in the United States (but about 50% of that used on the West Coast). See Figure 12, p. S69, in Supplement 7 for a graph of the growth of hydropower between 1965 and 2011.

According to the United Nations, only about 13% of the world's potential for hydropower has been developed. Much of the untapped potential is in China, India, South America, Central Africa, and parts of the former Soviet Union. China has plans to more than double its hydropower output during the next decade and is also building or funding more than 200 dams around the world.

However, some analysts expect that use of large-scale hydropower plants will fall slowly over the next several decades as many existing reservoirs fill with silt and become useless faster than new systems are built. Also, there is growing concern over emissions of methane, a potent greenhouse gas, from the decomposition of submerged vegetation in hydropower plant reservoirs, especially in warm climates. In fact, scientists at Brazil's National Institute for Space Research estimate that the world's largest dams altogether are the single largest human-caused source of methane.

In addition, if atmospheric temperatures continue to rise and change the world's climate as projected, many mountain glaciers will melt. Several of these melting glaciers are the primary sources of water for some hydropower plants, so the electrical outputs of these plants are likely to drop. Figure 16-22 lists the major advantages and disadvantages of using large-scale hydropower plants to produce electricity (Concept 16-3).

The use of *microhydropower generators* may become an increasingly important way to produce electricity. These are floating turbines, each about the size of an overnight suitcase. They can be placed in any stream or river without altering its course to provide electricity at a very low cost with very low environmental impact.

We Can Use Tides and Waves to Produce Electricity

We can also produce electricity from flowing water by tapping into the energy from *ocean tides* and *waves*. In some coastal bays and estuaries, water levels can rise or fall by 6 meters (20 feet) or more between daily high and low tides. Dams can be built across the mouths of such bays and estuaries to capture the energy in these flows for hydropower, but sites with large tidal flows are rare. Only two large tidal energy dams are currently operating, one at La Rance on the northern coast of France, and the other in Nova Scotia's Bay of Fundy.

Several turbines, resembling underwater wind turbines, have also been installed to tap the tidal flow of the

Trade-Offs

Large-Scale Hydropower

Advantages	Disadvantages
High net energy yield	Large land disturbance and displacement of people
Large untapped potential	High CH_4 emissions from rapid biomass decay in shallow tropical reservoirs
Low-cost electricity	Disrupts downstream aquatic ecosystems
Low emissions of CO_2 and other air pollutants in temperate areas	

Figure 16-22 Using large dams and reservoirs to produce electricity has advantages and disadvantages (Concept 16-3). ***Questions:*** Which single advantage and which single disadvantage do you think are the most important? Why? Do you think that the advantages of this technology outweigh its disadvantages? Why?

East River near New York City. They swivel to face the incoming and outgoing tides, and they have produced electricity efficiently. Such a system powers a town in Norway. However, these systems are limited to the small number of rivers that have adequate tidal flows.

For decades, scientists and engineers have also been trying to produce electricity by tapping wave energy along seacoasts where there are almost continuous waves. Large, snakelike chains of floating steel tubes have been installed off the coast of Portugal. The up-and-down motion of these chains driven by wave action generates electricity. Scientists estimate that learning how to tap into the world's wave power at an affordable cost could provide more than twice the amount of electricity that the world uses.

Production of electricity from tidal and wave systems is limited because of a lack of suitable sites, citizen opposition at some sites, high costs, and equipment damage from saltwater corrosion and storms. Proponents hope that improved technology can greatly increase the production of electricity from tides and waves sometime during this century.

16-4 What Are the Advantages and Disadvantages of Using Wind Power?

CONCEPT 16-4

When we include the environmental costs of using energy resources in their market prices, wind power is the least expensive and least polluting way to produce electricity.

Using Wind to Produce Electricity Is an Important Step toward Sustainability

Because today's wind turbines are so tall (**Core Case Study**) and have very long blades (Figure 16-23), they can extract more energy from the wind more efficiently than smaller turbines nearer the ground can. According to the DOE's National Renewable Energy Laboratory (NREL), wind farms with these giant turbines can generate more energy with fewer turbines and a lower land and ecological footprint than those with the smaller turbines, and this will help to lower the price of wind energy.

So far, most of the world's rapidly growing number of wind farms have been built on land in parts of Europe, China, and the United States. However, the frontier for wind energy is offshore wind farms (Figure 16-1, right), which have been built off the coasts of ten European countries as well as China and Japan. South Korea is planning a number of offshore wind farms.

Since 1990, wind power has been the world's second fastest-growing source of energy after solar cells (see the graph in Figure 10, p. S68, in Supplement 7), with more than 80 countries now harnessing energy from the wind. GOOD NEWS In order, the countries with the largest installed wind-power capacity in 2012 were China, the United States (**Core Case Study**), Germany, Spain, and India. However, many of China's wind farms are in remote areas and produce no electricity because some of the electrical transmission lines for China's proposed national smart grid (Case Study, p. 405) have yet to be built. Wind power produced about 3.5% of the world's electricity in 2011 and by 2050, could produce about 31% of the world's electricity, according to a 2013 analysis by the market research firm iSUPPLi and the Global Wind Energy Council.

Denmark, the world's most energy-efficient country, got 30% of its electricity from wind in 2012 and is aiming for 50% by 2020. Around the world, more than 400,000 people are employed in the production, installation, and maintenance of wind turbines (Figure 16-24). These job numbers are likely to rise rapidly in coming years.

In 2009, a Harvard University study led by Xi Lu estimated that wind power has the potential to produce 40 times the world's current use of electricity. Even though offshore wind farms are more costly to install, analysts expect to see increasing use of them because wind speeds over water are often stronger and steadier than those over land. Also, locating them offshore lessens the need for negotiations among multiple land owners over the locations of the turbines.

A 2009 study published in the *Proceedings of the U.S. National Academy of Sciences* estimated that the United States has enough wind potential to meet an estimated 16 to 22 times its current electricity needs (**Core Case Study**). Texas leads the nation in wind energy production, followed by California, Iowa, and South Dakota, with Iowa and South Dakota each getting about 24% of their electricity from wind power in 2012. Texas wind farms now supply 8% of the state's electricity and, by 2025, could supply about 90% of the state's current residential electricity needs.

According to a 2009 study by the U.S. Department of the Interior, with expanded and sustained subsidies, wind farms off of the Atlantic and Gulf coasts could generate enough electricity to more than replace all of the country's coal-fired power plants. Many Atlantic and Gulf coast states are making plans to tap into this vast source of energy and create jobs and income in the process. See Figure 37, p. S62, in Supplement 6 for a map of the potential supply of land-based and offshore wind energy in the United States.

Unlike fossil fuels, wind is abundant, widely distributed, and inexhaustible. A wind farm can be built within 9 to 12 months, can be expanded as needed, and has a

GREG GIRARD/National Geographic Creative

Figure 16-23 Tractor trailer hauling two wind turbine blades from Beijing to a wind farm in the Inner Mongolia region of China.

small land footprint. Although wind farms can cover large areas of land, the turbines occupy only about 1% of the land, which can be used for other purposes such as growing crops or raising cattle. Wind farms do not need water for cooling, unlike coal, natural gas, and nuclear power plants. Also, home owners and neighborhoods can use small, quiet wind turbines that operate at low speeds to produce electricity.

In addition, wind power has a high net energy yield. GOOD NEWS The DOE and the Worldwatch Institute estimate that, if we were to apply the full-cost pricing **principle of sustainability** by including the harmful environmental and health costs of various energy resources in comparative cost estimates, wind energy would be the cheapest way to produce electricity (**Concept 16-4**). In making the transition from climate-disrupting fossil fuels—especially coal—to a mix of renewable energy resources, wind has a wide lead. By 2012, wind turbines in more than 80 countries generated about 3.5 times more electricity than solar cells and about 22 times more electricity than geothermal power plants.

Like any energy source, wind power has some drawbacks. For example, areas with the greatest wind power potential are often sparsely populated and located far from cities. Thus, to take advantage of the potential for electricity from wind energy, the United States and other countries will have to invest in upgrading and expanding their outdated electrical grids (Figure 16-5), converting them to smart grid systems (Case Study, p. 405). However, even if we continue to rely on coal-fired and nuclear power plants, such an upgrade and expansion will still have to take place. Many of the new lines could be run along state-owned interstate highway corridors to avoid legal conflicts and further degradation of natural areas.

Another problem is that winds can die down and thus require a backup source of power, such as natural gas, for generating electricity. However, analysts calculate that a large number of wind farms in different areas connected to a smart electrical grid could take up the slack when winds die down in any one area and thus could make wind power a very stable source of electricity.

Suzlon Energy Ltd

Figure 16-24 Maintenance workers get a long-distance view from atop a wind turbine, somewhere in North America.

Trade-Offs

Wind Power

Advantages	Disadvantages
High net energy yield	Needs backup or storage system when winds die down
Widely available	Visual pollution for some people
Low electricity cost	Low-level noise bothers some people
Little or no direct emissions of CO_2 and other air pollutants	Can kill birds if not properly designed and located
Easy to build and expand	

© Cengage Learning

Figure 16-25 Using wind to produce electricity has advantages and disadvantages (Concept 16-4). ***Questions:*** Which single advantage and which single disadvantage do you think are the most important? Why? Do you think the advantages outweigh the disadvantages? Why?

Top: ©TebNad/Shutterstock.com. Bottom: Yegor Korzh/Shutterstock.com.

CONSIDER THIS. . .

CONNECTIONS Bird and Bat Deaths and Wind Turbines

Wildlife ecologists and bird experts have estimated that collisions with wind turbines kill as many as 440,000 birds each year in the United States, although more recent estimates put the figure at 70,000 to 100,000. Compare this to much larger numbers reported by Defenders of Wildlife and in 2013 by the U.S. Fish and Wildlife Service: domestic and feral cats kill 1.4 billion to 3.7 billion birds a year; hunters, more than 100 million; cars and trucks, about 80 million; pesticide poisoning, 67 million; and collisions with tall structures and windows, at least 10 million every year. Most of the wind turbines involved in bird and bat deaths were built with the use of outdated designs, and some were built in bird migration corridors. Wind power developers now avoid such corridors, as well as areas with large bat colonies, when building wind farms. Newer turbine designs reduce bird deaths considerably by using slower blade rotation speeds and by not providing places for birds to perch or nest.

In windy parts of the U.S. Midwest and in Canada, many farmers and ranchers welcome wind farms and some have become wind developers themselves. For each wind turbine located on a farmer's or rancher's land, the land owner typically receives $3,000 to $10,000 a year in royalties. And the land can still be used for growing crops or grazing cattle. An acre of land in northern Iowa planted in corn can produce about $1,000 worth of ethanol car fuel. The same site used for a single wind turbine can produce $300,000 worth of electricity per year, which is why many land owners are investing in wind farms.

Figure 16-25 lists the major advantages and disadvantages of using wind to produce electricity. According to energy analysts, wind power has more benefits and fewer serious drawbacks than any other energy resource, except for improvements in energy efficiency (Figure 16-4). With sufficient and consistent government incentives, wind power could supply more than 10% of the world's electricity and 20% of the electricity used in the United States by 2030. According to the DOE, this would reduce U.S. greenhouse gas emissions by an amount equivalent to that of taking 140 million motor vehicles off the road. It could also create hundreds of thousands of jobs. **GREEN CAREER:** wind-energy engineering

16-5 What Are the Advantages and Disadvantages of Using Biomass as an Energy Source?

CONCEPT 16-5A
Solid biomass is a renewable resource for much of the world's population, but burning it faster than it is replenished produces a net gain in atmospheric greenhouse gases.

CONCEPT 16-5B
We can use liquid biofuels derived from biomass to lessen our dependence on oil-based fuels, but creating biofuel plantations can degrade soil and biodiversity, increase greenhouse gas emissions, and lead to higher food prices.

We Can Produce Energy by Burning Solid Biomass

Biomass consists of plant materials (such as wood and agricultural waste) that we can burn directly as a solid fuel or convert into gaseous or liquid biofuels. Biomass is another indirect form of solar energy because it consists of combustible organic (carbon-containing) compounds in plant matter produced mainly by photosynthesis.

Solid biomass is burned mostly for heating and cooking, but also for industrial processes and for generating electricity. Wood, wood wastes, charcoal made from wood, and other forms of biomass used for heating and cooking supply 10% of the world's energy, 35% of the energy used in less-developed countries, and 95% of the energy used in the poorest countries. In agricultural areas, *crop residues* (such as sugarcane and cotton stalks, rice husks, straw, and coconut shells) and *animal manure* are collected and burned.

Wood is a renewable resource only if it is not harvested faster than it is replenished. The problem is that about 2.7 billion people in 77 less-developed countries face a *fuelwood crisis* and are often forced to meet their fuel needs by cutting forests faster than they can grow to replace the trees that are removed. One way to deal with this problem is to plant fast-growing trees, shrubs, or perennial grasses in *biomass plantations*. But repeated cycles of growing and harvesting these plantations can deplete the soil of key nutrients. Also, the clearing of forests and grasslands for such plantations destroys or degrades biodiversity. And some plantation tree species such as European poplar and American mesquite are invasive species that can spread from plantations and take over nearby areas.

Another problem with depending on solid biomass as a fuel is that clearing forests to provide the fuel reduces the amount of vegetation that would otherwise capture CO_2, and burning biomass produces CO_2. If the rate of use of biomass does not exceed the rate at which it is replenished by new plant growth, there is no net increase in CO_2 emissions. However, monitoring and managing this balance, globally or throughout any one country, is very difficult.

Burning solid biomass to produce electricity has a moderate net energy yield, but burning it in a cogeneration system almost doubles its net energy yield. Denmark gets almost half of its electricity by burning wood and agricultural wastes for cogeneration.

Figure 16-26 lists the general advantages and disadvantages of burning solid biomass as a fuel (Concept 16-5A).

We Can Convert Plants and Plant Wastes to Liquid Biofuels

Liquid biofuels such as *ethanol* (ethyl alcohol produced from plants and plant wastes) and *biodiesel* (produced from vegetable oils) are being increasingly used to fuel motor vehicles. The biggest producers of liquid biofuels are, in order, the United States (producing mostly ethanol from corn), Brazil (producing mostly ethanol from sugarcane residues), the European Union (producing mostly biodiesel from vegetable oils), and China (producing mostly ethanol from non-grain plant sources to avoid diverting grains from its food supply).

Biofuels have three major advantages over gasoline and diesel fuel produced from oil. *First,* biofuel crops can be grown throughout much of the world, and thus they can help countries to reduce their dependence on imported oil. *Second,* if these crops are not used faster than they are

Trade-Offs

Solid Biomass

Advantages

Widely available in some areas

Moderate costs

Medium net energy yield

No net CO_2 increase if harvested, burned, and replanted sustainably

Plantations can help restore degraded lands

Disadvantages

Contributes to deforestation

Clear-cutting can cause soil erosion, water pollution, and loss of wildlife habitat

Can open ecosystems to invasive species

Increases CO_2 emissions if harvested and burned unsustainably

Figure 16-26 Burning solid biomass as a fuel has advantages and disadvantages (Concept 16-5A). ***Questions:*** Which single advantage and which single disadvantage do you think are the most important? Why? Do you think the advantages outweigh the disadvantages? Why?

replenished by new plant growth, there is no net increase in CO_2 emissions, unless existing grasslands or forests are cleared to plant biofuel crops. *Third,* biofuels are easy to store and transport through existing fuel networks and can be used in motor vehicles at little or no additional cost.

However, in a 2007 UN report on bioenergy, and in another study by R. Zahn and his colleagues, scientists warned that large-scale biofuel-crop farming could degrade biodiversity by leading to more clearing of natural forests and grasslands; increase soil degradation, erosion, and nutrient leaching; push small farmers off their land; and raise food prices if farmers can make more money by growing corn and other crops to fuel cars rather than to feed livestock and people (**Concept 16-5B**).

CONSIDER THIS. . .

CONNECTIONS Biofuels and Climate Change

In 2007, Nobel Prize–winning chemist Paul Crutzen warned that intensive farming of biofuel crops could speed up atmospheric warming and projected climate change by producing more greenhouse gases than would be produced by burning fossil fuels instead of biofuels. This would happen if nitrogen fertilizers were used to grow corn and other biofuel crops. Such fertilizers, when applied to the soil, release large amounts of the potent greenhouse gas nitrous oxide. A 2008 study by Finn Danielsen and a team of other scientists concluded that keeping tropical rain forests intact is a better way to slow projected climate change than burning and clearing such forests and replacing them with biofuel plantations.

Another problem with biofuel production is that by growing corn and soybeans in dry climates that require irrigation, farmers reduce water supplies in these arid regions. In fact, the two most water-intensive ways to produce a unit of energy are by irrigating soybean crops to produce biodiesel fuel and by irrigating corn to produce ethanol.

The challenge is to grow crops for food and biofuels by using more sustainable agriculture (see Figure 12-33, p. 309) with less irrigation, land degradation, air and water pollution, greenhouse gas emissions, and degradation of biodiversity. Also, any system used to produce a biofuel should have at least a moderate net energy yield so that it can compete in the energy marketplace without large government subsidies.

CASE STUDY

Is Biodiesel the Answer?

If a truck or bus whizzes by and leaves a scent of fast food, it is probably running on *biodiesel.* This biofuel, designed to run in diesel engines, is produced from vegetable oil extracted from soybeans, rapeseeds, sunflowers, oil palms, and jatropha shrubs. It can also be made from used vegetable oils from restaurants. European Union countries (primarily Germany, France, and Italy) produce about 95% of the world's biodiesel, mostly from rapeseeds and sunflower seeds, and these countries hope to get 20% of their diesel fuel from this source within a decade. In Europe, more than half of all cars have diesel engines, primarily because they are as much as 40% more efficient than gasoline engines.

Aided by government subsidies, biodiesel production is growing rapidly in the United States. However, soybean and canola oil crops (rapeseeds or field mustard) grown for biodiesel production require large areas of land. In addition, using industrialized agriculture to produce these crops results in topsoil losses and fertilizer runoff. Biodiesel production also requires energy (mostly from crude oil and natural gas), which reduces its net energy yield and increases emissions of the greenhouse gases nitrous oxide (N_2O) and CO_2.

Brazil, Malaysia, and Indonesia produce biodiesel from palm oil, extracted from large plantations of African oil palm (see Figure 12-5, p. 282), and export much of it to Europe. The net energy yield for biodiesel from oil palm is 5 times that from rapeseeds used in Europe and about 8 to 9 times higher than the yield from soybeans used to produce biodiesel in the United States. However, increased burning and clearing of tropical forests and other wooded lands to establish oil palm plantations in these countries poses a serious threat to their biodiversity and releases CO_2 into the atmosphere. It also reduces CO_2 uptake by eliminating rain forests that store large amounts of carbon and replacing them with crops that store much less carbon. Also, African oil palm is an invasive plant that has taken over adjacent farms and forest areas in parts of Brazil, further degrading their biodiversity.

CASE STUDY

Is Ethanol the Answer?

Ethanol can be made from plants such as sugarcane, corn, and switchgrass, and from agricultural, forestry, and municipal wastes. This process involves converting plant starches or other plant materials into simple sugars, which are processed to produce ethanol.

Brazil, by far the world's largest sugarcane producer, is the world's second largest ethanol producer after the United States. Brazil makes its ethanol from *bagasse,* a residue produced when sugarcane is crushed. This sugarcane ethanol has a medium net energy yield that is about 8 times higher than that of ethanol produced from corn. About 45% of Brazil's motor vehicles run on ethanol or ethanol–gasoline mixtures produced from sugarcane grown on only 1% of the country's arable land.

Within a decade, Brazil could expand its sugarcane production, eliminate oil imports, and greatly increase ethanol exports to other countries. To do this, Brazil plans to clear large areas of its rapidly disappearing Cerrado, a wooded savanna region, to make way for sugarcane plantations. This will further stress biodiversity in this biodiversity hotspot (see Figure 10-28, p. 238). Other harmful effects of producing ethanol from sugarcane include CO_2 emissions from the burning of oil and gasoline during sugarcane production, heavy soil erosion and resulting water pollution after sugarcane plantations are harvested, and stresses on water supplies. Producing 1 liter (0.27 gallons) of ethanol from sugarcane requires the equivalent of about 174 bathtubs full of water.

National Renewable Energy Laboratory

Figure 16-27 Natural capital: The cellulose in this rapidly growing switchgrass can be converted into ethanol, but further research is needed to develop affordable production methods.

In the United States, most ethanol is made from corn (see Figure 11, p. S68, in Supplement 7). However, studies indicate that corn-based ethanol has a low net energy yield because fossil fuels are used heavily for growing the corn and converting it to ethanol. This is one reason why it has received big government subsidies. It also helps to explain why scientists calculate that using this ethanol results in a net increase in greenhouse gas emissions. Also, raising corn requires a great deal of water and thus adds to stressed water supplies in many areas.

According to a 2007 study by environmental economist Stephen Polasky, processing all of the corn grown in the United States into ethanol each year would meet only about 30 days' worth of the country's current demand for gasoline. This would leave no corn for other uses and would cause sharp increases in the prices of corn-based foods, including cereals, tortillas, and meats, with a devastating impact on the poor in several countries. Replacing all gasoline with corn-based ethanol in the United States would provide no major environmental benefits, and it would likely lead to higher emissions of climate-changing CO_2 and higher food prices for consumers. Also, energy expert Vaclav Smil has calculated that in order to replace all of the gasoline used in the United States with ethanol, about 6 times the country's total area of farmable land would have to be planted in corn.

Trade-Offs

Liquid Biofuels

Advantages	Disadvantages
Reduced CO_2 emissions for some crops	Fuel crops can compete with food crops for land and raise food prices
Medium net energy yield for biodiesel from oil palms	Fuel crops can be invasive species
Medium net energy yield for ethanol from sugarcane	Low net energy yield for corn ethanol and for biodiesel from soybeans
	Higher CO_2 emissions from corn ethanol

Figure 16-28 There are advantages and disadvantages to using liquid biofuels. (Concept 16-5B). ***Questions:*** Which single advantage and which single disadvantage do you think are the most important? Why?

An alternative to corn-based ethanol is *cellulosic ethanol,* which is produced from the inedible cellulose that makes up most of the biomass of plants—material such as leaves, stalks, and wood chips. A plant that could be used for cellulosic ethanol production is *switchgrass* (Figure 16-27), a tall perennial grass native to North American prairies that grows faster than corn. It is disease resistant and drought tolerant, and can be grown without the use of nitrogen fertilizers on land unfit for other crops. According to a 2008 study led by agronomist Ken Vogel, switchgrass-based ethanol has a medium net energy yield compared to a low net energy yield from corn-based ethanol.

One major problem is that affordable chemical processes for converting cellulosic material to ethanol are still being developed. Also, according to cellulosic-ethanol expert Robert Ranier, growing enough switchgrass to meet half of the U.S. demand for gasoline would require about 7 times the land area currently used for all U.S. corn production. In addition, we need more research to learn more about CO_2 emissions, invasive species effects, and net energy yields related to switchgrass and other potential cellulosic ethanol crops such as arundo and miscanthus plants.

Scientists are also evaluating the use of algae and bacteria to produce biofuels (Science Focus 16.2). Figure 16-28 compares the advantages and disadvantages of using biodiesel and ethanol liquid biofuels.

SCIENCE FOCUS 16.2

MAKING GASOLINE AND DIESEL FUEL FROM ALGAE AND BACTERIA

Scientists are looking for ways to produce biofuels that are almost identical to gasoline and biodiesel from various types of existing or genetically engineered oil-rich algae that can grow rapidly at any time of the year in various aquatic environments.

As they grow, the algae remove CO_2 from the atmosphere and convert it to oil, proteins, and other useful products. They also require much less land, water, and other resources than biofuel plantations do and would not affect food prices by competing for cropland. In addition, the algae could be grown in wastewater from sewage treatment plants where they would help to clean up the wastewater while producing biofuel. Another possibility would be to transfer CO_2 produced by coal-burning power plants into nearby algae ponds or bioreactors for use in making biofuel.

There are four major challenges in converting such technological dreams into reality: **(1)** cutting the very high cost of producing oil by such methods; **(2)** learning whether it is economically better to grow the algae in open ponds or in enclosed bioreactors; **(3)** finding an economically affordable way to get the oil out of the algae (for example, by drying and crushing the algae or removing it with a solvent); and **(4)** progressing from small-scale pilot projects to large-scale production systems.

A different approach is to make gasoline or diesel fuel from rapidly multiplying bacteria by using techniques developed in the new field of *synthetic biology.* Producing gasoline and diesel fuels from algae and bacteria could be done almost anywhere. The resulting fuels could be distributed by the world's current gasoline and diesel fuel distribution system. Thus, it is not surprising that some of the world's major oil companies are investing heavily in research on producing oil-like compounds from algae and bacteria, hoping to lock in numerous patents for such processes.

Critical Thinking

What might be some environmental risks of producing biofuels from algae or bacteria?

16-6 What Are the Advantages and Disadvantages of Using Geothermal Energy?

CONCEPT 16-6

Geothermal energy has great potential for supplying many areas with heat and electricity, and has a generally low environmental impact, but the sites where it can be produced economically are limited.

We Can Get Energy by Tapping the Earth's Internal Heat

Geothermal energy is heat stored in soil, underground rocks, and fluids in the earth's mantle. We can tap into this stored energy to heat and cool buildings and to produce electricity. Scientists estimate that using just 1% of the heat stored in the uppermost 5 kilometers (8 miles) of the earth's crust would provide 250 times more energy than that stored in all the earth's proven crude oil and natural gas reserves combined. GOOD NEWS

One way to capture geothermal energy is by using a *geothermal heat pump* system (Figure 16-29). It can heat and cool a house by exploiting the temperature difference, almost anywhere in the world, between the earth's surface and underground at a depth of 3–6 meters (10–20 feet), where the temperature typically is 10–20°C (50–60°F) year-round. In winter, a closed loop of buried pipes circulates a fluid, which extracts heat from the ground and carries it to a heat pump, which transfers the heat to a home's heat distribution system. In summer, this system works in reverse, removing heat from a home's interior and storing it in the ground. The EPA estimates that such a system can heat or cool a 190-square-meter (2,000-square-foot) house for as little as $1 a day.

According to the EPA, a well-designed geothermal heat pump system is the most energy-efficient, reliable, environmentally clean, and cost-effective way to heat or cool a space. Once installed, it produces no air pollutants and emits no CO_2. Installation costs can be high but are generally recouped within 3–5 years. Thereafter, such systems save money for their owners. Initial costs can be added to a home mortgage to spread the financial burden over two or more decades.

We can also tap into deeper, more concentrated *hydrothermal reservoirs* of geothermal energy (Figure 16-30). This is done by drilling wells into the reservoirs to extract their dry steam (with a low water content), wet steam (with a high water content), or hot water, which are then used to heat homes and buildings, provide hot water, grow vegetables in greenhouses, raise fish in aquaculture ponds, and spin turbines to produce electricity.

The United States is the world's largest producer of geothermal electricity from hydrothermal reservoirs. Most of it is produced in California, Nevada, Utah, and Hawaii (see Figure 39, p. S63, in Supplement 6 for a map of the

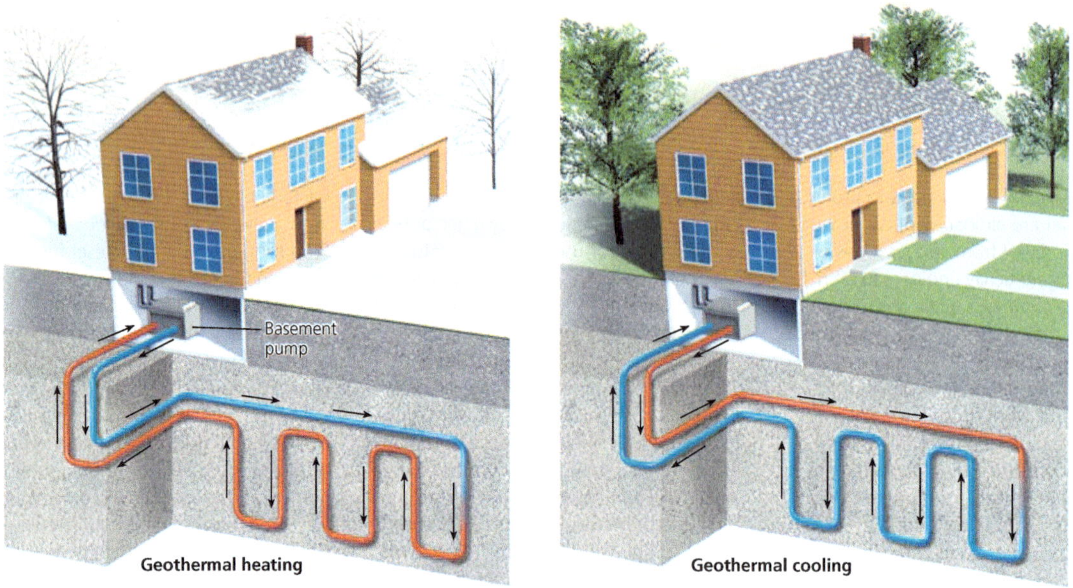

© Cengage Learning

Figure 16-29 Natural capital: A geothermal heat pump system can heat or cool a house almost anywhere.

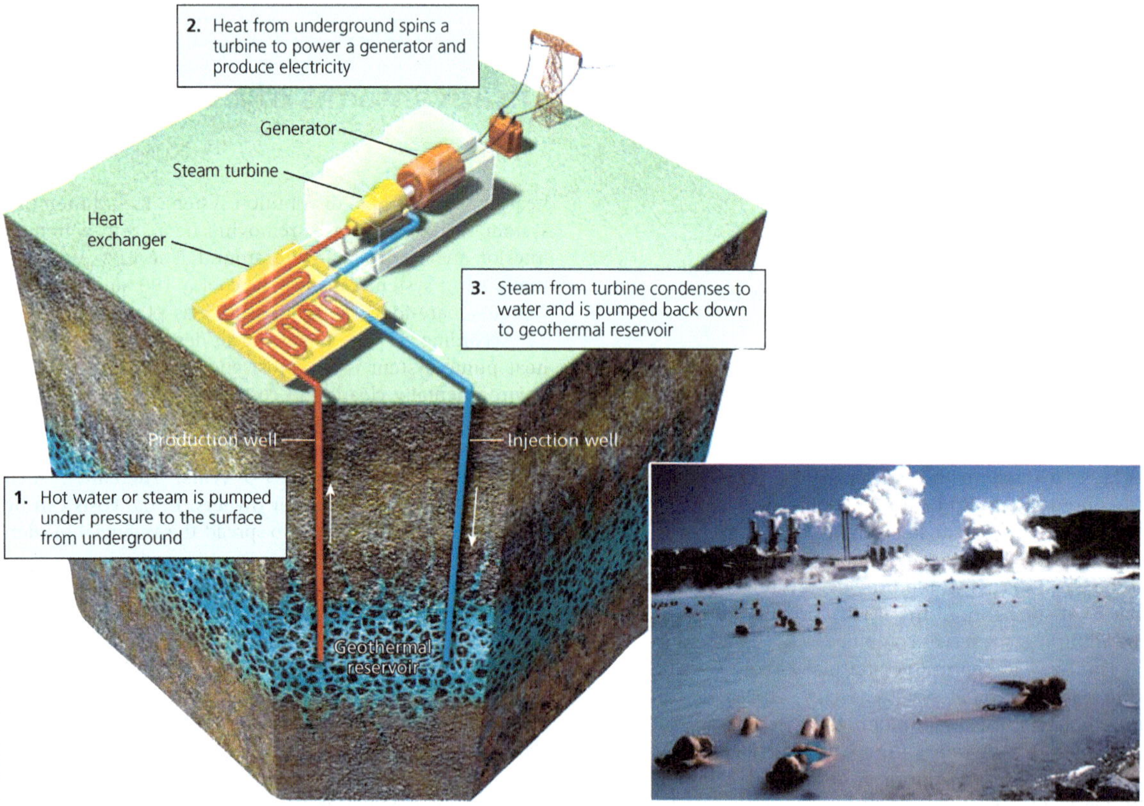

© Cengage Learning 2015

Richard Nowitz/National Geographic Creative

Figure 16-30 Power plants can produce electricity from heat extracted from underground geothermal reservoirs. The photo shows a geothermal power plant in Iceland that produces electricity and heats a nearby spa called the Blue Lagoon.

individuals matter 16.2

Andrés Ruzo—Geothermal Energy Sleuth and National Geographic Young Explorer

©Sofia Ruzo

Andrés Ruzo is a geophysicist with a driving passion to learn about geothermal energy and to show how this renewable and clean energy source can help us to solve some of the world's energy problems. As a boy, he spent his summers on the family farm in Nicaragua. Because the farm rests on top of the Casita Volcano, he was able to experience firsthand the power of the earth's internal heat.

As an undergraduate student at Southern Methodist University (SMU) in Dallas, Texas (USA), because of his boyhood experience, he took a course in volcanology. The course awakened his passion for geology along with a desire to learn more about the earth's heat as a source of energy. This led him to pursue a PhD in geophysics at SMU's Geothermal Laboratory.

This National Geographic Young Explorer and his wife and field assistant, Sofia, spent most of 2012 gathering data to develop the first detailed map of the geothermal energy resources of northern Peru. Their field work involved lowering temperature measuring equipment down into abandoned water wells and oil wells in the desert area of northern Peru where daytime temperatures can exceed 54°C (130°F). These data will provide him with a geothermal map showing how heat flows through the upper crust of the earth and describing its potential to be tapped as a source of energy.

Ruzo believes that geothermal energy is a hidden sleeping giant that can be tapped as an important renewable source of heat and electricity. He says that his goal in life is "to be a force of positive change in the world."

Background photo: N.Minton/Shutterstock.com

best geothermal sites in the continental United States). It meets the electricity needs of about 6 million Americans—an amount roughly equal to the combined populations of Los Angeles, California, and Houston, Texas—and supplies almost 6% of California's electricity. By 2012, there were 130 new geothermal power plants under construction or in development in the United States, enough to triple the country's electrical output from geothermal energy.

Iceland gets almost all of its electricity from hydroelectric and geothermal power plants (Figure 16-30, photo). The Philippines gets 18% of its electricity from geothermal plants—enough electricity for 19 million people. China has a large potential for geothermal power, which could help to reduce its dependence on coal-fired power plants (see Figure 38, p. S63, in Supplement 6 for a map of the world's best geothermal sites). In Peru, a National Geographic Young Explorer is carrying out research to develop that country's geothermal resources (Individuals Matter 16.2).

Another potential source of geothermal energy is *hot, dry rock* found 5 or more kilometers (3 or more miles) underground almost everywhere. It involves drilling a well into the rock and injecting water into the well. Some of this water can absorb enough heat to produce steam that is brought to the surface and used to spin turbines to generate electricity. According to the U.S. Geological Survey, tapping just 2% of this source of geothermal energy in the United States could produce more than 2,000 times as much electricity as is used every year in the country. The limiting factor is cost, which could be brought down by more research and improved technology. There is also concern that tapping this deep source of geothermal energy could increase the frequency of earthquakes in the drilled areas. **GREEN CAREER:** geothermal engineer

Figure 16-31 lists the major advantages and disadvantages of using geothermal energy (Concept 16-6). The biggest problem limiting the widespread use of geothermal energy is the lack of sites with high enough concentrations of heat needed to make this energy resource affordable for widespread use.

Trade-Offs

Geothermal Energy

Advantages	Disadvantages
Medium net energy yield and high efficiency at accessible sites	High cost except at concentrated and accessible sources
Lower CO_2 emissions than fossil fuels	Scarcity of suitable sites
Low cost at favorable sites	Noise and some CO_2 emissions

© Cengage Learning

Figure 16-31 Using geothermal energy for space heating and for producing electricity or high-temperature heat for industrial processes has advantages and disadvantages. (Concept 16-6). ***Questions:*** Which single advantage and which single disadvantage do you think are the most important? Why? Do you think the advantages of this energy resource outweigh its disadvantages? Why?

Photo: ©N.Minton/Shutterstock.com

16-7 What Are the Advantages and Disadvantages of Using Hydrogen as an Energy Source?

CONCEPT 16-7

Hydrogen is a clean energy source as long as it is not produced with the use of fossil fuels, but it has a negative net energy yield.

Will Hydrogen Save Us?

Hydrogen (H) is the simplest and most abundant chemical element in the universe. The sun produces its energy, which sustains life on the earth, through the nuclear fusion of hydrogen atoms (see Figure 2-9, bottom, p. 40).

Some scientists say that the fuel of the future is hydrogen gas (H_2). Most of their research has been focused on using fuel cells (Figure 16-32) that combine H_2 and oxygen gas (O_2) to produce electricity and water vapor ($2\ H_2 + O_2 \rightarrow 2\ H_2O$ + energy), which is emitted into the atmosphere.

Widespread use of hydrogen as a fuel would eliminate most of the outdoor air pollution that comes from motor vehicles and coal-burning power plants. It would also greatly reduce the threat of projected climate change, because in using it, we emit no CO_2—as long as the H_2 is not produced with the use of fossil fuels or nuclear power. Hydrogen also provides more energy per gram than does any other fuel, making it a lightweight fuel ideal for aviation.

So what's the catch? There are three challenges in turning the vision of hydrogen as a fuel into reality. *First,* there is hardly any hydrogen gas (H_2) in the earth's atmosphere, so it must be produced from hydrogen, which is chemically locked up in water and in organic compounds

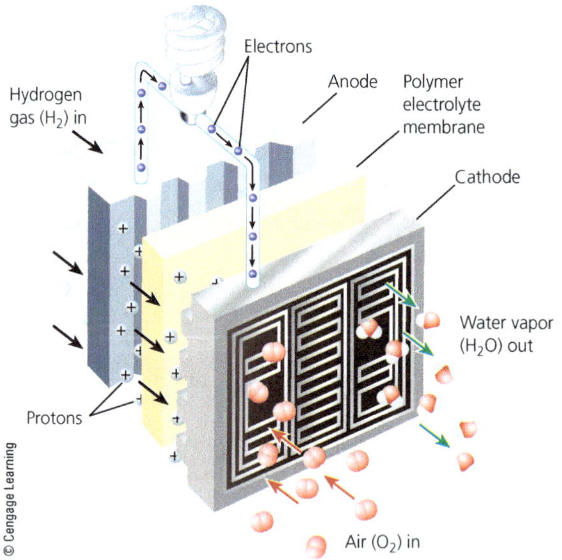

© Cengage Learning

Figure 16-32 A *fuel cell* takes in hydrogen gas and separates the hydrogen atoms' electrons from their protons. The electrons flow through wires to provide electricity, while the protons pass through a membrane and combine with oxygen gas to form water vapor. Note that this process is the reverse of *electrolysis,* the process of passing electricity through water to produce hydrogen fuel.

SCIENCE FOCUS 16.3

THE QUEST TO MAKE HYDROGEN WORKABLE

Scientists are proposing various ideas for producing and storing hydrogen gas (H_2) as fuel. For example, naturally occurring bacteria and algae can produce H_2 by biodegrading almost any organic material in a microbiological fuel cell.

The most likely production methods will involve using electricity generated by solar-cell power plants, wind farms, and geothermal energy. In 2008, MIT scientists Daniel Nocera and Matthew Kanan developed a catalyst made from inexpensive cobalt and phosphate salts that can be used to split water into hydrogen and oxygen using a relatively small amount of electricity. This breakthrough could make it affordable to use electricity produced by wind turbines or solar cells to produce H_2 and thus to store the energy produced by the wind and sun in the H_2. Researchers are also working on using green algae to produce H_2 from water (Figure 16-A).

Once produced, H_2 can be stored in a pressurized tank as liquid hydrogen. It can also be stored in solid metal hydride compounds and in sodium borohydride, which releases H_2 when heated. Scientists are also evaluating ways to store H_2 by coating the surfaces of activated charcoal or carbon nanofibers with it. When heated, these coated surfaces release the H_2. Another possibility is to store H_2 inside nano-sized glass microspheres that can easily be filled and refilled. More research is needed to transform these possibilities into realities.

Metal hydrides, sodium borohydride, charcoal powders, ammonia borane, carbon nanotubes, and glass microspheres containing H_2 will not explode or burn if a vehicle's fuel tank is ruptured in an accident. Thus, H_2 stored in such ways is a much safer fuel than gasoline, diesel fuel, natural gas, and concentrated ethanol. Also, the use of ultralight car bodies made of composites with an energy-efficient aerodynamic design would improve vehicle fuel efficiency so that large hydrogen fuel tanks would not be needed.

In 2007, engineering professor Jerry Woodall invented a new way to produce hydrogen on demand, without producing toxic fumes, by exposing pellets of an aluminum–gallium alloy to water. If this process is perfected and proves economically feasible, H_2 could be generated as needed inside a tank about the same size as an average car's gasoline tank, and thus the fuel would not have to be transported or stored.

Warren Gretz/National Renewable Energy Laboratory

Figure 16-A Scientists are evaluating the use of green algae that can use sunlight to produce hydrogen gas from water.

Critical Thinking

Do you think that governments should subsidize research and development of these and other technologies in order to help make H_2 a workable fuel? Explain.

---

such as methane and gasoline. We can produce H_2 by heating water or passing electricity through it; by stripping it from the methane (CH_4) found in natural gas and from gasoline molecules; and through a chemical reaction involving coal, oxygen, and steam. The problem is that it takes energy and money to produce H_2 using these methods. In other words, H_2 must be produced by using other forms of energy, and therefore has a *negative net energy yield.* The amount of energy it takes to make this fuel will always be more than the amount we can get by burning it.

Second, fuel cells are the best way to use H_2 to produce electricity, but current versions of fuel cells are expensive. However, progress in the development of nanotechnology (see Science Focus 14.1, p. 362) could lead to less expensive and more energy-efficient fuel cells.

Third, whether or not a hydrogen-based energy system produces less outdoor air pollution and CO_2 than a fossil fuel system depends on how the H_2 is produced. We could use electricity from coal-burning and conventional nuclear power plants to decompose water into H_2 and O_2. But this approach does not avoid the harmful environmental effects associated with using coal and the nuclear fuel cycle. Also, making H_2 from coal or stripping it from methane or gasoline adds much more CO_2 to the atmosphere per unit of heat generated than does burning these carbon-containing fuels directly.

If renewable energy sources such as wind farms and solar-cell power plants were used to make H_2, these CO_2 emissions would be sharply reduced. Making H_2 by decomposing the methane in natural gas could serve as a transition to producing H_2 by using renewable energy, as long as there were a sharp reduction of the methane leaks currently resulting from natural gas production (see Science Focus 15.1, p. 378).

Hydrogen's negative net energy yield is a serious limitation and means that this fuel will have to be subsidized in order for it to compete in the open marketplace with fuels that have medium-to-high net energy yields. However, widespread use of H_2 would greatly reduce air pollution and projected climate change. Some analysts contend that the potential for reducing the huge economic costs and environmental harm related to air pollution and projected climate change could make it worthwhile to subsidize the use of hydrogen (Science Focus 16.3).

In the 1990s, Amory Lovins (Individuals Matter 16.1) and his colleagues at the Rocky Mountain Institute designed a very light, safe, extremely efficient hydrogen-powered car. It is the basis of most prototype hydrogen fuel-cell cars now being tested by major automobile companies. Some analysts project that fuel-cell cars, running on affordable H_2 produced from natural gas, could be in widespread use by 2030 to 2050.

Larger, stationary fuel cells could provide electricity and heat for commercial and industrial users. A 45-story office building in New York City gets much of its heat from two large fuel-cell stacks. Japan has built a large fuel cell that produces enough electricity to run a small town.

A Canadian company is developing a unit about the size of a dishwasher that will allow consumers to use electricity to produce their own H_2 from tap water. The unit could be installed in a garage and used to fuel a hydrogen-powered vehicle overnight, when electricity rates are sometimes lower. In sunny areas, people could install rooftop panels of solar cells to produce and store H_2 for their cars. In Japan, some home owners get their electricity and hot water from refrigerator-sized fuel-cell units, which produce H_2 from the methane in natural gas. Figure 16-33 lists the major advantages and disadvantages of using hydrogen as an energy resource (Concept 16-7). **GREEN CAREER:** hydrogen energy

Trade-Offs

Hydrogen

Advantages	Disadvantages
Can be produced from plentiful water at some sites	Negative net energy yield
No CO_2 emissions if produced with use of renewables	CO_2 emissions if produced from carbon-containing compounds
Good substitute for oil	High costs create need for subsidies
High efficiency in fuel cells	Needs H_2 storage and distribution system

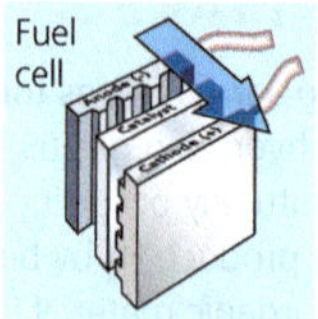

Figure 16-33 Using hydrogen as a fuel for vehicles and for providing heat and electricity has advantages and disadvantages. (Concept 16-7). ***Questions:*** Which single advantage and which single disadvantage do you think are the most important? Why? Do you think that the advantages of hydrogen fuel outweigh its disadvantages? Why?

Photo: LovelaceMedia/Shutterstock.com

16-8 How Can We Make the Transition to a More Sustainable Energy Future?

CONCEPT 16-8

We can make the transition to a more sustainable energy future by greatly improving energy efficiency, using a mix of renewable energy resources, and including the environmental and health costs of energy resources in their market prices.

Choosing Energy Paths

We need to develop energy policies with the future in mind, because experience shows that it usually takes at least 50 years and huge financial investments to phase in new energy alternatives. Creating energy policy involves trying to answer the following questions for *each* energy alternative:

- How much of the energy resource is likely to be available in the near future (the next 25 years) and in the long term (the next 50 years)?
- What is the estimated net energy yield (p. 375) for the resource?
- What are the estimated costs for developing, phasing in, and using the resource?
- What kinds of government research and development subsidies and tax breaks will be needed to help develop the resource?
- How will dependence on the resource affect national and global economic and military security?
- How vulnerable is the resource to terrorism?
- How will extracting, transporting, and using the resource likely affect the environment, the earth's climate, and human health?
- Does use of the resource produce hazardous, toxic, or radioactive substances that we must safely store for very long periods of time?

In considering possible energy futures, scientists and energy experts who have evaluated energy alternatives have come to three general conclusions. First, *during this century, there will likely be a gradual shift from large, centralized macropower systems to smaller, decentralized micropower systems* (Figure 16-34) such as wind turbines, household solar-cell panels, rooftop solar water heaters, and small natural gas turbines. There will also be a shift from gasoline-powered motor vehicles to hybrid and plug-in electric cars. This may be followed by a shift to fuel cells for cars and to stationary fuel cells for houses and commercial buildings.

A shift to a more decentralized energy system would improve national and economic security, because countries would rely on diverse, dispersed, domestic, renewable energy resources instead of on a smaller number of large power plants that are vulnerable to storm dam-

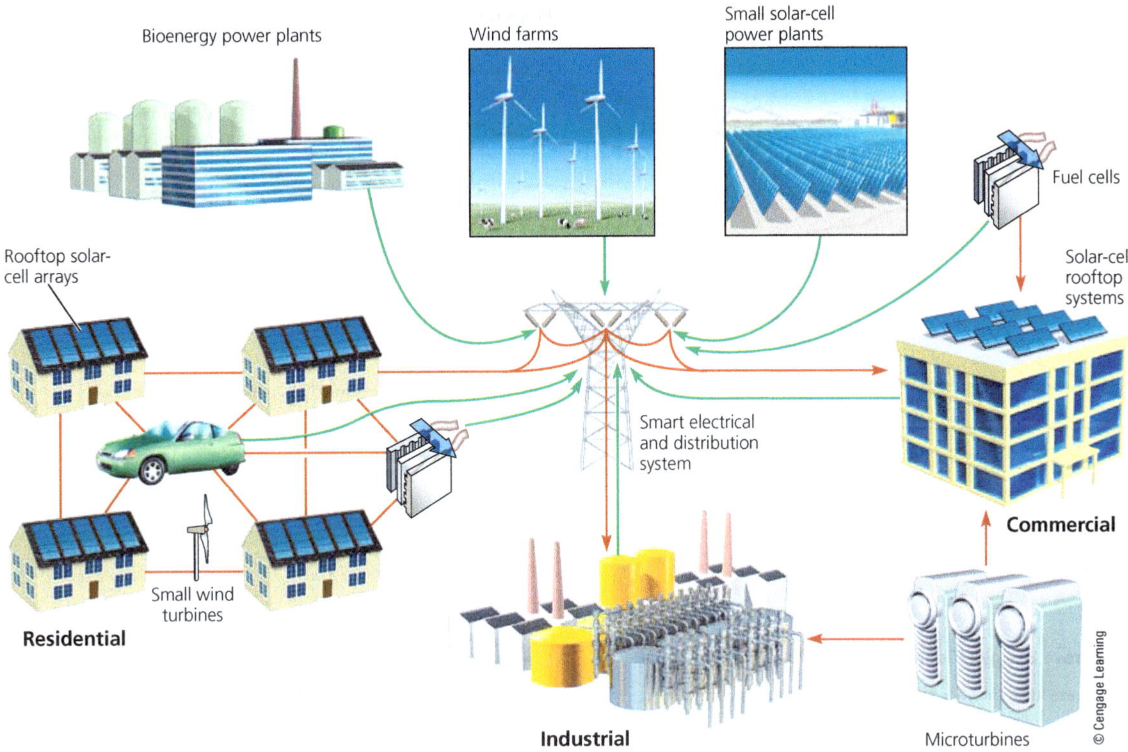

Figure 16-34 Solutions: During the next few decades, we will probably shift from dependence on a *centralized electric power system,* based on a few hundred large coal-burning and nuclear power plants, to a *decentralized power system,* in which electricity is produced by a large number of dispersed, small-scale, local power-generating systems that depend primarily on a variety of locally available, low-carbon renewable energy resources.

age and sabotage. Similarly, states and communities within countries could become more energy-independent and secure by making this shift. GOOD NEWS

The second general conclusion of experts about the future of energy use is that *a combination of improved energy efficiency and carefully regulated use of natural gas will be the best way to make the transition to using mostly renewable energy resources during this century* (**Concept 16-8**). By using a variety of locally available renewable energy resources, we could apply the biodiversity **principle of sustainability**.

The third general conclusion is that *because fossil fuels are still abundant and artificially cheap, we will continue to use them in large quantities* (Figure 16-35). This presents two major challenges. One is to find ways to reduce the harmful environmental impacts of widespread use of fossil fuels, with special emphasis on reducing outdoor air pollution and greenhouse gas emissions. The second is to find ways to include more of the harmful environmental and health costs of using fossil fuels in their market prices and thus to implement the full-cost pricing **principle of sustainability**.

China (which uses 20% of the world's energy) and the United States (which uses 19% of the world's energy) are the key players in making the shift away from fossil fuels and thus in slowing the rate of projected climate change during this century.

Figure 16-36 summarizes these and other strategies for making the transition to a more sustainable energy future over the next 50 years (**Concept 16-8**). Making such a transition presents exciting challenges as well as big rewards, including major cost savings, stimulating and fulfilling careers for young people, and improvements to the environment that could benefit many generations to come (see Chapter 1, Core Case Study, p. 4).

Economics, Politics, and Education Can Help Us Shift to More Sustainable Energy Resources

To most analysts, economics, politics, and consumer education hold the keys to making a shift to a mix of improved energy efficiency and more sustainable energy resources. Governments can use three strategies to help stimulate or reduce the short-term and long-term use of a particular energy resource.

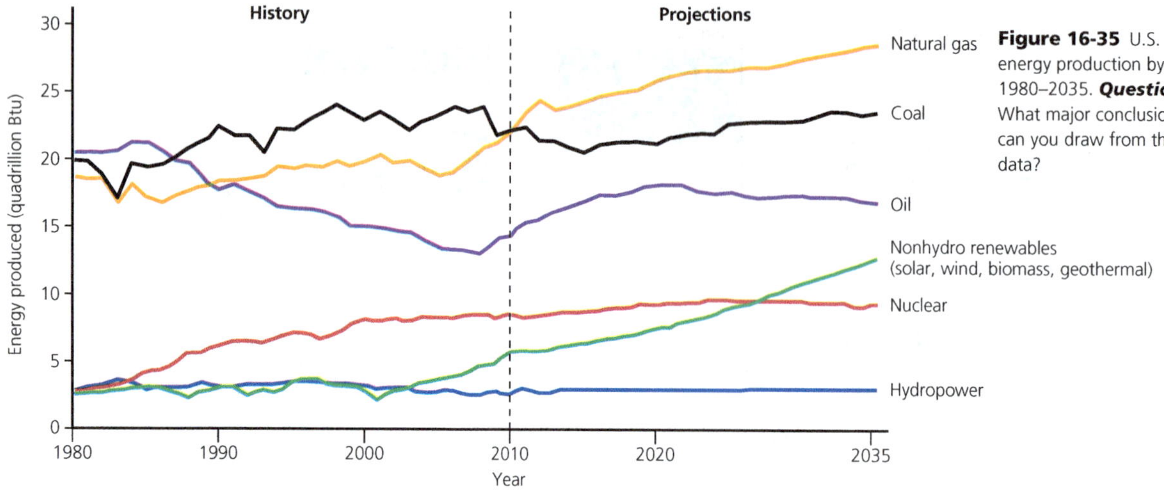

Figure 16-35 U.S. energy production by fuel, 1980–2035. ***Question:*** What major conclusions can you draw from these data?

Solutions

Making the Transition to a More Sustainable Energy Future

Improve Energy Efficiency

Increase fuel-efficiency standards for vehicles, buildings, and appliances

Provide large tax credits or fee-bates for buying efficient cars, houses, and appliances

Reward utilities for reducing demand for electricity

Greatly increase energy efficiency research and development

More Renewable Energy

Greatly increase use of renewable energy

Provide large subsidies and tax credits for use of renewable energy

Greatly increase renewable energy research and development

Reduce Pollution and Health Risk

Phase out coal subsidies and tax breaks

Levy taxes on coal and oil use

Phase out nuclear power subsidies, tax breaks, and loan guarantees

Figure 16-36 Energy analysts have made a number of suggestions for helping us make the transition to a more sustainable energy future (Concept 16-8). ***Questions:*** Which five of these solutions do you think are the best ones? Why?

First, they can *keep the prices of selected energy resources artificially low to encourage use of those resources.* They do this by providing research and development subsidies, tax breaks, and loan guarantees to encourage the development of those resources, and by enacting regulations that favor them. For decades, this approach has been employed to stimulate the development and use of fossil fuels and nuclear power in the United States, as well as in most other more-developed countries. This has created an uneven economic playing field that has *encouraged* energy waste and increased use of these nonrenewable energy resources, while it has *discouraged* improvements in energy efficiency and the development of a variety of renewable energy resources.

Many energy analysts argue that one of the most important steps that governments can take to level the economic playing field is to phase out the $250 billion to $300 billion in annual subsidies and tax breaks now provided worldwide for fossil fuels and nuclear energy—both of which are mature industries that could be left to stand on their own economically. Critics say that solar and wind energy receive too many government subsidies compared to fossil fuels and nuclear power. However, in 2010, fossil fuels and nuclear energy together received $6.7 billion in direct government subsidies compared to $6.1 billion for solar and wind combined, according to the Department of Energy. Energy analysts also call for greatly increasing subsidies and tax breaks for developing and phasing in affordable renewable energy and energy-efficiency technologies.

However, making such a shift in energy subsidies is difficult because of the immense political and financial power of the fossil fuel and nuclear power industries. It is not surprising that they vigorously oppose the loss of their subsidies and tax breaks. Generally, they have also opposed significant increases in subsidies and tax breaks for energy efficiency and for competing renewable energy sources.

A *second* major strategy that governments can use is to *allow the prices of selected energy resources to rise to their natural market levels by removing government subsidies, thereby discouraging the use of these resources.* They can do this by eliminating existing tax breaks and other subsidies that favor the use of the targeted resource. They can also tax and regulate the use of such resources as a way of helping to control and pay for the harmful environmental costs of using them. Such measures can increase government revenues, encourage improvements in energy efficiency, and reduce dependence on environmentally harmful energy resources. And they would represent an application of the full-cost pricing **principle of sustainability**.

Third, governments can *emphasize consumer education.* Even if governments offer generous financial incentives for energy efficiency and renewable energy use, people will not make such investments if they are uninformed—or misinformed—about the availability, advantages, disadvantages, and hidden environmental costs of various energy resources.

An excellent example of what a government can do to bring about a more sustainable energy mix is the case of Germany. It is the world's most solar-powered nation, with half of the world's installed capacity. Why does cloudy Germany have more solar water heaters and solar-cell panels than sunny France and Spain have? One reason is that the German government made the public aware of the environmental benefits of these technologies. The other is that the government provided consumers with substantial economic incentives for using the technologies.

We have the creativity, wealth, and most of the technologies we need to make the transition to a more sustainable energy future within your lifetime. GOOD NEWS Making this transition depends primarily on *education, economics,* and *politics*—on how well individuals understand environmental and energy problems and their possible solutions, and on how they vote and thus influence their elected officials. People can also vote with their pocketbooks by refusing to buy energy-inefficient and environmentally harmful products and services, and by letting company executives know about their choices. In 2012, environmental engineer Mark Z. Jacobson wrote: "You could power America with renewables from a technical and economic standpoint. The biggest obstacles are social and political."

Figure 16-37 lists some ways in which you can contribute to making the transition to a more environmentally and economically sustainable energy future.

What Can You Do?

Shifting to More Sustainable Energy Use

- Walk, bike, or use mass transit or a car pool to get to work or school
- Drive only vehicles that get at least 17 kpl (40 mpg)
- Have an energy audit done in the place where you live
- Superinsulate the place where you live and plug all air leaks
- Use passive solar heating
- For cooling, open windows and use fans
- Use a programmable thermostat and energy-efficient heating and cooling systems, lights, and appliances
- Turn down your water heater's thermostat to 43–49°C (110–120°F) and insulate hot water pipes
- Turn off lights, TVs, computers, and other electronics when they are not in use
- Wash laundry in cold water and air dry it on racks

Figure 16-37 Individuals matter: You can make a shift in your own life toward using energy more sustainably. ***Questions:*** Which three of these measures do you think are the most important ones to take? Why? Which of these steps have you already taken and which do you plan to take?

Big Ideas

- We should evaluate energy resources on the basis of their potential supplies, their net energy yields, and the environmental and health impacts of using them.
- By using a mix of renewable energy sources—especially solar, wind, flowing water, sustainable biofuels, and geothermal energy—we could drastically reduce pollution, greenhouse gas emissions, and biodiversity losses.
- Making the transition to a more sustainable energy future will require sharply increasing energy efficiency, using a mix of environmentally friendly renewable energy resources, and including the harmful environmental and health costs of energy resources in their market prices.

TYING IT ALL TOGETHER Wind Power and Sustainability

Ulrich Mueller/Shutterstock.com

In the **Core Case Study** that opens this chapter, we considered the immense potential of wind—an indirect form of solar energy—as an energy resource, particularly in the United States. Relying mostly on nonrenewable fossil fuels and nuclear power violates the three scientific **principles of sustainability** (see Figure 1-2, p. 6 or back cover). By relying more on a diversity of direct and indirect forms of renewable solar energy coupled with improving energy efficiency to reduce unnecessary waste of energy and money, we could implement these three principles of sustainability.

Making this shift to a more sustainable energy future involves

- relying much more on direct and indirect forms of solar energy for our electricity, heating and cooling, and other needs;
- recycling and reusing more materials and thus reducing wasteful and excessive consumption of energy and matter; and
- mimicking nature's reliance on biodiversity by using a diverse mix of locally and regionally available renewable energy resources.

By making this shift, we would also be applying the three social science **principles of sustainability** (see Figure 1-5, p. 9 or back cover). By including the harmful environmental and health costs of fossil fuels and nuclear energy in their market prices, we would be applying the full-cost pricing principle. This could only be accomplished with the aid of compromise and trade-offs in the political arena by applying the win-win solutions principle. And many analysts argue that such solutions would greatly benefit life on earth immediately and in the long run—a classic application of the principle of responsibility to future generations.

Chapter Review

Core Case Study

1. Describe the potential for using renewable energy from the wind to produce most of the electricity used in the United States.

Section 16-1

2. What are the two key concepts for this section? What is **energy efficiency**? Explain why we can think of energy efficiency as an energy resource. What percentage of the energy used in the United States is unnecessarily wasted? List five widely used energy-inefficient technologies. What are the major advantages of reducing energy waste? List four ways to save energy and money in industry. What is **cogeneration**? How can electric utility companies help people reduce their energy waste? What is a smart grid and why is it important? What are the hidden costs of using gasoline?

3. What are four ways to save energy and money in transportation? Summarize the development of more energy-efficient vehicles. Explain the importance of developing better batteries and list some advances in this area. What are fuel cells and what are their advantages? What are some technologies applied by green architecture? Explain benefits of living roofs. List four ways to improve energy efficiency in new buildings and eight ways to improve energy efficiency in existing buildings. List six ways in which you can save energy where you live. Give three reasons why we waste so much energy. List five advantages of relying more on a variety of renewable energy sources and list three factors that are holding back such a transition.

Section 16-2

4. What is the key concept for this section? Distinguish between a **passive solar heating system** and an **active solar heating system** and discuss the major advantages and disadvantages of using such systems for heating buildings. What are three ways to cool houses naturally? What are **solar thermal systems**, how are they used, and what are the major advantages and disadvantages of using them? What is a **solar cell** (**photovoltaic** or **PV cell**) and what are the major advantages and disadvantages of using such devices to produce electricity?

Section 16-3

5. What is the key concept for this section? Define **hydropower** and summarize the potential for expanding it. What are the major advantages and disadvantages of using hydropower? What is the potential for using tides and waves to produce electricity?

Section 16-4

6. What is the key concept for this section? What are the advantages of using taller wind turbines? Summarize the global potential for wind power. What are the major advantages and disadvantages of using wind to produce electricity?

Section 16-5

7. What are the two key concepts for this section? What is **biomass** and what are the major advantages and disadvantages of using wood to provide heat and electricity? What are the major advantages and disadvantages of using biodiesel and ethanol to power motor vehicles? Explain how algae and bacteria can be used to produce fuels nearly identical to gasoline and diesel fuel.

Section 16-6

8. What is the key concept for this section? What is **geothermal energy** and what are three sources of such energy? What are the major advantages and disadvantages of using geothermal energy as a source of heat and to produce electricity?

Section 16-7

9. What is the key concept for this section? What are the major advantages and disadvantages of using hydrogen as a fuel to use in producing electricity and powering vehicles?

Section 16-8

10. What is the key concept for this section? List eight questions that policy makers should ask about each source of energy. List three general conclusions that energy experts have come to in considering possible energy futures. List five major strategies suggested by such experts for making the transition to a more sustainable energy future. Explain three strategies that governments can use to encourage or discourage the use of an energy resource. What are this chapter's *three big ideas*? Explain how we would be applying the six **principles of sustainability** by improving energy efficiency and shifting to renewable energy resources.

Note: Key terms are in bold type.

Critical Thinking

1. Suppose that a wind power developer has proposed building a wind farm near where you live (**Core Case Study**). Would you be in favor of the project or opposed to it? Write a letter to your local newspaper or a blog for a website explaining your position and your reasoning. As part of your research, determine how the electricity you use now is generated and where the nearest power plant is located, and include this information in your arguments.

2. List five ways in which you unnecessarily waste energy during a typical day, and explain how these actions violate the three scientific **principles of sustainability**.

3. Congratulations! You have won $500,000 to build a more sustainable house of your choice. With the goal of maximizing energy efficiency, what type of house would you build? How large would it be? Where would you locate it? What types of materials would you use? What types of materials would you *not* use? How would you heat and cool the house? How would you heat water? What types of lighting, stove, refrigerator, washer, and dryer would you use? Which, if any, of these appliances could you do without?

4. Suppose that a homebuilder installs electric baseboard heat and claims, "It is the cheapest and cleanest way to go." Apply your understanding of the second law of thermodynamics (see Chapter 2, p. 43) and net energy yield (see Figure 15-2, p. 375) to evaluate this claim. Write a response to the homebuilder summarizing your findings.

5. Should buyers of energy-efficient motor vehicles receive large rebates funded by fees levied on gas guzzlers? Explain.

6. Do you think that the estimated hidden costs of gasoline should be included in its price at the pump? Explain. Would you favor much higher gasoline taxes if payroll taxes and income taxes were reduced to balance gasoline tax increases, with no net additional cost to consumers? Explain.

7. Explain why you agree or disagree with the following proposals made by various energy analysts:
   a. Government subsidies for all energy alternatives should be eliminated so that all energy resources can compete on a level playing field in the marketplace.
   b. All government tax breaks and other subsidies for conventional fossil fuels (oil, natural gas, and coal), synthetic natural gas and oil, and nuclear power (fission and fusion) should be eliminated. They should be replaced with subsidies and tax breaks for improving energy efficiency and developing renewable energy alternatives.
   c. Development of renewable energy alternatives should be left to private enterprise and should receive little or no help from the federal government, but the nuclear power and fossil fuels industries should continue to receive large federal government subsidies.

8. What percentages of the U.S. Department of Energy research and development budget would you devote to fossil fuels, nuclear power, renewable energy, and improving energy efficiency? How would you distribute your funds among the various renewable energy options? Explain your thinking.

Doing Environmental Science

Do a survey of energy use at your school, based on the following questions: How is the electricity generated? How are most of the buildings heated? How is water heated? How are most of the vehicles powered? How is the computer network powered? How could energy efficiency be improved, if at all, in each of these areas? How could your school make use of solar, wind, biomass, and other forms of renewable energy if it does not already do so? Write up a proposal for using energy more efficiently and sustainably at your school and submit it to school officials.

Global Environment Watch Exercise

In the *Renewable Energy* portal, search for information on the renewable energy policy in Germany (you may also use the World Map). What are some of the challenges the German government is facing, and how are they being addressed? What challenges would arise if a U.S. government mandate required a similar transition to renewable energies? Write a report answering these questions and discussing the effects of the "German Experiment" on the United States as it looks to revise its energy policy and address climate change.

Data Analysis

Study the table below and then answer the questions that follow it by filling in the blank columns in the table.

Combined City/Highway Fuel Efficiency for 2012 Cars (mpg)

Model	Miles per Gallon (mpg)	Kilometers per Liter (kpl)	Annual Liters (Gallons) of Gasoline	Annual CO_2 Emissions
Toyota Prius Hybrid	50			
Ford Focus AM-6	31			
Chevrolet Cruze A-S6	30			
Honda Accord A-S	27			
Ford F150 A-S6 Pickup	19			
Jeep Grand Cherokee A-6 SUV	16			
Lamborghini Aventador Coupe	13			

(Compiled by the authors using data from the U.S. Environmental Protection Agency.)

1. Using Supplement 1 (Measurement Units, p. S1), convert the miles per gallon figures in the table to kilometers per liter (kpl).

2. How many liters (and how many gallons) of gasoline would be used annually by each car if it were driven 19,300 kilometers (12,000 miles) per year?

3. How many kilograms (and how many pounds) of carbon dioxide would be released into the atmosphere each year by each car, based on the fuel consumption calculated in question 2, assuming that the combustion of gasoline releases 2.3 kilograms of CO_2 per liter (19 pounds per gallon)?

19 Climate Disruption

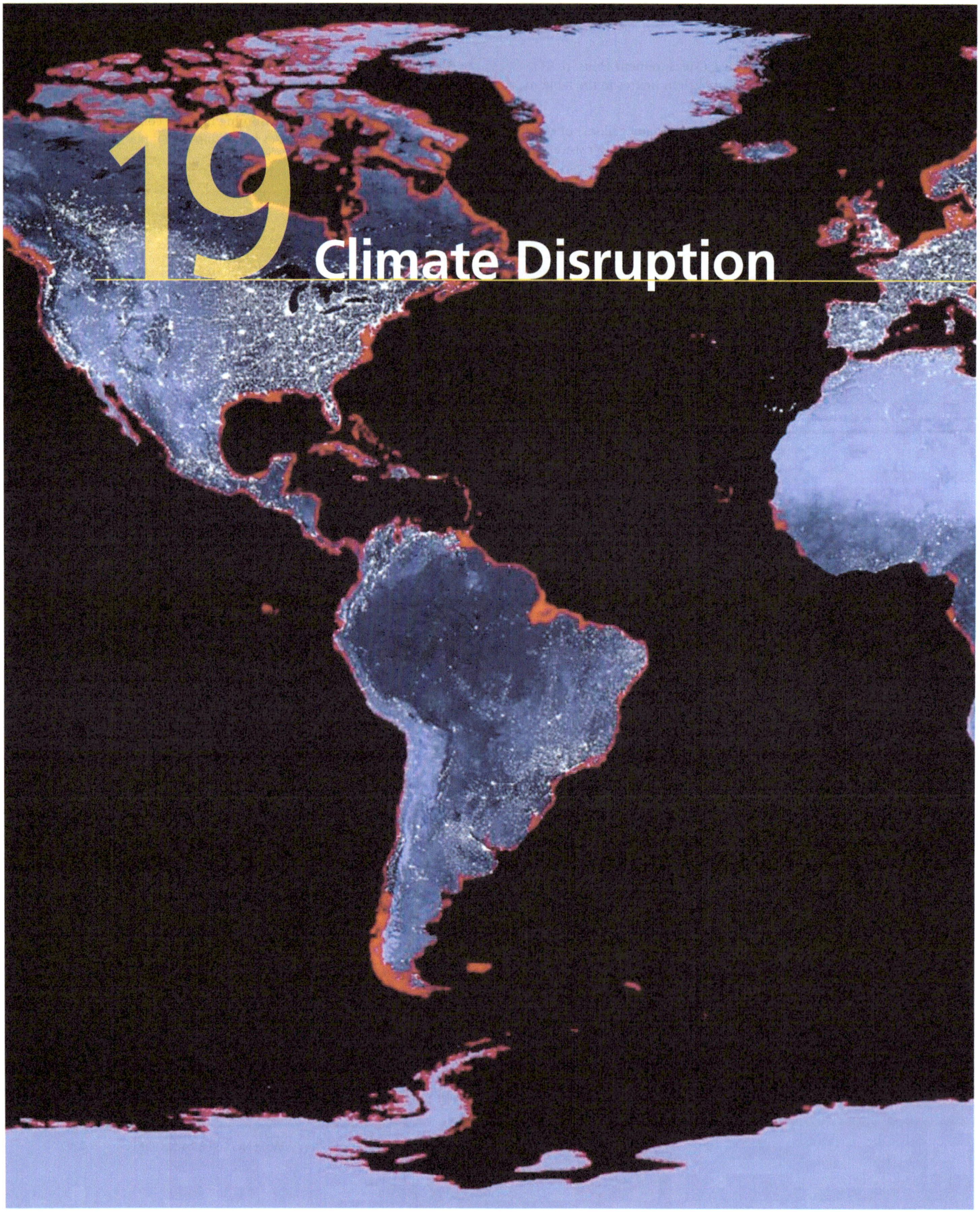

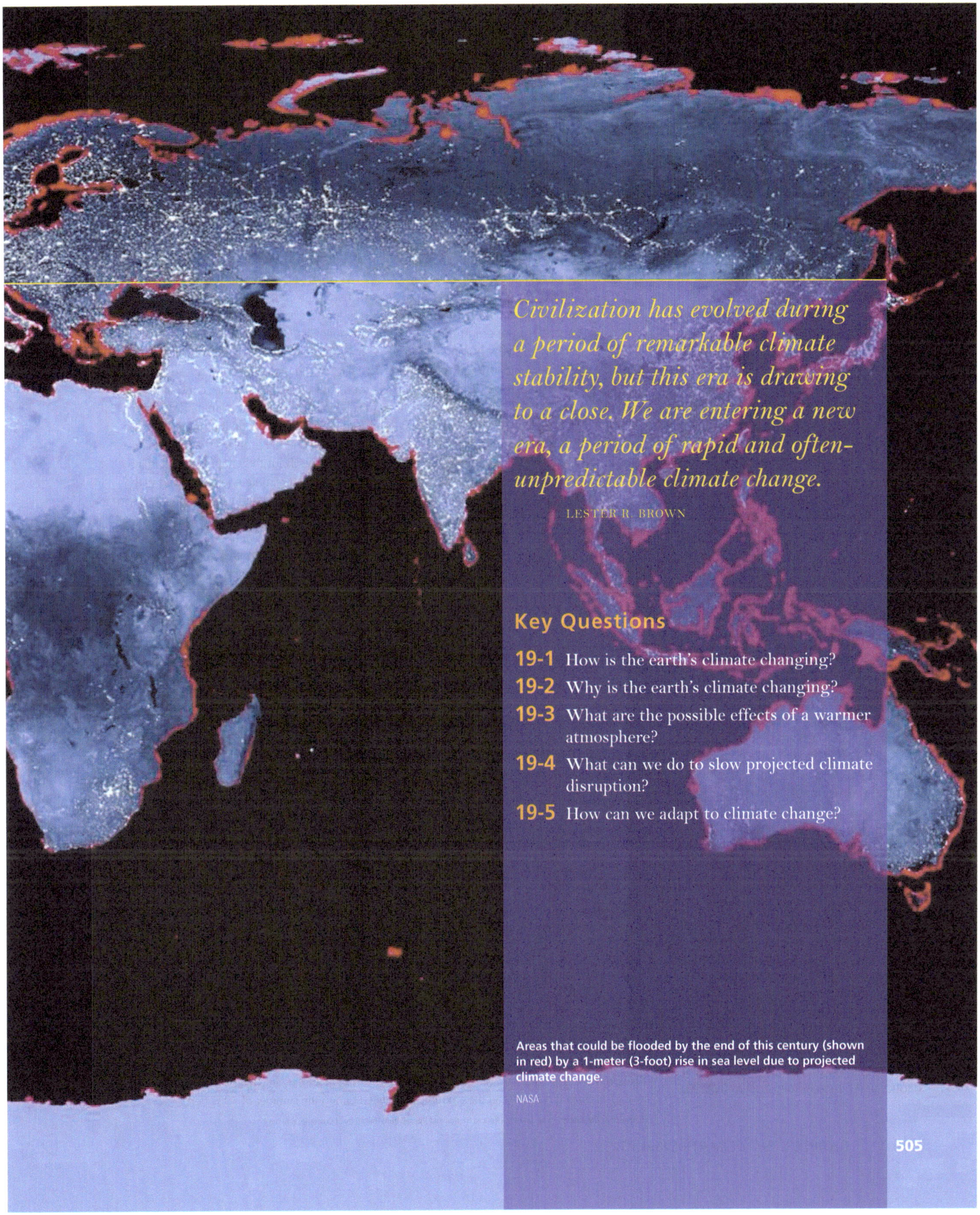

Civilization has evolved during a period of remarkable climate stability, but this era is drawing to a close. We are entering a new era, a period of rapid and often-unpredictable climate change.

LESTER R. BROWN

Key Questions

19-1 How is the earth's climate changing?

19-2 Why is the earth's climate changing?

19-3 What are the possible effects of a warmer atmosphere?

19-4 What can we do to slow projected climate disruption?

19-5 How can we adapt to climate change?

Areas that could be flooded by the end of this century (shown in red) by a 1-meter (3-foot) rise in sea level due to projected climate change.

NASA

CORE CASE STUDY Melting Ice in Greenland

Greenland is a largely ice-covered island in the Arctic Circle. It is the world's largest island, nearly the size of Mexico, and has a population of just under 58,000 people.

Greenland's ice, which formed during the last ice age, lies in glaciers that are as deep as 3.2 kilometers (2 miles) and cover about 80% of this mountainous island. Scientists are studying this ice closely because it appears to be melting at an accelerating rate during Greenland's summers (Figure 19-1). Some of this ice is replaced by snow during winter months, but the annual net loss of Greenland's ice has increased during recent years.

A number of climate scientists hypothesize that a key factor in this melting is *atmospheric warming*—the gradual rise of the average temperature of the atmosphere near the earth's surface. This warming is projected to increase significantly during this century. It could lead to rapid *climate disruption,* which is likely to have harmful effects on most of the earth's ecosystems, people, and economies.

Greenland's glaciers contain about 10% of the world's freshwater—enough water to raise the global sea level by as much as 7 meters (23 feet) if all of it were to melt and drain into the sea. It is highly unlikely that this will happen, but even a moderate loss of this ice over one or several centuries would raise sea levels considerably (see chapter-opening photo).

A glacier is a huge, moving ice mass that scrapes along extremely slowly, but scientists have demonstrated that it can pick up speed when meltwater on its surface, flowing down through its crevices, lubricates its bottom layer, which sits on bedrock. Some scientists hypothesize this is what is happening to some of northern Greenland's glaciers. In 2009, scientists found that some of Greenland's glaciers are also getting thinner at their ocean outlets. Such thinning can allow a glacier to float free of the ground at the shore, which has the effect of taking the brakes off and speeding the glacier's flow toward the sea. In some areas, the rate of Greenland's ice moving toward the sea has doubled since 1982, according to the U.S. National Aeronautics and Space Administration (NASA).

The Intergovernmental Panel on Climate Change (IPCC) is an international body of more than 2,500 scientists that was created in 1988 to study climate change. After reviewing tens of thousands of research studies, the IPCC and most of the world's major scientific bodies, including the National Academy of Sciences and the British Royal Society, have come to three major conclusions: **(1)** climate change is real and is happening now, **(2)** it is caused primarily by human activities such as the burning of fossil fuels and the clearing of forests, and **(3)** it is projected to accelerate during this century.

In its 2007 report, the IPCC made several projections. One of them was that the average global sea level is likely to rise by 18–59 centimeters (7–23 inches) by the end of this century. Since then, however, several scientists studying the melting of Greenland's ice have concluded that the rise in the average global sea level by 2100 will likely be closer to 100 centimeters (39 inches) and perhaps twice that. They estimate that Greenland's ice loss will make up half to two-thirds of that rise.

This change in projections about the rising average global sea level and Greenland's contribution to it is one example of how scientists are revising some IPCC projections upward due to the acceleration of rising temperatures at the earth's poles. It is an excellent example of how science advances its knowledge by taking into account new data and hypotheses.

Most climate scientists and researchers working in related disciplines consider the likely prospect of serious climate disruption during this century to be the single most serious environmental problem that humanity faces. Throughout this chapter, we examine the findings of climate scientists along with their projections related to atmospheric warming and climate disruption and their likely causes and effects. We also look at some possible ways to deal with this serious environmental, economic, and political challenge.

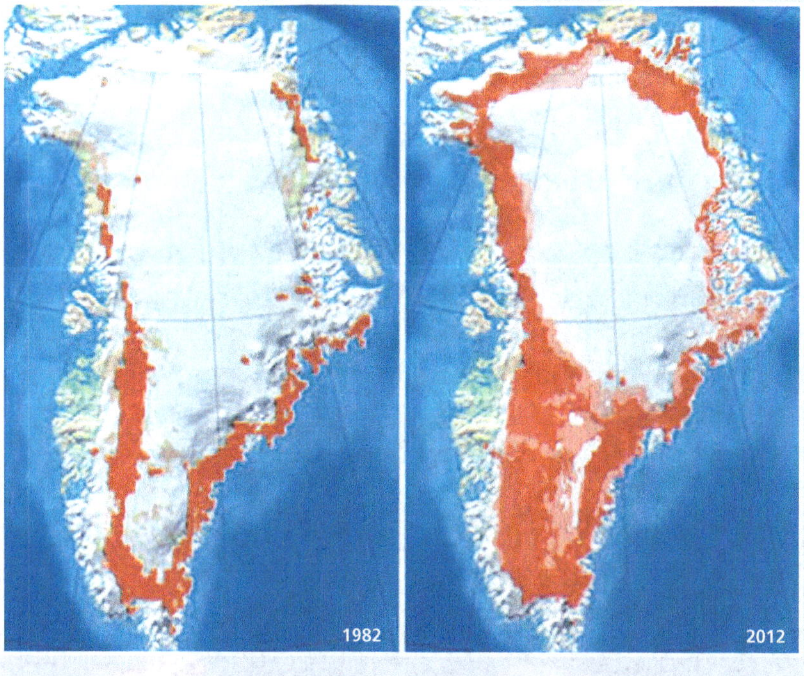

Figure 19-1 The total area of Greenland's glacial ice that melted by July 8, 2012 (red area in right image) was much greater than the amount that melted during the summer of 1982 (left).

(Compiled by the authors using data from Konrad Steffen and Russell Huff, University of Colorado, Boulder.)

19-1 How Is the Earth's Climate Changing?

CONCEPT 19-1
Considerable scientific evidence indicates that the earth's atmosphere is warming at a rapid rate that is likely to lead to significant climate disruption during this century.

Weather and Climate Are Not the Same

In thinking about climate change, it is very important to distinguish between weather and climate. *Weather* consists of short-term changes in atmospheric variables such as the temperature, precipitation, wind, and barometric pressure in a given area over a period of hours or days (see Supplement 5, p. S19).

By contrast, *climate* is determined by the *average* weather conditions of the earth or of a particular area, especially temperature and precipitation, over periods of at least three decades to thousands of years. According to the World Meteorological Organization, the minimum period considered for purposes of measuring climate is three decades. Scientists have used such long-term measurements to divide the earth into various climate zones (see Figure 7-2, p. 145).

During any period of 30 or more years, in a given area of the planet, there will often be hotter years and cooler years, and wetter years and drier years, as weather often fluctuates widely from day to day and from year to year. However, such short-term changes in weather do not necessarily tell us that the earth's climate or the climate of a particular area is warming or cooling. Climate scientists look at data on the normally fluctuating weather conditions for a particular area of the earth to see if there has been a general rise or fall in any measurements such as average temperature or precipitation over a period of at least 30 years. This is the only way they that they can determine whether and how the climate of an area has changed.

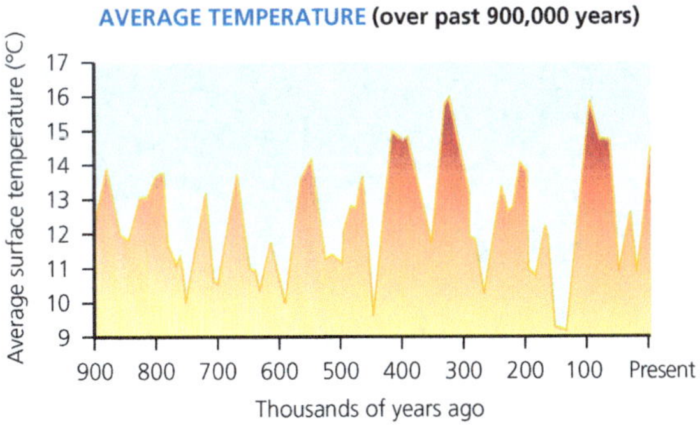

Figure 19-2 Over the past 900,000 years, the average global temperature of the atmosphere near the earth's surface has fluctuated widely. This graph is based on a body of scientific evidence that contains gaps, but the data clearly indicate general trends. ***Question:*** What are two conclusions you can draw from this graph?

(Compiled by the authors using data from Goddard Institute for Space Studies, Intergovernmental Panel on Climate Change, National Academy of Sciences, National Aeronautics and Space Administration, National Center for Atmospheric Research, and National Oceanic and Atmospheric Administration.)

Climate Change Is Not New but Has Accelerated

Climate change is neither new nor unusual. Over the past 3.5 billion years, the planet's climate has been altered many times by several factors, including volcanic emissions, changes in solar input, the shifting of tectonic plates (Figure 4-9, p. 86), and impacts by meteors and asteroids.

The earth's climate is also affected by global air circulation patterns (see Figure 7-3, p. 146), large areas of ice (**Core Case Study**) that reflect incoming solar energy and help cool the atmosphere, varying concentrations of the different gases that make up the atmosphere, and occasional slight changes in ocean currents (see Figure 7-5, p. 147). We look at these factors in this chapter. Average temperature and precipitation are the two main factors affecting global, regional, and local climates, but temperature is the key variable that climate specialists watch.

Over the past 900,000 years, the atmosphere has experienced prolonged periods of global cooling and global warming (Figure 19-2). These alternating cycles of freezing and thawing are known as *glacial* and *interglacial* (between ice ages) *periods*.

For roughly 10,000 years, we have had the good fortune to live in an interglacial period characterized by a fairly stable climate based mostly on a generally steady global average surface temperature. This allowed the human population to grow as agriculture developed, and later as cities grew. For the past 1,000 years, the average temperature of the atmosphere near the earth's surface has remained fairly stable but began to rise during the 19th and 20th centuries when a growing population began clearing large areas of forest and burning more fossil fuels. Most of the recent increase in the global average atmospheric temperature has taken place since 1975 (Figure 19-3).

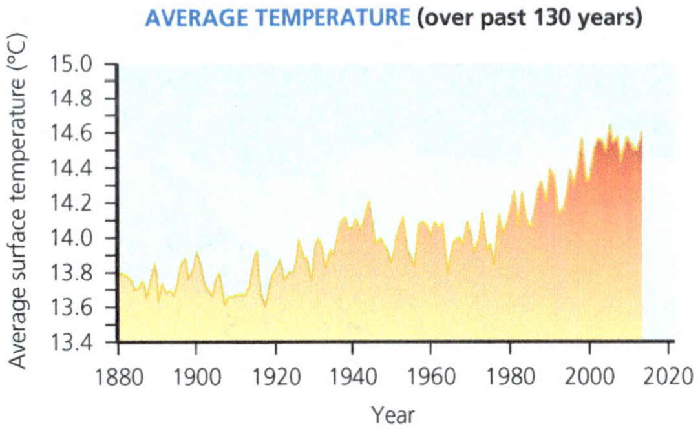

Figure 19-3 Atmospheric warming has trended upward since 1975 with a slight slowing of this trend between 2002 and 2012. ***Question:*** What are two conclusions you can draw from this graph?

(Compiled by the authors using data from Goddard Institute for Space Studies, Intergovernmental Panel on Climate Change, National Academy of Sciences, National Aeronautics and Space Administration, National Center for Atmospheric Research, and National Oceanic and Atmospheric Administration.)

U.S. Geological Survey

Figure 19-4 *Ice cores* are extracted by drilling deep holes into ancient glaciers at various sites such as this one near the South Pole in Antarctica. Analysis of ice cores yields information about the past composition of the lower atmosphere, temperature trends such as those shown in Figure 19-2, solar activity, snowfall, and forest fire frequency.

Scientists estimate past temperature changes such as those depicted in Figure 19-2 by analyzing different types of evidence. These include radioisotopes in rocks and fossils; plankton and radioisotopes in ocean sediments; tiny bubbles, layers of soot, and other materials trapped in different layers of ancient air found in ice cores from glaciers (Figure 19-4); pollen from the bottoms of lakes and bogs; tree rings; and temperature measurements taken regularly since 1861. These temperature measurements now include data from more than 40,000 measuring stations around the world. Such measurements have some limitations, but they show general changes in atmospheric temperature, which in turn can affect the earth's climate.

Figure 19-5 Melting of Himalayan glaciers between 1921 (top) and 2009 (bottom).

©Royal Geographical Society/Yale Environment 360

Royal Geographical Society/Yale Environment 360

The scientific evidence gathered in recent years indicates that atmospheric warming has been accelerating since about 1975 (Figure 19-3). In 1988, a number of nations, working through the World Meteorological Organization and the United Nations Environment Programme, established the Intergovernmental Panel on Climate Change (IPCC) to document past climate changes and project future changes. The IPCC network includes more than 2,500 climate scientists and scientists in disciplines related to climate studies from more than 130 countries.

In 2007, the IPCC issued a report based on more than 29,000 sets of data. This report, along with a set of reports issued in 2010 by the U.S. National Academy of Sciences, reports from the British Royal Society and the U.S. National Atmospheric and Oceanic Administration (NOAA), a 2011 analysis of global temperature measurements since 1750 by physicist Richard Muller and his colleagues, a 2012 report from NASA, and a 2012 report from the World Bank include the following findings:

- Between 1906 and 2012, the average global surface temperature has risen by about 0. 8C° (1.4F°), and over land, the average rise has been 1.4C° (2.5F°). (Land warms more than ocean waters do because heat can stay concentrated near the land surface while waves mix ocean water and transfer some of the heat to deeper water.) Most of this increase has taken place since 1975 (Figure 19-3).
- The first decade in this century (2000–2009) was the warmest decade since 1881, and 2012 was the warmest year on record. Of the 12 hottest years on record, 11 have occurred since 2000, and the twelfth was 1998.
- In some parts of the world, glaciers are melting at increasing rates (Figure 19-5 and **Core Case Study**).
- Floating arctic sea ice is shrinking. In 2012, it had shrunk to the smallest area ever recorded in 34

years of satellite measurements, according to NASA scientists.

- During the 20th century, the world's average sea level rose by 19 centimeters (7.6 inches)—an average of 1.9 cm (0.76 inches) per decade—mostly because of the expansion of ocean water as its temperature increased and to a lesser extent by runoff from melting land-based ice. By comparison, sea levels rose about 2 centimeters (three-quarters of an inch) in the 18th century and 6 centimeters (2 inches) in the 19th century.

Based on such evidence, the IPCC and other scientific bodies such as the U.S. National Academy of Sciences and the British Royal Society have concluded that the earth's lower atmosphere has warmed, especially since 1975, and that this warming is beginning to change the earth's climate.

19-2 Why Is the Earth's Climate Changing?

CONCEPT 19-2

Scientific evidence strongly indicates that the earth's atmosphere has been warming at a rapid rate since 1975 and that human activities, especially the burning of fossil fuels and deforestation, have played a major role in this warming.

The Natural Greenhouse Effect Plays a Key Role in Climate

The earth's climate is strongly influenced by changes in the amount of solar radiation reaching the earth. But a natural process called the *greenhouse effect* (see Figure 3-3, p. 54) also plays a key role in determining the earth's climate. It occurs when some of the solar energy absorbed by the earth radiates into the atmosphere as infrared radiation (heat) at various wavelengths. About 1% of the earth's lower atmosphere is composed of *greenhouse gases,* primarily water vapor (H_2O), carbon dioxide (CO_2), methane (CH_4), and nitrous oxide (N_2O). Heat radiated into the atmosphere by the earth causes molecules of these gases to vibrate and release infrared radiation with longer wavelengths into the lower atmosphere. As this radiation interacts with molecules in the air, it increases their kinetic energy and warms the lower atmosphere and the earth's surface.

There are several greenhouse gases, but we focus here on three major gases—CO_2, CH_4, and N_2O—that play varying roles in atmospheric warming because of their varying lifetimes in the atmosphere and varying warming potentials (Figure 19-6). Other greenhouse gases are hydrofluorocarbons (HFCs), chlorofluorocarbons (CFCs), and hydrochlorofluorocarbons (HCFCs), which are used as refrigerator coolants, cleaners, and fire retardants. Ozone (O_3) also has a warming effect.

Swedish chemist Svante Arrhenius first recognized the greenhouse effect in 1896. Since then, numerous laboratory experiments and measurements of temperatures at different altitudes have confirmed this effect—now one of the most widely accepted theories in the atmospheric sciences. Life on the earth and the world's economies are totally dependent on the greenhouse effect—one of the planet's most important forms of natural capital. Without it, the planet would be a frigid, uninhabitable place.

The atmospheric concentration of CO_2, as part of the carbon cycle (see Figure 3-17, p. 66), plays an important role in determining the average temperature of the atmosphere. Measurements of CO_2 in bubbles at various depths in ancient glacial ice indicate that changes in the levels of these gases in the lower atmosphere have correlated fairly closely with changes in the global average temperature near the earth's surface during the past 400,000 years (Figure 19-7). Scientists have noted a similar correlation between atmospheric temperatures and methane (CH_4) emissions.

Water vapor (H_2O), another greenhouse gas, accounts for about 66% of the earth's greenhouse effect. However, it stays in the atmosphere for only about 1 to 3 weeks,

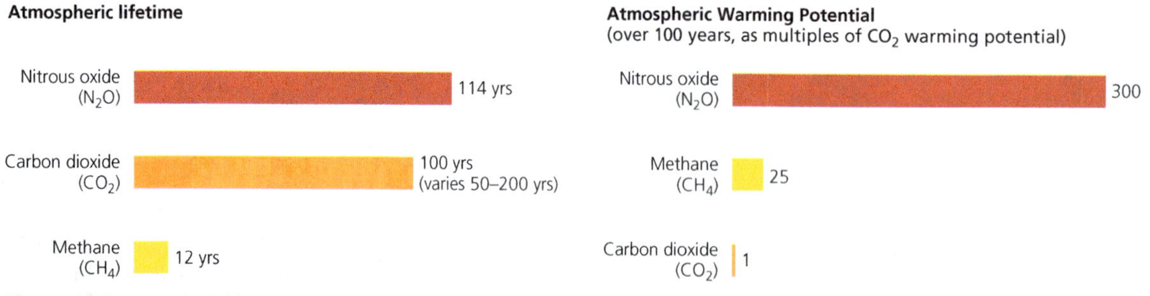

Figure 19-6 Atmospheric lifetimes and warming potentials for three major greenhouse gases.
(Compiled by the authors using data from National Oceanic and Atmospheric Administration and U.N. Framework Convention on Climate Change.)

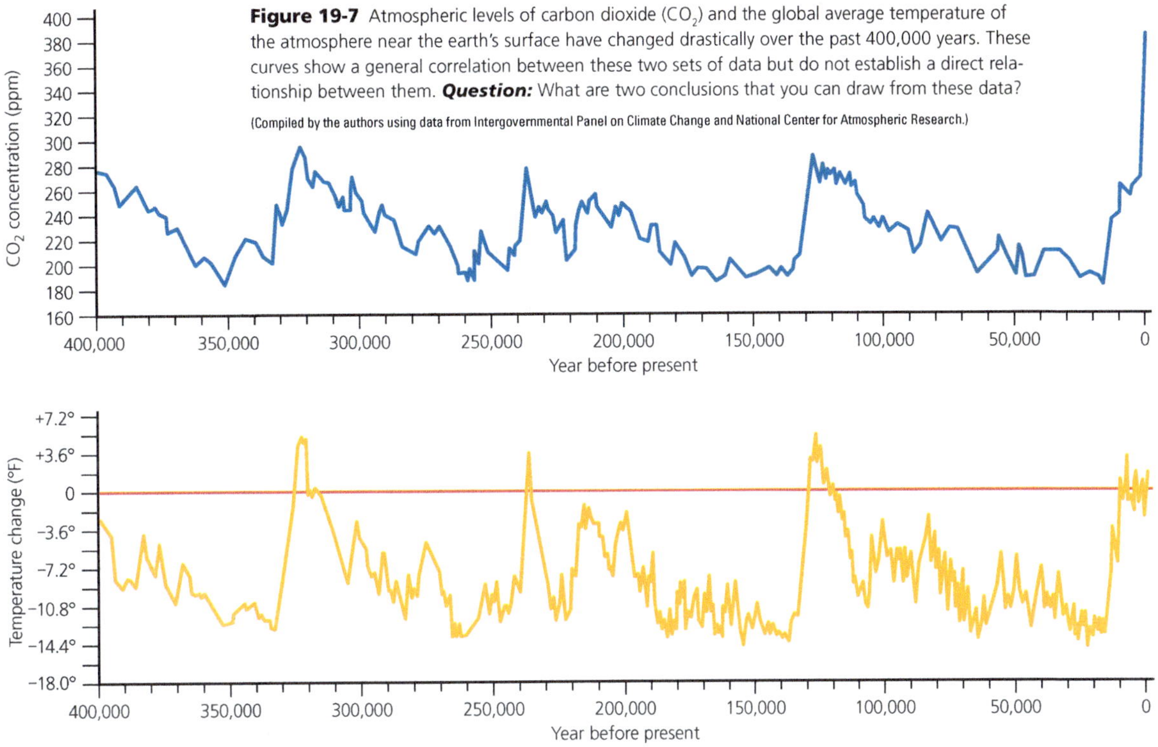

Figure 19-7 Atmospheric levels of carbon dioxide (CO_2) and the global average temperature of the atmosphere near the earth's surface have changed drastically over the past 400,000 years. These curves show a general correlation between these two sets of data but do not establish a direct relationship between them. ***Question:*** What are two conclusions that you can draw from these data?

(Compiled by the authors using data from Intergovernmental Panel on Climate Change and National Center for Atmospheric Research.)

on average, compared to years for the other gases (Figure 19-6). As the air warms, any given volume of air can hold more water vapor. Thus, the warming effect of other greenhouse gases can lead to a higher level of water vapor, which in turn amplifies the greenhouse effect. For this reason, scientists consider CO_2 to be the main engine of the atmospheric warming related to greenhouse gases, while water vapor plays a secondary role in response to CO_2 levels.

Human Activities Play a Key Role in Atmospheric Warming

Since the beginning of the Industrial Revolution in the mid-1700s, human activities—mainly the burning of fossil fuels, deforestation, and agriculture, in that order—have led to significant increases in the levels of greenhouse gases in the lower atmosphere, especially since 1960 (see Figure 14, p. S70, in Supplement 7).

Analyses of ice cores from Antarctica done in 2007 indicate that current levels of CO_2 in the atmosphere are higher than at any time during the past 800,000 years. Based on these studies, long-term changes in levels of CO_2 correlate fairly closely with changes in the global average temperature of the atmosphere near the earth's surface (Figures 19-7 and 19-8) (**Concept 19-2**). However, temperature increases generally lag behind increases in CO_2.

The burning of fossil fuels tops the list of human activities that emit CO_2, with coal being by far the highest emit-

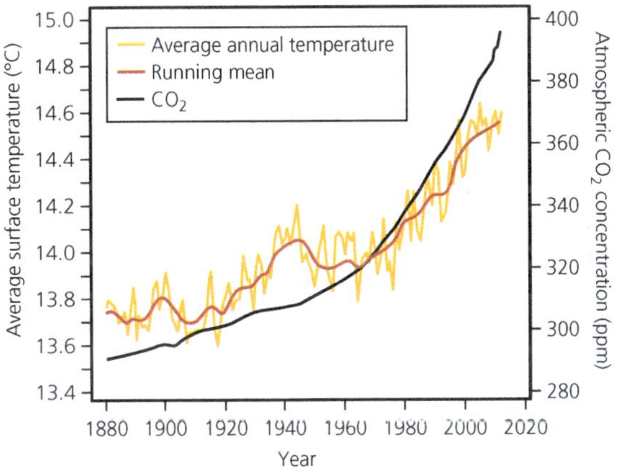

Figure 19-8 Comparison of atmospheric concentrations of carbon dioxide (CO_2) and the atmosphere's average temperature for the period 1880–2011. Temperature data show the fluctuating seasonal average temperatures (orange curve) as well as the *mean*, or average of these data (red curve). While the data show an apparent relationship between the mean temperature and CO_2 concentrations, they do not establish that these two variables are related.

(Compiled by the authors using data from NASA Goddard Institute of Space Studies, National Oceanic and Atmospheric Administration Earth System Research Laboratory, Earth Policy Institute, and Worldwatch Institute.)

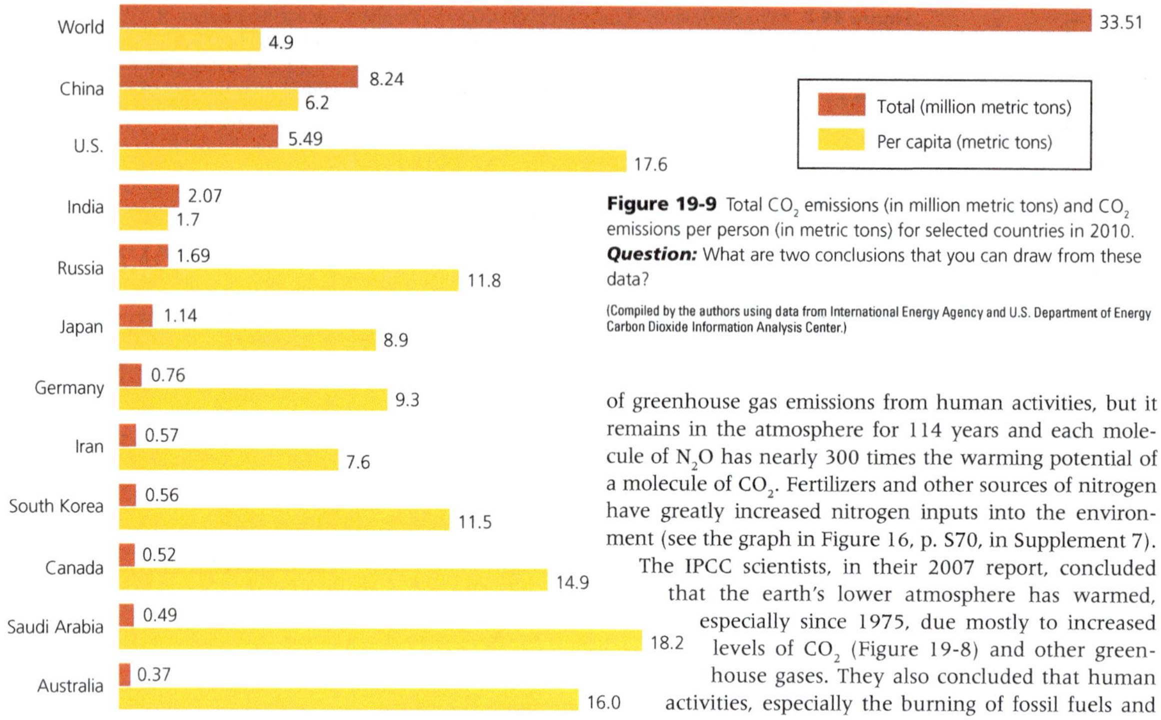

Figure 19-9 Total CO_2 emissions (in million metric tons) and CO_2 emissions per person (in metric tons) for selected countries in 2010. ***Question:*** What are two conclusions that you can draw from these data?

(Compiled by the authors using data from International Energy Agency and U.S. Department of Energy Carbon Dioxide Information Analysis Center.)

ter of all fossil fuels (see Figure 15-17, p. 388). The average atmospheric concentration of CO_2 rose from a level of 280 parts per million (ppm) at the start of the Industrial Revolution to 400 ppm in 2013, according to data from the NOAA (see Figure 14, p. S70, in Supplement 7). This amounts to a 43% increase in the amount of CO_2 in the atmosphere, and this is a long-term increase, because CO_2 typically remains in the atmosphere for about 100 years (Figure 19-6).

Ice core analysis reveals that about 70% of methane (CH_4) emissions during the last 275 years are the result of human activities, including livestock production, rice production, natural gas production, leaky coal mines, landfills, and the flooding of land behind large dams. In that time, methane levels in the atmosphere have tripled. Methane remains in the atmosphere for about 12 years and each molecule of methane has 25 times the warming potential of a molecule of carbon dioxide (Figure 19-6).

Methane emissions are expected to rise in the future if the atmosphere warms as projected and melts some of the permafrost in the arctic tundra and in the Arctic Ocean seabed, which will release a large volume of CH_4 into the atmosphere. Also, a 2012 study led by NASA's Eric Kort found surprising levels of methane being released through cracks in floating arctic ice. This methane is thought to be generated by marine bacteria, and emissions from this source will likely increase as arctic seawater warms and as more cracks appear in the melting sea ice.

Nitrous oxide (N_2O) levels have risen about 20% during the last 275 years, mostly as a result of the increased use of nitrogen fertilizers. This gas accounts for about 9% of greenhouse gas emissions from human activities, but it remains in the atmosphere for 114 years and each molecule of N_2O has nearly 300 times the warming potential of a molecule of CO_2. Fertilizers and other sources of nitrogen have greatly increased nitrogen inputs into the environment (see the graph in Figure 16, p. S70, in Supplement 7).

The IPCC scientists, in their 2007 report, concluded that the earth's lower atmosphere has warmed, especially since 1975, due mostly to increased levels of CO_2 (Figure 19-8) and other greenhouse gases. They also concluded that human activities, especially the burning of fossil fuels and deforestation, played an important role in this warming (Concept 19-2).

In 2010, China and the United States together accounted for about 41% of the world's annual greenhouse gas emissions (Figure 19-9). China has been the leading emitter of CO_2 since 2007. However, China's per capita CO_2 emissions are low because most members of its huge population are too poor to use large amounts of fossil fuels and other resources. The United States emits nearly 3 times as much CO_2 per person as China emits and about 10 times as much CO_2 per person as India emits. Figure 19-9 shows total and per capita CO_2 emissions for the world's highest-emitting countries.

In 2012, according to a survey by researchers at Stanford University, nearly 98% of climate scientists who were actively publishing their research agreed that the burning of fossil fuels has played a key role in atmospheric warming since 1975 (Figure 19-8). In 2013, the third National Climate Assessment written by 240 scientists, business leaders, and other climate experts concluded that "climate change, once considered an issue for a distant future, has moved firmly into the present."

Climate models (Science Focus 19.1) project that rising CO_2 emissions and atmospheric temperatures could bring about significant changes in the earth's climate that are likely to cause major ecological and economic disruption in the latter half of this century. IPCC scientists have identified several natural and human-influenced factors that might *amplify* or *dampen* the projected changes in the average atmospheric temperature. We now examine some of those factors.

What Role Does the Sun Play?

The energy output of the sun plays the key role in the earth's temperature, and this output has varied over millions of years. However, in 2007, climate researchers Claus Froehlich, Mike Lockwood, and Ben Santer all concluded in separate studies that most of the rise in global average atmospheric temperatures since 1975 (Figure 19-3) could not be the result of increased solar output. Instead, they determined that the energy output of the sun has dropped slightly during the past several decades. In 2011, a detailed data analysis by physicist Richard Muller and his colleagues confirmed this conclusion.

Froehlich noted that, according to satellite and weather balloon measurements since 1975, the troposphere has warmed while the stratosphere has cooled. This is the opposite of what a hotter sun would do. According to Santer, "If you increase output from the sun, you increase the amount of energy that arrives at the top of the Earth's atmosphere and get heating throughout the atmosphere." Instead, the data show that the atmosphere is now heating from the bottom up, which indicates that inputs at the earth's surface (most likely from human activities) play the more important role in atmospheric warming.

What Role Do Oceans Play?

The world's oceans absorb CO_2 from the atmosphere as part of the carbon cycle (see Figure 3-17, p. 66) and thus help to moderate the earth's average surface temperature and its climate. It is estimated that the oceans remove roughly a third of the CO_2 pumped into the lower atmosphere by human activities. The oceans also absorb heat from the lower atmosphere. Then, partly driven by this heat, ocean currents slowly transfer some of the CO_2 to the deep ocean (see Figure 7-5, p. 147), where it is buried in carbon compounds in bottom sediments for several hundred million years.

The ability of the oceans to absorb CO_2 decreases as water temperatures rise. As the oceans warm up, some of their dissolved CO_2 is released into the lower atmosphere, like CO_2 bubbling out of a warm carbonated soft drink. According to scientific measurements from a 2007 study, the upper portion of the oceans warmed by an average of 0.32–0.67°C (0.6–1.2°F) during the last century—an astounding increase considering the huge volume of water involved—most likely due to increasing atmospheric temperatures. A 2012 study by oceanographers Dean R. Roemmich and John Gibson found that the upper ocean warmed by about 0.2°C (0.1°F) between 1990 and 2010—a sharp increase for such a massive body of water in a short time.

As the oceans have absorbed more CO_2 and become warmer, they have also become more acidic, and this could result in some potentially harmful effects on marine ecosystems (see Science Focus 11.2, p. 252). We discuss this process and its effects in the next section.

There Is Uncertainty about the Effects of Cloud Cover on Atmospheric Warming

The role played by clouds in the warming of the atmosphere is uncertain. Warmer temperatures increase evaporation of surface water, which increases the relative humidity of the atmosphere in various parts of the world and creates more clouds that can either cool or warm the atmosphere.

An increase in thick and continuous *cumulus clouds* at low altitudes (Figure 19-10, left) could decrease surface warming by reflecting more sunlight back into space. But an increase in thin, wispy *cirrus clouds* at high altitudes (Figure 19-10, right) could cause more warming of the lower atmosphere by preventing some heat from escaping into space. Climate modelers have not been able to show which of these effects is more important and are working hard to understand more about the role of clouds in their climate models. One thing that current climate models do agree on is that any cooling effect that clouds might have will not be enough to offset the long-term heating effects of human-generated greenhouse gas emissions.

In 2010, Axel Lauer and other researchers developed a new model for evaluating the effects of cloud cover on atmospheric warming. According to their model, by the end of this century, higher CO_2 levels and a warmer atmosphere will likely shrink the size of low cumulus clouds. This will allow more solar radiation to reach the lower atmosphere, which will likely accelerate atmospheric warming toward the higher estimated temperatures shown in Figure 19-C (p. 515).

Tatiana Grozetskaya/Shutterstock.com

Cheryl Casey/Shutterstock.com

Figure 19-10 Cumulus clouds (left) are thick, relatively low-lying clouds that tend to decrease surface warming by reflecting some incoming solar radiation back into space. Cirrus clouds (right) are thin and float at high altitudes; they tend to warm the earth's surface by preventing some heat from flowing into space.

SCIENCE FOCUS 19.1

USING MODELS TO PROJECT FUTURE CHANGES IN ATMOSPHERIC TEMPERATURES

To project the effects of increasing levels of greenhouse gases on average global temperatures, scientists have developed complex *mathematical models,* which simulate interactions among incoming sunlight, clouds, landmasses, oceans, ocean currents, concentrations of greenhouse gases and air pollutants, and other factors within the earth's complex climate system. They run these continually improving models on supercomputers and compare the results to known past climate changes, from which they project future changes in the earth's average temperature. Figure 19-A gives a greatly simplified summary of some of the key interactions in the global climate system.

Such models provide *projections* of what is *likely* or *very likely* to happen to the average temperature of the lower atmosphere. How well these projections match what actually happens in the real world depends on the validity of the assumptions and variables built into the models and on the accuracy of the data used. Recall that no scientific research can give us absolute proof or certainty, but instead provides us with varying levels of certainty. However, when most experts in a particular scientific field generally agree that the level of certainty involved in a set of measurements or a model is at least 90%, they are very likely to be correct.

In 1990, 1995, 2001, and 2007, the IPCC published reports on how global temperatures have changed in the past (Figure 19-2) and made projections on how they are likely to change during this century and how such changes are likely to affect the earth's climate. According to the 2007 report, based on analysis of past climate data and the use of 19 climate models, it is very likely that human activities, especially the burning of fossil fuels, have played an important role in the observed atmospheric warming since 1975 (Figure 19-3 and Concept 19-2). The researchers based this conclusion on the fact that, after thousands of times running the models, the only way they can get the model results to match actual measurements is by including the human activities factor (Figure 19-B).

Similar to the 2007 IPCC report, more recent runs of several of the more than two dozen climate models now in use suggest that it is very likely that the earth's mean surface temperature will increase by 2–4.5°C (3.6–8.1°F) between now and 2100 (Figure 19-C) unless the world halts deforesta-

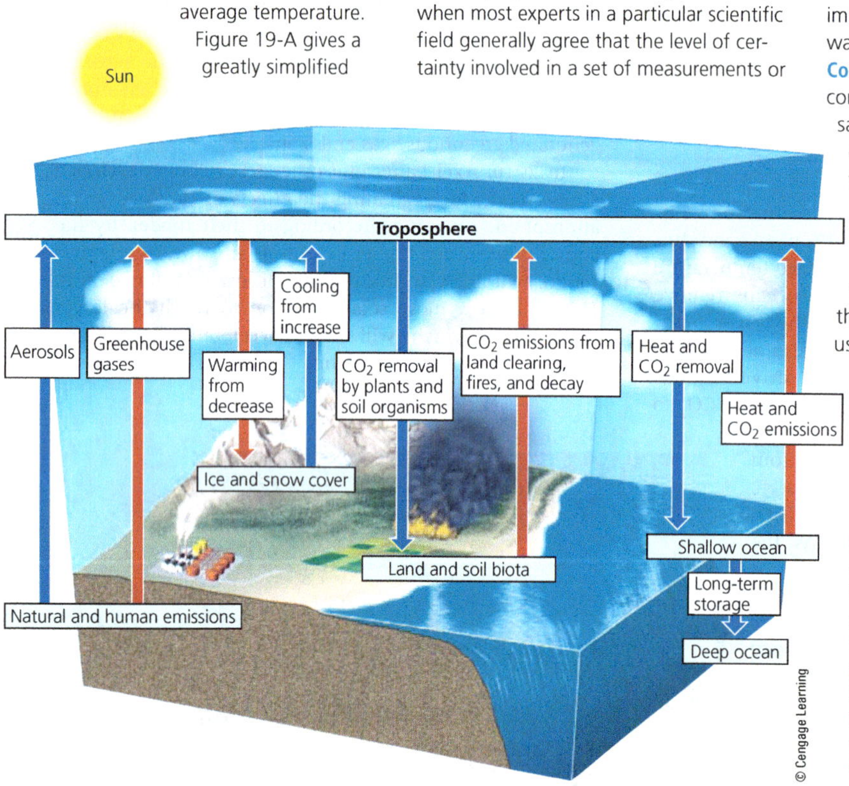

Figure 19-A A simplified model of some major processes that interact to determine the average temperature and greenhouse gas content of the lower atmosphere and thus the earth's climate. Red arrows show processes that warm the atmosphere and blue arrows show those that cool it. ***Question:*** Why do you think a decrease in snow and ice cover is adding to the warming of the atmosphere?

CONSIDER THIS. . .

CONNECTIONS Air Travel and Atmospheric Warming

Infrared satellite images indicate that wispy condensation trails (contrails) left behind by jet planes expand and turn into cirrus clouds, which tend to heat the atmosphere. A 2010 study led by Ulrike Burkhardt and Bernd Kärcher found that contrails are a significant contributor to atmospheric warming, more so than the total amount of CO_2 that is released by jet airplanes, at least in the short term.

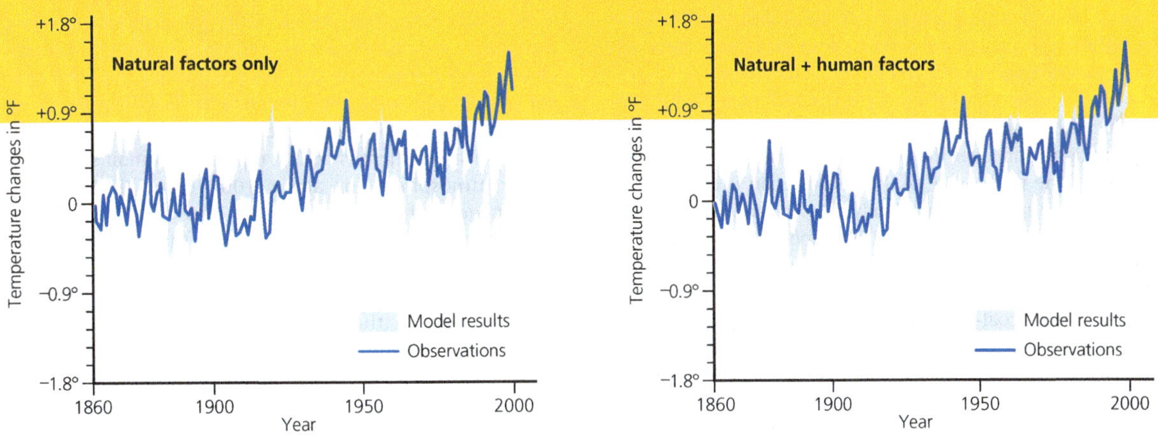

Figure 19-B Comparison of actual climate data with modeled projections for the period between 1860 and 2000 using natural factors only (left) and a combination of natural and human factors (right). Scientists have found that actual data match projections far more closely when human factors are included in the models.

(Compiled by the authors using data from Intergovernmental Panel on Climate Change, 2001 and 2007.)

tion, makes drastic cuts in greenhouse gas emissions from fossil fuel–burning power plants, factories, and motor vehicles, and switches to improved energy efficiency and low-carbon energy resources such as wind and solar energy (see Figure 16-36, p. 434).

In any scientific modeling, there is some degree of uncertainty, as indicated by the fairly wide range of projected temperature changes in Figure 19-C. Climate experts are working to improve these models and to narrow the projected range of temperatures. Also, current climate models are better at projecting likely general changes at the global level than at making projections for particular areas, which can have local climates that vary significantly from the average global climate. In other words, as the atmosphere warms globally, some areas will be hotter and some will be cooler. However, current climate models project that a global average temperature increase of 2–3°C (3.6–5.4°F) would be enough to overcome local cooling and lead to a warmer climate almost everywhere.

Critical Thinking

If the projected temperature increases in Figure 19-C take place, what are three major ways in which this will affect your lifestyle and that of any children or grandchildren you might have?

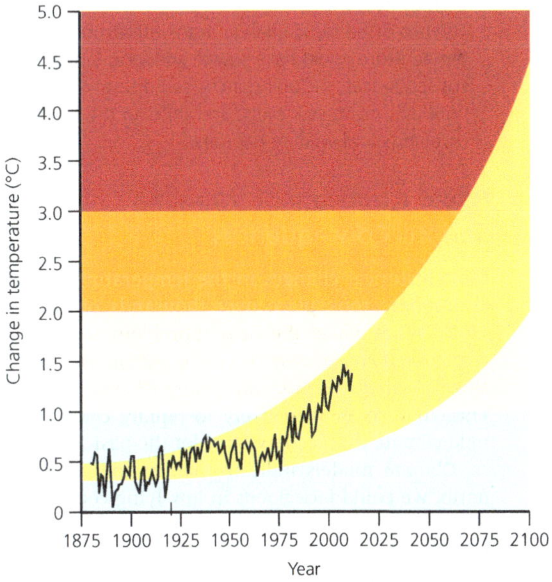

Figure 19-C Estimated changes (yellow area) and measured changes (black curve) in the average temperature of the atmosphere near the earth's surface between 1860 and 2011, and the projected range of temperature increase during the rest of this century.

(Compiled by the authors using data from U.S. National Academy of Sciences, National Center for Atmospheric Research, Intergovernmental Panel on Climate Change, and Hadley Center for Climate Prediction and Research.)

Outdoor Air Pollution Can Temporarily Slow Atmospheric Warming

There is considerable evidence that human activities have played a role in atmospheric warming, but there is also evidence that aerosol air pollution from human activities has slowed the rate of atmospheric warming. Aerosols (suspended microscopic droplets and solid particles) of various air pollutants are released or formed in the troposphere by volcanic eruptions and human activities (see

Figure 18-3, p. 477). They can hinder or enhance both the greenhouse warming effect and cloud formation, depending on factors such as their size and reflectivity.

Most aerosols, such as light-colored sulfate particles produced by fossil fuel combustion, tend to reflect incoming sunlight and cool the lower atmosphere as they seem to have done in the case of the South Asian Brown Clouds (see Chapter 18, Core Case Study, p. 474). A 2012 study led by Harvard University's Eric Leibensperger estimated that atmospheric temperatures in the eastern United States had dropped slightly between 1930 and 1990—due largely to pollution from coal-fired power and industrial plants—while worldwide average temperatures had increased slightly.

While sulfate particles reflect sunlight, black carbon particles, or *soot,* have the opposite effect. A 2008 study by atmospheric scientist Veerabhadran Ramanathan and his colleagues found that soot emitted into the air by coal-burning power and industrial plants, diesel exhaust, open cooking fires, and burning forests has a warming effect that is 4 times greater than was estimated earlier. These findings were supported by another study reported in 2009 by meteorologist Gunnar Myhre.

Climate scientists do not expect aerosols and soot pollutants to counteract or enhance projected climate change very much in the next 50 years for two reasons. *First,* aerosols and soot fall back to the earth or are washed out of the lower atmosphere within weeks or months, whereas CO_2 typically remains in the lower atmosphere for about 100 years (Figure 19-6). *Second,* aerosol and soot inputs into the lower atmosphere are being reduced—especially in more-developed countries—because of their harmful impacts on plants and human health.

According to the IPCC, the decline in atmospheric sulfate concentrations in some countries since 1970 has played a role in the warming of the atmosphere, especially since 1990, and this effect will grow as sulfate concentrations drop further due to improved air pollution regulations. We could possibly slow projected climate disruption by increasing outdoor sulfate air pollution. However, because such pollution kills more than 800,000 people a year (about 50,000 people in the United States), this would not be a desirable or ethically acceptable way to try to deal with projected climate disruption.

19-3 What Are the Possible Effects of a Warmer Atmosphere?

CONCEPT 19-3

The projected rapid change in the atmosphere's temperature could have severe and long-lasting consequences, including increased drought and flooding, rising sea levels, and shifts in the locations of croplands and wildlife habitats.

Rapid Atmospheric Warming Could Have Serious Consequences

Most historical changes in the temperature of the lower atmosphere took place over thousands of years (Figure 19-2). What makes the current problem urgent is that we face *a rapid projected increase in the average temperature of the lower atmosphere during this century* (Figure 19-C, p. 515). This, in turn, is very likely to rapidly change the fairly mild climate that we have had for the past 10,000 years.

Climate models indicate that, in the worst-case scenario, we could face floods in low-lying coastal cities, forests being consumed in vast wildfires, grasslands turning into dust bowls, rivers drying up, ecosystems collapsing, the extinction of up to half of the world's species, more intense and longer-lasting heat waves, more destructive storms and flooding, and the rapid spread of some infectious tropical diseases. If the models are correct, we will have to deal with many of these disruptive effects within this century—an incredibly short time span in terms of the earth's overall climate history.

Climate models also show that atmospheric warming and its harmful effects will be unevenly distributed, according to a 2012 World Bank report. Temperatures in the tropics could reach extremes that people there have not seen in over 10,000 years. The sea-level rise also would be 15–20% larger than the global average in tropical areas, and droughts could also be much worse in such areas. In our interconnected global economy, such effects would ripple strongly to other, less-affected areas of the world.

In anticipation of such possibilities, scientists have identified a number of *tipping elements,* or components of the climate system that could pass **climate change tipping points**—those thresholds beyond which natural systems could change for hundreds to thousands of years, with possibly catastrophic effects. One such tipping element is atmospheric CO_2 levels. A number of scientific studies and major climate models indicate that we need to prevent CO_2 levels from exceeding 450 ppm—an estimated tipping point beyond which we might experience large-scale climate changes that would last for hundreds to thousands of years. We have already reached 400 ppm, and without strong efforts to improve energy efficiency and replace fossil fuels with low-carbon energy resources, we could exceed 450 ppm within a couple of decades.

Another tipping element is global average atmospheric temperature. Several findings related to atmospheric warming are leading some scientists to suggest that we can no longer avoid a global temperature rise of more than 2C° (3.6F°), the threshold beyond which many say climate change will be very dangerous. Several of the effects of climate change that we will now examine could lead to these and other tipping points.

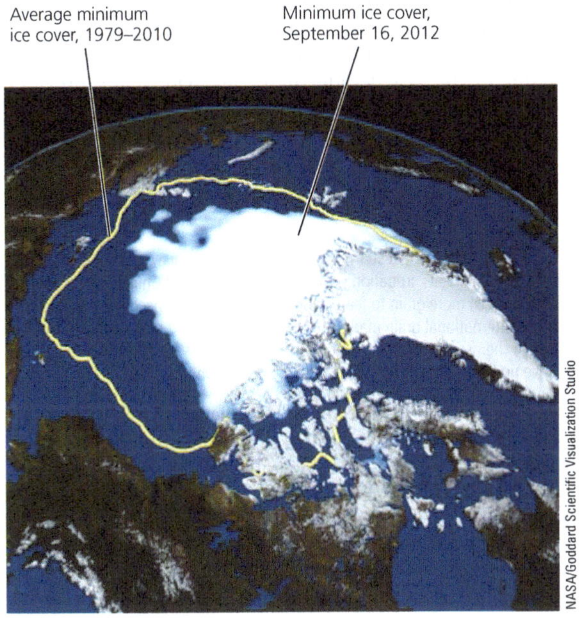

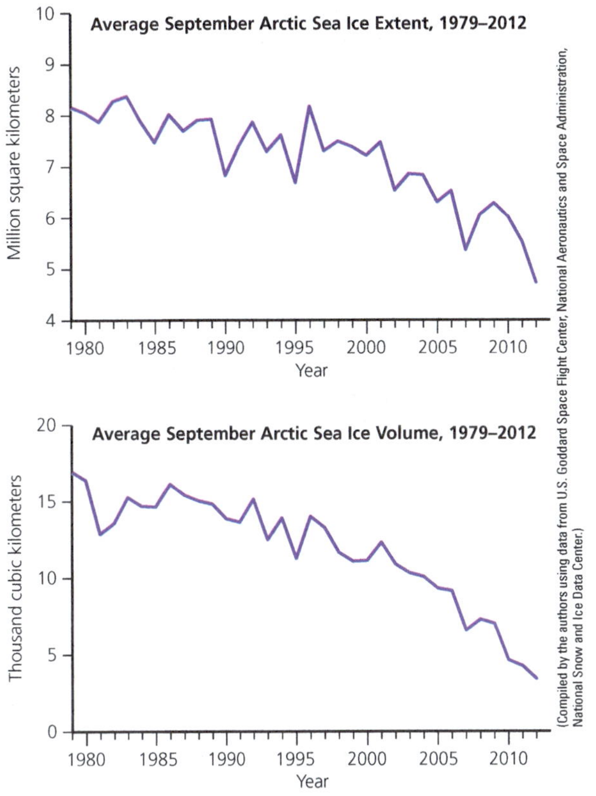

Figure 19-11 *The big melt:* Rising average atmospheric and ocean temperatures have caused more and more arctic sea ice to melt during the summer months. The yellow line added to this satellite image (left) shows the average summer minimum area of ice during the period 1979–2010, in contrast to the white, ice-covered summer minimum in 2012. The graphs (right) show that the sea ice melt has been accelerating. ***Question:*** Which is declining faster, the summer sea ice area, or the summer sea ice volume?

More Ice and Snow Are Likely to Melt

Each summer, some of the floating ice in the Arctic Sea melts and then refreezes during winter. But in recent years, rising average atmospheric and ocean temperatures have caused more and more ice to melt during the summer months. Satellite data show a 49% drop in the average area covered by summer arctic sea ice between 1979 and 2012 (Figure 19-11, left). In 2012, scientists measured the largest loss of arctic sea ice ever recorded in 34 years of satellite measurements—an area of sea ice larger than the continental United States.

According to National Snow and Ice Data scientist Julienne Stroeve, this ice area declined at an average of 6.5% per decade between 1979 and 2001 and since then has been falling twice as fast, on average (Figure 19-11, right, upper graph). Similarly, the volume of arctic sea ice has declined even more, due to dramatic thinning of the ice. In 2012, the total volume of arctic sea ice was less than one-fourth of what it was in 1979 (Figure 19-11, right, lower graph).

The melting of this ice is due to both warmer air above the ice and warmer water below the ice. Scientists are not sure what causes periodic rises and falls in the temperature of arctic seawater, occurring over 20- to 30-year cycles. These cycles are poorly understood and their impact on sea ice is largely unknown.

However, according to the 2007 IPCC report, arctic air temperatures have risen almost twice as fast as average temperatures in the rest of the world during the past 50 years. Historically, light-colored ice and snow in these regions has helped to cool the earth by reflecting incoming solar energy. However, the melting of ice and snow has exposed much darker land and sea areas, which reflect significantly less sunlight and absorb more solar energy. In addition, soot generated by North American, European, and Asian industries is darkening arctic ice and lessening its ability to reflect sunlight.

Recall from Chapter 2 our discussion of *positive feedback loops*—cycles of change that cause systems to change further in a certain direction (see Figure 2-16, p. 45). An important example of such a loop is the accelerating melting of arctic ice, which causes further warming, which, in turn, speeds the melting. As such a feedback loop kicks in, the rate of change and the resulting harmful effects can get out of control.

One of the climate-change tipping points that concern scientists is the complete melting of summer arctic sea ice. If the current trend continues, this summer ice may be gone by 2040, according to researchers Muyin Wang and

James Overland, and perhaps earlier according to others, including NASA climate scientist Jay Zwally and ocean physics professor Peter Wadhams. This would open these waters to shipping and would allow access to oil and mineral deposits in the arctic region. However, it could also lead to dramatic and long-lasting changes in weather and climate that could affect the whole planet.

For example, according to a 2012 study led by atmospheric scientists Jennifer A. Francis and Stephen J. Vavrus, changes in arctic seawater temperatures have already slowed the *jet stream*—a high, rapidly moving flow of air that affects global weather patterns (see Figure 2, p. S20, in Supplement 5). As a result, weather conditions can remain in certain areas for longer periods of time. Rainy weather can hold its position long enough to cause flooding, and dry weather can turn to drought. Another result, according to a 2010 report by NOAA scientist James Overland, could be colder and snowier winters in Europe, eastern North America, and eastern Asia because of a slower jet stream sinking southward and bringing cold air into those areas.

Another effect of this warming cycle is faster melting of polar land-based ice including that in Greenland (**Core Case Study** and Figure 19-12). This melting adds freshwater to the northern seas, as could an increase in rainfall, which is another projected effect of atmospheric warming. Some scientists hypothesize that at some point, this additional water could slow or stop ocean currents that move heat around the globe, which could ramp up the warming in Greenland. Some scientists are dedicating much of their time to studying Greenland's melting ice (Individuals Matter 19.1).

Another great storehouse of ice is the earth's mountain glaciers, which expand with added snowfall in winter (called the *snowpack*) and slowly melt during summer months. During the past 25 years, many of the world's mountain glaciers have been slowly shrinking (Figure 19-5) wherever summer melting exceeds the winter snowpack due to a warming atmosphere. For example, Glacier National Park in the U.S. state of Montana once had 150 glaciers, but by the end of 2010, only 25 remained, according to park officials.

Mountain glaciers play a vital role in the water cycle (see Figure 3-15, p. 63) by storing water as ice during cold seasons and releasing it slowly to streams during warmer seasons. A prime example of these high-elevation reservoirs is the glaciers of the Himalayan Mountains in Asia (Figure 19-5). They are a major source of water for large rivers such as the Ganges, which provides water for more than 400 million people in India and Bangladesh. They also feed China's Yangtze and Yellow Rivers, whose basins are home to more than 500 million people. In 2009, Swiss scientists reported that some Himalayan glaciers, as well as some glaciers in the European Alps, shrank farther and faster in the last decade than they had in the previous 150 years.

CONSIDER THIS. . .

CONNECTIONS Tibetan Plateau Glaciers and World Food Prices

Environmental analyst Lester Brown points out that as ice on the Tibetan Plateau in China melts, irrigation water will grow short, and China will need to import much more grain to feed its population. As China becomes more active in the international grain market, its 1.3 billion people, whose average incomes are on the rise, will be competing with U.S. consumers for grains, and this could drive up food prices in the United States and elsewhere.

About 80% of the mountain glaciers in South America's Andes range are also slowly shrinking. By 2010, Peru had lost 22% of its glaciers and about half of Bolivia's glacial ice had melted, according to the World Glacier Monitoring Service. If this continues, 53 million people in Bolivia, Peru, and Ecuador who rely on meltwater from the glaciers for irrigation and hydropower could face severe water, power, and food shortages at some point during the next 50 to 200 years. In the United States, people living in the Columbia, Sacramento, and Colorado River basins could face similar threats as the winter snowpack that feeds these rivers is projected to shrink by 70% by 2050, according to climate models.

Permafrost Is Likely to Melt: Another Dangerous Scenario

Permafrost occurs in soils found beneath 24% of the exposed land in the northern hemisphere. The amount of carbon locked up in permafrost soils is 50–60 times the

Figure 19-12 Accelerating melting of Greenland's ice has created flows such as this one, called *moulins*, which run from the surface to the base of the glaciers, leading to more dramatic melting and movement of the glacier toward the sea.

individuals matter 19.1

Konrad Steffen: Studying Melting Ice in Greenland

Photo by Konrad Steffen, University of Colorado/CIRES

In 2012, Konrad Steffen, formerly the Director of the Cooperative Institute for Research in Environmental Sciences (CIRES) at the University of Colorado, became Director of the Swiss Federal Institute for Forest, Snow and Landscape Research. He is one of the world's most respected experts on Greenland's melting ice (Core Case Study).

Since 1990, he and his colleagues have conducted extensive research on the Greenland ice sheet. This led to the building of the Greenland Climate Network—22 weather stations that use solar-powered monitoring equipment to automatically record the amounts of precipitation in Greenland and the melting and changes in the surface height of Greenland ice. Scientists correlate this data with satellite observations in order to estimate ice coverage in Greenland (Figure 19-1).

Steffen personally adapted and installed much of the equipment in these monitoring stations that provide scientists around the world with information on hourly conditions on Greenland's ice sheet and how they are changing. His research revealed that the melting of Greenland's ice accelerates when meltwater flows downward to bedrock (Figure 19-12). There it acts as a lubricant that accelerates the slide of ice into the ocean, an effect of atmospheric warming that is occurring more rapidly than was projected by the IPCC.

Steffen and his colleagues have also studied ice cores to look at the climate history of Greenland for the past 800 years. Their research has revealed that between 1979 and 2010, the area of ice that melted expanded by 30% because of rising summer air and water temperatures. This research also indicated that about 50% of Greenland's ice loss is due to melting, and the other 50% is the result of icebergs breaking off into the ocean. Using statistical modeling, Steffen has projected that the rise in sea level due to rising water temperatures and to the melting of polar ice might be 1 to 2 meters (3.3 to 6.6 feet) by the end of this century.

Background photo: ©Sofiaworld/Shutterstock.com

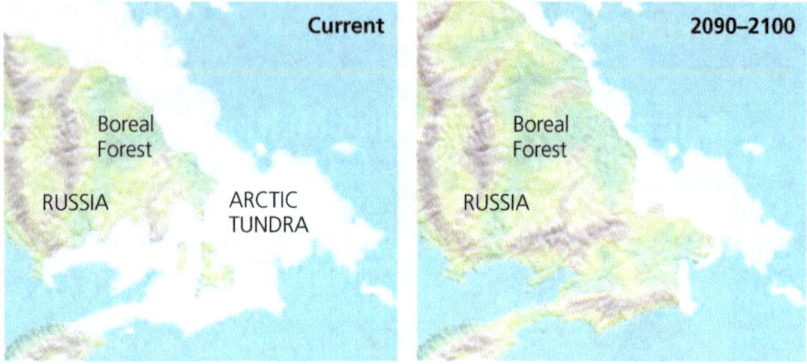

Figure 19-13 Projected thaw of permafrost in the arctic tundra of eastern Russia between 2004 and 2100 as a result of atmospheric warming. Tall shrubs and trees are likely to take over a growing area of the much warmer arctic tundra by the end of this century.

(Compiled by the authors using data from Intergovernmental Panel on Climate Change and 2004 Arctic Climate Impact Assessment.)

amount emitted as CO_2 each year from the burning of fossil fuels.

A 2012 report by the United Nations Environment Programme (UNEP) estimated that human-caused climate change is expected to thaw out significant amounts of permafrost. According to the 2004 Arctic Climate Impact Assessment, 10–20% of the Arctic's current permafrost might thaw during this century over a vast area of arctic tundra (Figure 19-13, left). If this happens, a great deal of stored organic material will rot and release large amounts of CH_4 and CO_2 into the atmosphere. This would accelerate projected warming, which would in turn melt more permafrost in yet another positive feedback loop.

Some scientists are concerned about another methane source—a layer of permafrost on the Arctic Sea floor and also on many arctic lake bottoms. In 2009, Natalia Shakhova and other researchers found areas of the seafloor to be leaking bubbles of methane. In 2012, aquatic ecologist Katey Walter Anthony (a National Geographic Emerging Explorer) found methane bubbling up from thousands of seeps on some arctic lake bottoms. In warmer parts of the world, yet another possible source of increasing methane emissions is tropical wetlands. If rainfall increases as projected in tropical areas, these wetlands will expand and produce more plants, which will eventually decay and release methane through anaerobic decomposition (see p. 57). Some scientists refer to such sources as "methane time bombs."

Sea Levels Are Rising

In 2007, the IPCC estimated that sea levels would rise by about 0.5 meters (20 inches) by the end of this century, primarily due to the expansion of seawater as it warms and to the melting of land-based ice as a result of atmospheric warming (**Core Case Study**). However, in 2011, the U.S. National Academy of Sciences (NAS) reported that, based on more recent studies, sea levels could rise by as much as 1.5 meters (5 feet) by 2100—about 3 times the increase estimated in the 2007 IPCC report. The NAS panel also warned that sea levels would probably keep rising for centuries after 2100.

The rise in sea level is projected to be more dramatic if the vast land-based glaciers in Greenland (**Core Case Study**) and western Antarctica, as well as many of the world's mountaintop glaciers, continue to melt at their current, or higher, rates as the atmosphere warms. Taking this into account, Konrad Steffen (Individuals Matter 19.1) projects a 1- to 2-meter (3.3- to 6.6-foot) rise in the average sea level by the end of this century. A 2012 analysis by Richard Alley and 46 other glaciologists concluded that ice losses from Greenland and West Antarctica due to a warming atmosphere and warming ocean have been accelerating.

According to the IPCC and the NAS study, a 1-meter (3.3-foot) rise in sea level during this century (excluding the additional effects of storm surges) could cause the following serious effects:

- Degradation or destruction of at least one-third of the world's coastal estuaries, wetlands, coral reefs, and deltas where much of the world's rice is grown.
- Disruption of many of the world's coastal fisheries.
- Flooding and erosion of low-lying barrier islands and gently sloping coastlines, especially in U.S. coastal states such as Texas, Louisiana, North and South Carolina, and Florida (Figure 19-14).
- Saltwater contamination of freshwater coastal aquifers resulting in degraded supplies of groundwater.
- Submersion of low-lying islands in the Indian Ocean (Figure 19-15), the Pacific Ocean, and the Caribbean Sea, which are home to 1 of every 20 of the world's people.

Figure 19-14 If the average sea level rises by 1 meter (3.3 feet), the areas shown here in red in the U.S. state of Florida will be flooded.

(Compiled by the authors using data from Jonathan Overpeck, Jeremy Weiss, and the U.S. Geological Survey.)

Figure 19-15 For a low-lying island nation like the Maldives in the Indian Ocean, even a small rise in sea level could spell disaster for most of its 295,000 people. About 80% of the 1,192 small islands making up this country lie less than 1 meter (3.2 feet) above sea level. Rising sea levels and higher storm surges during this century could flood most of these islands and their coral reefs.

- Flooding of some of the world's largest coastal cities, including Kolkata (India), Mumbai (India), Dhaka (Bangladesh), Guangzhou (China), Ho Chi Minh City (Vietnam), Shanghai (China), Bangkok (Thailand), Rangoon (Myanmar), Haiphong (Vietnam), and the U.S. cities of Miami, Florida, and New York City (see red areas in chapter-opening photo and Figure 19-14), and displacement of at least 100 million people.

The sea-level rise will not be uniform around the world, according to climate models, because of factors such as ocean currents and winds. This is already apparent, according to a 2012 study by the U.S. Geological Survey, which reported that the sea level is rising as much as 4 times faster than the global average along parts of the U.S. Atlantic coast.

The Oceans Are Becoming More Acidic

According to a 2005 report by the United Kingdom's Royal Society and a 2010 report by the U.S. National Academy of Sciences, rising levels of CO_2 in the ocean have increased the acidity of its surface waters by 30% since about 1800, and ocean acidity could reach dangerous levels before 2050. This is because much of the CO_2 absorbed by the ocean reacts with water to produce carbonic acid (H_2CO_3)—the same weak acid found in carbonated drinks—in a process called **ocean acidification** (see Science Focus 11.2, p. 252).

Scientists warn that this higher acidity threatens corals, snails, oysters, and other organisms with shells and body structures composed of calcium carbonate, because it hinders their ability to build and repair such structures. Oceanographer Carl Safina notes that acidification will slow the growth and repair of damaged coral reefs and could begin to dissolve some reefs by the end of the century.

Another problem is that increased acidity has been shown to decrease populations of phytoplankton that are the primary producer species of ocean food webs and that also remove CO_2 from the atmosphere. Phytoplankton are at the base of the ocean food webs, which include the human seafood supply, and could be severely degraded and might possibly even collapse if these producer organisms decline dramatically.

Severe Drought Is Likely to Become More Common

Recall that *drought* occurs when evaporation due to higher temperatures greatly exceeds precipitation for a prolonged period. A 2005 study by National Center for Atmospheric Research scientist Aiguo Dai and his colleagues found that severe and prolonged drought was affecting at least 30% of the earth's land (excluding Antarctica)—an area the size of Asia (Figure 19-16).

Some scientists hypothesize that higher temperatures due to atmospheric warming are intensifying the water cycle and adding more water vapor to the atmosphere. This would increase the warming effect (because water vapor is a greenhouse gas) and lead to more evaporation followed by greater warming in another example of a positive feedback loop. One effect of this intensified water cycle, according to the scientists, is that in some areas droughts will last much longer and will be more widespread. As a result, according to a 2007 study by climate researchers at NASA's Goddard Institute for Space Studies, by 2059, up to 45% of the world's land area could be experiencing extreme drought.

As droughts become worse and more widespread, the growth of trees and other plants in some areas has declined, which has reduced the removal of CO_2 from the atmosphere. Soils in some areas have dried out, making surface temperatures hotter. As some forests and grasslands are also drying out, wildfires are becoming more

Figure 19-16 These people were trying to get water from a well that was running dry in the drought-prone state of Gujarat in western India in 2010.

frequent in some areas, and this adds CO_2 to the atmosphere. Some streams and other surface waters are drying up. In some areas, water tables are falling, partly because farmers are irrigating more to make up for drier conditions. Climate scientists project that these combined effects from the projected worsening of prolonged droughts are likely to speed up atmospheric warming and lead to more prolonged droughts—one more example of a positive feedback loop.

Other Forms of Extreme Weather Will Be More Likely

Climate models show that along with increasing drought, a more intensified water cycle with more water vapor in the atmosphere will also bring more flooding to some areas due to heavy snowfall or rainfall (Figure 19-17).

In 2011, the IPCC issued a special report authored by Maarten van Aalst and other climate scientists on the increased likelihood of extreme weather events due to atmospheric warming. It suggested that heat waves that now occur once in a lifetime in areas of Latin America, Africa, and Asia, will become more frequent and hotter than any recorded so far. Another study projected that heavy rainstorms now occurring every 20 years in areas of North America will become 3 times more frequent by the end of the century and that floods in Southeast Asia will likely happen 4 times as often as they do now. And although there may be less snow in some areas, there could be more blizzards. Data from the U.S. Global Change Research Program indicate that the annual number of days of extreme heat in parts of the United States will rise dramatically by the end of the century.

There is controversy among scientists over whether projected atmospheric warming will increase the frequency and intensity of tropical storms and hurricanes. In 2008, climatologists Mark Saunders and Adam Lea analyzed data collected since 1950 and found that for every increase of about 0.8°C (1°F) in the water temperature of the Atlantic Ocean, the overall number of hurricanes and tropical storms increased by about a third, and the number of intense hurricanes increased by 45%. Other studies have supported these results.

However, some researchers blame the ups and downs in the intensity of tropical Atlantic hurricanes on natural climate cycles that can change ocean circulation patterns. It is important to note that we cannot conclusively tie any specific hurricane or severe winter storm, which are weather events, to atmospheric warming and climate change. However, in 2010, a World Meteorological Organization panel of experts on hurricanes and climate change from both sides of this scientific debate reached the following consensus view: projected atmospheric warming is likely to lead to fewer but stronger hurricanes that could cause more damage.

Climate Disruption Is a Threat to Biodiversity

According to the IPCC, projected climate disruption is likely to upset ecosystems and take a toll on biodiversity in areas of every continent. For example, the Amazon rain forest could be devastated, even by moderate warming. According to a 2009 British study led by Chris Jones, up to 85% of this forest could be lost if atmospheric warming increases as projected. The scientists noted that this would involve extensive losses in biodiversity and other ecosystem services as the rain forest dries out and becomes more prone to burning and the resulting conversion to tropical savanna (see Chapter 10, p. 226)—the result of exceeding an ecological tipping point.

According to the 1997 IPCC study, approximately 30% of the land-based plant and animal species assessed so far could disappear if the average global temperature change reaches 1.5–2.5°C (2.7–4.5°F). This percentage could grow to 70% if the temperature change exceeds 3.5°C (6.3°F) (**Concept 19-5**). The hardest hit will be plant and animal species in colder climates such as the polar bear in the Arctic (see Case Study, p. 203) and penguins in Antarctica; species that live at higher elevations; species with limited ranges such as some amphibians (see Chapter 4, Core Case Study, p. 78); and those with limited toler-

Figure 19-17 An elderly woman seeks a bag of rice in Dhaka, Bangladesh, after a 2007 flood that killed at least 198 people and drove 19 million people from their homes.

ance for temperature change. The primary cause of such extinctions would be loss of habitat.

Several species, including the comma butterfly of Great Britain, the golden-winged warbler in North America, and the mouse-like pica of Yosemite National park, are known to have migrated northward or upward within mountain ranges in recent decades to avoid the warmer conditions of climate change. A 2012 study by University of Washington scientists found that as the climate warms, about 9% of the western hemisphere's mammals (up to 40% in some areas) likely will not be able to move fast enough to find safe habitats that are cool enough for their survival. The researchers pointed out that while animals have moved to new ranges in the distant past to avoid climate change effects, today's landscape includes vast areas of farm fields, four-lane highways, and urban areas that will prevent some species from reaching new ranges.

One ecosystem that is already feeling the effects of climate change is the Southern Ocean near the Antarctic Peninsula. There, oceanographer Martin Montes-Hugo has been studying the phytoplankton that form the base of the Antarctic food web (see Figure 3-12, p. 60). His team compared phytoplankton production during two periods, 1978–1986 and 1998–2006, and found that it had declined nearly 90%. They said the main reason for this drop was the net shrinkage of sea ice during this period, which has resulted in less freshwater on which the phytoplankton depend. The researchers see this as part of the reason for the drop in numbers of Adélie penguins (top consumers in this food web; Figure 19-18) of more than 80% since 1975. Other vulnerable ecosystems are coral reefs (see Chapter 8, Core Case Study, p. 168), coastal wetlands, high-elevation mountaintops, and alpine and arctic tundra.

A warmer climate could also greatly increase populations of insects and fungi that damage trees. In the Canadian province of British Columbia, for example, warmer winters have led to surges in mountain pine beetle populations that have infected huge areas of lodgepole pine forests, which are now dying (Figure 19-19). Since 2000, pine beetles have also damaged about 60% of the lodgepole pines in the U.S. state of Colorado, which has been experiencing warmer winters that are more favorable to the beetles. Large numbers of dry and dead trees can increase the risk of severe forest fires that would further threaten biodiversity.

Because of such damage to trees in combination with drier conditions, forest fires might become more frequent and intense in some areas such as the southeastern and western United States. This would severely degrade some forest ecosystems, add more CO_2 to the atmosphere, reduce total CO_2 uptake by trees and plants, and help accelerate projected climate disruption through still another positive feedback loop.

Food Production Could Face an Overall Decline

Farmers will face dramatic changes due to shifting climates and a faster hydrologic cycle, if the earth warms as projected (Figure 19-C). In a warmer world, agricultural productivity will drop in some areas. This decrease is likely to be worse in areas where farmers depend on monocultures of plants bred to thrive at certain temperatures and with certain levels of rainfall.

Some scientists say that climate change is already affecting food production in certain areas. In Russia, the 2010 heat wave and drought destroyed one-third of the country's wheat crop. In the United States, the 2012 corn harvest was down 13% from 2011, even though the total area planted in corn was the largest since the 1930s, according to the U.S. Department of Agriculture. Some of these losses were related to intense heat that scientists in two 2012 studies argued was highly unlikely without the human influence on climate change.

On the other hand, crop productivity will likely increase in some areas in the short run. According to the IPCC, crop productivity is projected to increase slightly at middle to high latitudes in areas such as midwestern Canada, Russia, and Ukraine, with moderate warming, but decrease if warming goes too far. However, this effect may be limited because soils generally are more lacking in plant nutrients in these northern regions.

Figure 19-18 Declining populations of Adélie penguins in Antarctica have been linked to declines in floating sea ice and phytoplankton because of atmospheric warming.

Keith Szafranski/iStockphoto.com

Climate change models predict a decline in agricultural productivity in tropical and subtropical regions, especially in Southeast Asia and Central America. Also, flooding of river deltas due to rising sea levels could reduce crop production, partly because aquifers that supply irrigation water will be infiltrated by saltwater. Fish production in coastal aquaculture ponds could also be reduced.

Areas especially vulnerable to these effects will be the populous nations of Bangladesh, Egypt, and Vietnam, and parts of the African coast, according to a 2012 World Bank report. Food production could also drop in farm regions that are dependent on rivers fed by glaciers that are slowly disappearing, and in any arid and semiarid areas where droughts become more prolonged. Scientists warn that by 2050, some 200–600 million of the world's poorest and most vulnerable people could face starvation and malnutrition due to these projected effects of climate disruption.

Climate Disruption Threatens Human Health and Economies

According to IPCC and other reports, more frequent and prolonged heat waves in some areas will raise the numbers of deaths and illnesses, especially among older people, people in poor health, and the urban poor who cannot afford air conditioning. Fewer people will die from cold weather, but a 2007 study led by Mercedes Medina-Ramon of the Harvard University School of Public Health suggests that the projected rise in the number of heat-related deaths will exceed the projected drop in the number of cold-related deaths.

A 2012 World Bank report notes that climate change will likely hinder the fight against childhood mortality and malnutrition in many areas. While economic growth has helped areas of sub-Saharan Africa and South Asia in this regard, even the warming that now seems unavoidable is likely to reverse such gains, and warming by up to 4°C (7.2°F) could make these problems far worse than they are today.

A warmer, CO_2-rich world will likely favor rapidly multiplying insects, microbes, toxic molds, and fungi that make us sick, as well as plants that produce pollens that cause allergies and asthma attacks. Longer and more intense pollen seasons will mean more itchy eyes, runny noses, and asthma attacks. Insect pests and weeds will likely multiply, spread, and reduce crop yields.

Microbes that cause infectious tropical diseases such as dengue fever and yellow fever are likely to expand their ranges and numbers if mosquitoes that carry them spread to warmer temperate and higher-elevation areas, as they have begun to do (Figure 19-20). Latin America and other regions are seeing a rise in the incidence of dengue fever. The *Economist* reported in 2007 that Mexico's number of cases had grown fourfold since 2001.

Higher atmospheric temperatures and higher levels of water vapor in urban areas will also increase some forms of air pollution. The greatest effect will be to speed up the rate of the chemical reactions that produce ozone and other harmful chemicals in photochemical smog in urban areas (see Figure 18-9, p. 483). This in turn will likely cause more pollution-related deaths from heart ailments and respiratory problems.

Climate disruption is also likely to take a toll on human economies. Economist Nicholas Stern has estimated that the cost of projected climate change, if current trends continue, would be up to 20% per year of the world's economic activity as measured by gross domestic output. Other researchers say the cost could be higher, because most models used to calculate such costs have not accounted for ocean acidification, loss of glaciers, sea-level rise, or biodiversity losses.

For these reasons, projected climate disruption will very likely force many millions of people to move, in search of better conditions. Environmental scientist Norman Myers estimates that projected climate change during this century could produce at least 150 million, and perhaps 250 million, environmental refugees who will be forced to migrate because of greater hunger, flooding, and drought. (See the online Guest Essay on this topic by Myers.)

Considering the speed with which these and other effects of climate change are occurring, some scientists and analysts now argue that the IPCC and other climate science groups have been too conservative in their warnings of climate disruption (Science Focus 19.2).

Figure 19-19 With warmer winters, exploding populations of mountain pine beetles have munched their way through large areas of lodgepole pine forest in the Canadian province of British Columbia. In this photo, the orange-colored trees are those that are dead or dying—killed by these beetles.

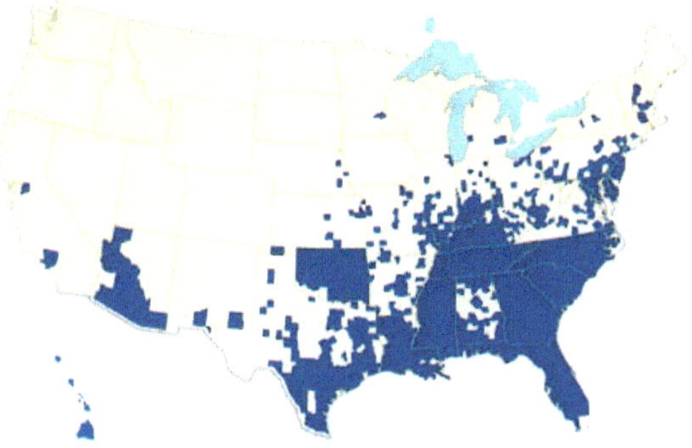

Figure 19-20 Areas in blue show counties in 28 U.S. states where one or both species of mosquitoes that transmit dengue fever had been found as of 2005. This supports the claim by some scientists that projected climate disruption will expand the range of mosquitoes that carry dengue fever, a debilitating and potentially deadly disease.
(Compiled by the authors using data from Natural Resources Defense Council, 2009.)

19-4 What Can We Do to Slow Projected Climate Disruption?

CONCEPT 19-4
We can reduce greenhouse gas emissions and the threat of climate disruption while saving money and improving human health if we cut energy waste and rely more on cleaner renewable energy resources.

Dealing with Climate Disruption Is Difficult

Many scientists argue that our most urgent priority is *to do all we can to avoid any and all estimated climate change tipping points*. Once we reach such a point, there is no going back. The problem is that we do not know how close we are to such tipping points, although there are no indications that we will cross any of them in the immediate future. However, it is estimated that if greenhouse gas emissions continue to rise, we could cross some of these thresholds within a few decades or by the end of this century. Figure 19-21 lists the tipping points that scientists are most concerned about.

Partly because of these possibilities, many analysts believe that addressing climate change is one of the most urgent scientific, political, economic, and ethical issues that humanity faces. But the following characteristics of this complex problem make it difficult to tackle:

- *The problem is global.* Dealing with this threat will require unprecedented and prolonged international cooperation.
- *The problem is a long-term political issue.* Voters and elected officials generally respond better to short-term problems than to long-term threats. Most of the people who could suffer the most serious harm from projected climate disruption during the latter half of this century have not been born.
- *The projected harmful and beneficial impacts of climate disruption are not spread evenly.* For example, higher-latitude nations such as Canada, Russia, and New Zealand could have higher crop yields, fewer deaths in winter, and lower heating bills. Other nations such as Bangladesh could see more flooding and higher death tolls.

- Atmospheric carbon level of 450 ppm
- Melting of all arctic summer sea ice
- Collapse and melting of the Greenland ice sheet
- Severe ocean acidification, collapse of phytoplankton populations, and a sharp drop in the ability of the oceans to absorb CO_2
- Massive release of methane from thawing arctic permafrost and from the arctic seafloor
- Collapse and melting of most of the western Antarctic ice sheet
- Severe shrinkage or collapse of Amazon rain forest

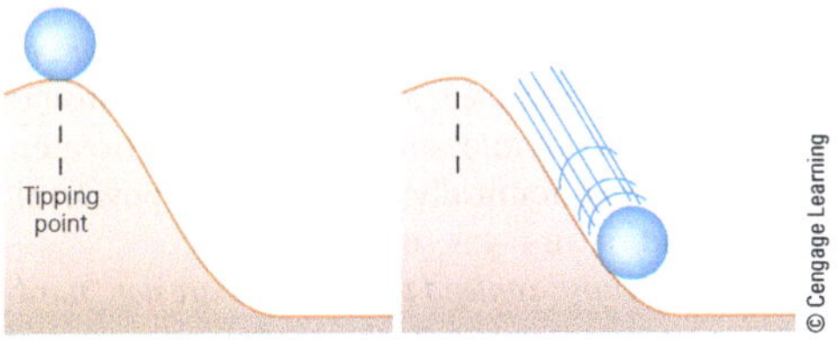

Figure 19-21 Climate scientists have come up with this list of possible climate change tipping points.

SCIENCE FOCUS 19.2

HAVE CLIMATE SCIENTISTS BEEN TOO CONSERVATIVE?

Some people who are skeptical about climate change have accused the IPCC scientists of being alarmists who are overstating the possible effects of atmospheric warming. However, a growing number of scientists are suggesting the opposite—that the IPCC has been too conservative in its projections.

One scientist who was once quite skeptical about climate change has become convinced not only that it is occurring but also that the IPCC projections are too conservative. Physicist Richard A. Muller founded the Berkeley Earth Surface Temperature project to monitor atmospheric temperatures. In 2011, he and his colleagues found that the average temperature over the earth's land had risen higher and faster than the IPCC had projected. Using 5 times the amount of temperature data available to the IPCC in 2007 along with newer climate models and statistical analysis tools, the team also concluded it is likely that "essentially all of this increase results from the human emission of greenhouse gases." The IPCC had concluded only that most of this warming was due to human activities.

Similarly, according to models used by scientists at the Massachusetts Institute of Technology and other institutions, the trend in atmospheric temperatures will likely take us to an average increase of 3.5 to 7.4C° (6.3 to 13.3F°) by 2100, whereas the IPCC in 2007 projected a rise of 2.4 to 6.4C° (4.3 to 11.5F°) (Figure 19-B). Other IPCC projections, compared with updated estimates produced by newer research and model runs, include the following:

- The IPCC projected that the Arctic would not be ice-free in summer until 2070; current projections estimate the ice-free summer could occur by around 2030.
- In 2007, the IPCC projected that in a worst-case scenario, humans would be emitting 30 billion tons of carbon per year by 2100. In fact, we have already exceeded that total, emitting about 31.6 billion tons in 2012, according to the International Energy Agency.
- The losses of ice in Greenland (**Core Case Study**) and Antarctica are at levels that the IPCC projected would be occurring about 100 years from now.
- Permafrost is melting faster in areas of Alaska and Siberia than models used by IPCC scientists had projected.
- IPCC projections on the rise in sea level, based mostly on expansion of the oceans' waters, estimated a rise of up to 0.6 meter (1.9 feet) by 2100. Current models that include land-based ice melt project that the sea-level rise will range between 0.7 and 1.9 meters (2.4 to 6.2 feet) by 2100.
- A 2011 study led by University of York researcher Chris Thomas found that plants and animals were migrating to higher elevations twice as fast as projected and that others were moving north 3 times as fast. University of Exeter scientists found in 2011 that extinction rates were more than twice as high as computer models had projected.

No model is perfect (Science Focus 19.1), and there are a number of reasons why the IPCC underestimated some effects of climate change. Generally, they have to do with unforeseen developments such as faster-than-expected melting of Greenland's ice and the sharp rise in emissions from coal-burning facilities, especially in China. Also, some scientists criticize the IPCC for not including a few important factors in their modeling, including the melting of arctic permafrost and the dynamics of melting glaciers.

Other scientists defend the IPCC, saying that science must necessarily be a painstaking process governed not by public opinion about dramatic events but by scientific principles. In other words, science is and must be conservative by its nature, according to these defenders.

Some critics argue that because of this fundamentally conservative approach to climate science, the consequences of climate change could be more severe because the public might not gain an important sense of urgency from the IPCC reports. Important decisions could be delayed until it is too late to succeed in slowing climate disruption.

Critical Thinking

Do you think that the need for scientists to be conservative in reporting their findings outweighs any need to urgently warn the public about possible threats indicated by their research? Explain.

- *Many proposed solutions, such as sharply reducing or phasing out the use of fossil fuels, are controversial* because they could disrupt economies and lifestyles and threaten the profits of economically and politically powerful oil, coal, and natural gas companies.
- *Some social scientists contend that humans are not "hard-wired" to respond to long-term threats.* Scientists and educators are frustrated by how difficult it is to convince the public that projected climate disruption could pose such a great threat to many people living on earth today, and even more so to their children, grandchildren, and great grandchildren.

For several years, many environmental analysts have been calling for urgent action at the national and international levels to curb greenhouse gas emissions, primarily from the top down, by regulating and taxing such emissions. Now a growing number of analysts are contending that this approach, by itself, will not work because, *first*, psychological research indicates that using fear and guilt and promoting sacrifice to change behavior rarely works. *Second*, people are primarily interested in the short-term, not long-term, benefits of changing their behavior. *Third*, for elected officials, spending their efforts on long-term problems is most often not in their best short-term political interests in getting reelected.

What Are Our Options?

There are two basic approaches to dealing with the projected harmful effects of global climate disruption. One, called *mitigation*, is to act to slow it and to avoid climate change tipping points. The other approach, called *adaptation*, is to recognize that some climate change is unavoidable and to try to reduce some of its harmful effects. (We discuss this option in Section 19-5.)

Regardless of the approach we take, many scientists and business leaders see the climate change problem as an urgent challenge, but also as one that is filled with economic opportunity. In 2009, a group of investors who collectively manage more than $13 trillion of the world's financial assets warned that climate disruption poses a threat to the global economy, but they projected that the challenge of making the transition to a low-carbon world presents an enormous global investment opportunity. For example, in the United States, electric cars are being made in Michigan, wind generators are being produced in Illinois, Oregon, and Iowa, and Texas produces more electricity from wind power than all but five of the world's countries.

We now look more specifically at mitigation options and at the opportunities they present.

We Could Control and Prevent Greenhouse Gas Emissions

One way to measure CO_2 emissions is by their global total weight. According to the International Energy Agency, the human population in 2012 emitted about 31.6 billion metric tons, or *gigatons*, of carbon into the atmosphere. Climate scientists have generally agreed that in order to avoid some of the severely harmful effects of climate change, we need to limit the global average temperature increase to 2°C (3.6°F) over the preindustrial global average temperature. To do that, we would have to limit our total carbon emissions to 565 gigatons between now and 2050, according to the latest climate models. However, at the current rate of emissions (which is on the rise), we will easily surpass that total by around 2030.

The problem is that the world's nations and energy companies all together hold reserves of fossil fuels that represent 2,795 gigatons—nearly 5 times what scientists say we can safely emit—according to the Carbon Tracker Initiative, a team of financial and environmental analysts. Currently, the survival of the politically powerful fossil fuel companies and most of the world's national economies depend on using all or most of these fossil fuel reserves. This is one of the major forces pushing climate change.

Between 2000 and 2008, CO_2 emissions in the United States leveled off, and between 2008 and 2012, they declined. This was due to the combination of an economic slowdown, improved fuel efficiency for new cars, and the partial shift from burning coal to using natural gas to produce the country's electricity. At the same time, CO_2 emissions in China were rising rapidly and by 2010 were twice as high as those in the United States. Analysts argue that unless China, by far the world's largest CO_2 emitter since 2007, can sharply cut its emissions, primarily by shifting from coal to a mixture of natural gas and low-carbon renewable energy sources, the world will likely experience a sharp rise in the global atmospheric temperature during this century.

Figure 19-22 lists a number of ways to slow the rate and degree of atmospheric warming and the resulting climate disruption caused by human activities (**Concept 19-4**). Among these solutions are four major *prevention* strategies that, by 2050, could reduce human greenhouse gas emissions by 57–83%, according to a 2010 study by the U.S. National Academy of Sciences:

1. Improve energy efficiency to reduce fossil fuel use, especially the use of coal.
2. Shift from carbon-based fossil fuels to low-carbon renewable energy resources based on local and regional availability.
3. Stop cutting down tropical forests and plant trees to help remove more CO_2 from the atmosphere.
4. Shift to more sustainable and climate-friendly agriculture (Figure 12-33, p. 309).

CONSIDER THIS. . .

THINKING ABOUT Climate Disruption Prevention Strategies

How are each of the four strategies listed above related to the three scientific **principles of sustainability** (see Figure 1-2, p. 6 or back cover)?

The solutions listed on the right side of Figure 19-22 are *output*, or *cleanup*, strategies, which focus on dealing with CO_2 after it has been produced. Many analysts argue that, rather than spending huge amounts of time and resources on such strategies, we should focus on reducing and preventing CO_2 emissions, as soon as possible.

In addition, some scientists urge us to ramp up efforts to reduce emissions of other greenhouse gases. For example, methane (CH_4) is emitted from landfills, rice paddies, some coal mines, older natural-gas pipelines, and the overall production of natural gas. In 2010, atmospheric scientists Veerabhadran Ramanathan and David Victor estimated that, just by collecting CH_4 from landfills, minimizing the rotting of rice in paddies, and replacing old leaky natural-gas pipelines, we could cut global CH_4 emissions by 40%.

Another climate-changing pollutant is soot generated by inefficient coal-burning power plants, diesel engines, brick-making kilns, and cook stoves, among many other sources. Soot is accumulating on Himalayan Mountain glaciers and arctic ice fields and contributing to the melting of this ice and to atmospheric warming. Ramanathan and Victor report that in China and India, programs to make power plants more efficient, to filter soot from diesel engines, to cut emissions from kilns, and to replace leaky stoves with efficient ones (see Figure 18-28, p. 496) could cut soot levels by almost two-thirds.

Some scientists argue that we could quickly slow atmospheric warming by focusing on reducing black carbon (a component of soot), methane, and hydrofluorocarbons (HFCs), partly because these chemicals are short-lived in the atmosphere, compared to CO_2 emissions, and because we have the technologies to accomplish such cuts fairly quickly. According to a 2011 study by the UNEP, major cuts in these greenhouse gas emissions could potentially reduce the rate of atmospheric warming in half by 2050, while also preventing up to 4 million deaths from air pollution and billions of dollars in crop losses every year.

In earlier chapters, we looked at several ways to prevent or reduce the release of greenhouse gases to the atmosphere in order to slow projected climate disruption. These have involved energy efficiency and the use of renewable energy (Chapter 16), sharply reducing tropical deforestation (Chapter 10), and shifting to more sustainable agriculture (Chapter 12).

Solutions

Slowing Climate Disruption

Prevention	Cleanup
Cut fossil fuel use (especially coal)	Sequester CO_2 by planting trees and preserving forests and wetlands
Shift from coal to natural gas	Sequester carbon in soil using biochar
Repair leaky natural gas pipelines and facilities	Sequester CO_2 deep underground (with no leaks allowed)
Improve energy efficiency	Sequester CO_2 in the deep ocean (with no leaks allowed)
Shift to renewable energy resources	Remove CO_2 from smokestack and vehicle emissions
Reduce deforestation	
Use more sustainable agriculture and forestry	
Put a price on greenhouse gas emissions	

Figure 19-22 Some ways to slow atmospheric warming and the resulting projected climate disruption during this century (Concept 19-3). ***Questions:*** Which five of these solutions do you think are the best ones? Why?

Top: Mark Smith/Shutterstock.com. Center: Yegor Korzh/Shutterstock.com. Bottom: ©pedrosala/Shutterstock.com.

We Could Try to Clean Up Greenhouse Gas Emissions

Analysts have proposed several output solutions—strategies for removing some of the CO_2 from the atmosphere or from smokestacks and storing (sequestering) it in other parts of the environment (Figure 19-22, right).

One way to increase the uptake of CO_2 is by quickly implementing a massive, global tree-planting program, especially on degraded land in the tropics. A similar approach is to restore wetlands that have been drained for farming. Wetlands are very efficient at taking up CO_2.

A third cleanup approach is to plant large areas of degraded land with fast-growing perennial plants such as switchgrass (see Figure 16-27, p. 426), which remove CO_2 from the air and store it in the soil. To be sustainable, however, this approach could not involve clearing existing forests.

Along these lines, a fourth carbon sequestration measure that is getting more attention is the use of *biochar*. Scientists have rediscovered this fertilizer that Amazon Basin natives used for growing crops several centuries ago. Producing biochar involves burning biomass such as chicken waste or wood in a low-oxygen environment to make a charcoal-like material—a process that produces no smoke and no odor. The resulting carbon-rich biochar makes an excellent organic fertilizer that helps to keep carbon in the soil for as long as 1,000 years. While making biochar does release CO_2, burying it stores considerably more CO_2 in the soil. Farmers could plant fast-growing trees on degraded land and use them to make biochar. They could then make money selling some of it as fertilizer.

A fifth cleanup approach is to help the natural uptake and storage of carbon by preserving and restoring natural forests. In 2007, biologist Renton Righelato and climate scientist Dominick Spracklen concluded that the amount of carbon that we could sequester by protecting and restoring forests is greater than the amount of carbon in CO_2 emissions that we would avoid by using biofuels such as biodiesel and ethanol (see Chapter 16 Case Studies, p. 425).

Another approach that has received a lot of attention is to remove CO_2 from the smokestacks of coal-burning power and industrial plants and to store it somewhere. This approach is generally referred to as **carbon capture and storage**, or **CCS** (Science Focus 19.3).

CONSIDER THIS. . .

CONNECTIONS **Sea Creatures, Carbon Dioxide, and Cement**

In 2010, earth science professor Brent Constantz developed a process for removing CO_2 from the smokestack emissions of a power plant and converting it to cement by spraying it with mineral-rich seawater. He developed this CCS method by mimicking sea creatures that convert CO_2 to calcium carbonate to form their skeletons and shells. He envisions converting CO_2 emissions into stone and locking them away essentially forever in the concrete foundations of our cities.

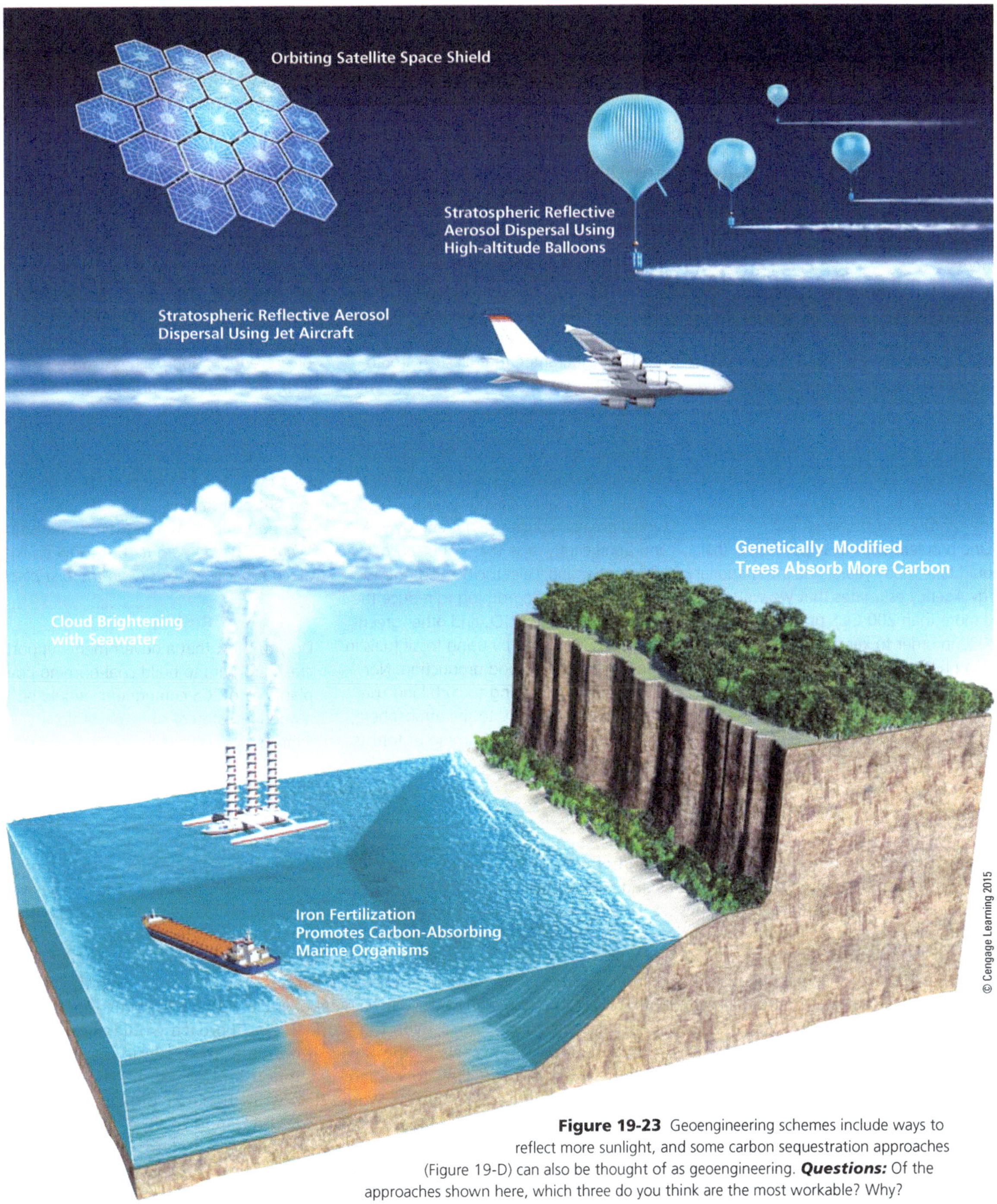

Figure 19-23 Geoengineering schemes include ways to reflect more sunlight, and some carbon sequestration approaches (Figure 19-D) can also be thought of as geoengineering. ***Questions:*** Of the approaches shown here, which three do you think are the most workable? Why?

Some Promote Geoengineering Schemes to Ward Off Climate Change

Many scientists and engineers want to use strategies that fall under the umbrella of *geoengineering,* or trying to manipulate certain natural conditions to help counter an enhanced greenhouse effect (Figure 19-23). In recent years, some of these scientists and engineers have become discouraged by the slow response of governments to what they see as the global emergency of projected climate dis-

SCIENCE FOCUS 19.3

IS CAPTURING AND STORING CO_2 THE ANSWER?

Figure 19-D shows several techniques for removing some of the CO_2 from the atmosphere and from smokestacks, and storing (sequestering) it in other parts of the environment.

Some researchers are studying ways to remove CO_2 from the exhaust coming out of smokestacks at coal-burning power and industrial plants. They hope to pump it deep underground into abandoned coal beds and oil and gas fields or to liquefy it and inject it into sediments under the seafloor (Figure 19-D). Another idea is to turn the collected CO_2 into charcoal or into carbonate rock. These schemes are generally referred to as *carbon capture and storage (CCS).*

One problem with this approach is that it would take many years. The International Energy Agency estimates that we would need more than 200 CCS power plants by 2030 in order to keep atmospheric warming below 3°C, and currently, no commercial-scale CCS plant has been built.

Also, it is becoming clear to scientists that stored CO_2 would have to remain sealed from the atmosphere forever. Any large-scale leaks caused by earthquakes or other shocks, as well as any number of smaller continuous leaks from CO_2 storage sites around the world, could dramatically increase atmospheric warming and climate disruption in a very short time. Some scientists doubt that we can develop the technology to guarantee that all such stored CO_2 would remain safely sequestered without some of it leaking out.

There are other serious problems. Removing all or most of the CO_2 from smokestack emissions and transporting and storing it takes a lot of energy. This would greatly reduce the net energy yield and significantly raise the cost of burning coal to supply electricity. Also, these measures would do nothing to reduce the massive amounts of CO_2 and other greenhouse gases emitted by using fossil fuels in transportation and food production. Nor would they do anything to curb land-use practices that add CO_2 to the atmosphere, such as the cutting and burning of forests. Most worrisome of all to some scientists, CCS could promote more use of coal, the world's most environmentally harmful fuel.

Environmental scientists including Peter Montague argue that CCS is a costly and extremely risky *output* solution to a serious problem that can be dealt with by using a variety of cheaper, quicker, and safer *input*, or *prevention*, approaches (Figure 19-22, left). They argue for producing the electricity we need by using low-carbon energy resources such as energy efficiency, solar cells, and wind power. To these scientists, when we face a problem such as CO_2 coming out of a smokestack or exhaust pipe, the most important question to ask is not *What do we do with it?* but, *How do we avoid producing the CO_2 in the first place?*

Critical Thinking

Do you think that a government-supported crash program to build coal-burning power plants with CCS components would be a good investment of taxpayer dollars? Explain.

---

ruption with its potentially serious harmful effects. They suggest that we look at the implications and costs of using large-scale geoengineering schemes as a last resort, if humanity fails to deal with the world's projected climate change emergency soon enough.

For example, some scientists have studied the 1991 eruption of Mount Pinatubo in the Philippines and the fact that the average global atmospheric temperature cooled for about 15 months after the eruption. The volcano pumped massive amounts of SO_2 into the atmosphere where chemical reactions converted it to sulfate particles, which were thought to have reflected some incoming sunlight into space, thus cooling the troposphere. These scientists have suggested using balloons, large jet planes, or giant cannons to inject sulfate particles into the stratosphere to get the same effect (Figure 19-23). Other scientists have called for placing a series of giant mirrors in orbit above the earth to reflect incoming sunlight for the same purpose.

Some scientists reject the idea of launching sulfates into the stratosphere as being too risky because of our limited knowledge about possible unknown effects. For example, if the sulfates reflected too much sunlight, they could reduce evaporation enough to alter the water cycle and worsen the already dangerous droughts in Asia and Africa. Also, a 2008 study by atmospheric scientist Simone Tilmes indicated that chlorine released by reactions involved in this scheme could speed up the thinning of the earth's vital ozone layer (see Chapter 18, p. 497).

Even if it worked, this approach would allow CO_2 levels in the lower atmosphere to continue rising and adding to climate disruption and ocean acidification. Some scientists would deal with the latter problem by building a global network of thousands of chemical plants that would remove acid from seawater to reduce its acidity. But this would be quite costly and also could have unpredictable and possibly harmful ecological effects on ocean ecosystems.

CONSIDER THIS. . .

THINKING ABOUT Tinkering with the Stratosphere

Do you think the projected benefits of injecting large quantities of sulfate particles into the stratosphere to help cool the troposphere would outweigh any projected negative results? Explain.

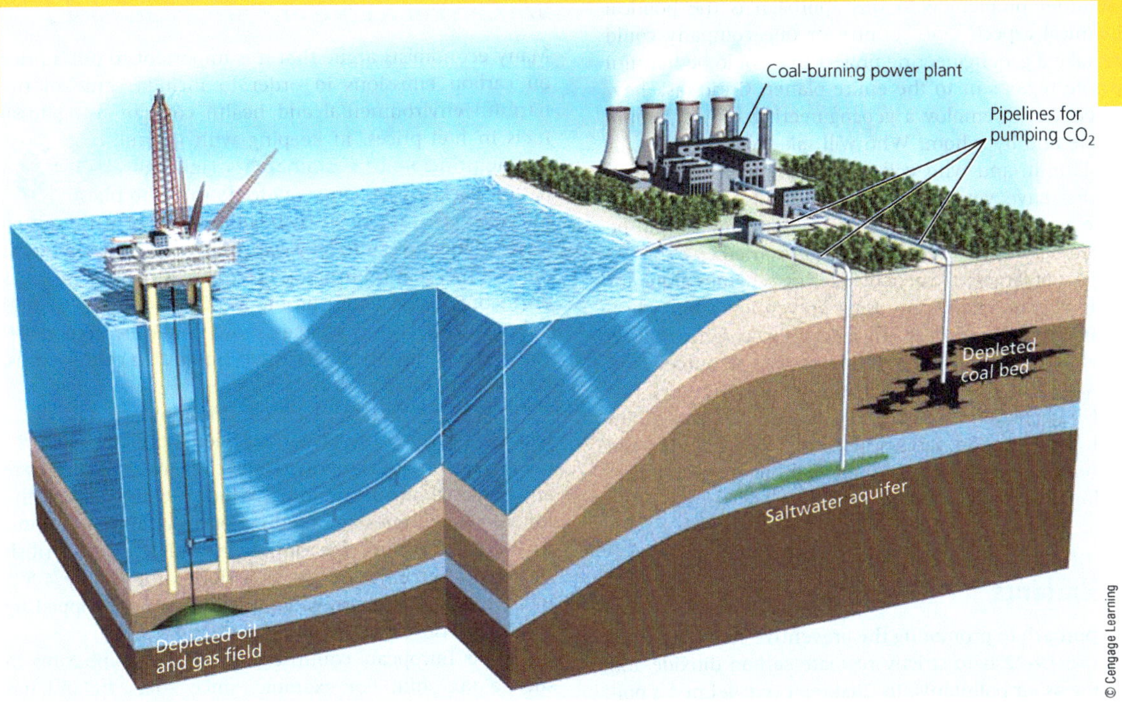

Figure 19-D Solutions: Several proposed methods for removing carbon dioxide from the atmosphere or from smokestacks and storing it in soil, plants, and deep underground reservoirs, as well as in sediments beneath the ocean floor. These proposed solutions are based on an *output approach*. ***Questions:*** Which two of these solutions do you think would work best? Which two would be the least effective? Why?

Another geoengineering proposal involves seeding the oceans with iron to promote the growth of more marine algae and other phytoplankton, which absorb huge amounts of CO_2 from the atmosphere as they grow (Figure 19-23). When they die, they sink to the seafloor, carrying the carbon with them. The proposal is controversial because the long-term effects of adding large amounts of iron to ocean ecosystems are unknown.

In a 2009 study involving an Atlantic Ocean test site seeded with iron, it was found that shrimp-like animals called copepods had quickly devoured the large swarm of algae produced by the seeding. If this were to happen regularly, more carbon would move up the food chain instead of sinking to the ocean floor, and eventually, it would be released by large organisms as CO_2, thus adding to the CO_2 problem. And according to a 2013 study by marine scientist Daniel Harrison, the amount of CO_2 removed from the atmosphere by this method is largely countered by the amount added to the atmosphere when fossil fuels are burned to produce the iron salts and to power the ships used in the seeding process. Thus, the net effect of this process is unknown.

Scientist James Lovelock has proposed another geoengineering idea. It would involve anchoring large vertical pipes in the oceans as part of a system that would allow wave motion to pump nutrient-rich water up from the deep-ocean layer to fertilize algae on the ocean surface. Lovelock contends that the resulting algal blooms would remove CO_2 from the atmosphere and emit dimethyl sulfide (see Figure 3-20, p. 69), which would contribute to the formation of low clouds that would reflect sunlight.

According to skeptical scientists, one major problem with most of these technological fixes and with some carbon capture and storage schemes (Science Focus 19.3) is that they require huge investments of energy and materials, and there is no guarantee that they will work. Geoengineering schemes all depend on complex machinery running constantly, flawlessly, and essentially forever, primarily to pump something from one place to another in the environment. If we rely on these systems and continue emitting greenhouse gases, and if the systems then fail, atmospheric temperatures will likely soar at a rapid rate and essentially ensure severe climate disruption.

Another problem with this approach is the political and ethical aspect. One country or one company could undertake a geoengineering project that could be harmful to whole regions or to the entire planet. Critics ask how any decision to employ a geoengineering project should be made and by whom. Who will take responsibility for controlling it, and who will pay for any damages? Such questions have yet to be answered, and critics such as climate scientist Ken Caldeira argue that they should be explored before any major projects get started.

Critics of large-scale geoengineering schemes argue for slowing climate change by using prevention approaches, such as improving energy efficiency, replacing fossil fuels with already available renewable energy resources, and drastically reducing tropical deforestation. They say this would be a better investment of limited resources than would gambling on large-scale, costly changes to the global environment that could have unknown and potentially long-lasting harmful effects.

Greenhouse Gases Can Be Regulated as Pollutants

One approach to promoting the preventive solutions listed in Figure 19-22 is to strictly regulate carbon dioxide and methane as air pollutants. In Chapter 1, we defined a pollutant as a chemical or other agent such as noise or heat in the environment that proves harmful to the health, survival, or activities of humans or other organisms.

Carbon dioxide is now classified as an air pollutant, but you might wonder how CO_2 can possibly be classified as such, because it is a natural and important part of the carbon cycle (see Figure 3-17, p. 66). Humans and most animal life-forms constantly emit CO_2 into the atmosphere, while plants remove it and use it for photosynthesis. However, one characteristic of any chemical that determines whether or not it should be classified as a pollutant is its *concentration* in the atmosphere, soil, water, or plant or animal tissues. For example, if we were to fill up a room with CO_2, everyone there would die very quickly from lack of oxygen. When we add CO_2 to the atmosphere faster than the carbon cycle can remove it, we help to warm the atmosphere and contribute to projected climate disruption. At some point in that process (at some atmospheric concentration), CO_2 becomes an air pollutant.

This is why, in 2009, the U.S. Environmental Protection Agency found that CO_2, CH_4, N_2O, and three other greenhouse gases are a danger to public health and welfare and should be listed as pollutants under the Clean Air Act. However, politically and economically powerful coal, oil, natural gas, and utility companies that make their money by producing and burning carbon-containing fossil fuels have successfully opposed regulating carbon dioxide for more than three decades and hope to overturn or greatly weaken any government efforts to regulate emissions of carbon dioxide and other greenhouse gases.

We Can Put a Price on Carbon Emissions

Many economists argue that it is important to put a price on carbon emissions in order to include some of the harmful environmental and health costs of using fossil fuels in fuel prices, in keeping with the full-cost pricing **principle of sustainability** (see Figure 1-5, p. 9 or back cover). One way to do this is to phase in *carbon taxes* or *fees* on each unit of CO_2 or CH_4 emitted by fossil fuel use, or to levy *energy taxes* or *fees* on each unit of fossil fuel that is burned (Figure 19-24). Lowering taxes on income, wages, and profits to offset such taxes could help to make such a strategy more politically acceptable, and it might also stimulate the economy. In other words, *tax pollution, not payrolls and profits*.

GOOD NEWS

In British Columbia, Canada, this approach has had good results. The province levied a tax of $30 per metric ton of CO_2 emitted, and then lowered corporate and personal income tax rates, which are now among the lowest of such rates among the world's wealthier nations. In the 4 years after the carbon tax was created, British Columbia's greenhouse gas emissions dropped by 4.5% while its population grew, and gasoline sales dropped by 2% while rising by 5% overall in Canada.

Some European countries are gradually phasing in such a tax shift. For example, since 2008, Ireland has taxed carbon dioxide emissions and used the revenue to help reduce the country's staggering debt and by 2012 had seen a 15% drop in its CO_2 emissions. However, economists warn that the biggest global players in climate change—the United States, Europe, and China—would need to tax carbon emissions at basically the same rates in order for any carbon tax to be effective. A 2012 study by the Congressional Budget Office estimated that between 2012 and 2022, a modest carbon tax could generate about $1.25 trillion in U.S. tax revenues.

Trade-Offs

Carbon and Energy Taxes

Advantages	Disadvantages
Simple to administer	Tax laws can get complex
Clear price on carbon	Vulnerable to loopholes
Covers all emitters	Doesn't guarantee lower emissions
Predictable revenues	Politically unpopular

Figure 19-24 Using carbon and energy taxes or fees to help reduce greenhouse gas emissions has advantages and disadvantages. ***Question:*** Which two advantages and which two disadvantages do you think are the most important and why?

Trade-Offs

Cap-and-Trade Policies

Advantages	Disadvantages
Clear legal limit on emissions	Revenues not predictable
Rewards cuts in emissions	Vulnerable to cheating
Record of success	Rich polluters can keep polluting
Low expense for consumers	Puts variable price on carbon

Figure 19-25 Using a cap-and-trade policy to help reduce greenhouse gas emissions has advantages and disadvantages. ***Question:*** Which two advantages and which two disadvantages do you think are the most important and why?

Another way to include some costs of carbon emissions in market prices is to place a cap on total human-generated CO_2 and CH_4 emissions in a country or region, issue permits to emit these pollutants, and then let polluters trade their permits in the marketplace. This *cap-and-trade* or *pollution reduction approach* (Figure 19-25) has a political advantage over taxes, but it is difficult to manage because there are so many different emitters of greenhouse gases. In addition, this approach works only if the original caps are set low enough to encourage a serious reduction in greenhouse gas emissions and are lowered on a regular basis to promote further cuts. A related and more politically acceptable approach is to use a *cap-and-rebate* scheme in which all money made by any trading of permits would go back to the public in the form of cash payments.

Environmental economists argue that, regardless of whether governments use taxes or a cap-and-trade or cap-and-rebate policy, the most critical goal is to put a price on carbon emissions to get all emitters to come as close as possible to paying the total estimated harmful environmental and health costs of their carbon emissions. Economist Gary Yohe and others contend that the resulting higher costs for fossil fuels will spur innovation in finding ways to reduce carbon emissions, improve energy efficiency, and phase in a mix of low-carbon renewable energy alternatives.

Government Subsidies Can Be Used to Address Climate Disruption

Another way in which governments can address climate disruption is to level the economic playing field by greatly increasing *government subsidies* (direct payments and other forms of financial support) to businesses and individuals to encourage their use of energy-efficiency technologies, low-carbon renewable energy sources, and more sustainable food production methods. This would also include phasing out or sharply reducing subsidies and tax breaks that encourage the use of fossil fuels and nuclear power, unsustainable agriculture, and the clearing of forests. In other words, during the next 2 to 3 decades, we could shift from environmentally degrading to environmentally sustaining subsidies and tax breaks.

One important form of government support is funding of research and development, a great deal of which has gone to the fossil fuel and nuclear energy industries in decades past. Governments could shift this research and development focus to clean energy industries. With such help, these industries could develop innovations that would lower the costs of energy resources such as wind, solar power, and geothermal energy so that they could compete more favorably in the marketplace with fossil fuels, which have been heavily subsidized for decades.

Governments Can Cooperate Internationally

Governments have entered into international climate negotiations. In December 1997, delegates from 161 nations met in Kyoto, Japan, to negotiate a treaty to slow atmospheric warming and its projected climate disruption. The first phase of the resulting *Kyoto Protocol* went into effect in February 2005 with 187 of the world's 194 countries (not including the United States) ratifying the agreement by late 2009.

This agreement required the 36 participating more-developed countries to cut their emissions of CO_2, CH_4, and N_2O to certain levels by 2012, when the treaty was to expire. Less-developed countries were excluded from this requirement, because such reductions would curb their economic growth. In 2005, participating countries began negotiating a second phase of the treaty, which was supposed to go into effect after 2012. That year, at the annual UN climate conference, held in Doha, Qatar, participating nations agreed to keep the Kyoto Protocol in effect until at least 2015 and to commit to more intensive but unspecified actions to reduce greenhouse gases, which were to be explored in 2013.

Some analysts see current and proposed international climate agreements as weak and slow responses to an urgent global problem, pointing out that atmospheric carbon dioxide levels have increased 35% faster than was expected under the Kyoto agreement. Many have also concluded that a top-down international agreement will be too slow and ineffective. Some call for the nations that produce most of the world's greenhouse gas emissions—especially the United States, China, India, Brazil, and the European Union nations (Figure 19-9)—to get together and work out an agreement to sharply lower their emissions.

Another way in which governments can cooperate is through *technology transfer*. Governments of more-developed countries could help to fund the transfer of the latest green technologies to less-developed countries so that they could bypass older, energy-inefficient and polluting technologies.

Some analysts argue that helping poorer countries to deal with the harmful effects of projected climate disruption makes sense, because many of these countries will suffer the most from such effects, which have been caused mostly by more-developed countries. They also point out that even if all countries do not suffer the worst effects of climate disruption, the whole world will eventually suffer very harmful effects due to rampant widespread poverty and homelessness, mass migrations of environmental refugees, and sharply decreased international security. Environmental analyst Lester Brown has suggested that increasing the current tax on each international currency transaction by a quarter of a penny could finance effective technology transfer.

Another important form of international cooperation involves efforts to protect large forests, especially tropical forests, which are important players in the carbon cycle. The effects of deforestation account for 12–17% of global greenhouse gas emissions, according to the Environmental Policy Institute. One such effort is the UN's Reducing Emissions from Deforestation and Forest Degradation (REDD) program, which helps to fund efforts by farmers and others who depend on forests and who seek to use them more sustainably. Since it was founded in 2008, the program has helped 44 countries in Africa, Asia, and Latin America to reduce deforestation. Brazil has slowed deforestation by 40% since 2008 with the help of the REDD program, and is projected to reach its reduction goal of 80% by 2020. GOOD NEWS

Some Countries, States, and Localities Are Leading the Way

Some nations are leading others in facing the challenges of projected climate disruption. Costa Rica aims to be the first country to become *carbon neutral* by cutting its net carbon emissions to zero by 2030. The country generates 78% of its electricity with renewable hydroelectric power and another 18% from renewable wind and geothermal energy.

Some analysts are urging rapidly developing countries such as China and India to shift toward more sustainable, low-carbon economic development by leapfrogging over the traditional forms of economic development that have led to the threat of climate disruption. If China and India continue their extensive coal burning and other high-carbon activities, the world will not be able to avoid the projected harmful effects of climate disruption.

While China emits more CO_2 into the atmosphere than any other country, it also has one of the world's most intensive energy efficiency programs, according to ClimateWorks, a global energy consulting firm. Chinese automobile fuel-efficiency standards are higher than U.S. standards, but enforcement of the standards is still weak. China's government is working with the country's top 1,000 industries

scottjay/iStockphoto.com

Figure 19-26 Portland, Oregon, has reduced car use by developing an efficient and widely used light-rail mass transit system. For more than 40 years, Portland has been consistently near the top of U.S. green city lists.

to implement tough energy efficiency goals. China is also rapidly becoming the world's leader in developing and selling solar cells, solar water heaters, wind turbines, advanced batteries, and plug-in hybrid electric cars.

By 2010, at least 30 U.S. states had set goals for reducing greenhouse gas emissions. California plans to get 33% of its electricity from low-carbon renewable energy sources by 2030. Since 1990, local governments in more than 650 cities around the world (including more than 450 U.S. cities) have established programs to reduce their greenhouse gas emissions.

The first major U.S. city to do this was Portland, Oregon. Between 1993 and 2012, the city cut its per capita greenhouse gas emissions below 1990 levels by 26%—more than any other U.S. city. Portland promotes energy-efficient buildings and the use of electricity from wind and solar sources. It has also built a network of bicycle trails and has greatly expanded its mass transit system (Figure 19-26). In 2009, the city implemented a Climate Action Plan with the goal of cutting its greenhouse gas emissions to 40% below 1990 levels by 2030 and to 80% below those levels by 2050. Far from hurting Portland's economy, these strategies have produced an economic boom and have saved the city $2 million a year on its energy bills.

Some Companies Are Reducing Their Carbon Footprints

Leaders of some of the most prominent U.S. companies, including Alcoa, DuPont, Ford Motor Company, General Electric, Johnson & Johnson, PepsiCo, and Shell Oil, have joined with leading environmental organizations to form the U.S. Climate Action Partnership. They have called on the government to enact strong national climate change legislation, realizing that significantly reducing their own greenhouse gas emissions will save them money.

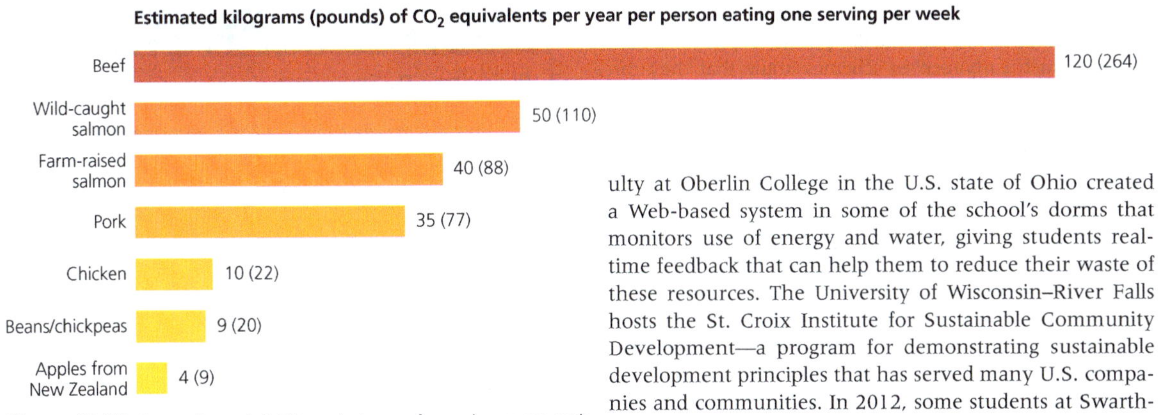

Figure 19-27 Comparison of diet items in terms of greenhouse gas emissions per year by a person eating one serving per week, in CO_2 equivalents. (Compiled by the authors using data from the Swedish Environmental Protection Agency.)

For example, between 1990 and 2006, DuPont slashed its energy usage and cut its greenhouse gas emissions by 72% and saved $3 billion, while the company grew its business by almost a third.

Leaders of these and many other major companies see an enormous profit opportunity in developing or using energy-efficient and cleaner technologies. For example, the production of building materials is energy-intensive and produces 10–12% of all U.S. CO_2 emissions. Companies such as the brick maker Calstar Products have found ways to produce building materials with much lower carbon outputs. Calstar recycles fly ash from coal-burning facilities and it has reduced the carbon footprint of brick-making by 80%. Many companies now recognize that "there is gold in going green."

Colleges and Universities Are Going Green

Many colleges and universities are also taking action. The Tempe campus of Arizona State University (ASU) boasts the largest collection of solar panels of any American university. ASU was the first university in the country to establish an entire academic program with its School of Sustainability.

The College of the Atlantic in Bar Harbor, Maine (USA) has been carbon neutral since 2007. It gets all of its electricity from renewable hydropower, and many of its buildings are heated with the use of renewable wood pellets. Some students there built a wind turbine that powers a nearby organic farm, which offers organic produce to the campus, to local schools, and to food banks.

EARTH University in Costa Rica has a mission to promote sustainable development in tropical countries. Its sustainable agriculture degree program has attracted students from more than 20 different nations.

The number of examples such as these is large and growing. Here are just a few more: Students and faculty at Oberlin College in the U.S. state of Ohio created a Web-based system in some of the school's dorms that monitors use of energy and water, giving students real-time feedback that can help them to reduce their waste of these resources. The University of Wisconsin–River Falls hosts the St. Croix Institute for Sustainable Community Development—a program for demonstrating sustainable development principles that has served many U.S. companies and communities. In 2012, some students at Swarthmore College asked their administrators to sell the college endowment's holdings in large fossil fuel companies, and students on dozens of other campuses have followed their lead by making similar requests. (See the Campus Sustainability features that appear in several other chapters of this book.)

Every Individual Choice Makes a Difference

Each of us will play a part in the projected acceleration of atmospheric warming and climate disruption during this century. Whenever we use energy generated by fossil fuels, for example, we add a certain amount of CO_2 to the atmosphere. Each use of energy adds to an individual's *carbon footprint*, the amount of CO_2 generated directly and indirectly by any one person or group. You can calculate your own carbon footprint by using a footprint calculator, several of which are available online.

One important aspect of your carbon footprint is your diet. Foods vary greatly in the greenhouse gases that result from their production and delivery (Figure 19-27). For example, processed foods require much more energy to produce than do whole foods such as fresh fruit and vegetables. Meat production, especially as it is done on factory farms, involves far higher greenhouse gas emissions than production of grains and vegetables. By choosing foods carefully, you can reduce your carbon footprint.

Another major factor in the average carbon footprint is nonfood purchases of common goods such as clothing and electronic devices. The Union of Concerned Scientists has estimated that such purchases represent 26% of the average American's carbon footprint, noting that this is an area where there is a great opportunity for cutting carbon emissions. There are many sources of information on the Internet and elsewhere comparing the carbon emissions involved with the production of various goods and services.

Some websites suggest ways for you to reduce, or *offset*, some of your carbon dioxide emissions by helping to fund or participate in projects such as forest restoration and renewable energy development. However, buyers need to do their homework, because some carbon-offset programs are not real and do not actually produce emissions cuts. Critics of such carbon-offset schemes argue that they can encourage people to continue producing greenhouse gases instead of making changes to reduce their carbon footprints by, for example, driving less, using mass transit, or consuming less electricity.

Figure 19-28 lists some ways in which you can cut your CO_2 emissions. One person taking each of these steps makes a small contribution to reducing global greenhouse gas emissions, but when millions of people take such steps, the cumulative result is quite large. In a 2009 study, human ecologist Thomas Dietz estimated that if each American took steps such as those listed in Figure 19-28, the cumulative impact of such individual actions could cut U.S. emissions of CO_2 by more than 7% within a decade.

What Can You Do?

Reducing CO_2 Emissions

- Calculate your carbon footprint (there are several helpful websites)
- Drive a fuel-efficient car, walk, bike, carpool, and use mass transit
- Reduce garbage by reducing consumption, recycling, and reusing more items
- Use energy-efficient appliances and compact fluorescent or LED lightbulbs
- Wash clothes in warm or cold water and hang them up to dry
- Close window curtains to keep heat in or out
- Use a low-flow showerhead
- Eat less meat or no meat
- Heavily insulate your house and seal all air leaks
- Use energy-efficient windows
- Set your hot-water heater to 49°C (120°F)
- Plant trees
- Buy from businesses working to reduce their emissions

Figure 19-28 Individuals matter: You can reduce your annual emissions of CO_2. ***Question:*** Which of these steps, if any, do you take now or plan to take in the future?

19-5 How Can We Adapt to Climate Change?

CONCEPT 19-5

While we can prepare for some climate change that is now inevitable, we could realize important economic, ecological, and health benefits by drastically reducing greenhouse gas emissions with the goal of slowing projected climate disruption.

We Can Prepare for Climate Disruption

According to global climate models, the world will have to make a 50–85% cut in emissions of greenhouse gases by 2050 to stabilize concentrations of these gases in the atmosphere. This would help to prevent the planet from heating up by more than 2C° (3.6F°) and to head off rapid changes in the world's climate and the projected harmful effects that would result.

However, because of the difficulty of making such large reductions, many analysts believe that while we work to slash emissions, we should also begin to prepare for the likely harmful effects of projected climate disruption. Figure 19-29 shows some ways to do so.

For example, relief organizations such as the International Red Cross are turning their attention to projects such as expanding mangrove forests as buffers against storm surges, building shelters on high ground, and planting trees on slopes to help prevent landslides. In the face of projected higher levels of precipitation and rising sea levels, seawall design and construction will likely be a major growth industry. And low-lying countries such as Bangladesh are trying to figure out what to do with millions of environmental refugees who will be flooded out by rising sea levels and more intense storms. Some cities plan to establish cooling centers to shelter residents during increasingly intense heat waves.

Several cities have developed adaptation plans. In London, England, a flood-control barrier is being strengthened at a cost of about $1 billion (Figure 19-30). Other cities, including the U.S. cities of New York City and Seattle, Washington, and several states have also developed adaptation plans. California is beefing up its forest firefighting capabilities and is proposing desalination plants to help relieve projected water shortages, which will worsen as some of its mountain glaciers melt. In the aftermath of flooding caused by Hurricane Sandy in 2012, New York City is considering new floodwalls and flood gates for its subway system. In anticipation of rising sea levels, Boston has elevated one of its sewage treatment plants.

Some people fear that emphasizing these adaptation approaches will distract us from the more urgent need to reduce greenhouse gas emissions. However, to some analysts, projected climate change is already such a serious threat that we have no alternative but to implement both prevention and adaptation strategies, and we have no time to lose in doing this on an emergency basis.

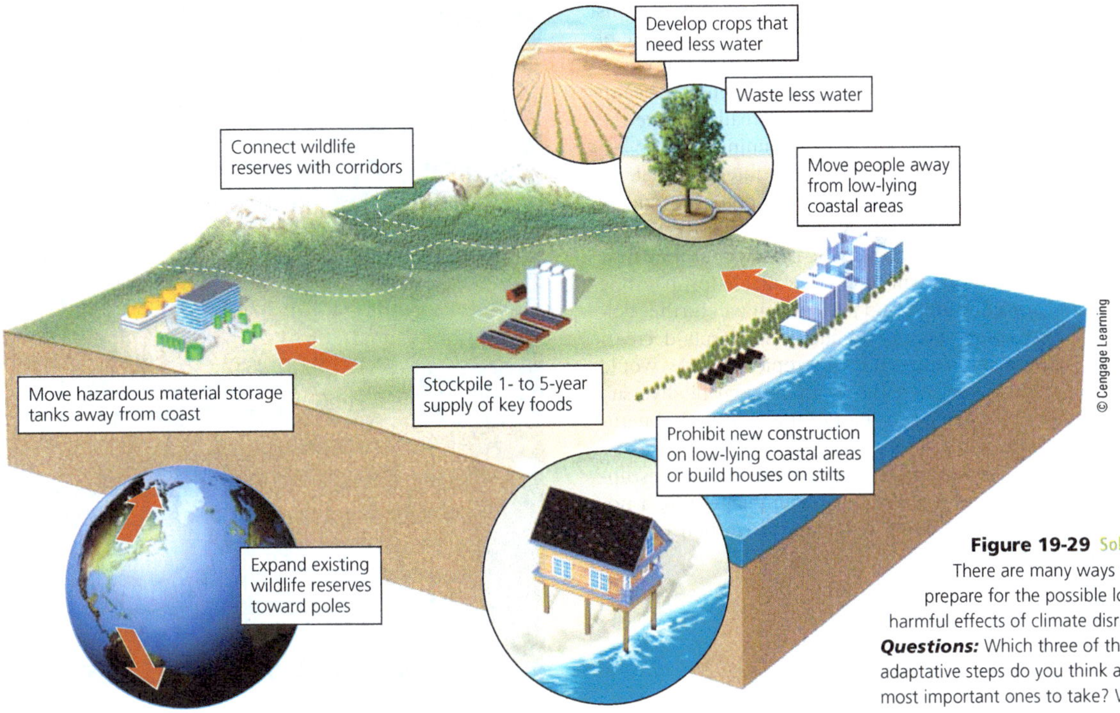

Figure 19-29 Solutions: There are many ways for us to prepare for the possible long-term harmful effects of climate disruption. ***Questions:*** Which three of these adaptative steps do you think are the most important ones to take? Why?

A No-Regrets Strategy

The threat of climate disruption pushes us to look for preventive solutions such as those listed in Figure 19-22. But suppose we had the ability to look into the future and, in doing so, we found that the climate models were wrong and that climate disruption was not a serious threat after all. Should we then return to the present and abandon the search for preventive solutions?

A number of climate and environmental scientists, economists, and business leaders say *no* to this question, and they call for us to begin implementing such changes now as a *no-regrets strategy.* They argue that actions such as those listed in Figure 19-22 should be implemented in any case because they will lead to very important environmental, health, and economic benefits.

For example, improving energy efficiency has numerous economic and environmental advantages (see Figure 16-4, p. 404). Individuals, companies, and local governments can save large amounts of money and improve their economic security by cutting their energy bills. In addition, the use of fossil fuels, especially coal, causes a

Figure 19-30 This high-tech, movable flood-control barrier—the second largest in the world—was built across the Thames River to prevent flooding from very high tides and storm surges in London, England.

great deal of air pollution, which harms or kills millions of people every year. Thus, a sharp decrease in the use of these fuels will improve the overall health of the population, save lives, and save money now spent on health care. In addition, cutting coal use will greatly reduce the land disruption resulting from surface mining.

Another benefit of the no-regrets strategy is that, by relying more on a mix of renewable, domestic energy resources, many countries will cut their costly dependence on imports of fossil fuels. The money that they now spend for fuel imports will then be available for investing in health care, education, and jobs, which will further bolster their economic security. In addition, sharply decreasing or halting the destruction of tropical forests would help to preserve the increasingly threatened biodiversity that makes up an important part of the earth's vital natural capital (see Figure 1-3, p. 7).

Thus, leaving out any concerns about climate disruption, there are plenty of reasons for implementing a no-regrets strategy for sharply reducing our carbon outputs on an urgent basis.

Big Ideas

- Considerable scientific evidence indicates that the earth's atmosphere is warming, mostly because of human activities, and that this is likely to lead to significant climate disruption during this century that could have severe and long-lasting harmful consequences.
- Reducing the projected harmful effects of rapid climate disruption during this century requires emergency action to increase energy efficiency, sharply reduce greenhouse gas emissions, and rely more on renewable energy resources.
- While we can prepare for some climate change that is now inevitable, we could realize important economic, ecological, and health benefits by drastically reducing greenhouse gas emissions with the goal of slowing climate change.

TYING IT ALL TOGETHER Melting Ice in Greenland and Sustainability

Jan Martin Will/Shutterstock.com

In this chapter, we have seen that human activities such as the burning of fossil fuels and the widespread clearing and burning of forests for planting crops have contributed to higher levels of greenhouse gases in the atmosphere. This has contributed to increased atmospheric warming, which in turn is projected to result in climate disruption during this century.

The effects of climate disruption include rapid, increased melting of land-based ice (**Core Case Study**) and sea ice, worsening drought, rising sea levels, ocean acidification, declining biodiversity, and severe threats to human health and economies. Many of these effects could further accelerate climate disruption in worsening spirals of change known as positive feedback loops.

We can apply the six **principles of sustainability** (see Figure 1-2, p. 6, Figure 1-5, p. 9, or back cover) to help reduce the harmful effects of projected climate disruption. We can reduce inputs of pollutants and greenhouse gases into the atmosphere by relying more on direct and indirect forms of solar energy than on fossil fuels; recycling and reusing matter resources much more widely than we do now; and mimicking biodiversity by using a variety of low-carbon renewable energy resources in place of fossil fuels and nuclear power. We can advance toward these goals by including the harmful environmental and health costs of fossil fuel use in market prices; seeking win-win solutions that will benefit both the economy and the environment; and giving high priority to passing along to future generations an environment in which they too can thrive.

Chapter Review

Core Case Study

1. Summarize the story of Greenland's melting glaciers (**Core Case Study**) and the possible effects of this process. Explain how it fits into the IPCC projections about climate change.

Section 19-1

2. What is the key concept for this section? Explain why weather and climate are not the same. Summarize the trends in atmospheric warming and cooling over the past 900,000 years and during the last century. How do scientists get information about past temperatures and climates? List eight pieces of evidence that indicate that the atmosphere has been warming since 1975.

Section 19-2

3. What is the key concept for this section? What is the greenhouse effect and why is it so important to life on the earth? What are greenhouse gases and how do they affect the atmosphere? What are three key greenhouse gases and how do they differ in atmospheric lifetimes and warming potentials? What is the role of water vapor in the atmosphere and how does it relate to other greenhouse gases in the warming process?

4. How have human activities affected atmospheric greenhouse gas levels during the last 275 years and especially in the last 30 years? List the major human activities that add CO_2, CH_4, and N_2O to the atmosphere. How do scientists use models to make projections about future temperature changes? What do a number of climate models project about temperature changes during this century? Explain how each of the following might contribute to projected atmospheric warming and resulting global climate disruption: **(a)** a hotter sun, **(b)** the oceans, **(c)** cloud cover, and **(d)** air pollution.

Section 19-3

5. What is the key concept for this section? What makes the current atmospheric warming problem urgent? Define **climate change tipping points** and give three examples. Briefly describe the possible effects of a warmer atmosphere on arctic sea ice and land-based ice, permafrost, and sea levels. Define **ocean acidification** and explain how and why it is occurring. Explain how atmospheric warming, global drought, extreme weather, and an intensified water cycle are related. What are the possible effects of climate disruption on biodiversity, food production, and human health and economies? Pick three of these effects and describe how each can become part of a positive feedback loop leading to increased climate disruption. Summarize the science community's discussion of whether or not the IPCC scientists were too conservative in their projections.

Section 19-4

6. What is the key concept for this section? What are five factors that make it difficult to deal with the problem of projected climate disruption? What are two basic approaches to dealing with the projected harmful effects of global climate disruption? List four major prevention strategies for reducing greenhouse gas emissions. Why do some scientists argue for focusing on reducing greenhouse gases other than carbon dioxide? List seven strategies for trying to clean up carbon emissions. What is **carbon capture and storage (CCS)**? Describe three problems associated with capturing and storing carbon dioxide emissions. What are geoengineering schemes (give two examples) and what are two major problems with most of them?

7. Summarize the argument for regulating greenhouse gases as pollutants. List two ways to put a price on carbon emissions. What are the advantages and disadvantages of using taxes on carbon emissions or energy use to help reduce greenhouse gas emissions? What is cap and trade and what are the advantages and disadvantages of using it to help reduce greenhouse gas emissions? What is the Kyoto Protocol and how has it progressed? What are three ways in which governments can cooperate internationally to deal with projected climate change?

8. Give two examples of what some countries are doing on their own to deal with climate change. Give two examples of what some companies are doing and two examples of what some colleges and universities have done to reduce their carbon footprints. List five ways in which you can reduce your carbon footprint.

Section 19-5

9. What is the key concept for this section? List five ways in which we can prepare for the possible long-term harmful effects of climate disruption. Describe the no-regrets strategy for dealing with the problems associated with energy waste and fossil fuel use, regardless of climate change.

10. What are this chapter's *three big ideas*? Explain how we can apply the six **principles of sustainability** (see Figure 1-2, p. 6, Figure 1-5, p. 9, or back cover) to the problem of projected climate disruption.

Note: Key terms are in bold type.

Critical Thinking

1. If you had convincing evidence that at least half of Greenland's glaciers (**Core Case Study**) were sure to melt during this century, would you argue for taking serious actions now to slow projected climate disruption? If so, summarize the arguments you would use. If not, explain why you would be opposed to such actions.
2. A top U.S. economic adviser to a president once gave a speech in Williamsburg, Virginia (USA), to representatives of governments from a number of countries. He told his audience not to worry about atmospheric warming because the average global temperature increases predicted by scientists were much less than the temperature increase he had experienced that day in traveling from Washington, DC, to nearby Williamsburg. What was the flaw in his reasoning? Write an argument that you could use to counter his claim.
3. China relies on coal for two-thirds of its commercial energy usage, partly because the country has abundant supplies of this resource. As a result, China's use of coal is playing a large role in the warming of the global atmosphere. Do you think China is justified in expanding its use of this resource as other countries, including the United States, have done with their coal resources? Explain. What are China's alternatives?
4. Explain why you would support or oppose each of the strategies listed in Figure 19-22 for slowing projected climate disruption caused by atmospheric warming.
5. Suppose someone tells you that carbon dioxide (CO_2) should not be classified as an air pollutant because it is a natural chemical that is part of the carbon cycle (see Figure 3-17, p. 66) and a chemical that we add to the atmosphere every time we exhale. Explain the faulty reasoning in this statement about CO_2.
6. One way to slow the rate of CO_2 emissions is to reduce the clearing of forests—especially in less-developed tropical countries where intense deforestation is taking place. Should the United States and other more-developed countries pay poorer countries to stop cutting their forests? Explain.
7. Some scientists have suggested that, in order to help cool the warming atmosphere, we could annually inject huge quantities of sulfate particles into the stratosphere. This might have the effect of reflecting some incoming sunlight back into space. Explain why you would support or oppose this geoengineering scheme.
8. What are three consumption patterns or other aspects of your lifestyle that directly add greenhouse gases to the atmosphere? Which, if any, of these habits would you be willing to give up in order to help slow projected climate disruption? Write up an argument that you could use to convince others to do the same. Compare your answers to those of your classmates.

Doing Environmental Science

Gather data on the average annual temperature and average annual precipitation for the area where you live over the last 30 years. Try to find data for as many of these years as possible, and plot these data to determine whether the average temperature and precipitation during this period has increased, decreased, or stayed about the same. Write a report summarizing your search for data, your results, and your conclusions.

Global Environment Watch Exercise

Within the GREENR database, use the World Map feature. Under "Browse" select *Climate Change* and click on the pin for Greenland. Find the latest information on the melting of glaciers in Greenland (**Core Case Study**). What is the rate of melting of the glaciers and how might this affect sea levels during this century? Briefly summarize the evidence used by scientists to support their statements about melting ice in Greenland. Summarize any information you find about ongoing and future studies of Greenland's ice.

Ecological Footprint Analysis

According to the International Energy Agency, the average American adds 19.6 metric tons (21.6 tons) of CO_2 per year to the atmosphere, compared with a world average of 4.23 metric tons (4.65 tons). The table on the next page is designed to help you understand the sources of your personal inputs of CO_2 into the atmosphere. You will be making calculations to fill in the blanks in this table.

Some typical numbers are provided in the "Typical Quantity per Year" column of the table. However, your calculations will be more accurate if you can use information based on your own personal lifestyle, which you can enter in the blank "Personal Quantity per Year" column. For example, you could add up your monthly utility bills for a year and divide the total by the number of persons in your household to get a rough estimate of your own utility use.

	Units per Year	Typical Quantity per Year	Personal Quantity per Year	Multiplier	Emissions per Year (lb. CO_2)
Residential Utilities					
Electricity	kwh	4,500		1.5	
Heating oil	gallons	37		22	
Natural gas	hundreds of cubic feet (ccf)	400		12	
Propane	gallons	8		13	
Coal	tons	—		4,200	
Transportation					
Automobiles	gallons	600		19	
Air travel	miles	2,000		0.6	
Bus, urban	miles	12		0.07	
Bus, intercity	miles	0		0.2	
Rail or subway	miles	28		0.6	
Taxi or limousine	miles	2		1	
Other motor fuel	gallons	9		22	
Household Waste					
Trash	pounds	780		0.75	
Recycled items	pounds	337		−2	
				Total (pounds)	
				Total (tons)	
				Total (kilograms)	
				Total (metric tons)	

Source: Thomas B. Cobb at Ohio's Bowling Green State University developed this CO_2 calculator.

1. Calculate your carbon footprint. To calculate your emissions, first complete the blank "Personal Quantity per Year" column as described above. Wherever you cannot provide personal data, use that listed in the "Typical Quantity per Year" column. Enter each number in the "Personal Quantity per Year" column to represent your annual consumption (using the units specified in the "Units per Year" column). Now multiply your annual consumption for each activity by the associated number in the "Multiplier" column to obtain an estimate of the pounds of CO_2 resulting from that activity, which you will enter in the "Emissions per Year" column. Finally, add the numbers in that column to find your carbon footprint, and express the final CO_2 result in both pounds and tons (1 ton = 2,000 pounds) and in kilograms and metric tons (1 kilogram = 2.2 pounds; 1 metric ton = 1.1 tons).
2. Compare your emissions with those of your classmates and with the per capita U.S. average of 19.6 metric tons (21.6 tons) of CO_2 per person per year. Actually, your answer should be considerably less—roughly about half the per capita value—because this computation only accounts for direct emissions. For instance, CO_2 resulting from driving a car is included, but the CO_2 emitted in manufacturing and disposing of the car is not. You can find more complete carbon footprint calculators at several sites on the Web.
3. Consider and list actions you could take to reduce your carbon footprint by 20%.

20 Water Pollution

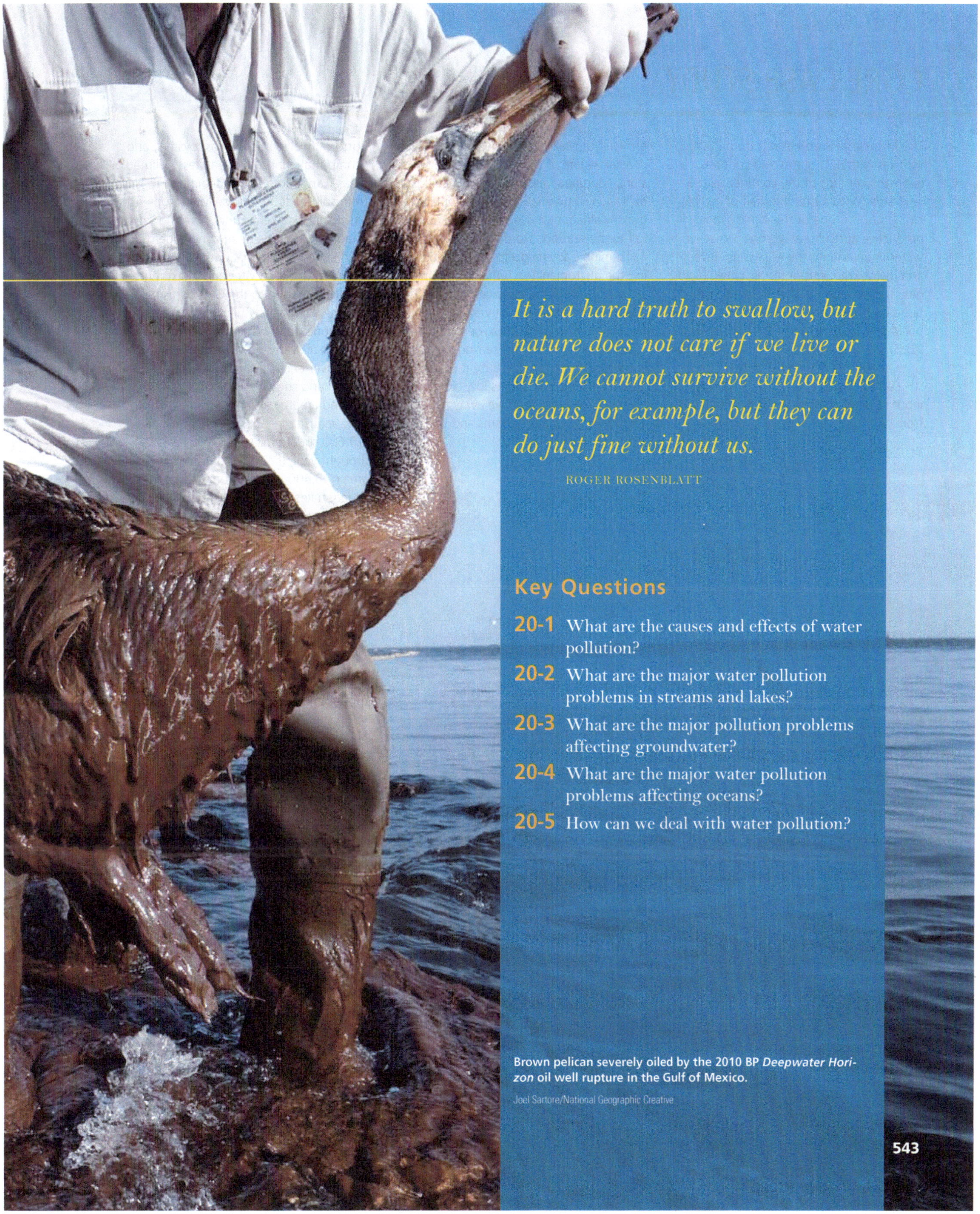

It is a hard truth to swallow, but nature does not care if we live or die. We cannot survive without the oceans, for example, but they can do just fine without us.

ROGER ROSENBLATT

Key Questions

20-1 What are the causes and effects of water pollution?

20-2 What are the major water pollution problems in streams and lakes?

20-3 What are the major pollution problems affecting groundwater?

20-4 What are the major water pollution problems affecting oceans?

20-5 How can we deal with water pollution?

Brown pelican severely oiled by the 2010 BP *Deepwater Horizon* oil well rupture in the Gulf of Mexico.

Joel Sartore/National Geographic Creative

CORE CASE STUDY The Gulf of Mexico's Annual Dead Zone

The Mississippi River basin (Figure 20-1, top) lies within 31 states and contains almost two-thirds of the continental U.S. land area. With more than half of all U.S. croplands, it is one of the world's most productive agricultural regions. However, water draining into the Mississippi River and its tributaries from farms, cities, factories, and sewage treatment plants in this huge basin contains sediments and other pollutants that end up in the Gulf of Mexico (Figure 20-1, bottom)—a major supplier of the country's fish and shellfish.

Each spring and summer, this huge input of plant nutrients, mostly nitrates from crop fertilizers, enters the northern Gulf of Mexico and overfertilizes the coastal waters of the U.S. states of Mississippi, Louisiana, and Texas. One result is an explosion of populations of phytoplankton (mostly algae) that eventually die and fall to the seafloor. Hordes of oxygen-consuming bacteria go to work decomposing the phytoplankton remains and, in the process, deplete the dissolved oxygen in the Gulf's bottom layer of water.

The huge volume of *oxygen-depleted water* resulting from this seasonal event is called a *dead zone* because it contains little animal marine life. Its low oxygen levels (called *hypoxia*) drive away faster-swimming marine organisms and suffocate bottom-dwelling fish, crabs, oysters, and shrimp that cannot move to less polluted areas. Large amounts of sediment, mostly from soil eroded from the Mississippi River basin, can also kill bottom-dwelling forms of aquatic life. The dead zone appears each spring and grows until fall when storms churn the water and redistribute dissolved oxygen to the Gulf bottom.

The size of the Gulf of Mexico's annual dead zone depends primarily on the amount of water flowing into the Mississippi River each year. In years with ample rainfall and snowmelt, such as 2003, it has covered an area as large as the state of Massachusetts—27,300 square kilometers (10,600 square miles). In 2012, a severe drought year, it covered a much smaller area of 6,400 square kilometers (2,500 square miles)—about the size of the state of Delaware.

The annual Gulf of Mexico dead zone is one of about 400 oxygen-depleted zones scattered around the world. The largest of these zones lies in the Baltic Sea to the south of Sweden in Northern Europe. Others have appeared in coastal waters of Australia, New Zealand, Japan, China, and South America.

Oxygen-depleted zones represent a disruption of the nitrogen cycle (Figure 3-18, p. 67) caused primarily by human activities. This is because huge quantities of nitrogen from nitrate fertilizers are added to ecosystems such as the Mississippi River and Gulf of Mexico much faster than the nitrogen cycle can remove them. As a result, by producing crops for food and ethanol fuel in the vast Mississippi basin, we end up disrupting coastal aquatic life and seafood production in the Gulf of Mexico.

This overfertilization of coastal waters is one of the world's many forms of water pollution. Some good news is that we know how to reduce the sizes of dead zones and how to reduce other forms of water pollution that we explore in this chapter.

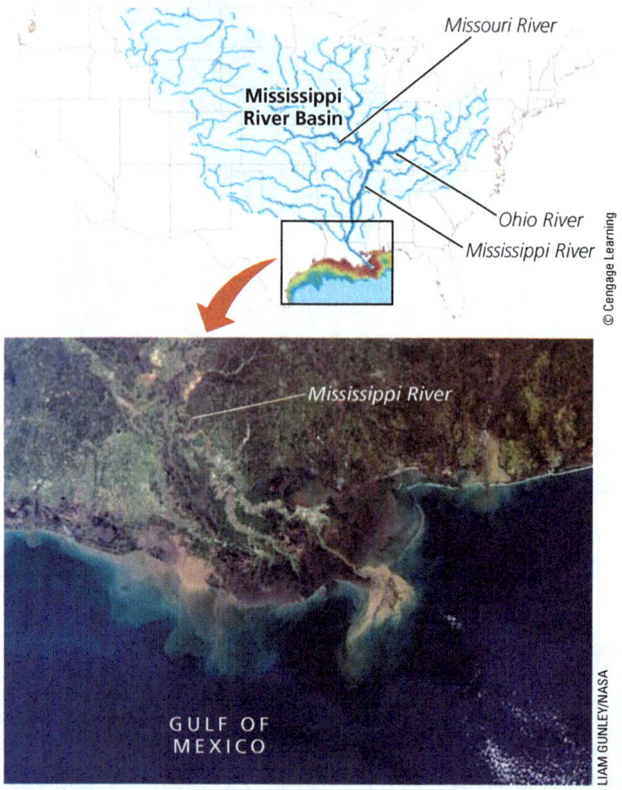

Figure 20-1 Water containing sediments, dissolved nitrate fertilizers, and other pollutants drains from the Mississippi River basin (top) into the Mississippi River and from there into the northern Gulf of Mexico (bottom). The boxed area at the bottom of the map shows the dead zone with the red area having the lowest oxygen level.

20-1 What Are the Causes and Effects of Water Pollution?

CONCEPT 20-1A
Water pollution causes illness and death in humans and other species, and disrupts ecosystems.

CONCEPT 20-1B
The chief sources of water pollution are agricultural activities, industrial facilities, and mining, but the growth of both the human population and our rate of resource use makes it increasingly worse.

Water Pollution Comes from Point and Nonpoint Sources

Water pollution is any change in water quality that can harm living organisms or make the water unfit for human uses such as drinking, irrigation, and recreation. It can come from a single (point) source, or from a larger, more dispersed (nonpoint) source. **Point sources** discharge pollutants into bodies of surface or underground water at specific locations through drain pipes (Figure 20-2), ditches, or sewer lines. Examples include factories, sewage treatment plants (which remove some, but not all, pollutants), underground mines (Figure 14-15, p. 360), and oil tankers.

Because point sources are located at specific places, they are fairly easy to identify, monitor, and regulate. Most of the world's more-developed countries have laws that help control point-source discharges of harmful chemicals and disease organisms into aquatic systems. However, in most of the less-developed countries, there is little control of such discharges.

Nonpoint sources are broad and diffuse areas where rainfall or snowmelt washes pollutants off the land into bodies of surface water. Examples include runoff of eroded soil and chemicals such as fertilizers and pesticides from cropland, livestock feedlots, logged forests (Figure 10-12, p. 224), urban streets, parking lots, lawns, and golf courses. We have made little progress in controlling water pollution from nonpoint sources because of the difficulty and expense of identifying and controlling discharges from so many diffuse sources. According to the U.S. Environmental Protection Agency (EPA), nonpoint source pollution is the main reason why 40% of all U.S. rivers, lakes, and estuaries are still not clean enough for uses such as fishing and swimming, despite the enactment of major water pollution control laws more than 35 years ago.

Agricultural activities are by far the leading cause of water pollution. Sediment eroded from agricultural lands (Figure 20-3) is the most common pollutant. Other major agricultural pollutants include fertilizers and pesticides, bacteria from livestock and food-processing wastes, and excess salts from the soils of irrigated cropland. *Industrial facilities*, which emit a variety of harmful inorganic and organic chemicals, are a second major source of water pollution.

CONSIDER THIS. . .

CONNECTIONS Cleaning Up the Air and Polluting the Water

Stricter air pollution control laws (see Chapter 18, p. 493) have forced coal-burning power plants in more-developed countries to remove many of the harmful gases and particles from their smokestack emissions. But this results in hazardous ash, which is typically placed in slurry ponds (see Figure 15-18, p. 388) that can rupture and cause water pollution. Thus, the problem of coal ash illustrates one way in which air and water pollution are connected.

Mining is the third biggest source of water pollution (Figure 20-4). Surface mining disturbs the land (see Figure 14-11, p. 358; Figure 14-12, p. 359; and 14-14, p. 359), creating major erosion of sediments and runoff of toxic chemicals into surface waters.

Another form of water pollution is caused by the widespread use of human-made materials such as plastics that are used to make millions of products. Much of the plastic that is improperly discarded eventually ends up in waterways and in the oceans (Figure 20-5). Such discarded plastic products can harm various forms of wildlife (see Figure 11-4, p. 253).

CONSIDER THIS. . .

CONNECTIONS Atmospheric Warming and Water Pollution

Projected climate change from atmospheric warming will likely contribute to water pollution in some areas of the globe. In a warmer world, some regions

Age Fotostock/SuperStock

Figure 20-2 This is a point source of water pollution in Gargas, France.

Figure 20-3 Nonpoint sediment from farmland runoff flows into streams and sometimes changes their courses or dams them up. As measured by weight, it is the largest source of water pollution. ***Question:*** What do you think the owner of this farm could have done to prevent such sediment pollution?

Figure 20-4 Water pollution flows from the wastes of the abandoned Climax Mine, a molybdenum mining site in Colorado that is one of the major U.S. waste sites yet to be cleaned up.

will get more precipitation and other areas will get less. More intense downpours will flush more harmful chemicals, plant nutrients, and disease-causing microorganisms into some waterways, while prolonged drought will reduce river flows that dilute wastes in other areas.

Major Water Pollutants Have Harmful Effects

Table 20-1 lists the major types of water pollutants along with examples of each and their harmful effects and sources (**Concept 20-1A**).

One of the major water pollution problems that we face is exposure to infectious disease organisms (pathogens) primarily through contaminated drinking water. Scientists have identified more than 500 types of disease-causing bacteria, viruses, and parasites that can be transferred into water from the wastes of humans and animals. They can get into drinking water and cause painful, debilitating, and often life-threatening diseases, including typhoid fever, cholera, hepatitis B, giardiasis, and cryptosporidium.

Figure 20-5 Plastics and other forms of waste pollute this mountain lake, as well as many other bodies of water around the world, and can release harmful chemicals into the water.

The World Health Organization (WHO) estimates that about 1 billion people—one of every seven in the world—do not have access to clean drinking water. The WHO estimates that each year, more than 1.6 million people die from largely preventable waterborne infectious diseases that they get by drinking contaminated water or by not having enough clean water for adequate hygiene. This adds up to an average of nearly 4,400 premature deaths a day, 90% of them among children younger than age 5. This death toll could be reduced with more widespread use of well-developed methods for detecting pollutants (Science Focus 20.1).

Table 20-1 Major Water Pollutants and Their Sources

Type/*Effects*	Examples	Major Sources
Infectious agents (pathogens) *Cause diseases*	Bacteria, viruses, protozoa, parasites	Human and animal wastes
Oxygen-demanding wastes *Deplete dissolved oxygen needed by aquatic species*	Biodegradable animal wastes and plant debris	Sewage, animal feedlots, food-processing facilities, paper mills
Plant nutrients *Cause excessive growth of algae and other species*	Nitrates (NO_3^-) and phosphates (PO_4^{3-})	Sewage, animal wastes, inorganic fertilizers
Organic chemicals *Add toxins to aquatic systems*	Oil, gasoline, plastics, pesticides, cleaning solvents	Industry, farms, households
Inorganic chemicals *Add toxins to aquatic systems*	Acids, bases, salts, metal compounds	Industry, households, surface runoff, mining sites
Sediments *Disrupt photosynthesis, food webs, other processes*	Soil, silt	Land erosion
Heavy metals *Cause cancer, disrupt immune and endocrine systems*	Lead, mercury, arsenic	Unlined landfills, household chemicals, mining refuse, industrial discharges
Thermal *Make some species vulnerable to disease*	Heat	Electric power and industrial plants

SCIENCE FOCUS 20.1

TESTING WATER FOR POLLUTANTS

Scientists use a variety of methods to measure water quality. For example, they test samples of water for the *presence of various infectious agents* such as certain strains of the coliform bacteria *Escherichia coli*, or *E. coli* (Figure 20-A), which live in the colons and intestines of humans and other animals and thus are present in their fecal wastes. Although most strains of coliform bacteria do not cause disease, their presence indicates that water has been exposed to human or animal wastes that are likely to contain disease-causing agents.

To be considered safe for drinking, a 100-milliliter (about 1/2-cup) sample of water should contain no colonies of coliform bacteria. To be considered safe for swimming, such a water sample should contain no more than 200 colonies of coliform bacteria. By contrast, a similar sample of raw sewage may contain several million coliform bacterial colonies.

Another indicator of water quality is its *level of dissolved oxygen (DO)*. Excessive inputs of oxygen-demanding wastes can deplete DO levels in water. Figure 20-B shows the relationship between dissolved oxygen content and water quality. In the dead zone formed in the Gulf of Mexico (**Core Case Study**), dissolved oxygen levels near the seafloor typically drop below 3 ppm.

Scientists can use *chemical analysis* to determine the presence and concentrations of specific organic chemicals in polluted water. They can also monitor water pollution by using living organisms as *indicator species*. For example, they remove aquatic plants such as cattails from areas contaminated with fuels, solvents, and other organic chemicals, and analyze them to determine the exact pollutants contained in their tissues. Scientists also determine water quality by analyzing bottom-dwelling species such as mussels, which feed by filtering water through their bodies.

Genetic engineers are working to develop bacteria and yeasts (single-celled fungi) that glow in the presence of specific pollutants such as toxic heavy metals in the ocean, toxins in the air, and carcinogens in food.

Scientists measure the amount of sediment in polluted water by evaporating the water in a sample and weighing the resulting sediment. They also use instruments called colorimeters, which measure specific wavelengths of light shined through a water sample to determine the concentrations of pollutants in the water. Another tool is the turbidimeter, used to measure the *turbidity*, or cloudiness, of water samples containing sediment.

The technology for testing waters is evolving. For example, in 2010, researchers in Spain used five robot fish, each one about as long as a seal, to detect potentially hazardous pollutants in a river.

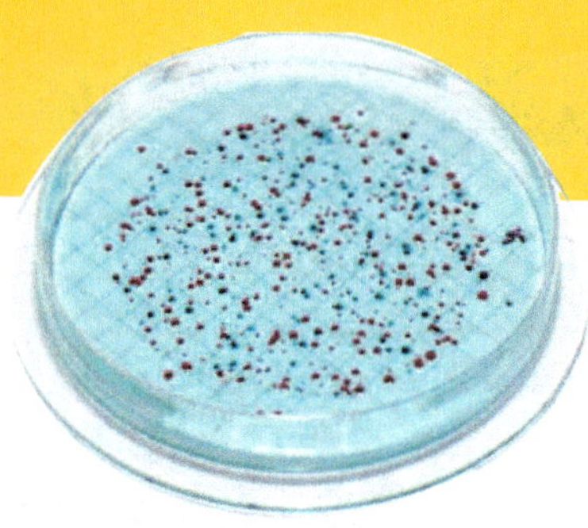

Figure 20-A Colonies of coliform bacteria growing on a petri dish.

Critical Thinking

Runoff of fertilizer into a lake from farmland, lawns, and sewage treatment plants can overload the water with nitrogen and phosphorus—plant nutrients that can cause algae population explosions. How could this lower the dissolved oxygen level of the water and lead to fish kills?

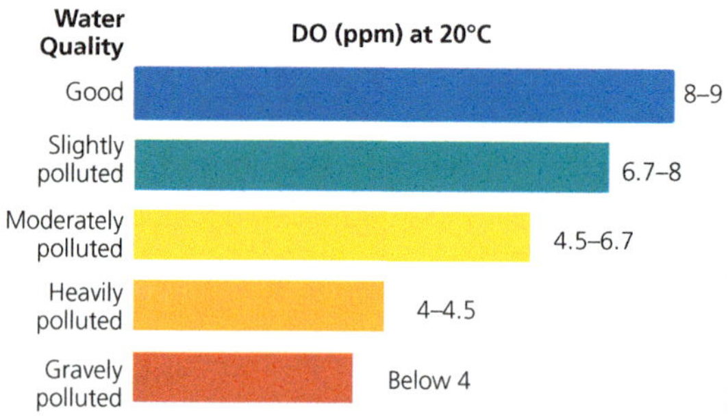

Figure 20-B Scientists measure dissolved oxygen (DO) content in parts per million (ppm) at 20°C (68°F) as an indicator of water quality. Only a few fish species can survive in water with less than 3 ppm of dissolved oxygen at this temperature. ***Question:*** Would you expect the dissolved oxygen content of polluted water to increase or decrease if the water were heated? Explain.

20-2 What Are the Major Water Pollution Problems in Streams and Lakes?

CONCEPT 20-2A

Streams and rivers around the world are extensively polluted, but they can cleanse themselves of many pollutants if we do not overload them or reduce their flows.

CONCEPT 20-2B

Adding excessive nutrients to lakes from human activities can disrupt their ecosystems, and prevention of such pollution is more effective and less costly than cleaning it up.

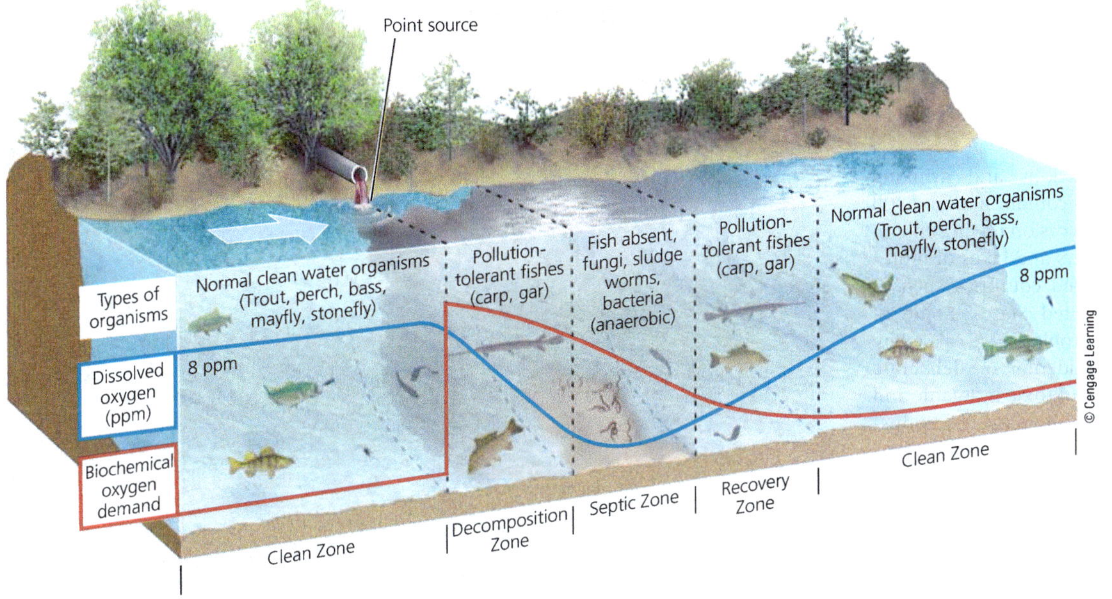

Animated Figure 20-6 **Natural capital:** A stream can dilute and decay degradable, oxygen-demanding wastes, and it can also dilute heated water. This figure shows the oxygen sag curve (blue) and the curve of oxygen demand (red). Streams recover from oxygen-demanding wastes and from the injection of heated water if they are given enough time and are not overloaded (**Concept 20-2A**). ***Question:*** What would be the effect of putting another discharge pipe emitting biodegradable waste to the right of the one in this picture?

Streams Can Cleanse Themselves, If We Do Not Overload Them

Flowing rivers and streams can recover rapidly from moderate levels of degradable, oxygen-demanding wastes through a combination of dilution and bacterial biodegradation of such wastes. But this natural recovery process does not work when streams become overloaded with such pollutants or when drought, damming, or water diversion reduce their flows (**Concept 20-2A**). Also, while this process can remove biodegradable wastes, it does not eliminate slowly degradable and nondegradable pollutants. In a flowing stream, the breakdown of biodegradable wastes by bacteria depletes dissolved oxygen and creates an *oxygen sag curve* (Figure 20-6).

Stream Pollution in More-Developed Countries

Laws enacted in the 1970s to control water pollution have greatly increased the number and quality of wastewater treatment plants in the United States and in most other more-developed countries. Such laws also require industries to reduce or eliminate their point-source discharges of harmful chemicals into surface waters. This is an impressive accomplishment given the country's growth in population, economic activity, and resource consumption since the passage of these laws.

GOOD NEWS

One success story is the cleanup of the U.S. state of Ohio's Cuyahoga River. It was so polluted that it caught fire several times and, in 1969, was photographed while burning as it flowed through the city of Cleveland toward Lake Erie. The highly publicized image of this combustible river prompted elected officials to enact laws to limit the discharge of industrial wastes into the river and into local sewage systems and to provide funds for upgrading sewage treatment facilities. Today, the river is cleaner, is no longer flammable, and is widely used by boaters and anglers. This accomplishment illustrates the power of bottom-up pressure by citizens, who prodded elected officials to change a severely polluted river into an economically and ecologically valuable public resource.

Another spectacular cleanup occurred in Great Britain. In the 1950s, the Thames River was little more than a flowing, smelly sewer. Now, after 50 years of effort and large inputs of money from British taxpayers and private industry, the Thames has made a remarkable recovery. Commercial fishing is thriving and the number of fish species in the river has increased 20-fold since 1960. In addition, many species of waterfowl and wading birds have returned to their former feeding grounds along the banks of the Thames.

© CHINA DAILY/Reuters/Corbis

Figure 20-7 Natural capital degradation: This portion of China's Yangtze River near the Three Gorges dam is heavily polluted with chemicals, sediment, and trash.

Fish kills and drinking water contamination still occur occasionally in some of the rivers and lakes of more-developed countries such as the United States. Some of these problems are caused by the accidental or deliberate release of toxic inorganic and organic chemicals by industries and mining operations (see Figure 14-15, p. 360, and Figure 20-4). Another cause is malfunctioning sewage treatment plants. A third cause is nonpoint runoff of pesticides and excess plant nutrients from cropland and animal feedlots.

Stream Pollution in Less-Developed Countries

In most less-developed countries, stream pollution from discharges of untreated sewage and industrial wastes is a serious and growing problem. According to the Global Water Policy Project, most cities in less-developed countries discharge 80–90% of their untreated sewage directly into rivers, streams, and lakes whose waters are often used also for drinking, bathing, and washing clothes.

According to the World Commission on Water for the 21st Century, half of the world's 500 major rivers are heavily polluted, and most of these polluted waterways run through less-developed countries. A majority of these countries cannot afford to build waste treatment plants and do not have, or do not enforce, laws for controlling water pollution.

In Latin America and Africa, most streams passing through urban or industrial areas suffer from severe pollution. Industrial wastes and sewage pollute more than two-thirds of India's water resources and 54 of the 78 rivers and streams monitored in China. In some parts of China, river water is too toxic to touch, much less drink. Liver and stomach cancer, linked in some cases to water pollution, are among the leading causes of death in the Chinese countryside where many industries have been relocated.

In 2010, Chinese officials reported that huge islands of garbage are threatening to jam the flow of water over the country's massive Three Gorges Dam on the Yangtze River (Figure 20-7). As it decays, this garbage will pollute the water.

Too Little Mixing and Low Flow Rates Make Lakes and Reservoirs Vulnerable to Water Pollution

Lakes and reservoirs are generally less effective at diluting pollutants than streams are, for two reasons. *First,* lakes and reservoirs often contain stratified layers (see Figure 8-16, p. 179) that undergo little vertical mixing. *Second,* they have low flow rates or no flow at all. The flushing and changing of water in lakes and large artificial reservoirs can take from 1 to 100 years, compared with several days to several weeks for streams.

As a result, lakes and reservoirs are more vulnerable than streams are to contamination by runoff or discharge of plant nutrients, oil, pesticides, and nondegradable toxic substances, such as lead, mercury, and arsenic. These contaminants can kill bottom-dwelling organisms and fish, as well as birds that feed on contaminated aquatic organisms. Many toxic chemicals and acids also enter lakes and reservoirs from the atmosphere (see Figure 18-12, p. 485).

Due to the stratified layers and reduced flows in lakes and reservoirs, the concentrations of some harmful chemicals are biologically magnified as they pass through food webs in these waters. Examples include DDT (see Figure 9-13, p. 202), PCBs (see Chapter 17, Case Study, p. 453), some radioactive isotopes, and some mercury compounds.

Cultural Eutrophication Is Too Much of a Good Thing

Eutrophication is the name given to the natural nutrient enrichment of a body of water such as a lake, coastal areas at the mouth of a river (**Core Case Study**), or a slow-

Figure 20-8 In 2010, severe cultural eutrophication affected Chaohu Lake near Hefei City in China's Anhui Province.

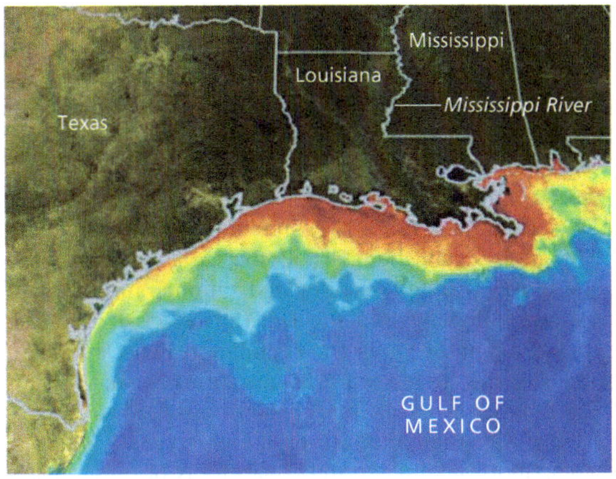

Figure 20-9 The colorized area in this satellite image of the northern Gulf of Mexico represents the seasonal dead zone of 2012 with the red area having the lowest oxygen level.

moving stream. It is caused mostly by runoff of plant nutrients such as nitrates and phosphates from land bordering such bodies of water.

An *oligotrophic lake* is low in nutrients and its water is clear (see Figure 8-15, p. 179). Over time, some lakes become more eutrophic (see Figure 8-17, p. 180) as nutrients are added from natural and human sources in the surrounding watersheds. Near urban or agricultural areas, human activities can greatly accelerate the input of plant nutrients to a lake—a process called **cultural eutrophication**. Such inputs involve mostly nitrate- and phosphate-containing effluents from various sources, including farmland, feedlots, urban streets and parking lots, chemically fertilized suburban yards, mining sites, and municipal sewage treatment plants. Some nitrogen also reaches lakes by deposition from the atmosphere (Figure 18-12, p. 485).

During hot weather or drought, this nutrient overload can produce dense growths, or "blooms," of organisms such as algae and cyanobacteria in slow-moving surface waters (see Figure 8-17, p. 180). When the algae die, they are decomposed by swelling populations of aerobic bacteria, which deplete the dissolved oxygen in the surface layer of water near the shore, as well as in the bottom layer of a lake or coastal area. This can kill fish, shellfish, and other aerobic aquatic animals that cannot move to safer waters. If excess nutrients continue to flow into a lake, anaerobic bacteria take over and produce gaseous products such as smelly, highly toxic hydrogen sulfide and methane.

According to the EPA, about one-third of the 100,000 medium to large lakes and 85% of the large lakes near major U.S. population centers have some degree of cultural eutrophication. The International Water Association estimates that more than half of the lakes in China, along with some of its coastal zones, suffer from cultural eutrophication (Figure 20-8).

There are several ways to *prevent* or *reduce* cultural eutrophication. We can use advanced (but expensive) waste treatment processes to remove nitrates and phosphates from wastewater before it enters a body of water. We can also use a preventive approach by banning or limiting the use of phosphates in household detergents and other cleaning agents, and by employing soil conservation (see Chapter 12, p. 304) and other controls to reduce nutrient runoff (**Concept 20-2B**).

There are several ways to *clean up* waters suffering from cultural eutrophication. They include mechanically removing excess weeds, controlling undesirable plant growth with herbicides and algaecides, and pumping air into lakes and reservoirs to prevent oxygen depletion, all of which are expensive and energy-intensive methods.

As usual, pollution prevention is more effective and quite often cheaper in the long run than cleanup. Most lakes and other surface waters can recover from cultural eutrophication, if excessive inputs of plant nutrients are stopped. In addition to cultural eutrophication, pollution causes other problems that plague many lakes (Case Study that follows).

Revisiting the Gulf of Mexico: An Extreme Case of Cultural Eutrophication

Since the 1950s, the level of nitrates discharged from the Mississippi River into the northern Gulf of Mexico has nearly tripled (**Core Case Study**). This has disrupted the nitrogen cycle and animal life in the area's affected coastal waters. Figure 20-9 provides a closer look at the severe oxygen depletion (shown in red) in 2012 in the dead zone of the northern Gulf of Mexico, where dissolved oxygen levels were typically less than 2 parts per million (Figure 20-B)—low enough to kill bottom-dwelling shellfish.

Excess nutrients fueling an explosive growth of algae (eutrophication) is a natural process that often occurs in some areas of the ocean where upwellings bring nutrients from the ocean bottom to surface waters (see Figure 7-2, p. 145). However, excessive inputs of plant nutrients, especially nitrates, from human activities (cultural eutrophication) is the primary cause of the annual dead zone in the northern Gulf of Mexico and in many other hypoxic zones around the world, including parts of Chesapeake Bay (see Figure 8-13, p. 177) and the San Francisco Bay.

In more detail, here is how the Gulf's seasonal dead zone forms. During the spring and summer, nitrate-laden freshwater flowing into the Gulf forms an oxygen-rich layer on top of the Gulf's cooler and more dense saltwater. Because there are few storms at this time of year, this sun-heated upper layer of water remains fairly calm and does not mix with the bottom layer of low-oxygen water. The combination of sunlight and large inputs of nitrate plant nutrients from fertilizer and sewage into the freshwater layer leads to massive blooms of phytoplankton, mostly blue-green algae.

When these algae die, they sink into the saltier water below where they are decomposed by oxygen-consuming bacteria, which use up nearly all of the dissolved oxygen, leaving less than 2 parts per million in the deeper water. Mobile species can survive this lack of oxygen by migrating to oxygen-rich waters, but certain species of fish, shellfish, and other organisms are not able to escape, and they die off.

Thus, the oxygen-depleted bottom layer of water becomes a dead zone for certain species. This disrupts the Gulf's food web, because die-offs of such species lead to the deaths of seabird and marine mammal species that depend on the dying fish and shellfish for their survival. The dead zone breaks up, beginning in early fall when cooler weather, storms, and hurricanes mix the top and lower layers of water and distribute dissolved oxygen throughout the layers.

In addition to greatly increased nitrate levels from human inputs, other human factors have contributed to the formation of the dead zone. In efforts to control flooding along the upper Mississippi River, engineers have dredged and straightened parts of the river and raised its banks with levees in many places. This has had the effect of speeding the river's flow of nutrients and sediment pollution into the Gulf. Also, most of the river basin's original freshwater wetlands, which acted as natural filters that helped to remove excess nutrients and sediments from flood water, have been drained for farming and urban development.

The seasonal formation of dead zones in the northern Gulf of Mexico, mostly resulting from human activities, is a reminder that in nature, everything is connected. Plant nutrients flowing into the Mississippi from a farm in Iowa or a sewage treatment plant in Wisconsin help to cause fish kills (Figure 20-10) a thousand miles away on the gulf coast of Texas. Researchers fear that if the size of the Gulf's annual dead zone is not sharply reduced, its long-term effects could permanently alter the ecological makeup of these coastal waters.

AP Photo/Houston Chronicle, Thomas B. Shea

Figure 20-10 On August 13, 2012, hundreds of thousands of dead shad fish washed ashore on Jamaica Beach near Galveston, Texas. The likely cause of this die-off was low oxygen levels in the Gulf of Mexico.

CASE STUDY

Pollution in the Great Lakes

The five interconnected Great Lakes of North America (Figure 20-11) contain about 95% of the fresh surface water in the United States and one-fifth of the world's fresh surface water. At least 38 million people in the United States and Canada obtain their drinking water from these lakes.

Despite their enormous size, these lakes are vulnerable to pollution from point and nonpoint sources. One reason is that each year, less than 1% of the water entering these lakes flows east into the St. Lawrence River and then out to the Atlantic Ocean, meaning that pollutants can take as long as 100 years to be flushed out to sea.

By the 1960s, many areas of the Great Lakes were suffering from severe cultural eutrophication, huge fish kills, and contamination from bacteria and a variety of toxic industrial wastes. The impact on Lake Erie was particularly intense because it is the shallowest of the Great Lakes and has the highest concentrations of people and industrial activity along its shores.

In 1972, the United States and Canada signed the Great Lakes Water Quality Agreement, which is considered a model of international cooperation. They agreed to spend more than $20 billion to maintain and restore the chemical, physical, and biological integrity of the Great Lakes basin ecosystem. This program has helped to cut the number and sizes of algal blooms, raised dissolved oxygen levels, boosted sport and commercial fishing catches in Lake Erie, and allowed most swimming beaches to reopen—making this one of the nations' greatest environmental success stories. These improvements resulted mainly from the use of new or upgraded GOOD NEWS

Figure 20-11 The five Great Lakes of North America make up the world's largest freshwater system. Dozens of major cities in the United States and Canada are located on their shores, and water pollution in the lakes is a growing problem.

sewage treatment plants, better treatment of industrial wastes, and bans on the use of detergents, household cleaners, and water conditioners that contain phosphates. Most of these measures were instituted largely as a result of bottom-up citizen pressure.

Despite this important progress, many problems remain. Increasing nonpoint runoff of pesticides and fertilizers resulting from urban sprawl, fueled by population growth, now surpasses industrial pollution as the greatest threat to the lakes. Bottom sediments in 26 toxic hotspots remain heavily polluted. In addition, *biological pollution* in the form of growing populations of zebra mussels (Figure 11-20, p. 268) and more than 180 other invasive species threaten some native aquatic species and cause at least $200 million a year in damages (see Chapter 11, Case Study, p. 268).

Air quality over the Great Lakes has generally improved, but about half of the toxic compounds entering the lakes still come from atmospheric deposition of pesticides, mercury from coal-burning plants, and other toxic chemicals from as far away as Mexico and Russia. A survey done by Wisconsin biologists found that one of every four fish taken from the Great Lakes was unsafe for human consumption.

Current efforts are aimed at improving the water quality of the lakes and other aquatic systems within the Great Lakes basin. To that end, the major goals are to clean up contaminated river bottoms, prevent erosion and runoff into lakes and streams, protect and restore wetlands, clean up toxic hotspots, and reduce the number of invasive species entering the lakes.

Some environmental and health scientists call for taking a prevention approach and banning the use of toxic chlorine compounds, such as bleach used in the pulp and paper industry, which is prominent around the Great Lakes. They would also ban new waste incinerators, which can release toxic chemicals into the atmosphere, and they would stop the discharge into the lakes of 70 toxic chemicals that threaten human health and wildlife. So far, officials in the industries involved have successfully opposed such bans.

20-3 What Are the Major Pollution Problems Affecting Groundwater?

CONCEPT 20-3A

Chemicals used in agriculture, industry, transportation, and homes can spill and leak into groundwater and make it undrinkable.

CONCEPT 20-3B

There are both simple ways and complex ways to purify groundwater used as a source of drinking water, but protecting it through pollution prevention is the least expensive and most effective strategy.

Groundwater Cannot Cleanse Itself Very Well

Aquifers provide drinking water for about half of the U.S. population and 95% of Americans who live in rural areas. According to many scientists, groundwater pollution is a serious threat. Common pollutants such as fertilizers, pesticides, gasoline, and organic solvents can seep into groundwater from numerous sources (Figure 20-12). People who dump or spill gasoline, oil, and paint thinners and other organic solvents onto the ground also help to contaminate groundwater (Concept 20-3A).

Scientists are now concerned about a new and growing potential threat to groundwater—the drilling of thousands of new natural gas wells in parts of the United States involving a process called *hydraulic fracturing*, or *fracking* (see Chapter 15, Case Study, p. 383). Groundwater contamination could result from leaky well pipes and pipe fittings and from contaminated wastewater brought to the surface during fracking operations. Without strict monitoring and enforcement of pollution regulations (which at this point do not exist), fracking could become a serious groundwater pollution threat.

Once a pollutant from a leaking underground storage tank or other source contaminates groundwater, it fills the aquifer's porous layers of sand, gravel, or bedrock like water saturates a sponge. This makes removal of the contaminant difficult and costly. The slowly flowing groundwater disperses the pollutant in a widening *plume* of contaminated water. If the plume reaches a well used to extract groundwater, the toxic pollutants can get into drinking water and into water that is used to irrigate crops.

When groundwater becomes contaminated, it cannot cleanse itself of degradable wastes as quickly as flowing surface water can. Groundwater flows so slowly—usually less than 0.3 meter (1 foot) per day—that contaminants

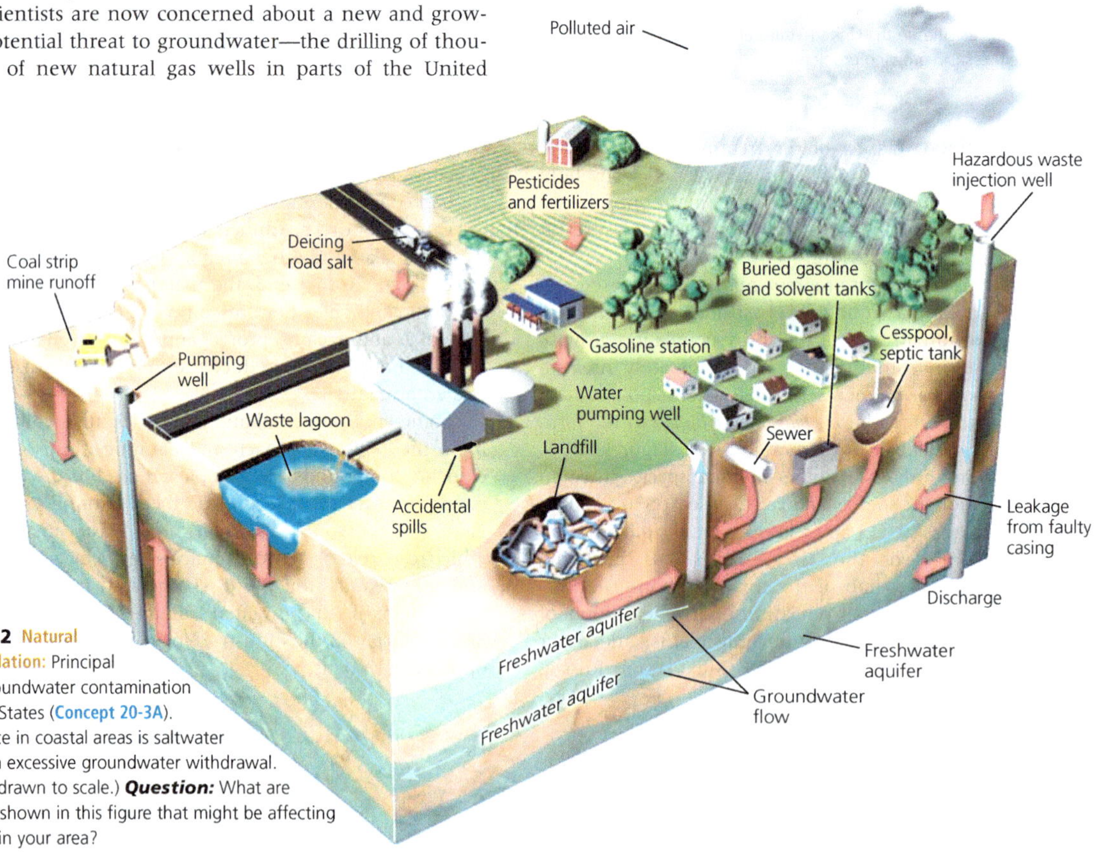

Figure 20-12 Natural capital degradation: Principal sources of groundwater contamination in the United States (Concept 20-3A). Another source in coastal areas is saltwater intrusion from excessive groundwater withdrawal. (Figure is not drawn to scale.) ***Question:*** What are three sources shown in this figure that might be affecting groundwater in your area?

are not diluted and dispersed effectively. In addition, groundwater usually has much lower concentrations of dissolved oxygen (which helps decompose many contaminants) and smaller populations of decomposing bacteria. The usually cold temperatures of groundwater also slow down chemical reactions that decompose wastes.

Groundwater Pollution Is a Serious Hidden Threat in Some Areas

On a global scale, we do not know much about groundwater pollution because few countries go to the great expense of locating, tracking, and testing aquifers. But the results of scientific studies in scattered parts of the world are alarming.

Groundwater provides about 70% of China's drinking water. According to the Chinese Ministry of Land and Resources, in 2010, about 90% of China's shallow groundwater was polluted with chemicals such as toxic heavy metals, organic solvents, nitrates, petrochemicals, and pesticides. About 37% of this groundwater is so polluted that it cannot be treated for use as drinking water. Every year, according to the WHO and the World Bank, water pollution causes an estimated 190 million Chinese to become ill and kills about 60,000.

In the United States, an EPA survey of 26,000 industrial waste ponds and lagoons found that one-third of them had no liners to prevent toxic liquid wastes from seeping into aquifers. One-third of these sites are within 1.6 kilometers (1 mile) of a drinking water well. In addition, almost two-thirds of America's liquid hazardous wastes are injected into the ground in disposal wells (Figure 20-12), some of which leak their contents into aquifers that are used as sources of drinking water.

By 2008, the EPA had cleaned up about 357,000 of the more than 479,000 underground tanks in the United States that were leaking gasoline, diesel fuel, home heating oil, or toxic solvents into groundwater. During this century, scientists expect many of the millions of such tanks, which have been installed around the world, to become corroded and leaky, possibly contaminating groundwater and becoming a major global health problem. Determining the extent of a leak from a single underground tank can cost $25,000–$250,000, and cleanup costs range from $10,000 to more than $250,000. If the chemical reaches an aquifer, effective cleanup is often not possible or is too costly.

Groundwater that is used as a source of drinking water can also be contaminated with *nitrate ions* (NO_3^-), especially in agricultural areas where nitrates in fertilizer can leach into groundwater. Nitrite ions (NO_2^-) in the stomach, colon, and bladder can convert some of the nitrate ions in drinking water to organic compounds that have been shown in tests to cause cancer in more than 40 animal species. The conversion of nitrates in tap water to nitrites in infants under 6 months old can cause a potentially fatal condition known as "blue baby syndrome," in which blood lacks the ability to carry sufficient oxygen to body cells.

It can take decades to thousands of years for contaminated groundwater to cleanse itself of *slowly degradable wastes* (such as DDT). On a human time scale, *nondegradable wastes* (such as toxic lead and arsenic, see the following Case Study) remain in the water permanently. Although there are ways to clean up contaminated groundwater (Figure 20-13, right), such methods are very expensive. Cleaning up a single contaminated aquifer can cost anywhere from $10 million to $10 billion. Thus, preventing groundwater contamination (Figure 20-13, left) is the only effective and affordable way to deal with this serious water pollution problem.

CASE STUDY

Arsenic in Drinking Water

Arsenic contaminates drinking water when a well is drilled into an aquifer where the soil and rock are naturally rich in arsenic, or when human activities such as mining and ore processing release arsenic into drinking water supplies. Some rivers used for drinking water also are contaminated naturally, having originated in springs that have high levels of arsenic.

The World Health Organization (WHO) estimates that more than 140 million people in 70 countries are drinking water with arsenic concentrations of 5–100 times the accepted safe level of 10 parts per billion (ppb). Some scientists from the WHO and other organizations argue that even the 10 ppb standard is not safe.

Solutions

Groundwater Pollution

Prevention	Cleanup
Find substitutes for toxic chemicals	Pump to surface, clean, and return to aquifer (very expensive)
Keep toxic chemicals out of the environment	Inject microorganisms to clean up contamination (less expensive but still costly)
Require leak detectors on underground tanks	Pump nanoparticles of inorganic compounds to remove pollutants (still being developed)
Ban hazardous waste disposal in landfills and injection wells	
Store harmful liquids in aboveground tanks with leak detection and collection systems	

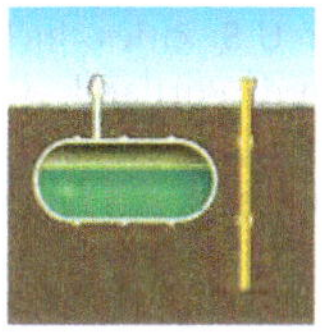

Figure 20-13 There are ways to prevent and ways to clean up contaminated groundwater, but prevention is the only effective approach (Concept 20-3B). ***Question:*** Which two of these preventive solutions do you think are the most important? Why?

Arsenic levels are especially high in Bangladesh, China, India's state of West Bengal, and parts of northern Chile. The WHO estimates that long-term exposure to nondegradable arsenic in drinking water is likely to cause hundreds of thousands of premature deaths from cancer of the skin, bladder, and lungs.

There is also concern over arsenic levels in drinking water in parts of the United States. According to the EPA, some 13 million people in several thousand communities, mostly in the western half of the country, are exposed to arsenic levels of 3–10 ppb in their drinking water.

In 2006, researchers from Rice University in Houston, Texas (USA), reported that by suspending nanoparticles of rust in arsenic-contaminated water and then drawing them out with handheld magnets, they had removed enough arsenic from the water to make it safe to drink. Use of this technique could greatly reduce the threat of arsenic in drinking water for many families at a cost of a few cents a day.

GOOD NEWS

There Are Many Ways to Purify Drinking Water

Most of the more-developed countries have laws establishing drinking water standards. But most of the less-developed countries do not have such laws or, if they do have them, they do not enforce them. There are both simple ways and complex ways to purify drinking water obtained from surface and groundwater sources (**Concept 20-3B**).

In more-developed countries, wherever people depend on surface water sources, water is usually stored in a reservoir for several days. This improves its clarity and taste by increasing its dissolved oxygen content and allowing suspended matter to settle. The water is then pumped to a purification plant and treated to meet government drinking water standards. In areas with very pure groundwater or surface water sources, little treatment is necessary. Several major U.S. cities, including New York City, Boston, Seattle, and Portland, Oregon, have avoided building expensive water treatment facilities by investing in protection of the forests and wetlands in the watersheds that provide their water supplies.

CONSIDER THIS. . .

THINKING ABOUT Protecting the Sources of Drinking Water

Where does the community in which you live get its drinking water? If the source is surface water, could the community save money and help to preserve biodiversity by finding ways to protect the watershed for this source?

We have the technology to convert sewer water into pure drinking water, which could help to reduce depletion of some water supplies. However, recycling wastewater is expensive and it faces opposition from citizens and from some health officials who are unaware of the advances in this technology, much of it developed to recycle wastewater on the International Space Station. In a world where we will face deepening shortages of drinking water, wastewater purification is likely to become a major growth business and is much cheaper than using desalination to turn ocean water into drinking water (p. 335). Wastewater is being recycled to provide freshwater in Israel, Singapore, Orange County and San Diego in California, and El Paso, Texas, among a growing number of places.

In many areas of both more-developed and less-developed countries, people have turned to bottled water, hoping that it is a safe alternative to local drinking water sources (see the following Case Study). However, there are several simple ways to purify drinking water that are far less costly than using bottled water. For example, in tropical countries that lack centralized water treatment systems, the WHO urges people to purify drinking water by exposing a clear plastic bottle filled with contaminated water to intense sunlight. The sun's heat and ultraviolet (UV) rays can kill infectious microbes in as little as 3 hours. Painting one side of the bottle black improves heat absorption in this simple solar disinfection method, which applies the solar energy **principle of sustainability**.

GOOD NEWS

SUSTAINABILITY

Where this measure has been used, the incidence of dangerous childhood diarrhea has decreased by 30–40%. In 2012, researchers found they could speed up the disinfection process by adding lime juice to the bottles of water. In 2011, scientists in Brazil found that adding dried bits of banana peels is an inexpensive way to remove toxic metals such as mercury, arsenic, and lead from drinking water polluted by mining and industrial wastes.

In 2007, Danish inventor Torben Vestergaard Frandsen developed the LifeStraw™, an inexpensive, portable water filter that eliminates many viruses and parasites from water that is drawn through it (Figure 20-14). This filter has been particularly useful in Africa, where aid agencies are distributing it.

Another product that could be helpful for the 1 billion people who do not have access to safe drinking water is PUR, a powder containing chlorine and iron sulfate. A user empties a packet of PUR into a large container of dirty water, then stirs the mixture and lets it sit for about 20 minutes. After the water is poured through a piece of clean cloth to filter out the particles, it is ready to drink.

CASE STUDY

Is Bottled Water a Good Option?

Bottled water can be a useful (but expensive) option in countries and areas where people do not have access to safe and clean drinking water. However, despite some problems, experts say the United States has some of the world's cleanest drinking water. Municipal water systems in the United States are required to test their water regularly for a number of pollutants and to make the results available to citizens.

Yet about half of all Americans worry about getting sick from tap water contaminants, and many drink high-

Figure 20-14 The *LifeStraw™* is a personal water purification device that gives many poor people access to safe drinking water. Here, four young men in Uganda demonstrate its use. ***Question:*** Do you think the development of such devices should make prevention of water pollution less of a priority? Explain.

priced bottled water or install expensive water purification systems. Americans are the world's largest consumers of bottled water, followed by Mexico, China, and Brazil. In 2011, Americans spent about $15 billion on bottled water—enough to meet the annual drinking water needs of the roughly 1 billion people in the world who routinely lack access to safe, clean drinking water.

Studies by the Natural Resources Defense Council (NRDC) reveal that in the United States, a bottle of water costs between 240 and 10,000 times as much as the same volume of tap water. In 2008, water expert Peter Gleick estimated that more than 40% of the expensive bottled water that Americans drink is really bottled tap water. And a 4-year study by the NRDC concluded that most bottled water is of good quality. However, they found traces of bacteria and synthetic organic chemicals in 23 of the 123 brands tested. Bottled water is less regulated than tap water and the EPA contamination standards that apply to public water supplies do not apply to bottled water.

About a third of all bottled water suppliers do not disclose anything about the treatment or purity of the water they sell. However, the International Bottled Water Association and the nonprofit National Sanitation Foundation International certify bottled water companies that meet their standards. Cautious buyers use only bottled water that is certified by such organizations.

Use of bottled water also causes environmental problems. In the United States, according to the Container Recycling Institute, more than 67 million plastic water bottles are discarded every day. Each year, this amounts to enough bottles to, if lined up end-to-end, wrap around the planet 149 times. Most water bottles are made of recyclable polyethylene terephthalate (PET) plastic, but in the United States, only about 29% of these bottles get recycled. Many of the rest end up in landfills, where they can remain for hundreds of years, or in incinerators, which release some of their harmful chemicals into the atmosphere. Also, many millions of discarded bottles get scattered on the land and wind up in rivers, lakes (Figure 20-5), and oceans.

It takes huge amounts of energy to manufacture bottled water and to transport it across countries and around the world, as well as to refrigerate much of it in stores. Toxic gases and liquids are released during the manufacture of plastic water bottles, and greenhouse gases and other air pollutants are emitted by the fossil fuels burned to make them and to deliver bottled water to suppliers. In addition, withdrawing water for bottling is helping to deplete some aquifers. There is also concern about health risks from chemicals such as bisphenol A (BPA, see Science Focus 17.3, p. 456) that can leach into the water from the plastic in some water bottles, especially if they are exposed to the hot sun.

Because of these harmful environmental impacts and the high cost of bottled water, there is a growing *back-to-the-tap* movement. From San Francisco to New York to Paris, city governments, restaurants, schools, religious groups, and many consumers are refusing to buy bottled water.

Also, people are increasingly using refillable glass and metal bottles, filling them with tap water and using simple filters to improve the taste and color of water where necessary. In Germany and other countries, most bottled water is sold in returnable and reusable glass bottles. Some health officials suggest that before drinking expensive bottled water or buying costly home water purifiers, consumers have their water tested by local health departments or private labs (but not by companies trying to sell water purification equipment).

Using Laws to Protect Drinking Water Quality

About 54 countries, most of them in North America and Europe, have legal standards for safe drinking water. For example, the U.S. Safe Drinking Water Act of 1974 requires the EPA to establish national drinking water standards, called *maximum contaminant levels,* for any pollutants that could have adverse effects on human health. Currently, this act strictly limits the levels of 91 potential contaminants in U.S. tap water. However, in most of the less-developed countries, such laws do not exist or are not enforced.

Health scientists call for strengthening the U.S. Safe Drinking Water Act in several ways. Here are three of their recommendations:

- Combine many of the drinking water treatment systems that serve fewer than 3,300 people with nearby larger systems to make it easier for these smaller systems to meet federal standards.
- Strengthen and enforce requirements concerning public notification of violations of drinking water standards.
- Ban the use of any toxic lead in new plumbing pipes, faucets, and fixtures. Current law allows for fixtures with up to 10% lead content to be sold as lead-free.

According to the Natural Resources Defense Council (NRDC), making these three improvements would cost U.S. taxpayers an average of about $30 a year per household.

However, certain industries are pressuring elected officials to weaken the Safe Drinking Water Act, because complying with it increases their costs. One proposal is to eliminate national testing of drinking water and requirements for public notification about violations of drinking water standards. Another proposal would allow states to give providers of drinking water a permanent right to violate the standard for a given contaminant if they claim they cannot afford to meet the standard. Also, some critics call for greatly reducing the EPA's already-low budget for enforcing the Safe Drinking Water Act.

20-4 What Are the Major Water Pollution Problems Affecting Oceans?

CONCEPT 20-4A

Most ocean pollution originates on land and includes oil and other toxic chemicals, as well as solid waste, which threaten fish and wildlife and disrupt marine ecosystems.

CONCEPT 20-4B

The key to protecting the oceans is to reduce the flow of pollution from land and air and from streams emptying into ocean waters.

Ocean Pollution Is a Growing and Poorly Understood Problem

Coastal areas—especially wetlands, estuaries, coral reefs, and mangrove swamps—bear the brunt of our enormous inputs of pollutants and wastes into the ocean (Figure 20-15). Roughly 40% of the world's people (53% in the United States) live on or near coastlines, which helps to explain why 80% of marine pollution originates on land (**Concept 20-4A**), and coastal populations are projected to double by 2050.

According to a study by the U.N. Environment Programme (UNEP), 80–90% of the municipal sewage from coastal areas of less-developed countries is dumped into oceans without treatment. This often overwhelms the ability of the coastal waters to degrade these biodegradable wastes. Many areas of China's coastline, for example, are so choked with algae growing on the nutrients provided by sewage that some scientists have concluded that large areas of China's coastal waters can no longer sustain marine ecosystems.

In deeper waters, the oceans can dilute, disperse, and degrade large amounts of raw sewage and other types of degradable pollutants. Some scientists suggest that it is safer to dump sewage sludge, toxic mining wastes, and most other harmful wastes into the deep ocean than to bury them on land or burn them in incinerators. Other scientists disagree, pointing out that we know less about the deep ocean than we do about the moon. They add that dumping harmful wastes into the ocean would delay urgently needed pollution prevention measures and promote further degradation of this vital part of the earth's life-support system.

Recent studies of some U.S. coastal waters have found vast colonies of viruses thriving in raw sewage and in effluents from sewage treatment plants (which do not remove viruses) and leaking septic tanks. According to one study, on average, one-fourth of the people using coastal beaches in the United States develop ear infections, sore throats, eye irritations, respiratory disease, or gastrointestinal disease from swimming in seawater containing infectious viruses and bacteria.

CONSIDER THIS. . .

CONNECTIONS Cruise Ships and Water Pollution

A cruise liner can carry as many as 6,300 passengers and 2,400 crewmembers, and it can generate as much waste (toxic chemicals, garbage, sewage, and waste oil) as a small city. Many cruise ships dump these wastes at sea. In U.S. waters, such dumping is illegal, but some ships continue dumping secretively, usually at night. Some environmentally aware vacationers are refusing to go on cruise ships that do not have sophisticated systems for dealing with the wastes they produce.

Runoff of sewage and agricultural wastes into coastal waters can carry large quantities of nitrate (NO_3^-) and phosphate (PO_4^{3-}) plant nutrients, which can cause explosive growths of harmful algae and lead to dead zones (**Core Case Study**) during summer months. These *harmful algal blooms* are called red, brown, or green toxic tides (Figure 20-16). They can release waterborne and airborne toxins that poison seafood, kill fish and some fish-eating birds, and discourage tourism in the affected coastal areas. Each year, harmful algal blooms lead to the poisoning of about 60,000 Americans who eat shellfish contaminated by the algae.

Harmful algal blooms occur annually in about 400 oxygen-depleted zones around the world, mostly in temperate coastal waters and in large bodies of water with

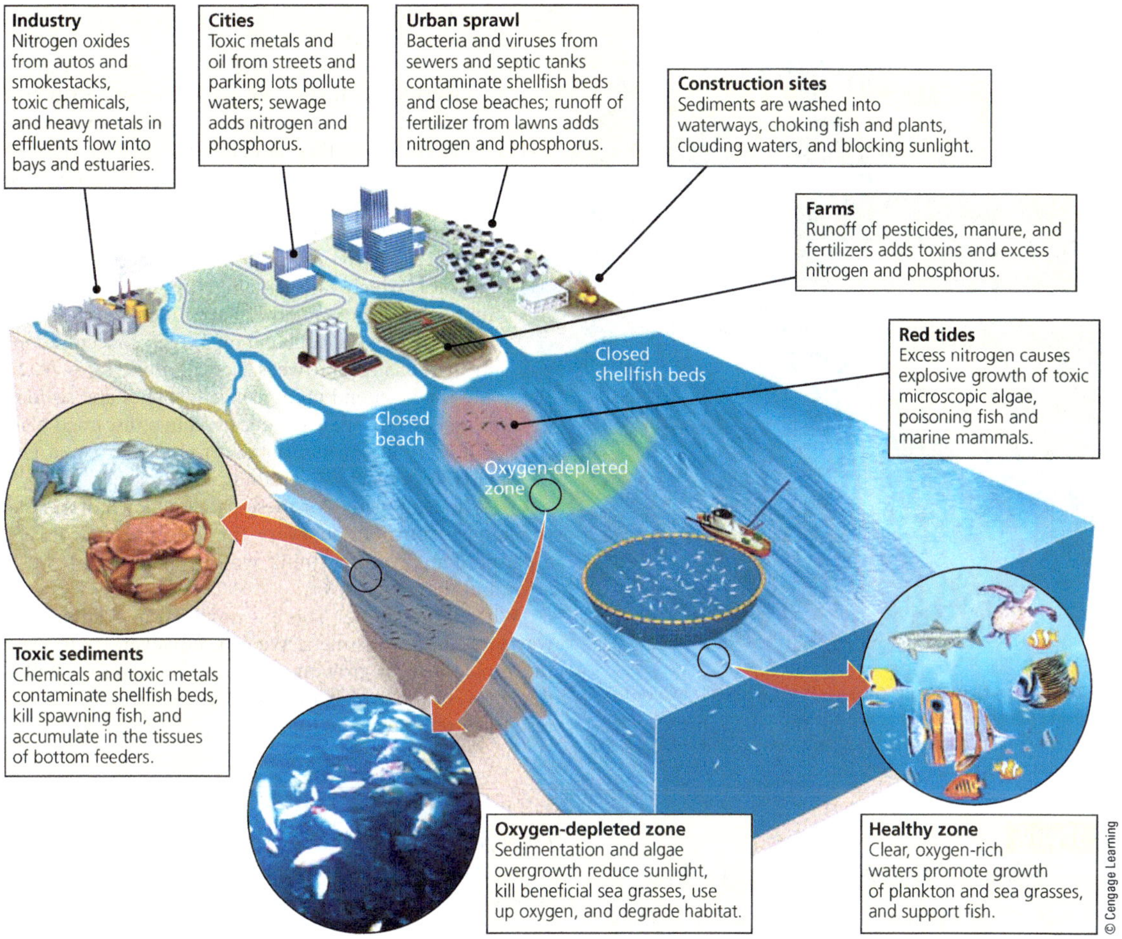

Figure 20-15 Natural capital degradation: Residential areas, factories, and farms all contribute to the pollution of coastal waters. ***Question:*** What do you think are the three worst pollution problems shown here? For each one, how does it affect two or more of the ecosystem and economic services listed in Figure 8-5 (p. 170)?

restricted outflows, such as the Baltic and Black seas. The largest of these zones in U.S. coastal waters forms each year in the northern Gulf of Mexico (**Core Case Study**). A 2008 study by Luan Weixin, of China's Dalain Maritime University, found that water pollutants such as nitrates and phosphates seriously contaminated about half of China's shallow coastal waters. Warmer ocean water temperatures, thought by most scientists to be due to climate change, are extending the size and duration of dead zones in the world's oceans.

Figure 20-16 *A red tide:* This harmful algal bloom in Washington state's Puget sound contains organisms that give the water a red tint and can be toxic to fish, wildlife, and people. ***Question:*** What are two ways in which this sort of pollution could be prevented?

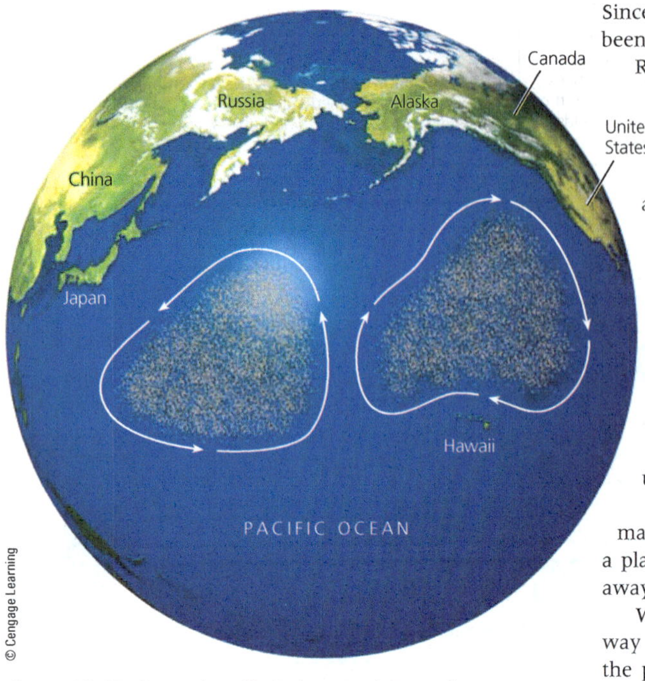

Figure 20-17 The North Pacific Garbage Patch is actually two vast, slowly swirling masses of small plastic particles floating just under the water. Four other huge garbage patches have been discovered in the world's other major oceans.

CASE STUDY

Ocean Garbage Patches: There Is No Away

In 1997, ocean researcher and sea captain Charles Moore accidentally discovered two gigantic, slowly rotating masses of small pieces of plastic and other solid wastes in the middle of the North Pacific Ocean near the Hawaiian Islands. These wastes, floating on or just beneath the ocean's surface and known as the North Pacific Garbage Patch, are trapped there by a vortex where rotating ocean currents called *gyres* meet (Figure 20-17).

Roughly 80% of this trash comes from the land—washed or blown off beaches, pouring out of storm drains, and floating down streams and rivers that empty into the sea from the west coast of North America and the east coast of Asia (**Concept 20-4A**). Most of the rest is dumped into the ocean from cargo and cruise ships.

The North Pacific Garbage Patch is estimated by some scientists to occupy an area at least the size of Texas. However, such estimates are difficult to verify because this plastic-laden soup consists mostly of small rice-grain- to pencil-eraser-sized particles of plastic suspended slightly underwater, and thus it is difficult to see and measure. Since it was discovered, other huge garbage patches have been discovered in other oceans.

Research shows that the tiny plastic particles can be harmful to marine mammals, to some seabirds such as albatrosses (Figure 20-18), and to some fishes and other aquatic species that mistake them for food and swallow them. Because these animals cannot digest the plastic, it can cause them to die from starvation, poisoning, or choking.

Fish that feed on plankton ingest these tiny plastic particles, which can contain PCBs, DDT, hormone-mimicking BPA, and other harmful chemicals. These long-lived toxins can build up to high concentrations in food chains and webs (see Figure 9-13, p. 202), and they can end up in fish sandwiches and seafood dinners. Thus, toxic chemicals from a discarded plastic grocery bag, water bottle, or chips bag could end up in our stomachs. Everything is connected.

Who is to blame for these patches? Each of us risks making a contribution every time we use and discard a plastic item. We may think we have simply thrown it away, but there is no away.

What can be done? There is no practical or affordable way to clean up this mess. We can, however, try to keep the problem from growing worse by focusing on pollution prevention, which means practicing the four Rs of resource use: Refuse, Reduce, Reuse, and Recycle.

Ocean Pollution from Oil

Crude petroleum (oil as it comes out of the ground) and *refined petroleum* (fuel oil, diesel, gasoline, and other processed petroleum products; see Figure 15-4, p. 377) reach

Figure 20-18 It is likely that this plastic debris found in the stomach of this albatross caused it to starve to death.

U.S. Coast Guard

Figure 20-19 The *Deepwater Horizon* drilling platform burned and sank in the Gulf of Mexico after exploding on April 20, 2010.

the ocean from a number of sources and become highly disruptive pollutants (**Concept 20-4A**).

The most visible sources are tanker accidents, such as the *Exxon Valdez* oil spill in the U.S. state of Alaska in 1989, and blowouts at offshore oil drilling rigs, such as that of the BP *Deepwater Horizon* rig in the Gulf of Mexico in 2010 (see the Case Study that follows). However, studies show that the largest source of ocean oil pollution is urban and industrial runoff from land. Much of this waste oil comes from leaks in pipelines, refineries, and other oil-handling and storage facilities. An estimated one-third to one-half of all ocean oil pollution comes from oil and oil products that are intentionally dumped or accidentally spilled or leaked onto the land or into sewers by home owners and industries.

Volatile organic hydrocarbons in oil kill many aquatic organisms immediately upon contact, especially if these animals are in their vulnerable larval forms. Other chemicals in oil form tar-like globs that float on the surface and coat the feathers of seabirds (see chapter-opening photo) and the fur of marine mammals. This oil coating destroys their natural heat insulation and buoyancy, causing many of them to drown or die of exposure from loss of body heat.

Heavy oil components that sink to the ocean floor or wash into estuaries and coastal wetlands can smother bottom-dwelling organisms such as crabs, oysters, mussels, and clams, or make them unfit for human consumption. Some oil spills have killed coral reefs.

Research shows that, within about 3 years, populations of many forms of marine life can recover from exposure to large amounts of crude oil in warm waters with fairly rapid currents. But in cold and calm waters, full recovery can take decades. In 2009, some 20 years after the *Exxon Valdez* spill, researchers found patches of oil remaining on some stretches of the shoreline of the U.S. state of Alaska's Prince William Sound. In addition, recovery from exposure to refined oil, especially in estuaries and salt marshes, can take 10–20 years or longer. Oil slicks that wash onto beaches can have a serious economic impact on coastal residents, who lose income normally gained from fishing and tourism.

Some oil spills that are not too large can be partially cleaned up by mechanical means including floating booms, skimmer boats, and absorbent devices such as large pillows filled with feathers or hair. But scientists estimate that current cleanup methods can recover no more than 15% of the oil from a major spill.

Thus, *preventing* oil pollution is the most effective and, in the long run, the least costly approach (**Concept 20-4B**). One of the best ways to prevent tanker spills is to use oil tankers with double hulls. Stricter safety standards and inspections could help to reduce oil well blowouts at sea. Most importantly, businesses, institutions, and citizens living in coastal areas must take care to prevent leaks and spillage of even the smallest amounts of oil and oil products such as paint thinners and gasoline.

CONSIDER THIS. . .

THINKING ABOUT Ocean Oil Pollution

What are three ways in which you might be contributing to ocean oil pollution? How could you reduce your contribution to this environmental problem?

CASE STUDY

The BP *Deepwater Horizon* Oil-Rig Spill

On April 20, 2010, the world learned a harsh lesson about the possible environmental impacts of deep-sea oil drilling when the BP Company's *Deepwater Horizon* offshore oil-drilling rig exploded (Figure 20-19). The accident occurred in the Gulf of Mexico, 64 kilometers (40 miles) off the Louisiana coast, after the wellhead on the ocean bottom, almost 1.6 kilometers (1 mile) below the surface, ruptured and released natural gas. The resulting fire and explosion killed 11 of the rig's crewmates and injured 17 more. After burning and belching oil smoke into the air for 36 hours, the rig sank.

During the 3 months following this accident, the ruptured wellhead on the ocean floor released about 4.9 million barrels (206 million gallons) of crude oil before it was capped. It was the largest accidental oil spill that ever occurred in U.S. waters. After the rupture, engineers worked frantically, trying to cap the gushing well from almost a mile above the wellhead, using robot-controlled submarines. It took them several attempts before they succeeded in stopping the flow of oil.

The oil contaminated some ecologically vital coastal marshes, mangrove forests, sea-grass beds, fish nurseries, and deep coral reefs along Louisiana's coast and along the barrier island coastlines of Mississippi, Alabama, and parts of Florida. Scientists suspect that deep-ocean aquatic organisms were affected, but the true extent of the ecological damage will not be known for years. According to estimates from the U.S. Coast Guard, the National Oceanic and Atmospheric Administration (NOAA) and BP, the oil spill killed at least 6,100 seabirds and oiled another 2,000 (see chapter-opening photo) and killed more than 600 sea turtles and oiled another 460.

The spill also disrupted the livelihoods of people who depend on the Gulf Coast's fisheries, and it caused large economic losses for the area's tourism businesses (Figure 20-20). By 2012, BP had spent or pledged to spend roughly $40 billion for cleanup costs, damages, restoration costs, and fines for felony misconduct and neglect. In 2013, the company was in U.S. federal court facing the possibility of being held accountable for several billion dollars more in damages and additional penalties.

Several studies, including that of a Presidential commission, have concluded that the main causes of the accident were failure of equipment that could have detected the leak earlier, a faulty blowout preventer, failure of several safety valves, and a number of poor decisions made by workers and managers. This event also revealed flaws in federal oversight of offshore drilling, which many argue has included too cozy a relationship between the oil industry and its Department of Interior regulators.

Since the accident, the U.S. government has developed new standards for each step in the offshore drilling process and agreed to give a higher priority to environmental concerns in the leasing process. However, the Presidential oil spill commission called for close oversight by Congress to ensure that such reforms are put in place and carefully monitored.

Tyrone Turner/National Geographic Creative

Figure 20-20 This cleanup crew in protective suits was combing the beach in Gulf Shores, Alabama after the 2010 BP oil spill, looking for deposits of toxic crude oil.

The 2012 report of the Presidential oil spill commission faulted the U.S. Congress for not providing funding to help prevent another catastrophe and to implement a major program of ecosystem restoration of the Gulf of Mexico's wetlands and barrier islands. The report also included a strong warning about the risks of drilling in arctic waters, where harsh conditions could greatly increase the risk of oil spills that would likely cause much longer-lasting damage than that experienced in the Gulf of Mexico.

The widespread global publicity following the BP accident helped to educate the public about the dangers of accidents that can release large quantities of oil into the oceans. However, there has been little attention focused on the numerous smaller oil spills that have been occurring more-or-less continually for almost 50 years in Nigeria's oil-producing Niger Delta, where there is little or no government oversight of oil company operations. Each year, an amount of oil roughly equal to that of the Exxon Valdez disaster leaks and spills from rusted and aging pipes and other facilities in the delta region. Decades of such leaks have caused widespread ecological destruction and threats to human health. Some scientists argue that these spills altogether are more damaging than the large, well-known spills.

20-5 How Can We Deal with Water Pollution?

CONCEPT 20-5

Reducing water pollution requires that we prevent it, work with nature to treat sewage, and use natural resources far more efficiently.

Reducing Ocean Water Pollution

Most ocean pollution occurs in coastal waters and comes from human activities on land (**Concept 20-4A**). Figure 20-21 lists ways to prevent pollution of coastal waters and ways to reduce it.

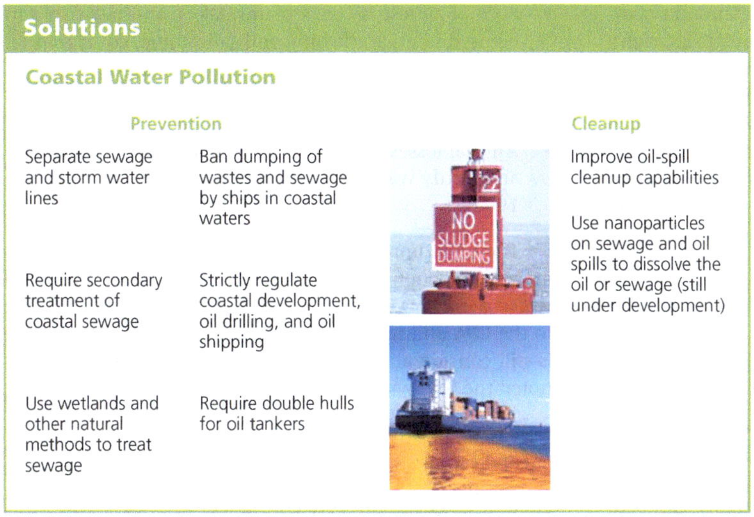

Figure 20-21 Methods for preventing excessive pollution of coastal waters and methods for cleaning it up (Concept 20-4B). ***Question:*** Which two of these solutions do you think are the best ones? Why?

The key to protecting the oceans is to reduce the flow of pollution from land and air and from streams emptying into these waters (Concept 20-4B). Thus, ocean pollution control must be linked with land-use and air pollution control policies, which in turn are linked to energy policies (see Figure 16-36, p. 434) and climate policies (Figure 19-22, p. 528).

Reducing Surface Water Pollution from Nonpoint Sources

There are a number of ways to reduce nonpoint sources of water pollution, most of which come from agricultural practices. Examples include the following: GOOD NEWS

- Reducing soil erosion and fertilizer runoff by keeping cropland covered with vegetation and using conservation tillage (see Chapter 12, p. 305) and other soil conservation methods (see Figure 12-28, p. 305).
- Using fertilizers that release plant nutrients slowly.
- Using no fertilizers on steeply sloped land.
- Relying more on organic farming (see Chapter 12, Core Case Study, p. 278) and more sustainable food production (see Figure 12-33, p. 309) to reduce the use and runoff of plant nutrients and pesticides.
- Planting buffer zones of vegetation between cultivated fields (Figure 20-22) and between animal waste storage sites and nearby surface waters.
- Setting discharge standards for nitrate chemicals from sewage treatment and industrial plants.

The annual formation of the dead zone in the Gulf of Mexico (Core Case Study and Figure 20-C) will be difficult to prevent because of the importance of the Mississippi River basin for growing crops. However, nutrient inputs can be reduced with the widespread use of the fertilizer management practices listed above.

Experience shows that we can sharply reduce the formation of dead zones. For example, a vast dead zone in the Black Sea largely disappeared in the 1990s when the breakup of the Soviet Union led to a sharp increase in fertilizer prices. Reducing nutrient inputs has also shrunk the sizes of dead zones in New York's Hudson River and in California's San Francisco Bay. Also, countries along Europe's Rhine River have reduced nitrate levels in the North Sea's dead zone by about 35%, primarily by reducing nitrate inputs from sewage treatment and industrial plants.

In addition to reducing nitrate inputs, it would be helpful to capture nutrients after they leave farm fields. This would involve two steps. The first would be to protect the

Figure 20-22 This buffer zone of vegetation on an Iowa farm helps to reduce runoff of fertilizer and pesticides into waterways from nearby crop fields. However, such zones are rare.

remaining inland and coastal wetlands, vegetation in *riparian zones* (zones along river banks; see Figure 10-23, right, p. 232), and floodplains, all of which act as natural filters to pull nutrients from waters that are running off the land and spilling out of streams. However, many of these natural features have been replaced by cropland. For example, in Ohio, Indiana, and Iowa, landowners have drained more than 80% of these states' natural wetlands, mostly to plant crops. Also, many wetlands and floodplains have been cut off from their sources of freshwater by flood-control levees along the rivers that feed them.

The second step would involve restoring key wetlands that have been destroyed or degraded and, where feasible, restoring and reconnecting rivers to their natural floodplains. In addition to reducing water pollution, this would restore natural habitats for a variety of species in keeping with the biodiversity **principle of sustainability** (see Figure 1-2, p. 6 or back cover).

Point-source water pollution is considerably easier to control than non-point-source pollution, and there have been a number of success stories around the world in this area. One of the best examples is that of U.S. efforts to deal with this problem (see the following Case Study).

CASE STUDY

The U.S. Experience with Reducing Point-Source Pollution

Efforts to control pollution of surface waters in the United States are based on the Federal Water Pollution Control Act of 1972 (renamed the Clean Water Act when it was amended in 1977) and the 1987 Water Quality Act. The Clean Water Act sets standards for allowed levels of 100 key water pollutants and requires polluters to get permits that limit the amounts of these various pollutants that they can discharge into aquatic systems.

The EPA has also been experimenting with a *discharge trading policy,* which uses market forces to reduce water pollution as has been done with sulfur dioxide for air pollution control (see Chapter 18, p. 495). Under this program, a permit holder can pollute at higher levels than allowed in its permit by buying credits from permit holders who are polluting below their allowed levels. Environmental scientists warn that the effectiveness of such a system depends on how low the cap on total pollution levels in any given area is set and on how regularly the cap is lowered. They also warn that discharge trading could allow water pollutants to build up to dangerous levels in areas where credits are bought.

According to the EPA, the Clean Water Act of 1972 led to numerous improvements in U.S. water quality, including the following:

- 95% of all Americans are served by public drinking water systems that must meet federal health standards.
- 60% of all tested U.S. streams, lakes, and estuaries can be used safely for fishing and swimming, compared to 33% in 1972.
- 75% of the U.S. population is served by sewage treatment plants.
- Annual losses of U.S. wetlands that naturally absorb and purify water have been reduced by 80% since 1992.

These are impressive achievements, given the increases in the U.S. population and its per capita consumption of water and other resources since 1972. However, according to the EPA and recent studies by the *New York Times,* there is still considerable room for improvement. Those reports noted the following:

- About 40% of the nation's surveyed streams, lakes, and estuaries are still too polluted for swimming or fishing.
- Runoff of animal wastes from hog, poultry, and cattle feedlots and meat processing facilities pollutes seven of every ten U.S. rivers.
- Tens of thousands of gasoline storage tanks in 43 states are leaking.
- One in five U.S. water treatment systems violated the Safe Drinking Water Act between 2003 and 2008, releasing sewage and chemicals such as arsenic and radioactive uranium.
- There are currently no federal drinking water standards for hexavalent chromium, which studies indicate is a potent carcinogen. In 2011, residents of Midland, Texas, sued Dow Chemical for allegedly putting dangerous levels of this chemical into their drinking water.
- About 45% of the country's largest water polluters have declared that the Clean Water Act no longer applies to waters that they are polluting. This results from a U.S. Supreme Court decision that created uncertainty over which waterways are protected by the law. About 117 million Americans (32% of the population) get some or all of their drinking water from sources that are being polluted in this way.

Some environmental scientists call for strengthening the Clean Water Act. Suggested improvements include shifting the focus of the law to water pollution prevention instead of focusing mostly on end-of-pipe removal of specific pollutants; greatly increased monitoring for violations of the law and much larger mandatory fines for violators; and regulating irrigation water quality (for which there is no federal regulation). Another suggestion is to expand the rights of citizens to bring lawsuits to ensure that water pollution laws are enforced. Still another suggestion is to rewrite the Clean Water Act to clarify that it covers all waterways (as Congress originally intended). This would eliminate confusion about which waterways are covered and stop major polluters from using this confusion to keep polluting in many areas.

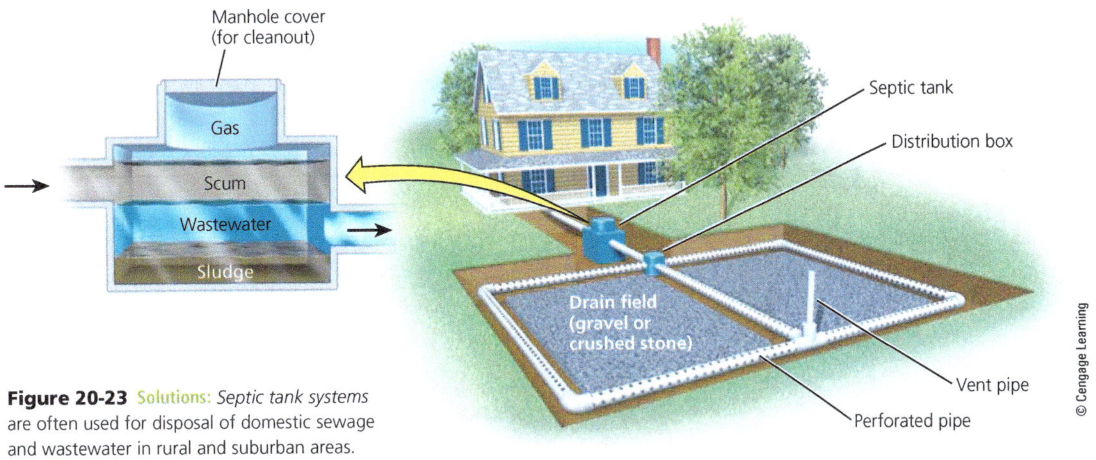

Figure 20-23 Solutions: *Septic tank systems are often used for disposal of domestic sewage and wastewater in rural and suburban areas.*

Many businesses and legislators oppose these proposals, contending that the Clean Water Act's regulations are already too restrictive and costly. Some state and local officials argue that in many communities, it is unnecessary and too expensive to test all the water for pollutants as required by federal law.

Some members of Congress, under pressure from regulated industries, go further and suggest seriously weakening or repealing the Clean Water Act. They contend that such regulations hinder economic growth and prevent job growth. In 2012, William K. Reilly, who headed the EPA from 1989 to 1993 and served as a co-chairman of the Presidential commission on offshore drilling, said: "If we buy into the misguided notion that reducing protection of our waters will somehow ignite the economy, we will shortchange our, health, environment, and economy."

Sewage Treatment Reduces Water Pollution

In rural and suburban areas of more-developed countries, sewage from each house usually is discharged into a **septic tank** with a large drainage field (Figure 20-23). In such a system, household sewage and wastewater is pumped into a settling tank, where grease and oil rise to the top and solids fall to the bottom and are decomposed by bacteria. The partially treated wastewater that results is discharged in a large drainage field through small holes in perforated pipes embedded in porous gravel or crushed stone just below the soil's surface. As these wastes drain from the pipes and percolate downward, the soil filters out some potential pollutants and soil bacteria decompose biodegradable materials.

About one-fourth of all homes in the United States are served by septic tanks. If these systems are not installed correctly and pumped out regularly, they can cause sewage to back up into homes or to pollute nearby groundwater and surface water. Chlorine bleaches, drain cleaners, and antibacterial soaps should not be used in these systems, because they can kill the bacteria that decompose the wastes. Kitchen sink garbage disposals should not be used either, because they can overload septic systems.

In urban areas in the United States and other more-developed countries, most waterborne wastes from homes, businesses, and storm runoff flow through a network of sewer pipes to *wastewater* or *sewage treatment plants*. Raw sewage reaching a treatment plant typically undergoes one or two levels of wastewater treatment. The first is **primary sewage treatment**—a *physical* process that uses screens and a grit tank to remove large floating objects and to allow solids such as sand and rock to settle out. Then the waste stream flows into a primary settling tank where suspended solids settle out as sludge (Figure 20-24, left).

By itself, primary treatment removes about 60% of the suspended solids and 30–40% of the oxygen-demanding organic wastes from sewage. It removes no phosphates, nitrates, salts, radioisotopes, trace amounts of medications, pesticides, or pathogens such as viruses.

The second level is **secondary sewage treatment**—a *biological* process in which aerobic bacteria remove as much as 90% of dissolved and biodegradable, oxygen-demanding organic wastes (Figure 20-24, right). A combination of primary and secondary treatment removes 95–97% of the suspended solids and oxygen demanding organic wastes, 70% of most toxic metal compounds and nonpersistent synthetic organic chemicals, 70% of the phosphorus, and 50% of the nitrogen. However, this process removes only a tiny fraction of the persistent and potentially toxic organic substances found in some pesticides and in discarded medicines that people put into sewage systems, and it does not kill pathogens.

A third level of cleanup, *advanced* or *tertiary sewage treatment*, uses a series of specialized chemical and physical processes to remove specific pollutants left in the water after primary and secondary treatment. In its most common form, advanced sewage treatment uses special filters to remove phosphates and nitrates from wastewater before it is discharged into surface waters. This third

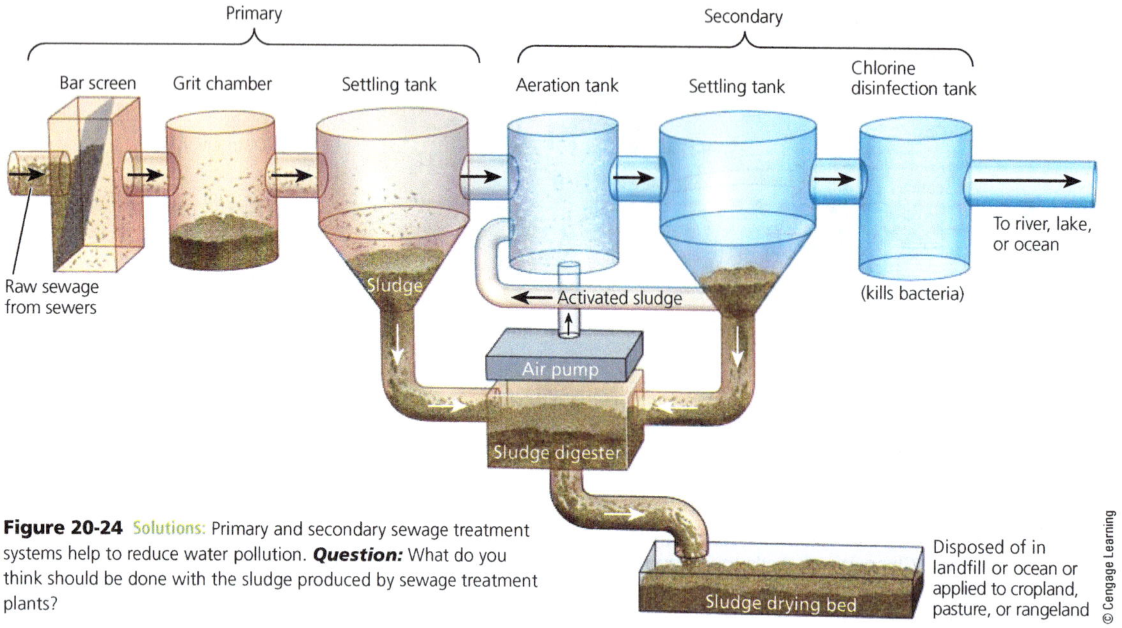

Figure 20-24 Solutions: Primary and secondary sewage treatment systems help to reduce water pollution. ***Question:*** What do you think should be done with the sludge produced by sewage treatment plants?

stage would help a great deal in reducing nutrient overload from nitrates and phosphates, but because of its high costs, it is not widely used.

Before discharge, water from sewage treatment plants usually undergoes *bleaching*, to remove water coloration, and *disinfection*, to kill disease-carrying bacteria and some (but not all) viruses. The usual method for accomplishing this is *chlorination*. But chlorine can react with organic materials in water to form small amounts of chlorinated hydrocarbons. Some of these chemicals cause cancers in test animals, can increase the risk of miscarriages, and can damage the human nervous, immune, and endocrine systems. The use of other disinfectants, such as ozone and ultraviolet light, is increasing, but they cost more and their effects do not last as long as those of chlorination.

Federal law in the United States requires primary and secondary treatment in all municipal sewage treatment plants, but exemptions from secondary treatment are possible for towns and cities that can show that the cost of installing such systems poses an excessive financial burden. In addition, the EPA says that at least two-thirds of the country's sewage treatment plants have sometimes violated water pollution regulations. At least 500 U.S. cities have failed to meet federal standards for sewage treatment plants, and 34 East Coast cities simply screen out large floating objects from their sewage before discharging it into the coastal waters of the Atlantic Ocean.

Some cities have two separate networks of pipes, one for carrying storm-water runoff from streets and parking lots, and the other for carrying sewage. But about 1,200 U.S. cities have combined these two systems into one set of pipes to cut costs. Heavy rains or a high volume of sewage from too many users hooked up to such combined systems can cause them to overflow and discharge untreated sewage (called *effluents*) directly into surface waters such as the Great Lakes (see Case Study, p. 552).

According to the EPA, at least 40,000 of these overflows occur each year in the United States, and about 7.1 million people get sick every year from swimming in waters contaminated by overflows that contain sewage. The best way to deal with this overflow problem is to prevent it by using separate systems to carry storm water and sewage (**Concept 20-5**).

We Can Improve Conventional Sewage Treatment

Some scientists call for redesigning the conventional sewage treatment system shown in Figure 20-24. The idea is to prevent toxic and hazardous chemicals from reaching sewage treatment plants and thus from getting into sludge and water discharged from such plants.

They suggest several ways to implement this pollution prevention approach. One is to require industries and businesses to remove toxic and hazardous wastes from water sent to municipal sewage treatment plants. Another is to encourage industries to reduce or eliminate their use and waste of toxic chemicals.

A major problem is that building and maintaining sewage treatment plants costs too much for governments in less-developed countries where untreated or poorly treated wastewater causes environmental and human health problems. Some individuals are trying to address this problem by showing that wastewater and sewage sludge can be thought of as valuable resources to be recycled (Individuals Matter 20.1).

individuals matter 20.1

Ashley Murray: Wastewater Engineer, Entrepreneur, and National Geographic Emerging Explorer

© Matt Muspratt

Ashley Murray grew up in a family of small business owners and earned MS and PhD degrees in engineering from the University of California. She has worked in several countries, including China, India, Senegal, Kenya, Uganda, and Ghana, where she now lives. She has used her background in business, science, and engineering to found and run Waste Enterprisers, a company devoted to revolutionizing the way we deal with wastewater and fecal sludge, especially in less-developed countries.

Her approach is to promote cleaning up and reusing wastewater by viewing the energy it contains as a financially valuable resource that can be harnessed and sold. Her business relies on using human waste as a raw material and converting it into a renewable solid fuel that can be used by industries such as cement plants. Revenue from the sale of fuel more than covers the processing costs, essentially turning sanitation into the byproduct of manufacturing fuel.

After 18 months of research to validate the concept and conduct bench-scale experiments, Waste Enterprisers has built a demonstration plant in Accra, Ghana, where they have the capacity to produce 2 tons of fuel per day. This is produced from approximately 100 cubic meters of raw human waste from the city's pit latrines and septic tanks (see background photo). Waste Enterprisers are conducting burning trials with local industries and plan to build a commercial scale plant in Kenya next year.

Waste Enterprisers' fuel can be used as a replacement for coal in kilns and boilers. When cement industries burn the fuel they can even incorporate the residual ash into their cement product, making the process a zero-waste, closed loop solution to sanitation.

Murray hopes to build human waste-to-fuel plants across sub-Saharan Africa. She says, "the real goal is improving basic sanitation, health, and environmental conditions for some of the world's poorest populations."

Background photo: ©Ashley Murray

©Peepoople/Camilla Wirseen

Figure 20-25 The Peepoo is a single-use biodegradable plastic bag that can be used as a toilet in the urban slums of less-developed countries. Each bag costs 3¢ and the company buys back the used bags for 1¢ and turns the waste into fertilizer.

Another suggestion is to require or encourage more households, apartment buildings, and offices to eliminate sewage outputs by switching to waterless, odorless *composting toilet systems,* to be installed, maintained, and managed by professionals. These systems, pioneered several decades ago in Sweden, convert nutrient-rich human fecal matter into a soil-like humus that can be used as a fertilizer supplement.

By using such systems, we can return the plant nutrients in human waste to the soil, thus mimicking the chemical cycling **principle of sustainability.** This would reduce the need for energy-intensive, water-polluting commercial fertilizers. Composting toilets also save large amounts of water, reduce water bills, and decrease the amount of energy used to pump and purify water along with the resulting greenhouse gas emissions. For example, in the United States the collection, distribution, and treatment of drinking water and wastewater (Figure 20-24) releases as much CO_2 each year as 10 million cars, and this could be reduced if more people used waterless composting toilets. The downside of these systems is the initial cost for home owners, which is much higher than that of conventional toilets.

One of the authors of this book (Miller) used a waterless compositing toilet for over a decade with no problems in his office-home facility. This more environmentally sustainable replacement for the conventional toilet is now being used in parts of more than a dozen countries, including China, India, Mexico, Syria, and South Africa.

In 2009, a Swedish architect and professor developed a biodegradable single-use plastic bag, called a Peepoo (Figure 20-25), that can be used as a toilet in urban slums and in other areas where 40% of the world's people do not have access to toilets. After it is used, the bag is knotted and buried. A thin layer of urea in this bag kills the disease-producing pathogens in the feces and helps break down the waste into plant nutrients that are then recycled—a simple and inexpensive, low-tech application of the chemical cycling **principle of sustainability**.

Solutions

Water Pollution

- Prevent groundwater contamination
- Reduce nonpoint runoff
- Work with nature to treat sewage and reuse treated wastewater
- Find substitutes for toxic pollutants
- Practice the four Rs of resource use (refuse, reduce, reuse, recycle)
- Reduce air pollution
- Reduce poverty
- Slow population growth

Figure 20-26 Ways to prevent or reduce water pollution (Concept 20-5). ***Question:*** Which two of these solutions do you think are the best ones? Why?

Many communities are using unconventional, but highly effective, *wetland-based sewage treatment systems,* which work with nature (Science Focus 20.2). Such systems would be ideal in some less-developed countries where water is short and disease is rampant.

CONSIDER THIS. . .

CAMPUS SUSTAINABILITY

Oberlin College

Robb Williamson/NREL

Oberlin College in Ohio is consistently listed as one of the greenest colleges in the United States. A group of Oberlin students worked with faculty members and architects to design a more sustainable environmental studies building powered by solar panels, which produce 30% more electricity than the building uses. A living machine (Science Focus 20.2) in the building's greenhouse purifies all of its wastewater. The building collects rainwater that is used to irrigate the campus's grasses, gardens, and meadow, which contain a diversity of plant and animal species. The college also gets half of its electricity from renewable energy resources and has a car-sharing program.

There Are Sustainable Ways to Reduce or Prevent Water Pollution

It is encouraging that since 1970, most of the world's more-developed countries have enacted laws and regulations that have significantly reduced point-source water

SCIENCE FOCUS 20.2

TREATING SEWAGE BY WORKING WITH NATURE

Some communities and individuals are seeking better ways to purify sewage by working with nature (**Concept 20-5**). Biologist John Todd has developed an ecological approach to treating sewage, which he calls *living machines* (Figure 20-C).

This purification process begins when sewage flows into a passive solar greenhouse or outdoor site containing rows of large open tanks populated by an increasingly complex series of organisms. In the first set of tanks, algae and microorganisms decompose organic wastes, with sunlight speeding up the process. Water hyacinths, cattails, bulrushes, and other aquatic plants growing in the tanks take up the resulting nutrients.

After flowing though several of these natural purification tanks, the water passes through an artificial marsh made of sand, gravel, and bulrushes, which filters out algae and remaining organic waste. Some of the plants also absorb, or *sequester,* toxic metals such as lead and mercury and secrete natural antibiotic compounds that kill pathogens.

Next, the water flows into aquarium tanks, where snails and zooplankton consume microorganisms and are in turn consumed by crayfish, tilapia, and other fish that can be eaten or sold as bait. After 10 days, the clear water flows into a second artificial marsh for final filtering and cleansing. The water can be made pure enough to drink by treating it with ultraviolet light or by passing the water through an ozone generator, usually immersed out of sight in an attractive pond or wetland habitat. Operating costs are about the same as those of a conventional sewage treatment plant. These systems are widely used on a small scale. However, they have been difficult to maintain on a scale large enough to handle the typical variety of chemicals in the sewage wastes from more-developed urban areas.

More than 800 cities and towns around the world (150 in the United States) use natural or artificially created wetlands to treat sewage as a lower-cost alternative to expensive waste treatment plants. For example, in Arcata, California—a coastal town of about 17,000 people—scientists and workers created some 65 hectares (160 acres) of wetlands between the town and the adjacent Humboldt Bay. The marshes and ponds, developed on land that was once a dump, act as a natural waste treatment plant. The project cost was less than half the estimated price of a conventional treatment plant.

This system returns purified water to Humboldt Bay, and the sludge that is removed is processed for use as fertilizer. The marshes and ponds also serve as an Audubon Society bird sanctuary, which provides habitats for thousands of seabirds, otters, and other marine animals.

Figure 20-C Solutions: Ecological wastewater purification system called a *living machine* at the Solar Sewage Treatment Plant in the U.S. city of Providence, Rhode Island. Biologist John Todd is demonstrating this ecological process that he invented for purifying wastewater by using the sun and a series of tanks containing living organisms.

The town even celebrates its natural sewage treatment system with an annual "Flush with Pride" festival.

This approach and the living machine system developed by John Todd apply all three scientific **principles of sustainability**: using solar energy, employing natural processes to remove and recycle nutrients and other chemicals, and relying on a diversity of organisms and natural processes.

Critical Thinking

Can you think of any disadvantages of using such a nature-based system instead of a conventional sewage treatment plant? Do you think any such disadvantages outweigh the advantages? Why or why not?

pollution. These improvements were largely the result of *bottom-up* political pressure on elected officials from individuals and groups. GOOD NEWS

On the other hand, little has been done to reduce water pollution in most of the less-developed countries. However, China plans to provide all of its cities with small sewage treatment plants that will make wastewater clean enough to be recycled back into the urban water supply systems. If China is successful in this ambitious plan, it could become a world leader in developing ways to tackle both water pollution and water scarcity through systems that recycle water.

Many environmental and health scientists argue that the next step is to increase efforts to reduce and prevent water pollution in both more- and less-developed countries. They would begin by asking the question: *How can we avoid producing water pollutants in the first place?* (**Concept 20-5**). Figure 20-26 lists ways to achieve this goal over the next several decades.

This shift to pollution prevention will not take place unless citizens put political pressure on elected officials and also take actions to reduce their own daily contributions to water pollution. Figure 20-27 lists some actions you can take to help reduce water pollution.

What Can You Do?

Reducing Water Pollution

- Fertilize garden and yard plants with manure or compost instead of commercial inorganic fertilizer
- Minimize use of pesticides, especially near bodies of water
- Prevent yard wastes from entering storm drains
- Do not use water fresheners in toilets
- Do not flush unwanted medicines down the toilet
- Do not pour pesticides, paints, solvents, oil, antifreeze, or other harmful chemicals down the drain or onto the ground

Figure 20-27 Individuals matter: You can help reduce water pollution. ***Question:*** Which three of these steps do you think are the most important ones to take? Why?

Big Ideas

- There are a number of ways to purify drinking water, but the most effective and least costly strategy is pollution prevention.
- The key to protecting the oceans is to reduce the flow of pollution from land and air, and from streams emptying into ocean waters.
- Reducing water pollution requires that we prevent it, work with nature in treating sewage, and use natural resources far more efficiently.

TYING IT ALL TOGETHER Dead Zones and Sustainability

The **Core Case Study** that opens this chapter explains how certain farming practices can violate the chemical cycling **principle of sustainability** (see Figure 1-2, p. 6 or back cover) by disrupting the nitrogen cycle. This occurs every year when nitrate plant nutrients in fertilizers flow in runoff from farmland into streams in the Mississippi River basin. Much of this flow goes into the Mississippi River and ends up overfertilizing an area of the Gulf of Mexico, depleting the dissolved oxygen and reducing seafood production in these coastal waters. This process also violates the biodiversity **principle of sustainability** by disrupting ecological interactions among species in river and coastal ecosystems.

The three scientific **principles of sustainability** can guide us in reducing and preventing water pollution. We can use solar energy to purify much more of the water we use, and this will help us to reduce water waste. We can also use natural nutrient cycles (chemical cycling) to treat our wastes in wetland-based sewage treatment systems (Science Focus 20.2). In addition, we can view human waste as an important and potentially profitable source of nutrients. Finally, by preventing the pollution and degradation of aquatic systems and their bordering terrestrial systems, we will help to preserve biodiversity—a key component of the life-support system that we depend on for maintaining water supplies and water quality.

Chapter Review

Core Case Study

1. Describe the nature and causes of the annual dead (oxygen-depleted) zone in the Gulf of Mexico.

Section 20-1

2. What are the two key concepts for this section? What is **water pollution**? Distinguish between **point sources** and **nonpoint sources** of water pollution and give two examples of each. Explain how certain efforts to clean up the air can contribute to water pollution. Explain how atmospheric warming will likely increase water pollution.

3. List seven major types of water pollutants and three diseases that can be transmitted to humans through polluted water. Describe some of the chemical and biological methods that scientists use to measure water quality.

Section 20-2

4. What are the two key concepts for this section? Explain how streams can cleanse themselves and how this cleansing process can be overwhelmed. Describe the varying states of stream pollution in more- and less-developed countries. Give two reasons why lakes cannot cleanse themselves as readily as streams can. Distinguish between **eutrophication** and **cultural eutrophication**. List ways to prevent or reduce cultural eutrophication. Explain how seasonal dead zones form in the Gulf of Mexico. Describe the pollution of the Great Lakes and the progress being made in reducing this pollution.

Section 20-3

5. What are the two key concepts for this section? Explain why groundwater cannot cleanse itself very well. What are the major sources of groundwater contamination in the United States? List three harmful effects of groundwater pollution. Describe the threat from arsenic in groundwater. List three ways to prevent groundwater contamination and three ways to clean up groundwater contamination. Explain why prevention is the most important approach.

6. Explain how drinking water is purified typically in more-developed countries. List three ways to provide safe drinking water in poor countries. Describe the environmental problems caused by the widespread use of bottled water. Summarize the U.S. laws for protecting drinking water quality. List three ways to strengthen the U.S. Safe Drinking Water Act.

Section 20-4

7. What are the two key concepts for this section? How are coastal waters and deeper ocean waters most often polluted? What causes algal blooms and what are their harmful effects? Describe an ocean garbage patch and explain how it can harm marine life and how the growth of such patches could be prevented. How serious is oil pollution of the oceans? What are its effects and what can be done to reduce such pollution? Describe the 2010 BP *Deepwater Horizon* oil blowout and its causes and effects. How might such accidents be prevented in the future?

Section 20-5

8. What is the key concept for this section? List four ways to reduce water pollution from nonpoint sources. Describe the U.S. successes and failures with using laws and regulations to reduce point-source water pollution. What is a **septic tank** and how does it work? Define **primary sewage treatment** and **secondary sewage treatment** and explain how they are used to treat wastewater.

9. What are three ways to improve conventional sewage treatment? Summarize Ashley Murray's plans for improving sanitation in less-developed countries. What is a waterless composting toilet system? Explain how we can use wetlands to treat sewage. Describe John Todd's use of living machines to treat sewage.

10. List six ways to prevent or reduce water pollution. List five things you can do to help reduce water pollution. What are this chapter's *three big ideas*? Explain how we can use the three scientific **principles of sustainability** (see Figure 1-2, p. 6 or back cover) to prevent and reduce water pollution.

Note: Key terms are in bold type.

Critical Thinking

1. How might you be contributing directly or indirectly to the annual dead zone that forms in the Gulf of Mexico (**Core Case Study**)? What are three things you could do to reduce your impact?

2. Suppose a large number of dead fish are found floating in a lake. How would you determine whether they died from cultural eutrophication or from exposure to toxic chemicals?

3. Assume that you are a regulator charged with drawing up plans for controlling water pollution, and briefly describe your strategy for controlling water pollution from each of the following sources: **(a)** a pipe from a factory discharging effluent into a stream, **(b)** a parking lot at a shopping mall bordered by a stream, and **(c)** a farmer's field on a slope next to a stream.
4. How might you be contributing directly or indirectly to groundwater pollution? What are three things you could do to reduce your impact?
5. When you flush your toilet, where does the wastewater go? Trace the actual flow of this water in your community from your toilet through sewers to a wastewater treatment plant (or to a septic system) and from there to the environment. Try to visit a local sewage treatment plant to see what it does with wastewater. Compare the processes it uses with those shown in Figure 20-24. What happens to the sludge produced by this plant? What improvements, if any, would you suggest for this plant?
6. In your community, **(a)** what is the source of drinking water? **(b)** How is drinking water treated? **(c)** What problems related to drinking water, if any, have arisen in your community? **(d)** What actions, if any, has your local government taken to solve them?
7. List three ways in which you could apply **Concept 20-5** in order to make your lifestyle more environmentally sustainable.
8. Congratulations! You are in charge of the world. What are three actions you would take to **(a)** sharply reduce point-source water pollution in more-developed countries, **(b)** sharply reduce nonpoint-source water pollution throughout the world, **(c)** sharply reduce groundwater pollution throughout the world, and **(d)** provide safe drinking water for the poor and for other people in less-developed countries?

Doing Environmental Science

Do some research using the library and/or the Internet and evaluate the relative effectiveness of three different home water purification devices. Try to determine the costs of these systems, including purchase, installation, and maintenance costs, and record these costs for each system. Determine the type or types of water pollutants that each device is supposed to remove and the effectiveness of each such device. You might also consider looking up reviews of each product and finding some users of these products to interview. Write a report summarizing your findings.

Global Environment Watch Exercise

Visit the *Gulf of Mexico Oil Spill 2010* topic portal and research the ecological effects of the oil well blowout and spill. Write a report that answers the following questions:

1. Is oil from the spill in the Gulf still considered to be a problem?
2. How has it affected the coastal wetlands along the gulf?
3. How has it affected the gulf's bottom-dwelling species?
4. How has it affected bird life in the gulf?
5. How has it affected the fishing industry there?

Data Analysis

In 2006, scientists assessed the overall condition of the estuaries on the western coasts of the U.S. states of Oregon and Washington. To do so, they took measurements of various characteristics of the water, including dissolved oxygen (DO), in selected locations within the estuaries. The concentration of DO for each site was measured in terms of milligrams (mg) of oxygen per liter (L) of water sampled. The scientists used the following DO concentration ranges and quality categories to rate their water samples: water with greater than 5 mg/L of DO was considered *good* for supporting aquatic life; water with 2 to 5 mg/L of DO was rated as *fair;* and water with less than 2 mg/L of DO was rated as *poor.*

The following graph shows measurements taken in bottom water at 242 locations. Each triangle mark represents one or more measurements. The x-axis on this graph represents DO concentrations in mg/L. The y-axis represents percentages of the total area of estuaries studied (estuarine area).

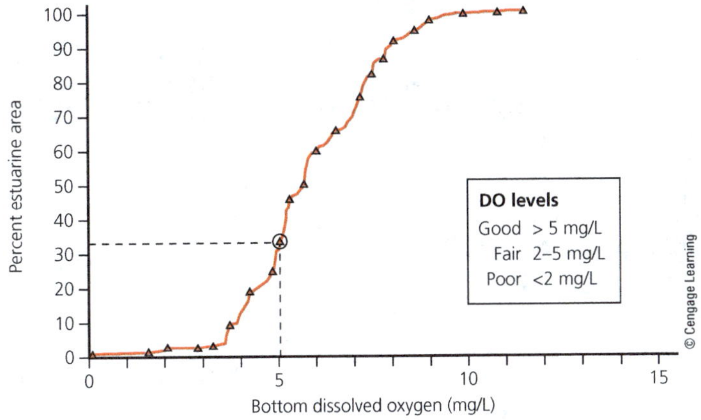

Concentrations of dissolved oxygen in bottom waters of estuaries in Washington and Oregon.

To read this graph, pick one of the triangles and observe the values on the x- and y-axes. For example, note that the circled triangle lines up approximately with the 5-mg/L mark on the x-axis and with a value of about 34% on the y-axis. This means that waters at this particular measurement station (or stations), along with about 34% of the total area being studied, are estimated to have a 5% or lower DO concentration.

Use this information, along with the graph, to answer the following questions:

1. Half of the estuarine area has waters falling below a certain DO concentration level, and the other half has levels above that level. What is that level, in mg/L?
2. Give your estimate of the highest DO concentration measured and your estimate of the lowest concentration.
3. Approximately what percentage of the estuarine area studied is considered to have poor DO levels? About what percentage has fair DO levels, and about what percentage has good DO levels?

21 Solid and Hazardous Waste

Follow nature's example, realize waste's potential.

GUNTER PAULI

Key Questions

21-1 What are solid waste and hazardous waste, and why are they problems?

21-2 How should we deal with solid waste?

21-3 Why are refusing, reducing, reusing, and recycling so important?

21-4 What are the advantages and disadvantages of burning or burying solid waste?

21-5 How should we deal with hazardous waste?

21-6 How can we make the transition to a more sustainable low-waste society?

Recycling computer monitors in Grand Prairie, Texas.

Peter Essick/National Geographic Creative

CORE CASE STUDY E-Waste—An Exploding Problem

Electronic waste, or *e-waste,* consists of discarded television sets, cell phones, computers, and other electronic devices (see chapter-opening photo). It is the fastest-growing solid waste problem in the United States and in the world. Each year, Americans discard several hundred million cell phones, personal computers, television sets, iPods, Blackberries, fax machines, printers, and other electronic products.

Only about 14% of all U.S. e-waste is recycled, according to the Consumer Electronics Association. Most of the rest of it ends up in landfills and incinerators, even though many of the components in electronic devices contain valuable materials that could be recycled or reused. These wasted resources include high-quality plastics and valuable metals such as aluminum, copper, platinum, silver, gold, and rare-earth metals that are used to build most of the world's electronic devices and military weapons systems (see Chapter 14, Core Case Study, p. 350). E-waste is also a source of toxic and hazardous chemicals that can contaminate air, surface water, groundwater, and soil and cause serious and even life-threatening health problems for e-waste workers.

Much of the e-waste in the United States that is not buried or incinerated is shipped to China (Figure 21-1), India, and other Asian and African countries where labor is cheap and environmental regulations are weak. Workers there—many of them children—dismantle, burn, and treat e-waste products with acids to recover valuable metals and reusable parts. As they do this, they are exposed to toxic metals such as lead and mercury and other harmful chemicals. The remaining scrap is dumped into waterways and fields or burned in open fires, exposing many people to highly toxic chemicals called dioxins.

Transfer of such hazardous waste from more-developed to less-developed countries was banned by the International Basel Convention, a treaty now signed by 179 countries to reduce and control the movement of hazardous waste across international borders. Despite this ban, much of the world's e-waste is not officially classified as hazardous waste or is illegally smuggled out of some countries. The United States can export its e-waste legally because it is the only industrialized nation that has not ratified the Basel Convention.

The European Union (EU) has led the way in dealing with e-waste, taking a *cradle-to-grave* approach, which requires manufacturers to take back electronic products at the end of their useful lives and repair, remanufacture, or recycle them. In the EU, e-waste is banned from landfills and incinerators. Japan is also taking this approach. To cover the costs of these programs, consumers pay a recycling tax on electronic products, an example of implementing the full-cost pricing **principle of sustainability** (see Figure 1-5, p. 9 or back cover). However, because of the high costs involved in complying with these laws, much of this waste that is supposed to be recycled domestically is shipped illegally to other countries.

Electronic waste is just one of many types of solid and hazardous waste that we discuss in this chapter.

Peter Essick/National Geographic Creative

Figure 21-1 These workers in Taizhou City in China's Zhejiang Province are recovering valuable materials from scrap computers shipped from the United States. ***Question:*** Have you disposed of an electronic device lately? If so, how did you dispose of it?

21-1 What Are Solid Waste and Hazardous Waste, and Why Are They Problems?

CONCEPT 21-1A
Solid waste contributes to pollution and includes valuable resources that could be reused or recycled.

CONCEPT 21-1B
Hazardous waste contributes to pollution, as well as to natural capital degradation, health problems, and premature deaths.

We Throw Away Huge Amounts of Useful Things

In the natural world, wherever humans are not dominant, there is essentially no waste because the wastes of one organism become nutrients or raw materials for others (see Figure 3-10, p. 59). This natural cycling of nutrients is the basis for one of the three scientific **principles of sustainability** (see Figure 1-2, p. 6 or back cover).

Rechitan Sorin/Shutterstock.com

Modern humans violate this principle in a big way. We produce huge amounts of waste materials that go unused and that pollute the environment. For example, the manufacturing of a single desktop computer requires 700 or more different materials obtained from mines, oil wells, and chemical factories all over the world, along with large amounts of energy. For every 0.5 kilogram (1 pound) of electronics it contains, approximately 3,600 kilograms (8,000 pounds) of waste were created—an amount roughly equal to the weight of a large pickup truck.

Because of the law of conservation of matter (see Chapter 2, p. 40) and the nature of human lifestyles, we will always produce some waste. However, studies and experience indicate that we could reduce this waste of potential resources and the resulting environmental harm it causes by up to 80%.

One major category of waste is **solid waste**—any unwanted or discarded material we produce that is not a liquid or a gas. Solid waste can be divided into two types. One type is **industrial solid waste** produced by mines (see Figure 14-10, p. 358), farms, and industries that supply people with goods and services. The other is **municipal solid waste (MSW)**, often called *garbage* or *trash*, which consists of the combined solid waste produced by homes and workplaces other than factories. Examples include paper and cardboard, food wastes, cans, bottles, yard wastes, furniture, plastics, metals, glass, wood, and e-waste (**Core Case Study**). Some resource experts believe that we should change the name of the trash we produce from MSW of MWR—mostly wasted resources (**Concept 21-1A**).

Much of this waste ends up in rivers (see Figure 20-7, p. 550), lakes (Figure 20-5, p. 547), the ocean (Figure 20-17, p. 560), and natural landscapes (Figure 21-2). Discarded plastic items are a threat to many terrestrial animal species, as well as millions of seabirds (see Figure 20-18, p. 560), marine mammals (see Figure 11-4, p. 253), and sea turtles, which can mistake a floating plastic sandwich bag for a jellyfish or get caught in plastic fishing nets (see Figure 11-1, p. 248). About 80% of the plastics in the ocean are blown or washed in from beaches, rivers, storm drains, and other sources, and the rest get dumped into the ocean from ocean-going garbage barges, ships, and fishing boats.

In more-developed countries, most MSW is buried in landfills or burned in incinerators. In many less-developed countries, much of it ends up in open dumps, where poor people eke out a living finding items they can use or sell (Figure 21-3). The United States is the world's largest producer of MSW (see the following Case Study).

Figure 21-2 Various types of solid waste have been dumped in this isolated mountain area.

William Albert Allard/National Geographic Creative

Figure 21-3 This child and woman survive by searching this open trash dump in India for useful items.

Hazardous Waste Is a Serious and Growing Problem

Another major category of waste is **hazardous**, or **toxic**, **waste**—any discarded material or substance that threatens human health or the environment because it is poisonous, dangerously chemically reactive, corrosive, or flammable. Examples include industrial solvents, hospital medical waste, car batteries (containing lead and acids), household pesticide products, dry-cell batteries (containing mercury and cadmium), and ash and sludge from incinerators and coal-burning power and industrial plants (see Figure 15-18, p. 388, and Figure 15-19, p. 389). Improper handling of these wastes can lead to pollution of air and water, degradation of ecosystems, and severe health threats (Concept 21-1B).

What Harmful Chemical Are in Your Home?

Cleaning
Disinfectants
Drain, toilet, and window cleaners
Spot removers
Septic tank cleaners

Paint Products
Paints, stains, varnishes, and lacquers
Paint thinners, solvents, and strippers
Wood preservatives
Artist paints and inks

General
Dry-cell batteries (mercury and cadmium)
Glues and cements

Gardening
Pesticides
Weed killers
Ant and rodent killers
Flea powders

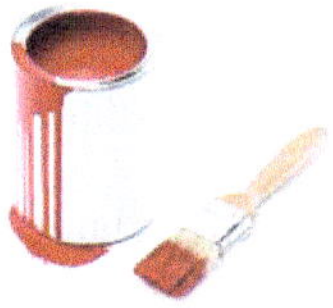

Automotive
Gasoline
Used motor oil
Antifreeze
Battery acid
Brake and transmission fluid

Figure 21-4 Harmful chemicals are found in many homes. The U.S. Congress has exempted the disposal of many of these household chemicals and other items from government regulation. ***Question:*** Which of these chemicals could you find in the place where you live?

Top: ©tuulijumala/Shutterstock.com. Center: Katrina Outland/Shutterstock.com. Bottom: Karramba Production/Shutterstock.com.

Figure 21-4 lists some of the harmful chemicals found in many homes. The two largest classes of hazardous wastes are *organic compounds* (such as various solvents, pesticides, PCBs, and dioxins) and nondegradable *toxic heavy metals* (such as lead, mercury, and arsenic).

Another form of extremely hazardous waste is the highly radioactive waste produced by nuclear power plants and nuclear weapons facilities (see Chapter 15, p. 391). Such wastes must be stored safely for 10,000 to 240,000 years, depending on what radioactive isotopes are present in them. After 60 years of research, scientists and governments have not found a scientifically and politically acceptable way to safely isolate these dangerous wastes for such long periods of time.

According to the U.N. Environment Programme (UNEP), more-developed countries produce 80–90% of the world's hazardous wastes, and the United States is the largest producer. However, as China continues to industrialize rapidly, largely without adequate pollution controls, it may soon take over the number-one spot. In order, the top three U.S. producers of hazardous waste are the military, the chemical industry, and the mining industry.

CASE STUDY

Solid Waste in the United States

The United States leads the world in total solid waste production and in solid waste per person. With only 4.6% of the world's people, the United States produces about 25% of the world's solid waste. According to the U.S. Environmental Protection Agency (EPA), about 98.5% of all solid waste produced in the United States is industrial

Figure 21-5 A year's worth of solid waste produced by family of four in San Diego, California. Recyclable items are on the left and the rest of their MSW is on the right.

solid waste from mining (76%), agriculture (13%), and industry (9.5%). The remaining 1.5% of U.S. solid waste is MSW.

While the EPA reports regularly on the amount and composition of U.S. MSW, Columbia University researchers and *BioCycle* magazine have gathered actual nationwide MSW data and found that it differs significantly from the EPA solid waste data, which is based on theoretical matter-flow models that are being reevaluated. According to this data-based analysis published in 2010, Americans produce much more waste and recycle and compost less waste than the EPA has estimated.

The Columbia and *BioCycle* researchers estimated that the total amount of MSW produced by the United States in 2008 (the latest year for which they had data) was about 354 million metric tons (389 million tons). This means that in 2008 each American threw away an average of about 3.2 kilograms (7 pounds) of trash every day (about a third more than what the EPA estimated), or about 1,168 kilograms (2,570 pounds) in a year (Figure 21-5). The researchers also found that about 24% of these wastes were recycled (which is lower than the EPA estimate of about 34%). These estimates do not include the much larger numbers for industrial solid waste—about 98.5% of all solid waste—generated to support the American economy and lifestyles.

In 2010, the largest categories of this MSW were paper and cardboard (29% of total), food waste (14%), yard waste (13%), plastics (12%), and metals (9%). According to the U.S. Energy Information Administration, packaging makes up 32% of all the country's MSW—the single greatest category of MSW in the United States. Every year, the United States generates enough MSW to fill a bumper-to-bumper convoy of garbage trucks long enough to encircle the globe almost 6 times, or to reach more than halfway to the moon. About 67% of it is dumped in landfills and about 9% is incinerated, but much of it ends up as litter.

Consider some of the solid wastes that consumers throw away in the high-waste economy of the United States:

- Enough tires each year to encircle the planet almost 3 times.
- An amount of disposable diapers each year that, if linked end to end, would reach to the moon and back 7 times.
- Enough carpet each year to cover the U.S. state of Delaware.
- About 2.8 million nonreturnable plastic bottles *every hour*. If laid end-to-end, the number of these bottles produced each year would reach from the earth to the moon and back about 6 times.
- About 274 million plastic shopping bags per day, an average of about 3,200 every second.
- Enough office paper each year to build a wall 3.5 meters (11 feet) high across the country from New York City to San Francisco, California.
- Enough plastic wrap each year to shrink-wrap Texas.
- 25 billion nonrecyclable plastic foam cups a year—enough, if lined up end to end, to circle the earth at its equator 436 times.

Most of these wastes break down very slowly, if at all. Lead, mercury, glass, plastic foam, and most plastic bottles essentially take forever to break down; an aluminum can takes 500 years; plastic bags, 400 to 1,000 years; and a plastic six-pack holder, 100 years (Science Focus 21.1).

SCIENCE FOCUS 21.1

GARBOLOGY AND TRACKING TRASH

How do we know about the composition of trash in landfills? Much of that information comes from research by *garbologists* such as William Rathje, an anthropologist who pioneered the field of garbology in the 1970s at the University of Arizona. These scientists work in the fashion of archaeologists, training their students to sort, weigh, and itemize people's trash, and to bore holes in garbage dumps and analyze what they find.

Many people think of landfills as huge compost piles where biodegradable wastes are decomposed within a few months. But garbologists looking at the contents of landfills found 50-year-old newspapers that were still readable and hot dogs and pork chops buried for decades that had not decayed. In landfills (as opposed to open dumps), trash can resist decomposition for perhaps centuries because it is tightly packed and protected from sunlight, water, and air, and from the bacteria that could digest and decompose much of these wastes in keeping with the chemical cycling **principle of sustainability**.

In 2009, a team of researchers, led by Carlo Riatti, at the Massachusetts Institute of Technology (MIT) conducted a project called "Trash Track." The project's goals were to find out where urban trash goes and to help New York City increase its recycling rate from the current 30% to 100% by 2030.

The researchers attached wireless transmitters about the size of a matchbook to several thousand different items of trash produced by volunteer participants in New York City, Seattle, Washington, and London, England. They are now using these battery-operated electronic tags to track the trash items and show their real-time movements whether they go to recycling plants, landfills, or incinerators. For example, in Seattle, Washington, 75% of the tracked waste items reached recycling plants.

Critical Thinking

Should landfills be exposed to more air and water to hasten decomposition of their wastes? Explain.

21-2 How Should We Deal with Solid Waste?

CONCEPT 21-2

A sustainable approach to solid waste is first to reduce it, then to reuse or recycle it, and finally to safely dispose of what is left.

We Can Burn, Bury, or Recycle Solid Waste or Produce Less of It

We can deal with the solid wastes we create in two ways. One method is **waste management**, in which we attempt to control wastes in ways that reduce their environmental harm without seriously trying to reduce the amount of waste produced. It typically involves mixing wastes together and then transferring them from one part of the environment to another, usually by burying them, burning them, or shipping them to another location. The second approach is **waste reduction**, in which we produce much less waste and pollution, and the wastes we do produce are considered to be potential resources that we can reuse, recycle, or compost (Concept 21-2).

There is no single solution to the solid waste problem. Most analysts call for using **integrated waste management**—a variety of coordinated strategies for both waste disposal and waste reduction (Figure 21-6).

Environmental scientists and the EPA call for much greater emphasis on waste prevention and reduction and pollution prevention than on waste disposal. Figure 21-7 compares the science-based waste-management goals of the EPA and National Academy of Sciences with actual waste-management trends based on analysis of waste data by Columbia University and *BioCycle* researchers.

Some scientists and economists estimate that we could eliminate up to 80% of the solid waste we produce by rigorously applying the strategies shown in Figure 21-7, left. Let us look more closely at these options in the order of priorities suggested by scientists.

We Can Cut Solid Wastes by Refusing, Reducing, Reusing, and Recycling

Waste reduction (Concept 21-2) is based on the four Rs of resource use:

- **Refuse**: Don't use it.
- **Reduce**: Use less.
- **Reuse**: Use it over and over.
- **Recycle**: Convert it to useful items and buy products made from recycled materials.

From an environmental standpoint, the first three Rs are preferred because they are *input,* or *waste prevention,* approaches that tackle the problem of waste production at the front end—before it occurs. Recycling is important

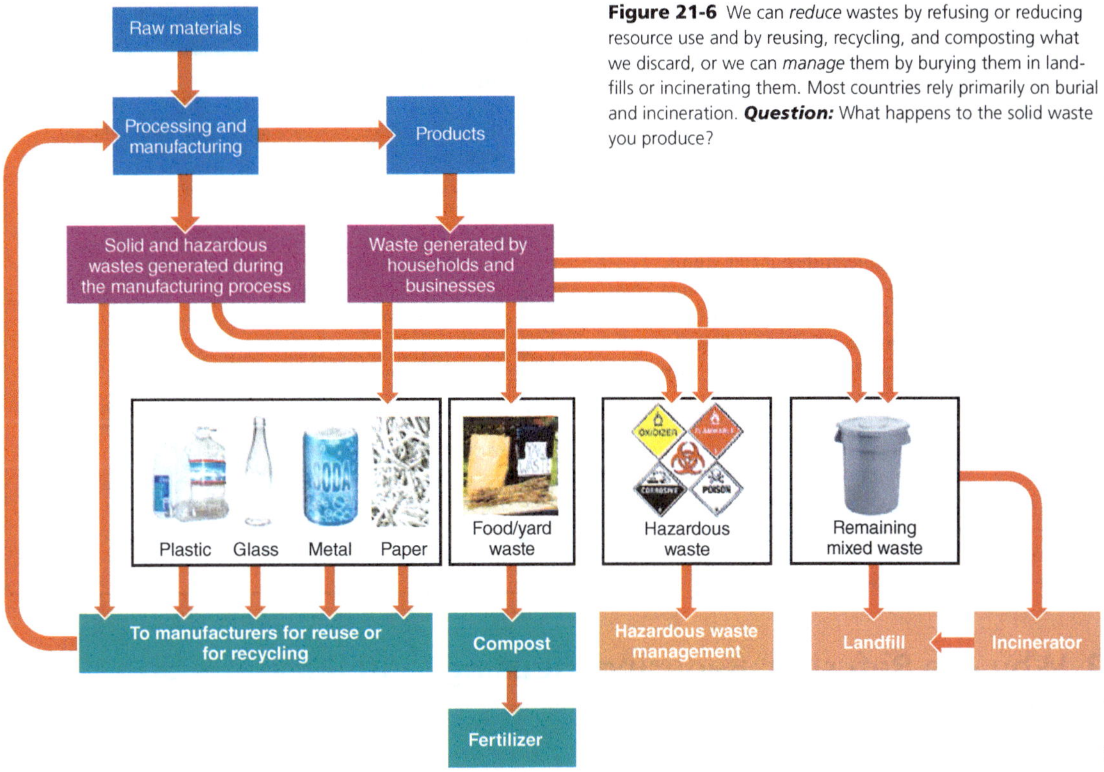

Figure 21-6 We can *reduce* wastes by refusing or reducing resource use and by reusing, recycling, and composting what we discard, or we can *manage* them by burying them in landfills or incinerating them. Most countries rely primarily on burial and incineration. ***Question:*** What happens to the solid waste you produce?

but it deals with wastes after they have been produced. An important form of recycling is **composting**, which involves using bacteria to decompose yard trimmings, vegetable food scraps, and other biodegradable organic wastes into materials than can increase soil fertility.

By refusing, reducing, reusing, recycling, and composting, we save matter and energy resources, reduce pollution (including greenhouse gas emissions), help protect biodiversity, and save money. Figure 21-8 lists some ways in which you can use the four Rs of waste reduction to reduce your output of solid waste.

Here are six strategies that industries and communities have used to reduce resource use, waste, and pollution.

First, *change industrial processes to eliminate or reduce the use of harmful chemicals.* Since 1975, the 3M company has taken this approach and in the process saved $1.2 billion (see Chapter 17, Case Study, p. 464).

Second, *redesign manufacturing processes and products to use less material and energy.* For example, the weight of a typical car has been reduced by about one-fourth since the 1960s through the use of lighter steel and lightweight plastics and composite materials.

Third, *develop products that are easy to repair, reuse, remanufacture, compost, or recycle.* For example, Xerox photocopiers that are leased by businesses are made of reusable or recyclable parts that allow for easy remanufacturing and

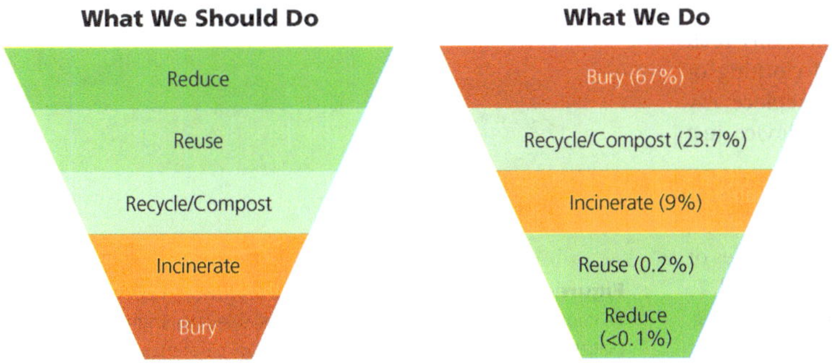

Figure 21-7 Priorities recommended by the U.S. National Academy of Sciences for dealing with municipal solid waste (left) compared with actual waste-handling practices in the United States based on data (right). ***Question:*** Why do you think most countries do not follow most of the scientific-based priorities listed on the left?

(Compiled by the authors using data from U.S. Environmental Protection Agency, U.S. National Academy of Sciences, Columbia University, and *BioCycle*.)

are projected to save the company $1 billion in manufacturing costs.

Fourth, *eliminate or reduce unnecessary packaging*. Use the following hierarchy for product packaging: no packaging, reusable packaging, and recyclable packaging. The 37 European Union countries require the recycling of 55–80% of all packaging waste.

Fifth, *use fee-per-bag waste collection systems* that charge consumers for the amount of waste they throw away but provide free pickup of recyclable and reusable items.

Sixth, *establish cradle-to-grave responsibility laws* that require companies to take back various consumer products such as electronic equipment (**Core Case Study**), appliances, and motor vehicles, as Japan and many European countries do.

What Can You Do?

Solid Waste

- Follow the four Rs of resource use: Refuse, Reduce, Reuse, Recycle
- Ask yourself whether you really need what you're buying and refuse packaging wherever possible
- Rent, borrow, or barter goods and services when you can, buy secondhand, and donate or sell unused items
- Buy things that are reusable, recyclable, or compostable, and be sure to reuse, recycle, and compost them
- Buy products with little or no packaging and recycle any packaging as much as possible
- Avoid disposables such as paper and plastic bags, plates, cups, and utensils, disposable diapers, and disposable razors whenever reusable versions are available
- Cook with whole, fresh foods, avoid heavily packaged processed foods, and buy products in bulk whenever possible
- Discontinue junk mail as much as possible and read online newspapers and magazines and e-books

Figure 21-8 Individuals matter: You can save resources by reducing your output of solid waste and pollution. ***Questions:*** Which three of these steps do you think are the most important ones to take? Why? Which of these things do you already do?

21-3 Why Are Refusing, Reducing, Reusing, and Recycling So Important?

CONCEPT 21-3

By refusing and reducing resource use and by reusing and recycling what we use, we decrease our consumption of matter and energy resources, reduce pollution and natural capital degradation, and save money.

There Are Alternatives to the Throwaway Economy

Today's more-developed economies are based largely on getting people to buy more and more stuff whether they need it or not. Such buying helps to support businesses that sell us these materials, and it supports jobs. However, there is growing evidence that in the long run, economies based on steadily increasing consumption of resources are not environmentally or economically sustainable.

Refusing to buy some things, especially those with a significant environmental impact, is one way to live more sustainably. It begins by asking important questions such as: "Do I really need this stuff—this piece of clothing or this electronic gadget? Do I really need to replace my mobile phone, computer, or other electronic device just because a newer model has come out?"

Reducing is the next level of living more sustainably. It means buying less, and involves asking questions such as: "How many pairs of shoes or pants or how many blouses or skirts do I need? How many electronic devices do I need?"

Reusing is the third level of more sustainable living. It involves cleaning and using items such as containers, dishes, utensils, and razor blades over and over, and thus increasing the typical life span of such products. This form of waste reduction decreases the use of matter and energy resources, cuts waste and pollution (including greenhouse gases), creates local jobs, and saves money (**Concept 21-3**).

The problem is that in today's industrialized societies, we have increasingly substituted throwaway items for reusable ones, which has resulted in more pollution and growing masses of solid waste. For example, if the 1 billion throwaway paper coffee cups (Figure 21-9, left) used by one famous chain of donut shops each year were lined

Levent Konuk/Shutterstock.com

ffolas/Shutterstock.com

Figure 21-9 *Reuse:* Instead of using throwaway paper (left) or plastic cups, many people take their own refillable cups (right) to workplaces, coffee shops, restaurants, and school cafeterias.

Brenda Carson/Shutterstock.com

Figure 21-10 *Reuse.* A growing number of people are using reusable cloth bags instead of throwaway plastic and paper bags.

What Can You Do?

Reuse

- Buy beverages in refillable glass containers
- Use reusable lunch containers
- Store refrigerated food in reusable containers
- Use rechargeable batteries and recycle them when their useful life is over
- When eating out, bring your own reusable container for leftovers
- Carry groceries and other items in a reusable basket or cloth bag
- Buy used furniture, cars, and other items, whenever possible

Figure 21-11 Individuals matter: There are many ways to reuse the items we purchase. ***Question:*** Which of these suggestions have you tried and how did they work for you?

up end-to-end, they would encircle the earth twice. Many coffee shops are reducing such waste and cutting their costs by discounting their hot drink prices for customers who bring their own refillable mugs (Figure 21-9, right).

In general, reuse is on the rise, although there is great potential for more of it. Denmark, Finland, and the Canadian province of Prince Edward Island have banned all beverage containers that cannot be reused. In Finland, 95% of the soft drink, beer, wine, and spirits containers are refillable. The latest rechargeable batteries come fully charged, can hold a charge for up to 2 years when they are not used, and can be recharged in about 15 minutes. We could greatly reduce toxic waste by replacing all conventional batteries with such rechargeables.

Instead of using throwaway paper or plastic bags to carry home groceries and other items they purchase, many people have switched to reusable cloth bags (Figure 21-10). Both paper and plastic bags are environmentally harmful, and the question of which is more damaging has no clear-cut answer. In many countries, the landscape is littered with plastic bags, as are many areas of the ocean. They can take 400 to 1,000 years to break down. Less than 1% of the estimated 102 billion plastic bags used each year in the United States are recycled. In addition to being an eyesore, they often block drains and sewage systems and can kill wildlife and livestock that try to eat them or become ensnared in them.

To encourage people to carry reusable bags, the governments of Ireland, Taiwan, and the Netherlands tax plastic shopping bags. In Ireland, a tax of about 25¢ per bag helped to cut plastic bag litter by 90% as people switched to reusable bags. Kenya, Uganda, Rwanda, South Africa, Bangladesh, Bhutan, parts of India, Taiwan, China, Australia, France, and Italy also have banned the use of all or most types of plastic shopping bags. By 2012, the U.S. cities of San Francisco and Los Angeles, California; Austin, Texas; and Portland, Oregon, along with the state of Hawaii, had instituted complete bans or phase-outs of plastic bags. At least 20 other U.S. communities had followed suit.

There are many other ways to reuse various items, some of which are listed in Figure 21-11. GOOD NEWS

There Is Great Potential for Recycling

Households and workplaces produce five major types of materials that can be recycled into new, useful products: paper products, glass, aluminum, steel, and some plastics. We can reprocess such materials in two ways. In **primary**, or **closed-loop**, **recycling**, materials such as aluminum cans are recycled into new products of the same type. In **secondary recycling**, waste materials are converted into different products.

Scientists also distinguish between two types of recyclable wastes: *preconsumer* or *internal waste* generated in a manufacturing process, and *postconsumer* or *external waste* generated by consumers' use of products. Preconsumer waste makes up more than three-fourths of the total.

Recycling involves three steps: collecting materials for recycling, converting the recycled materials to new products, and the selling and buying of products containing recycled material. Recycling does not work unless all three of these steps are taken consistently.

In 2010, the United States recycled or composted less than a quarter of its MSW. According to the EPA, this included the recycling of about 96% of all lead–acid batteries, 72% of all newspapers, directories, and newspaper inserts, 67% of all steel cans, and 50% of all aluminum cans discarded, according to EPA estimates. Recycling rates for some MSW categories have increased, partly because the number of curbside-pickup recycling pro-

Courtesy of Green Culture, Inc.

Figure 21-12 When the compost in this backyard composter drum is ready to be used, the device can be wheeled to a garden and emptied into the soil.

grams has risen to about 9,000, now serving about half of the U.S. population. Experts say that with education and proper incentives, the United States could recycle and compost 80% of its MSW, in keeping with the chemical cycling **principle of sustainability**.

Composting is another form of recycling that mimics nature's recycling of nutrients. It involves using bacteria to decompose yard trimmings, vegetable food scraps, and other biodegradable organic wastes. The resulting organic material can be added to soil to supply plant nutrients, slow soil erosion, retain water, and improve crop yields. Home owners can compost such wastes in simple backyard containers, in composting piles that must be turned over occasionally, or in small composting drums (Figure 21-12) that are rotated to mix the wastes and speed up the decomposition process.

Some cities in Canada and in many European Union countries collect and compost 85% or more of their biodegradable wastes in centralized community facilities. In the United States, about 3,000 municipal composting programs recycle about 60% of the yard wastes in the country's MSW (Figure 21-13), and in 2012, at least 25 states had some version of a ban on yard wastes in sanitary landfills. Compost from such programs can be used as organic soil fertilizer, topsoil, or landfill cover, and as a source of revenues for the communities.

To be successful, a large-scale composting program must be located carefully and odors must be controlled, especially near residential areas. Sometimes, composting takes place in huge indoor buildings. In the Canadian city of Edmonton, Alberta, an indoor facility the size of eight football fields composts 50% of the city's organic solid waste. Composting programs must also exclude toxic materials that can contaminate the compost and make it unsafe for fertilizing crops and lawns.

We Can Mix or Separate Household Solid Wastes for Recycling

One way to recycle is to send mixed household and business wastes to centralized *materials-recovery facilities* (MRFs, see Figure 21-6). There, machines or workers separate the mixed waste to recover valuable materials for sale to manufacturers as raw materials. The remaining paper, plastics, and other combustible wastes are recycled or burned to produce steam or electricity, which is used to run the recovery plant or is sold to nearby industries or homes.

Such plants are expensive to build, operate, and maintain. If not operated properly, they can emit CO_2 and toxic air pollutants, and they produce a toxic ash that must be disposed of safely, usually in landfills. Because MRFs

Figure 21-13 Large-scale municipal composting site.

imging/Shutterstock.com

require a steady diet of garbage to make them financially successful, their owners have a vested interest in increasing the flow of matter and energy resources. Thus, use of MRFs encourages people to produce more trash—the reverse of what many scientists urge us to do (Figure 21-7, left).

To many experts, it makes more environmental and economic sense for households and businesses to separate their trash into recyclable categories such as glass, paper, metals, certain types of plastics, and compostable materials. This *source separation* approach produces much less air and water pollution and costs less to implement than do MRFs. It also saves more energy, provides more jobs per unit of material, and yields cleaner and usually more valuable recyclables.

To promote separation of wastes for recycling, about 7,000 communities in the United States use a *pay-as-you-throw* or *fee-per-bag* waste collection system. They charge households and businesses for the amount of garbage that is picked up, but do not charge them for picking up materials separated for recycling or reuse. When the U.S. city of Ft. Worth, Texas, instituted such a program, the proportion of households recycling their trash went from 21% to 85%. In addition, the city went from losing $600,000 in its recycling program to making $1 million a year because of increased sales of recycled materials to industries.

GOOD NEWS

Recycling Paper

About 55% of the world's industrial tree harvest is used to make paper. However, according to the U.S. Department of Agriculture, we could make tree-free paper from straw and other agricultural residues and from the fibers of rapidly growing plants such as kenaf (see Figure 10-17, p. 228) and hemp.

The pulp and paper industry is the world's fifth largest energy consumer and uses more water to produce a metric ton of its product than any other industry. In both Canada and the United States, it is the third-largest industrial energy user and polluter, and paper is the dominant material in the MSW of both countries.

Paper (especially newspaper and cardboard) is easy to recycle. Recycling newspaper involves removing its ink, glue, and coating and then reconverting the paper to pulp, which is pressed into new paper. Making recycled paper uses 64% less energy and produces 35% less water pollution and 74% less air pollution than does making paper from wood pulp, and, of course, no trees are cut down.

In 2009, the Natural Resources Defense Council (NRDC) mounted a campaign to pressure paper companies to stop cutting down trees in some of North America's old-growth forests in order to make toilet paper. The NRDC estimates that nearly 425,000 trees would be saved each year if every U.S. household used just one 500-sheet roll of toilet paper made from 100% recycled paper in place of a roll made from virgin timber.

Recycling Plastics

Plastics consist of various types of large polymers, or *resins*—organic molecules made by chemically linking organic chemicals produced mostly from oil and natural gas. About 46 different types of plastics are used in consumer products, and some products contain several kinds of plastic.

Currently, only about 7% by weight of all plastic wastes in the United States (and 13% of plastic containers and packaging) is recycled. These percentages are low because there are many different types of plastic resins, which are difficult to separate from products that contain them. However, progress is being made in the recycling of plastics (Individuals Matter 21.1) and in the development of more degradable bioplastics (Science Focus 21.2).

Recycling Has Advantages and Disadvantages

Figure 21-14 lists the advantages and disadvantages of recycling (Concept 21-3). Whether recycling makes economic sense depends on how we look at its economic and environmental benefits and costs.

Critics of recycling programs argue that recycling is costly and adds to the taxpayer burden in communities where recycling is funded through taxation. They say that recycling may make economic sense for valuable and easy-to-recycle materials in MSW such as paper and paperboard, steel, and aluminum, but probably not for cheap or plentiful resources such as glass made from silica.

Proponents of recycling point to studies showing that the net economic, health, and environmental benefits of recycling (Figure 21-14, left) far outweigh the costs. For example, the EPA estimates that recycling and composting in the United States in 2010 reduced emissions of climate-changing carbon dioxide by an amount roughly equal to

Trade-Offs

Recycling

Advantages	Disadvantages
Reduces energy and mineral use and air and water pollution	Can cost more than burying in areas with ample landfill space
Reduces greenhouse gas emissions	Reduces profits for landfill and incinerator owners
Reduces solid waste	Inconvenient for some

Figure 21-14 Recycling solid waste has advantages and disadvantages (Concept 21-3). ***Questions:*** Which single advantage and which single disadvantage do you think are the most important? Why?

Photo: Jacqui Martin/Shutterstock.com

individuals matter 21.1

Mike Biddle's Contribution to Plastics Recycling

©World Economic Forum

In 1994, Mike Biddle, a former PhD engineer with Dow Chemical, and his business partner Trip Allen founded MBA Polymers, Inc. Their goal was to develop a commercial process for recycling high-value plastics from complex streams of manufactured goods such as computers, electronics, appliances, and automobiles. They succeeded by designing a 21-step automated process that separates plastics from nonplastic items in mixed waste streams, and then separates plastics from each other by type and grade. The process then converts them to pellets that can be used to make new products.

The pellets are cheaper than virgin plastics because the company's process uses 90% less energy than that needed to make a new plastic and because the raw material is discarded junk that is cheap or free. In addition, greenhouse gas emissions from this process are much lower than those from the process of making virgin plastics. And recycling plastic wastes reduces the need to incinerate them or bury them in landfills.

The company is considered a world leader in plastics recycling. It operates a large, state-of-the-art research and recycling plant in Richmond, California (USA), and recently opened the world's two most advanced plastics recycling plants in China and Austria. MBA Polymers has won many awards, including the Thomas Alva Edison Award for Innovation, and it was selected by *Inc.* magazine as one of "America's Most Innovative Companies."

Maybe you can become an environmental entrepreneur by using your brainpower to develop an environmentally beneficial and financially profitable process or business.

Background photo: justasc/Shutterstock.com

that emitted by 36 million passenger vehicles. According to the Steel Recycling Institute, recycling a metric ton of steel eliminates the need to mine 2.3 metric tons (2.5 tons) of iron ore and 1.3 metric tons (1.4 tons) of coal. Proponents also argue that the U.S. recycling industry employs about 1.1 million people—more than the number of people working in the U.S. auto industry—and that its annual revenues are much larger than those of the waste-management industry.

Cities that make money by recycling and that have higher recycling rates tend to use a *single-pickup system* for both recyclable and nonrecyclable materials, instead of a more expensive dual-pickup system. Successful systems also tend to use a pay-as-you-throw approach, and they have citizens and businesses sort their trash more carefully. San Francisco, California (USA) uses such a system, and in 2012, the city recycled, composted, or reused 78% of its MSW. GOOD NEWS

SCIENCE FOCUS 21.2

BIOPLASTICS

Most of today's plastics are made from organic polymers produced from petroleum-based chemicals. This may change as scientists shift to developing plastics made from biologically based chemicals.

Henry Ford, who developed the first Ford motorcar, supported research on the development of a bioplastic made from soybeans and another made from hemp. A 1914 photograph shows him using an ax to strike the body of a car made from soy bioplastic to demonstrate its strength and resistance to denting.

However, as oil became widely available, petrochemical plastics took over the market. Now, with projected climate change and other environmental problems associated with the use of oil, chemists are stepping up efforts to make biodegradable and more environmentally sustainable plastics. These *bioplastics* can be made from corn, soy, sugarcane, switchgrass (see Figure 16-27, p. 426), chicken feathers, and some components of garbage.

Compared to conventional oil-based plastics, properly designed bioplastics are lighter, stronger, and cheaper, and the process of making them requires less energy and produces less pollution per unit of weight. Instead of being sent to landfills, some packaging made from bioplastics (Figure 21-A) can be composted to produce a soil conditioner, in keeping with the chemical cycling **principle of sustainability**.

Some bioplastics are more ecofriendly than others. For example, some are made from corn raised by industrial agricultural methods, which require great amounts of energy, water, and petrochemical fertilizers and thus have a very large ecological footprint. In evaluating and choosing bioplastics, scientists urge consumers to learn how they were made and whether or not they are truly recyclable or compostable.

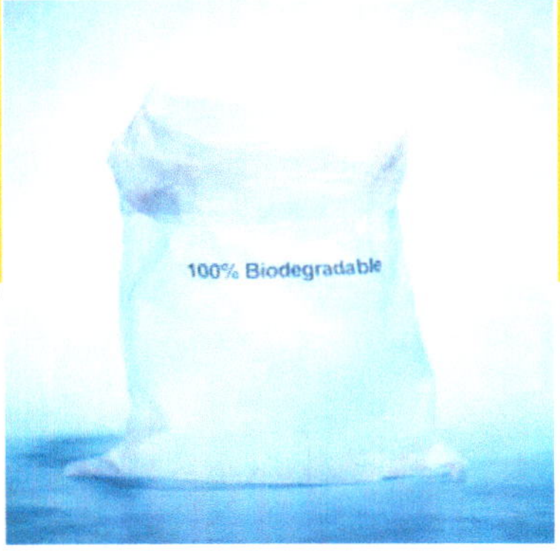

© Stefanie Timmermann/iStockphoto.com

Figure 21-A Biodegradable plastic bag.

Critical Thinking

What might be some disadvantages of bioplastics? Do you think they outweigh the advantages?

21-4 What Are the Advantages and Disadvantages of Burning or Burying Solid Waste?

CONCEPT 21-4

Technologies for burning and burying solid wastes are well developed, but burning contributes to air and water pollution and greenhouse gas emissions, and buried wastes eventually contribute to the pollution and degradation of land and water resources.

Burning Solid Waste Has Advantages and Disadvantages

Globally, MSW is burned in more than 600 large *waste-to-energy incinerators,* which use the heat they generate to boil water and make steam for heating water or interior spaces, or for producing electricity (Figure 21-15). They are intended to burn trash that cannot be recycled or reused. There are 115 of these incinerators in the United States, according to Columbia University and *BioCycle* researchers.

The United States incinerates about 9% of its MSW. One reason for this fairly low percentage is that incineration has a bad reputation stemming from past use of highly polluting and poorly regulated incinerators. Also, incineration competes with an abundance of low-cost landfills in many parts of the country.

By contrast, Denmark incinerates 54% of its MSW in waste-to-energy incinerators with newer designs. The country also recycles or composts 42% of its MSW (compared to 24% in the United States) and buries only 4% of it in landfills (compared to 67% in the United States). Denmark's incinerators use dozens of filters to remove pollutants such as toxic mercury and dioxins. They run so cleanly that they release just a fraction of the toxic dioxins released by typical home fireplaces and backyard barbecues, and they exceed European air pollution standards by a factor of 10. A 2009 EPA study concluded that landfills emit more air pollutants than modern waste-to-energy incinerators. On the other hand, the toxic chemicals that are filtered out still have to be stored somewhere.

Denmark and most other European nations burn their trash in small, less costly, community-based incinerators instead of hauling the trash long distances by truck to large incinerators, as is done in the United States. Germany incinerates 34% of its MSW and recycles and composts most of the rest.

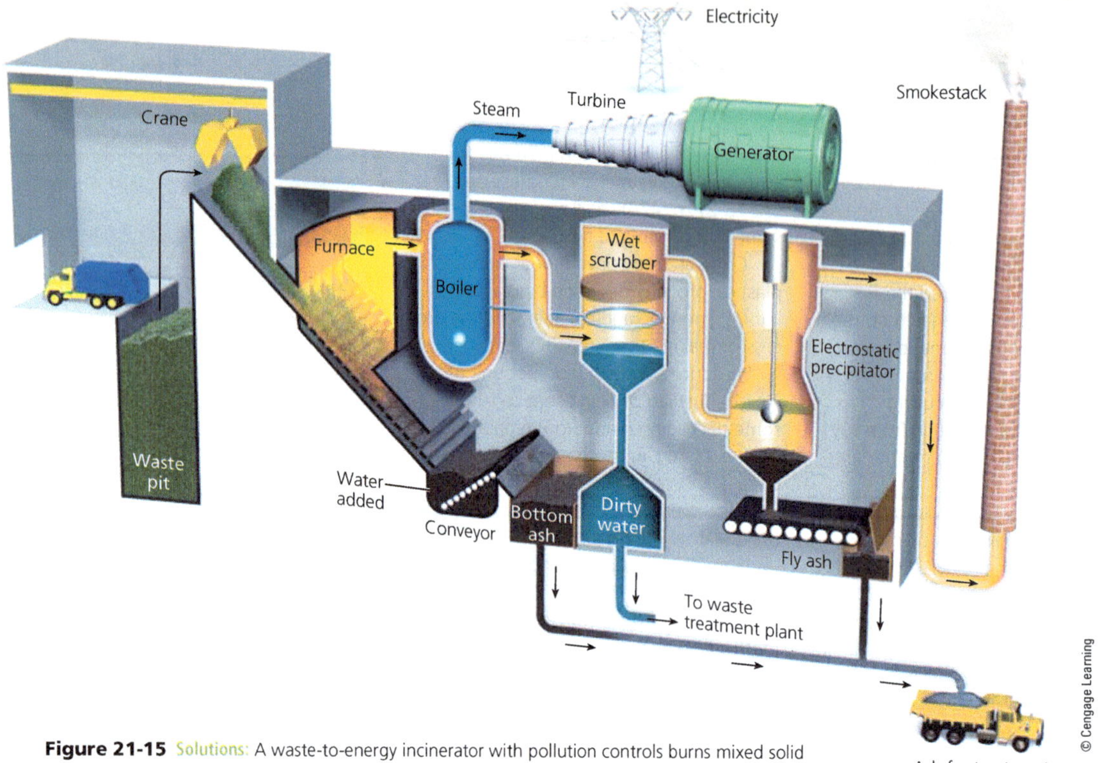

Figure 21-15 Solutions: A waste-to-energy incinerator with pollution controls burns mixed solid wastes and recovers some of the energy to produce steam to use for heating or producing electricity. ***Questions:*** Would you invest in such a project? Why or why not?

Trade-Offs

Waste-to-Energy Incineration

Advantages	Disadvantages
Reduces trash volume	Expensive to build
Produces energy	Produces a hazardous waste
Concentrates hazardous substances into ash for burial	Emits some CO_2 and other air pollutants
Sale of energy reduces cost	Encourages waste production

Figure 21-16 Incinerating solid waste has advantages and disadvantages (Concept 21-4). These trade-offs also apply to the incineration of hazardous waste. ***Questions:*** Which single advantage and which single disadvantage do you think are the most important? Why?

Top: Ulrich Mueller/Shutterstock.com. Bottom: Dmitry Kalinovsky/Shutterstock.com.

In the United States, incinerators are used in some areas, including the city of Fort Myers, Florida, which has closed its landfills, recycles or composts half of its trash, and burns the rest to provide electricity for 36,000 homes. However, many U.S. citizens, local governments, and environmental scientists oppose any use of waste incinerators because it undermines efforts to increase reuse and recycling. Once an incinerator is built, it must be fed large volumes of trash in order to be profitable for its owners. Figure 21-16 lists the advantages and disadvantages of using incinerators to burn solid waste.

Burying Solid Waste Has Advantages and Disadvantages

About 67%, by weight, of the MSW in the United States is buried in sanitary landfills, compared to 80% in Canada, 15% in Japan, and 4% in Denmark. There are two types of landfills. In newer landfills, called **sanitary landfills** (Figure 21-17), solid wastes are spread out in thin layers, compacted, and covered daily with a fresh layer of clay or plastic foam. This process helps keep the material dry and reduces leakage of contaminated water (leachate)

Figure 21-17 Solutions: A state-of-the-art *sanitary landfill* is designed to eliminate or minimize environmental problems that plague older landfills. ***Question:*** Some experts say that these landfills will eventually develop leaks and could emit toxic liquids. How do you think this could happen?

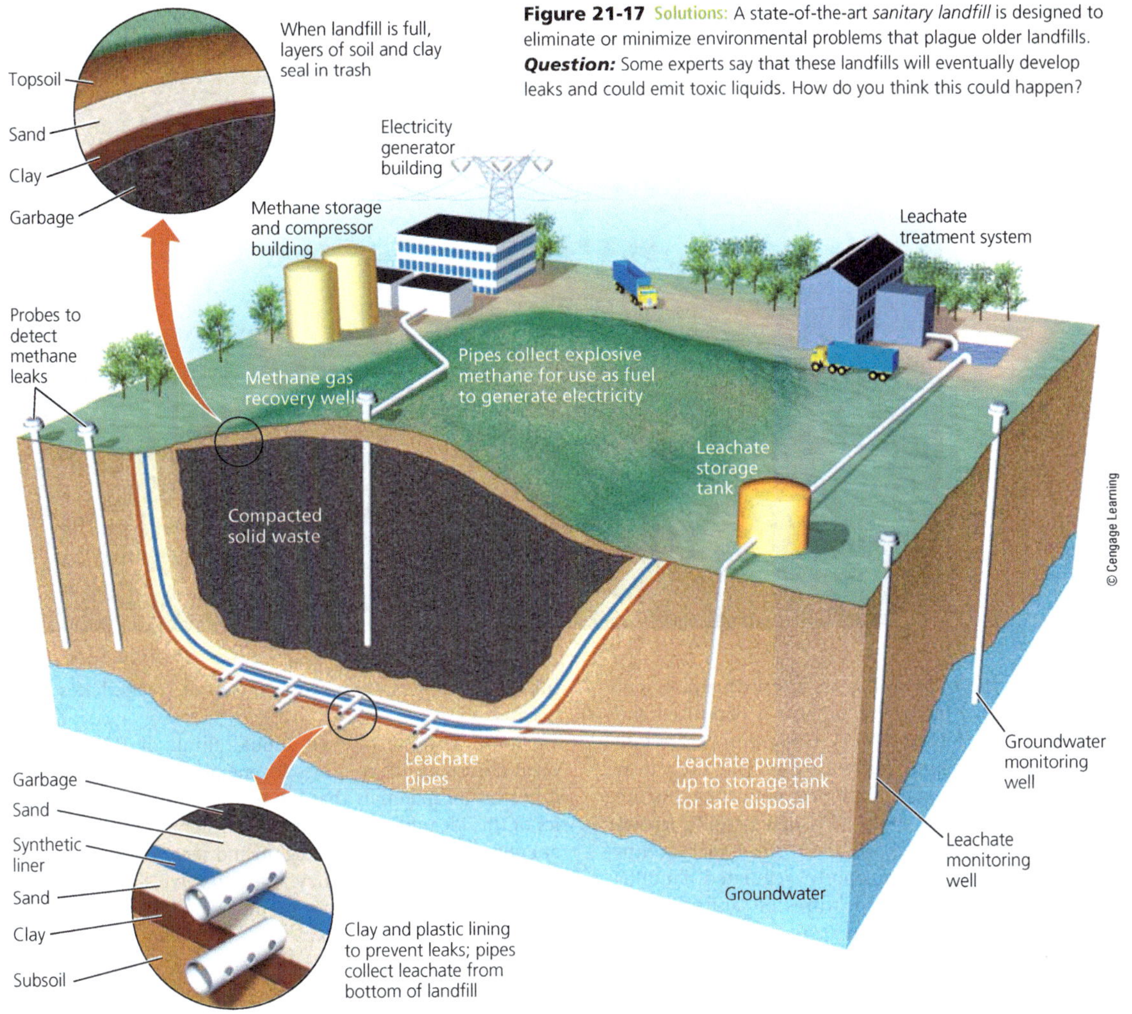

from the landfill. This covering also lessens the risk of fire, decreases odor, and reduces accessibility to vermin.

The bottoms and sides of these landfills have strong double liners and containment systems that collect the liquids leaching from them. Some landfills also have systems for collecting and burning methane, the potent greenhouse gas that is produced when the wastes decompose in the absence of oxygen. Figure 21-18 lists the advantages and disadvantages of using sanitary landfills to dispose of solid waste.

Trade-Offs

Sanitary Landfills

Advantages	Disadvantages
Low operating costs	Noise, traffic, and dust
Can handle large amounts of waste	Releases greenhouse gases (methane and CO_2) unless they are collected
Filled land can be used for other purposes	Output approach that encourages waste production
No shortage of landfill space in many areas	Eventually leaks and can contaminate groundwater

Figure 21-18 Using sanitary landfills to dispose of solid waste has advantages and disadvantages (Concept 21-4). ***Questions:*** Which single advantage and which single disadvantage do you think are the most important? Why?

Photo: © redbrickstock.com/Alamy

Joel Sartore/National Geographic Creative

Figure 21-19 Some open dumps like this one near the city of Matamoros, Mexico, are routinely burned, releasing pollutants into the atmosphere.

The second type of landfill is an **open dump**, essentially a field or large pit where garbage is deposited and sometimes burned (Figure 21-19). Open dumps are rare in more-developed countries, but are widely used near major cities in many less-developed countries (Figure 21-3). China disposes of about 85% of its rapidly growing mountains of solid waste mostly in rural open dumps or in poorly designed and poorly regulated landfills that do not have most of the features of sanitary landfills. Also, some solid wastes end up in rivers, lakes, and oceans.

Some analysts say we should think of landfills and open dumps as future mines for metals, plastics, paper, and other useful materials that we throw away. However, critics of this idea point out that it takes significant amounts of energy and is expensive to dig up and separate the materials in landfills and dumps. They argue that we should not mix different types of waste that we may want to separate later.

21-5 How Should We Deal with Hazardous Waste?

CONCEPT 21-5

A more sustainable approach to hazardous waste is first to produce less of it, then to reuse or recycle it, then to convert it to less-hazardous materials, and finally to safely store what is left.

We Can Use Integrated Management of Hazardous Waste

Figure 21-20 shows an integrated management approach suggested by the U.S. National Academy of Sciences that establishes three priority levels for dealing with hazardous waste: produce less; convert as much of it as possible to less-hazardous substances; and put the rest in long-term, safe storage (Concept 21-5). Denmark follows these priorities, but most countries do not.

As with solid waste, the top priority should be pollution prevention and waste reduction. With this approach, industries try to find substitutes for toxic or hazardous materials, reuse or recycle the hazardous materials within industrial processes, or use them as raw materials for making other products. (See the online Guest Essays on this subject by Lois Gibbs and Peter Montague.)

At least 33% of industrial hazardous wastes produced in the European Union are exchanged through clearinghouses where they are sold as raw materials for use by other industries. The producers of these wastes do not have to pay for their disposal and recipients get low-cost raw materials. About 10% of the hazardous waste in the United States is exchanged through such clearinghouses, a figure that could be raised significantly.

GOOD NEWS

We can also recycle e-waste (**Core Case Study**). However, most e-waste recycling efforts create further hazards, especially for workers in some less-developed countries (see the following Case Study).

Produce Less Hazardous Waste

- Change industrial processes to reduce or eliminate hazardous waste production
- Recycle and reuse hazardous waste

→

Convert to Less Hazardous or Nonhazardous Substances

- Natural decomposition
- Incineration
- Thermal treatment
- Chemical, physical, and biological treatment
- Dilution in air or water

→

Put in Perpetual Storage

- Landfill
- Underground injection wells
- Surface impoundments
- Underground salt formations

Figure 21-20 *Integrated hazardous waste management:* The U.S. National Academy of Sciences has suggested these priorities for dealing with hazardous waste (Concept 21-5). ***Question:*** Why do you think most countries do not follow these priorities?

CASE STUDY

Recycling E-waste

In some countries, workers in e-waste recycling operations—many of them children—are often exposed to toxic chemicals as they dismantle the electronic trash to extract its valuable metals or other parts that can be sold for reuse or recycling (Core Case Study and Figure 21-21).

According to the United Nations, more than 70% of the world's e-waste is shipped to China. A center for such waste is the small port city of Guiyu, where the air reeks of burning plastic and acid fumes. There, more than 5,500 small-scale e-waste businesses employ over 30,000 people (including some children) who work at very low wages in dangerous conditions to extract valuable metals like gold and copper and various rare-earth metals from millions of discarded computers, television sets, and cell phones.

These workers usually wear no masks or gloves, often work in rooms with no ventilation, and are usually exposed to a cocktail of toxic chemicals. They carry out dangerous activities such as smashing TV picture tubes with large hammers to recover certain components—a method that releases large amounts of toxic lead dust into the air. They also burn computer wires to expose copper, melt circuit boards in metal pots over coal fires to extract lead and other metals, and douse the boards with strong acid to extract gold.

After the valuable metals are removed, leftover parts are burned or dumped into rivers or onto the land. Atmospheric levels of deadly dioxin in Guiyu are up to 86 times higher than World Health Organization safety standards, and an estimated 82% of the area's children younger than age 6 suffer from lead poisoning.

The United States produces roughly 50% of the world's e-waste (Core Case Study) and recycles only about 14% of it, according to the Consumer Electronics Association (CEA). However, e-waste recycling is on the rise. In 2012, at least 20 states had banned the disposal of computers and TV sets in landfills and incinerators, and these measures set the stage for an emerging, highly profitable *e-cycling* industry. According to a CEA report, electronics recycling in the United States increased by 53% between 2010 and 2011 and the number of e-waste drop-off sites rose from 5,000 to 7,500. Also, 13 states as well as New York City make manufacturers responsible for recycling most electronic devices.

Figure 21-21 This young girl in Dhaka, Bangladesh, is recycling batteries by hammering them apart to extract tin and lead. The workers at this shop are mostly women and children.

Some U.S. electronics manufacturers have free recycling programs. A few will arrange to pick up discarded electronic products or will pay shipping costs to have them returned. Some have called for a standardized U.S. federal law that makes manufacturers responsible for taking back all electronic devices they produce and recycling them domestically with the process paid for by a recycling fee.

Recycling probably will not keep up with the explosive growth of e-waste, and there is money to be made from illegally sending such materials to other countries. According to Jim Puckett, coordinator of the Basel Action Network, an organization working to stem the flow of toxic waste, the only real long-term solution is a *prevention* approach through which electrical and electronic products are designed without the use of toxic materials and in such a way that they can be easily repaired, remanufactured, and recycled.

CONSIDER THIS. . .

THINKING ABOUT E-Waste Recycling

Would you support a recycling fee on all electronic devices? Why or why not?

We Can Detoxify Hazardous Wastes

The first step in dealing with hazardous wastes is to collect them. In Denmark, all hazardous and toxic waste from industries and households is delivered to any of 21 transfer stations throughout the country. From there it is taken to a large processing facility, where three-fourths of the waste is detoxified by physical, chemical, and biological methods. The rest is buried in a carefully designed and monitored landfill.

Physical methods for detoxifying hazardous wastes include using charcoal or resins to filter out harmful solids, distilling liquid wastes to separate out harmful chemicals, and precipitating (allowing natural processes to separate) such chemicals from solution. Especially deadly wastes can be encapsulated in glass, cement, or ceramics and then put in secure storage sites.

Chemical methods are used to convert hazardous chemicals to harmless or less harmful chemicals through chemical reactions. Currently, chemists are testing the use of *cyclodextrin*—a type of sugar made from cornstarch—to remove toxic materials such as solvents and pesticides from contaminated soil and groundwater. To clean up a site contaminated by hazardous chemicals, a solution of cyclodextrin is applied. After this molecular sponge-like material moves through the soil or groundwater, picking up various toxic chemicals, it is pumped out of the ground, stripped of its contaminants, and reused.

Another approach is the use of *nanomagnets,* magnetic nanoparticles coated with certain compounds that can remove various pollutants from water. Magnetic fields are used to remove the pollutant-coated nanomagnets. The pollutants can then be separated out and disposed of or recycled, and the magnetic nanoparticles can be reused.

Some scientists and engineers consider *biological methods* for treatment of hazardous waste to be the wave of the future. One such approach is *bioremediation,* in which bacteria and enzymes help to destroy toxic or hazardous substances, or convert them to harmless compounds. For example, microorganisms are used at some sites to break down hazardous chemicals, such as PCBs, pesticides, and oil, leaving behind harmless substances such as water and water-soluble chloride salts. Bioremediation takes a little longer to work than most physical and chemical methods, but it costs much less. (See the online Guest Essay by John Pichtel on this topic.)

Another approach is *phytoremediation,* which involves using natural or genetically engineered plants to absorb, filter, and remove contaminants from polluted soil and water (Figure 21-22). Various plants have been identified as "pollution sponges" that can help to clean up soil and water contaminated with chemicals such as pesticides, organic solvents, and radioactive or toxic metals. Phytoremediation is still being evaluated and is slow (it can take several growing seasons) compared to other alternatives. However, it can be used to help clean up toxic hotspots.

We can incinerate hazardous wastes to break them down and convert them to harmless or less harmful chemicals such as carbon dioxide and water. This has the same combination of advantages and disadvantages as does burning solid wastes (Figure 21-16). However, incinerating hazardous waste without effective and expensive air pollution controls can release air pollutants such as highly toxic dioxins, and it produces an extremely toxic ash that must be safely and permanently stored in a landfill or vault especially designed for this hazardous end product.

We can also detoxify hazardous wastes by using *plasma gasification,* a technology that uses arcs of electrical energy to produce very high temperatures in order to vaporize trash in the absence of oxygen. This process reduces the volume of a given amount of waste by 99%, produces a synthetic gaseous fuel, and encapsulates toxic metals and other materials in glassy lumps of rock. Figure 21-23 lists the major advantages and disadvantages of using this process.

So why are we not making widespread use of the plasma arc torch to detoxify hazardous wastes? The main reason is its high cost. Plasma arc companies are working to bring the cost down.

We Can Store Some Forms of Hazardous Waste

Ideally, we should use burial on land or long-term storage of hazardous and toxic wastes in secure vaults only as the third and last resort after the first two priorities have been exhausted (Figure 21-20 and Concept 21-5). Currently, however, burial on land is the most widely used method in the United States and in most countries, largely because it is the least expensive of all methods.

The most common form of burial is *deep-well disposal,* in which liquid hazardous wastes are pumped under pressure through a pipe into dry, porous rock formations far beneath aquifers that are tapped for drinking and irrigation water (see Figure 20-12, p. 554). Theoretically, these

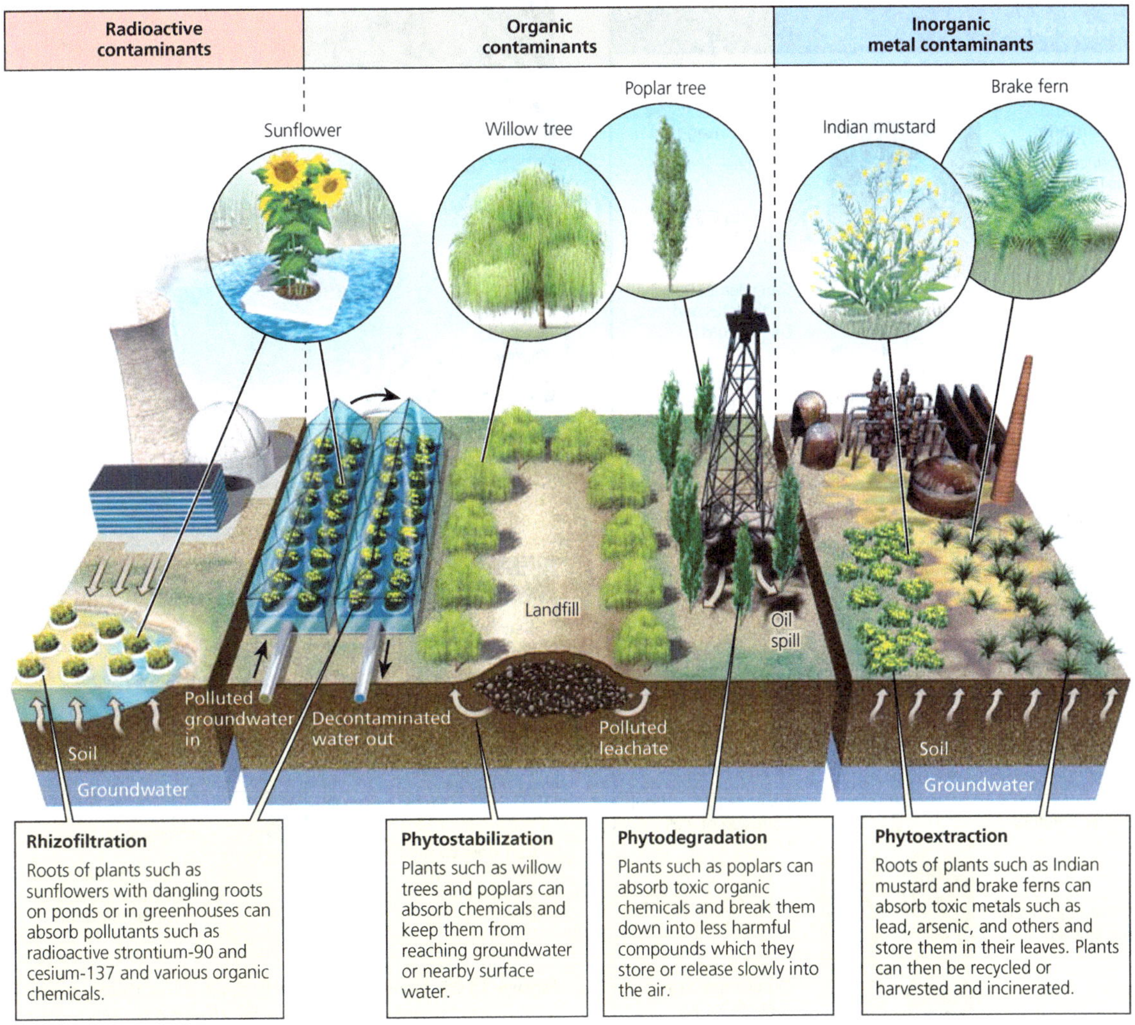

Figure 21-22 Solutions: Phytoremediation involves using various types of plants that function as pollution sponges to clean up contaminants such as radioactive substances (left), organic compounds (center), and toxic metals (right) from soil and water.

(Compiled by the authors using data from American Society of Plant Physiologists, U.S. Environmental Protection Agency, and Edenspace.)

liquids soak into the porous rock material and are isolated from overlying groundwater by essentially impermeable layers of clay and rock. The cost is low and the wastes can often be retrieved if problems develop.

However, this fairly inexpensive, out-of-sight and out-of-mind approach presents some problems. There are a limited number of such sites and limited space within them. Sometimes the wastes can leak into groundwater from the well shaft or migrate into groundwater in unexpected ways. Also, this is an output approach that encourages the production of hazardous wastes. In the United States, almost two-thirds of liquid hazardous wastes are injected into deep disposal wells, and this amount will increase sharply as the country relies more on fracking to produce natural gas and oil trapped in shale rock (see Figure 15-A, p. 378). Many scientists believe that current regulations for deep-well disposal in the United States are inadequate and should be improved. Figure 21-24 lists the advantages and disadvantages of deep-well disposal of liquid hazardous wastes.

Surface impoundments are lined ponds, pits, or lagoons in which liquid hazardous wastes are stored (Figure 21-25). Sometimes they include liners to help contain the waste. As the water evaporates, the waste settles and becomes more concentrated. But using no liner, using leaking single liners, and failing to use double liners can

Trade-Offs

Plasma Arc

Advantages	Disadvantages
Produces a mixture of CO and H_2 that can be used as a fuel	High cost
Mobile. Easy to move to different sites	Produces CO_2 and CO
Produces no toxic ash	Can release particulates and chlorine gas
	Can vaporize and release toxic metals and radioactive elements

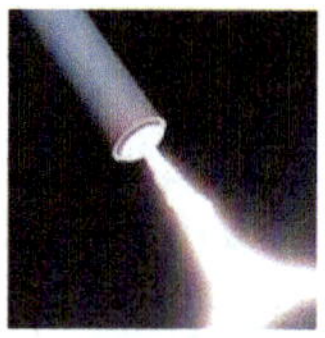

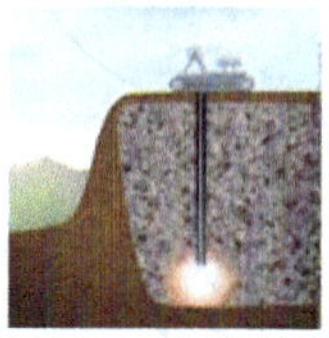

© Cengage Learning

Figure 21-23 Using *plasma gasification* to detoxify hazardous wastes has advantages and disadvantages. ***Questions:*** Which single advantage and which single disadvantage do you think are the most important? Why?

©Ken Sherman/Bruce Coleman USA, Inc.

Figure 21-25 A surface impoundment for storing liquid hazardous wastes located in Niagara Falls, New York (USA). Such sites can pollute the air and nearby groundwater and surface water.

Trade-Offs

Deep-Well Disposal

Advantages	Disadvantages
Safe if sites are chosen carefully	Leaks can occur from corrosion of well casing
Wastes can often be retrieved	Emits CO_2 and other air pollutants
Low cost	Output approach that encourages waste production

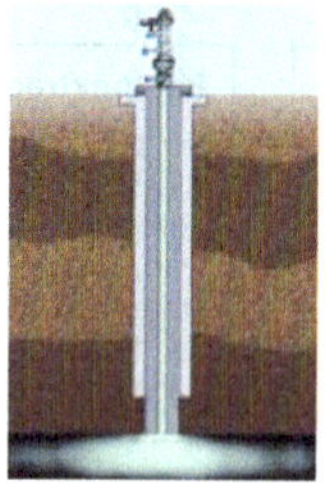

© Cengage Learning 2015

Figure 21-24 Injecting liquid hazardous wastes into deep underground wells has advantages and disadvantages. ***Questions:*** Which single advantage and which single disadvantage do you think are the most important? Why?

Trade-Offs

Surface Impoundments

Advantages	Disadvantages
Low cost	Water pollution from leaking liners and overflows
Wastes can often be retrieved	Air pollution from volatile organic compounds
Can store wastes indefinitely with secure double liners	Output approach that encourages waste production

© Cengage Learning 2015

Figure 21-26 Storing liquid hazardous wastes in surface impoundments has advantages and disadvantages. ***Questions:*** Which single advantage and which single disadvantage do you think are the most important? Why?

allow such wastes to percolate into the groundwater, and because these impoundments are not covered, volatile harmful chemicals can evaporate into the air. Also, powerful storms can cause them to overflow. Figure 21-26 lists the advantages and disadvantages of this method.

EPA studies have found that 70% of the storage ponds in the United States have no liners and could threaten groundwater supplies. According to the EPA, eventually all impoundment liners are likely to leak and could contaminate groundwater.

There are some highly toxic materials (such as mercury; see Chapter 17, Core Case Study, p. 442) that we cannot destroy, detoxify, or safely bury. The best way to deal with such materials is to prevent or reduce their use and store them in sealed containers. These containers could be placed above ground in specially designed storage buildings or underground in salt mines or bedrock caverns where they can be inspected on a regular basis and retrieved if necessary.

Sometimes both liquid and solid hazardous wastes are put into drums or other containers and buried in carefully designed and monitored *secure hazardous waste landfills* (Figure 21-27). This is the least-used method because of the expense involved.

Figure 21-28 lists some ways in which you can reduce your output of hazardous waste—the first step in dealing with it.

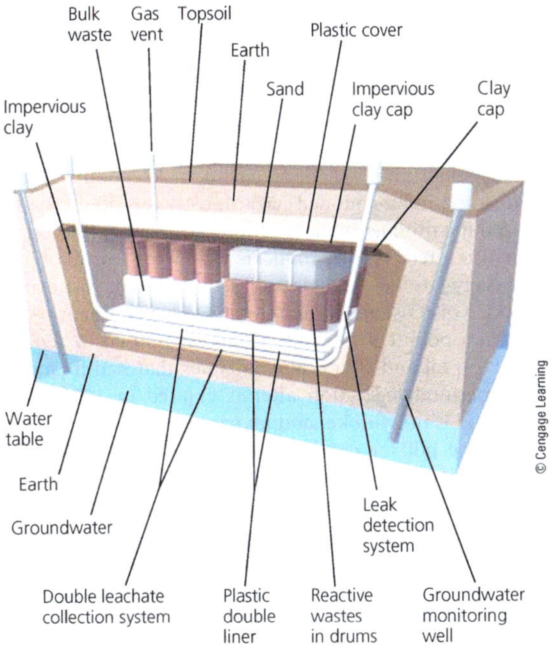

Figure 21-27 Solutions: Hazardous wastes can be isolated and stored in a secure hazardous waste landfill.

CASE STUDY

Hazardous Waste Regulation in the United States

About 5% of all hazardous waste produced in the United States is regulated under the Resource Conservation and Recovery Act (RCRA, pronounced "RICK-ra"), passed by the U.S. Congress in 1976 and amended in 1984. Under this act, the EPA sets standards for the management of several types of hazardous waste and issues permits to companies that allow them to produce and dispose of a certain amount of those wastes by approved methods. Permit holders must use a *cradle-to-grave* system to keep track of waste they transfer from a point of generation (cradle) to an approved off-site disposal facility (grave), and they must submit proof of this disposal to the EPA.

RCRA is a good start, but about 95% of the hazardous and toxic wastes, including e-waste, produced in the United States is not regulated. In most other countries, especially less-developed countries, the amount of regulated waste is even smaller.

CONSIDER THIS. . .

THINKING ABOUT **Hazardous Waste**

Why is it that 95% of the hazardous waste, including the growing amounts of e-waste (Core Case Study) produced in the United States, is not regulated? Do you favor regulating such wastes? What do you think would be the economic consequences of doing so?

What Can You Do?

Hazardous Waste

- Avoid using pesticides and other hazardous chemicals, or use them in the smallest amounts possible
- Use less harmful substances instead of commercial household cleaners. For example, use vinegar to polish metals, clean surfaces, and remove stains and mildew, and baking soda to clean utensils and to deodorize and remove stains.
- Do not dump pesticides, paints, solvents, oil, antifreeze, or other hazardous chemicals down the toilet, down the drain, into the ground, into the garbage, or down storm drains. Free hazardous waste disposal services are available in many cities.
- Do not throw old fluorescent lightbulbs (which contain mercury) into regular trash. Many communities and home product retailers offer free recycling of these bulbs.

Figure 21-28 Individuals matter: You can reduce your output of hazardous wastes (Concept 21-5). ***Questions:*** Which two of these measures do you think are the most important ones to take? Why?

In 1976, the U.S. Congress passed the Toxic Substances Control Act, which was intended to regulate and ensure the safety of the thousands of chemicals used in the manufacture of many products and contained in many products. Under this law, companies must notify the EPA before introducing a new chemical into the marketplace, but they are not required to provide any data about its safety. In other words, any new chemical is viewed as safe unless the EPA, with a very limited budget, can show that it is harmful.

Since 1976, the EPA has used this act to ban only 5 of the roughly 80,000 chemicals in use. Environmental and health scientists call for Congress to reform this almost-40-year-old law by requiring manufacturers to provide data showing that a chemical or product containing a certain chemical is safe before it can be sold in the marketplace.

In 1980, the U.S. Congress passed the *Comprehensive Environmental Response, Compensation, and Liability Act,* commonly known as the *CERCLA* or *Superfund* program, supervised by the EPA. Its goals are to identify sites, commonly called Superfund sites, where hazardous wastes have contaminated the environment (Figure 21-29) and to clean them up on a priority basis. The worst sites—those that represent an immediate and severe threat to human health—are put on a *National Priorities List* and scheduled for cleanup using EPA-approved methods.

By 2013, there were 1,320 sites on the Superfund list, along with 54 proposed new sites, and 365 sites had been cleaned up and removed from the list. The Waste Management Research Institute estimates that at least 10,000 sites should be on the priority list and that cleanup of these sites would cost about $1.7 trillion, not including legal fees. This is a glaring example of the economic and environmental value of emphasizing waste reduc-

Figure 21-29 Leaking barrels of toxic waste found at a Superfund site that has since been cleaned up.

tion and pollution prevention over the *end-of-pipe* cleanup approach that most countries rely on.

In 1984, Congress amended the Superfund Act to give citizens the right to know what toxic chemicals are being stored or released in their communities. This required 23,800 large manufacturing facilities to report their annual releases into the environment of any of nearly 650 toxic chemicals. If you live in the United States, you can find out what toxic chemicals are being stored and released in your neighborhood by going to the EPA's *Toxic Release Inventory* website.

The Superfund Act, designed to make polluters pay for cleaning up abandoned hazardous waste sites, greatly reduced the number of illegal dumpsites around the country. It also forced waste producers who were fearful of liability claims to reduce their production of such waste and to recycle or reuse much more of it. However, under pressure from polluters, the U.S. Congress refused to renew the tax on oil and chemical companies that had financed the Superfund legislation after it expired in 1995. The Superfund is now broke, and taxpayers, not polluters, are footing the bill for cleanups when the responsible parties cannot be found. As a result, the pace of cleanup has slowed.

The U.S. Congress and several state legislatures have also passed laws that encourage the cleanup of *brownfields*—abandoned industrial and commercial sites such as factories, junkyards, older landfills, and gas stations. In most cases, they are contaminated with hazardous wastes. Brownfields can be cleaned up and reborn as parks, nature reserves, athletic fields, ecoindustrial parks (see the Case Study at the end of this chapter), and residential neighborhoods. GOOD NEWS

21-6 How Can We Make the Transition to a More Sustainable Low-Waste Society?

CONCEPT 21-6

Shifting to a low-waste society requires individuals and businesses to reduce resource use and to reuse and recycle wastes at local, national, and global levels.

Grassroots Action Has Led to Better Solid and Hazardous Waste Management

In the United States, individuals have organized grassroots (bottom-up) citizen movements to prevent the construction of hundreds of incinerators, landfills, treatment plants for hazardous and radioactive wastes, and polluting chemical plants in or near their communities. Health risks from incinerators and landfills, when averaged over the entire country, are quite low, but the risks for people living near such facilities are much higher.

Manufacturers and waste industry officials point out that something must be done with the toxic and hazardous wastes created in the production of certain goods and services. They contend that even if local citizens adopt a "not in my back yard" (NIMBY) approach, the waste will always end up in someone's back yard.

Many citizens do not accept this argument. To them, the best way to deal with most toxic and hazardous waste is to produce much less of it, as suggested by the U.S. National Academy of Sciences (Figure 21-20). For such materials, they believe that the goal should be "not in anyone's back yard" (NIABY) or "not on planet Earth" (NOPE), which calls for drastically reducing production of such wastes by emphasizing pollution prevention and using the *precautionary principle* (see Chapter 9, p. 211).

Providing Environmental Justice for Everyone Is an Important Goal

Environmental justice is an ideal whereby every person is entitled to protection from environmental hazards regardless of race, gender, age, national origin, income, social class, or any political factor. (See the online Guest Essay on this subject by Robert Bullard.)

Studies have shown that a lopsided share of polluting factories, hazardous waste dumps, incinerators, and landfills in the United States are located in communities populated mostly by African Americans, Asian Americans, Latinos, and Native Americans. Studies have also shown that, in general, toxic waste sites in white communities have been cleaned up faster and more completely than similar sites in African American and Latino communities have.

Such environmental discrimination in the United States and in other parts of the world has led to a growing grassroots approach to this problem that is known as the *environmental justice movement.* Supporters of this group have pressured governments, businesses, and environmental organizations to become aware of environmental injustice and to act to prevent it. They have made some progress toward their goals, but have a long way to go.

CONSIDER THIS. . .

THINKING ABOUT Environmental Injustice

Have you or anyone in your family ever been a victim of environmental injustice? If so, what happened? What would you do to help prevent such environmental injustice?

We Can Encourage Reuse and Recycling

Three factors hinder reuse and recycling. *First,* the market prices of almost all products do not include the harmful environmental and health costs associated with producing, using, and discarding them.

Second, the economic playing field is uneven, because in most countries, resource-extracting industries receive more government tax breaks and subsidies than reuse and recycling industries get.

Third, the demand and thus the price paid for recycled materials fluctuates, mostly because buying goods made with recycled materials is not a priority for most governments, businesses, and individuals.

How can we encourage reuse and recycling? Proponents say that leveling the economic playing field is the best way to start. Governments can *increase* subsidies and tax breaks for reusing and recycling materials, and *decrease* subsidies and tax breaks for making items from virgin resources.

One way to include some of the harmful environmental costs of products in their prices, while encouraging recycling, is to attach a small deposit fee to the price of recyclable items, as is done in many European countries, several Canadian provinces, and ten U.S. states that have *bottle bills.* Such laws place a deposit fee of 5 or 10 cents on each beverage container, and consumers can recover that fee by returning their empty containers to the store. In 2012, these ten states recycled at least 70% of their bottles and cans, compared to 28% in states with no bottle bills. This approach is in keeping with the full-cost pricing **principle of sustainability**.

Another strategy is to greatly increase use of the fee-per-bag waste collection system. When the U.S. city of Fort Worth, Texas, instituted such a program, the proportion of households recycling their trash went from 21% to 85%. The city went from losing $600,000 in its recycling program to making $1 million a year because of increased sales of recycled materials to industries.

Governments can also pass laws requiring companies to take back and recycle or reuse packaging and electronic waste discarded by consumers (**Core Case Study**), as is done in Japan and some European Union countries. Another important strategy is to encourage or require government purchases of recycled products to help increase demand for and lower prices of these products. Also, citizens can pressure governments to require product labeling that lists the recycled content of products, as well as the types and amounts of any hazardous materials they contain. This would help consumers to make more informed choices about the environmental consequences of buying certain products. It would also help to expand the market for recycled materials by spurring demand for them.

Reuse, Recycling, and Composting Present Economic Opportunities

A growing number of people are saving money through reuse, regularly going to yard sales, flea markets, secondhand stores, and online sites such as eBay and craigslist. The Freecycle Network links people who want to give away household belongings to people who want or need them. Its nearly 7 million members reuse an average of 640 metric tons (700 tons) of items every day, which is about the amount of solid waste that arrives at a mid-size landfill every day.

For many, recycling has become a business opportunity. In particular, *upcycling,* or recycling materials into products of a higher value (Figure 21-30), is a growing field. For example, a British company called Worn Again is converting discarded textiles such as old hot-air balloons and worn-out seat covers into windbreaker jackets and other products. And Professor Na Lu in North Carolina has found a way to upcycle plastic bottles to make a building material that could outperform composite lumber and wood lumber. Entrepreneurs see upcycling as an area of great opportunity.

Rebecca Hale/National Geographic Creative

Figure 21-30 *Upcycling:* This handbag was made from an old airline seat.

Companies are also realizing that they can save money by minimizing packaging or by using *dual-use packaging*. For example, Hewlett-Packard has packaged some of its laptop computers in padded laptop bags that can be used indefinitely. In doing so, the company reduced packaging waste for those computers by 97%. As an added benefit, HP made these bags out of recycled materials.

International Treaties Have Reduced Hazardous Waste

Environmental justice also applies at the international level. For decades, some more-developed countries had been shipping hazardous wastes to less-developed countries. However, since 1992, an international treaty known as the Basel Convention has been in effect. It bans participating countries from shipping hazardous waste (including e-waste; see **Core Case Study**) to or through other countries without their permission. In 1995, the treaty was amended to outlaw all transfers of hazardous wastes from industrial countries to less-developed countries. By 2012, this agreement had been ratified (formally approved and implemented) by 179 countries. The United States, Afghanistan, and Haiti have signed but have not ratified the convention.

This ban will help, but it will not wipe out the very profitable illegal shipping of hazardous wastes. Hazardous waste smugglers evade the laws by using an array of tactics, including bribes, false permits, and mislabeling of hazardous wastes as recyclable materials.

In 2000, delegates from 122 countries completed a global treaty known as the Stockholm Convention on Persistent Organic Pollutants (POPs). It regulates the use of 12 widely used persistent organic pollutants that can accumulate in the fatty tissues of humans and other animals that occupy high trophic levels in food webs. At such levels, these hazardous chemicals can reach levels hundreds of thousands of times higher than their levels in the general environment (see Figure 9-13, p. 202). Because they persist in the environment, POPs can also be transported long distances by wind and water.

The original list of 12 hazardous chemicals, called the *dirty dozen*, includes DDT and 8 other chlorine-containing persistent pesticides, PCBs, dioxins, and furans. Using blood tests and statistical sampling, medical researchers at New York City's Mount Sinai School of Medicine found that it is likely that nearly every person on earth has detectable levels of POPs in their bodies. The long-term health effects of this involuntary global chemical experiment are largely unknown.

By 2012, 178 countries had ratified a strengthened version of the POPs treaty that seeks to ban or phase out the use of these hazardous chemicals and to detoxify or isolate existing stockpiles. It allows 25 countries to continue using DDT to combat malaria until safer alternatives are found. The United States has not yet ratified this treaty. The list of regulated POPs is expected to grow.

In 2000, the Swedish Parliament enacted a law that, by 2020, will ban all potentially hazardous chemicals that are persistent in the environment and can accumulate in living tissue. This law also requires industries to perform risk assessments on the chemicals they use and to show that these chemicals are safe to use, as opposed to requiring the government to show that they are dangerous. In other words, chemicals are assumed to be guilty until proven innocent—the reverse of the current policy in the United States and most other countries. There is strong opposition to this approach in the United States, especially from most of the industries that produce and use potentially hazardous chemicals.

We Can Make the Transition to Low-Waste Societies

According to physicist Albert Einstein, "A clever person solves a problem; a wise person avoids it." Many people are taking these words seriously. The governments of Norway, Austria, and the Netherlands have committed to reducing their resource waste by 75%. Many school cafeterias, restaurants, national parks, and corporations are participating in a rapidly growing "zero waste" movement to reduce, reuse, and recycle, and some have lowered their waste outputs by up to 80%, with the ultimate goal of eliminating their waste outputs.

GOOD NEWS

Many environmental scientists argue that we can prevent pollution, reduce waste, and make a transition to a low-waste society by understanding and following these key principles:

1. Everything is connected.
2. There is no *away*, as in *to throw away*, for the wastes we produce.
3. Producers and polluters should pay for the wastes they produce.
4. We can mimic nature by reusing, recycling, composting, or exchanging most of the municipal solid wastes we produce (see the following Case Study).

CASE STUDY

Industrial Ecosystems: Copying Nature

An important goal for a more sustainable society is to make its industrial manufacturing processes cleaner and more sustainable by redesigning them to mimic how nature deals with wastes—an approach called *biomimicry*. In nature, according to the chemical cycling **principle of sustainability**, the waste outputs of one organism become the nutrient inputs of another organism, so that all of the earth's nutrients are endlessly recycled. This explains why there is essentially no waste in undisturbed ecosystems.

One way for industries to mimic nature is to reuse or recycle most of the minerals and chemicals they use,

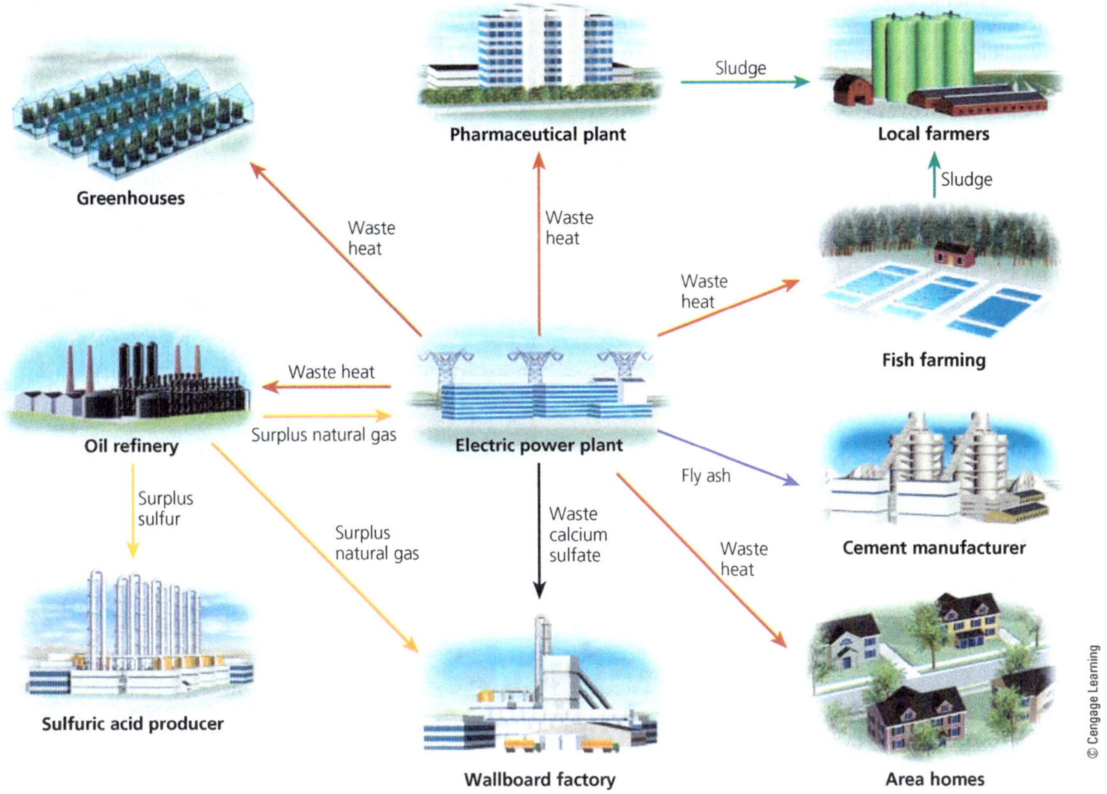

Figure 21-31 Solutions: This *industrial ecosystem* in Kalundborg, Denmark, reduces waste production by mimicking a natural ecosystem's food web. The wastes of one business become the raw materials for another, thus mimicking the way that nature recycles chemicals. ***Question:*** Is there an industrial ecosystem near where you live or go to school? If not, think about where and how such a system could be set up.

instead of burying or burning them or shipping them somewhere. Another method that industries can use to mimic nature is to interact with each other through *resource exchange webs* in which the wastes of one manufacturer become the raw materials for another—similar to food webs in natural ecosystems (see Figure 3-12, p. 60).

This is happening in Kalundborg, Denmark, where an electric power plant and nearby industries, farms, and homes are collaborating to save money and to reduce their outputs of waste and pollution within what is called an *ecoindustrial park*, or *industrial ecosystem*. They exchange waste outputs and convert them into resources, as shown in Figure 21-31. This cuts pollution and waste and reduces the flow of nonrenewable mineral and energy resources through the local economy. GOOD NEWS

Today, more than 40 ecoindustrial parks (18 of them in the United States) operate in various places around the world, and more are being built or planned—some of them on brownfield sites. A number of people who work in the rapidly growing field of industrial ecology are focusing on developing a global network of industrial ecosystems over the next few decades, and this could lead to an important *ecoindustrial revolution*. **GREEN CAREER:** industrial ecology

These and other industrial forms of biomimicry provide many economic benefits for businesses. By encouraging recycling and waste-reduction prevention, they reduce the costs of managing solid wastes, controlling pollution, and complying with pollution regulations. They also reduce a company's chances of being sued because of damages to people and the environment caused by their actions. In addition, companies improve the health and safety of workers by reducing their exposure to toxic and hazardous materials, thereby reducing company health insurance costs. Biomimicry also encourages companies to come up with new, environmentally beneficial, and less resource-intensive chemicals, processes, and products that they can sell worldwide.

Biomimicry involves two major steps. The first is to observe certain changes in nature and to study how natural systems have responded to such changing conditions over many millions of years. The second step is to try to copy or adapt these responses within human systems in order to help us deal with various environmental challenges. In the case of solid and hazardous wastes, the food web serves as a natural model for responding to the growing problem of these wastes. This is in keeping with the three scientific **principles of sustainability** (see Figure 1-2, p. 6 or back cover) that nature has used for billions of years.

Big Ideas

- The order of priorities for dealing with solid waste should be to produce less of it, reuse and recycle as much of it as possible, and safely burn or bury what is left.
- The order of priorities for dealing with hazardous waste should be to produce less of it, reuse or recycle it, convert it to less-hazardous material, and safely store what is left.
- We need to view solid wastes as wasted resources, and hazardous wastes as materials that we should not be producing in the first place.

TYING IT ALL TOGETHER E-Waste and Sustainability

Bakalusha/Shutterstock.com

One of the problems of maintaining a high-waste society is the growing mass of e-waste (**Core Case Study**) and other types of solid and hazardous waste discussed in this chapter. The challenge is to make the transition from an unsustainable high-waste, throwaway economy to a more sustainable low-waste, reducing–reusing–recycling economy as soon as possible.

Such a transition will require applying the six **principles of sustainability** (see Figure 1-2, p. 6, Figure 1-5, p. 9, or back cover). We can reduce our outputs of solid and hazardous waste by relying much less on fossil fuels and nuclear power (which produces long-lived, hazardous radioactive wastes) while relying much more on renewable energy from the sun, wind, and flowing water. We can mimic nature's chemical cycling processes by reusing and recycling materials as much as possible. Integrated waste management, which uses a diversity of approaches and emphasizes waste reduction and pollution prevention, is a way to mimic nature's use of biodiversity.

By including more of the harmful environmental and health costs of the consumer economy in market prices, we would be applying the full-cost pricing **principle of sustainability** while encouraging people to refuse, reduce, reuse, and recycle. In doing so, we would benefit the environment, create new jobs and businesses capitalizing on the four Rs, and gain health and environmental benefits for ourselves, thus finding win-win solutions. This could also lead to lower levels of resource use per person, and thus a lower demand for materials that eventually become solid and hazardous wastes. All these measures together would help us to pass along to future generations a world that is at least as livable as the one we have enjoyed.

Chapter Review

Core Case Study

1. Explain how and why electronic waste (e-waste) has become a growing solid waste problem.

Section 21-1

2. What are the two key concepts for this section? Distinguish among **solid waste**, **industrial solid waste**, **municipal solid waste (MSW)**, and **hazardous (toxic) waste**, and give an example of each. Summarize the types and sources of solid waste generated in the United States and explain what happens to it. What is garbology?

Section 21-2

3. What is the key concept for this section? Distinguish among **waste management**, **waste reduction**, and **integrated waste management**. Summarize the priorities that prominent scientists believe we should use for dealing with solid waste and compare them to actual practices in the United States. Distinguish among **refusing**, **reducing**, **reusing**, and **recycling** in dealing with the solid wastes we produce. Why are the first three Rs preferred from an environmental standpoint? What is **composting**? List six ways in which industries and communities can reduce resource use, waste, and pollution.

Section 21-3

4. What is the key concept for this section? Explain why refusing, reducing, reusing, and recycling are so important and give examples of each. List five ways to reuse various items.

5. Distinguish between **primary (closed-loop) recycling** and **secondary recycling**. What are three important steps that must occur for any recycling program to work? What are some benefits of composting? What is the benefit of having households and businesses separate their trash into recyclable categories? What is the fee-per-bag approach? Explain how paper and some plastics are being recycled. What are bioplastics? What are the major advantages and disadvantages of recycling?

Section 21-4

6. What is the key concept for this section? What are the major advantages and disadvantages of using incinerators to burn solid and hazardous waste? Distinguish between **sanitary landfills** and **open dumps**. What are the major advantages and disadvantages of burying solid waste in sanitary landfills?

Section 21-5

7. What is the key concept for this section? What are the priorities that scientists believe we should use in dealing with hazardous waste? Summarize the problems involved in sending e-wastes to less-developed countries for recycling. Describe three ways to detoxify hazardous wastes. What is bioremediation? What is phytoremediation? What are the major advantages and disadvantages of incinerating hazardous wastes? What are the major advantages and disadvantages of using plasma gasification to detoxify hazardous wastes?

8. What are the major advantages and disadvantages of disposing of liquid hazardous wastes **(a)** in deep underground wells and **(b)** in surface impoundments? What is a secure hazardous waste landfill? List four ways to reduce your output of hazardous waste. Summarize how laws in the United States regulate hazardous wastes. What is a brownfield?

Section 21-6

9. What is the key concept for this section? How has grassroots action improved solid and hazardous waste management in the United States? What is **environmental justice** and how well has it been applied in locating and cleaning up hazardous waste sites in the United States? What are three factors that discourage recycling? What are three ways to encourage recycling and reuse? Give three examples of how people are saving or making money through reuse, recycling, and composting. Describe regulation of hazardous wastes at the global level through the Basel Convention and the treaty to control persistent organic pollutants (POPs).What is biomimicry? What is an industrial ecosystem?

10. What are this chapter's three big ideas? Explain how we could deal with the growing problems of e-waste and other wastes (Core Case Study) by applying the six **principles of sustainability**.

Note: Key terms are in bold type.

Critical Thinking

1. Do you think that manufacturers of computers, television sets, and other forms of e-waste (**Core Case Study**) should be required to take their products back at the end of their useful lives for repair, remanufacture, or recycling in a manner that is environmentally responsible and that does not threaten the health of recycling workers? Explain. Would you be willing to pay more for these products to cover the costs of such a take-back program? If so, what percentage more per purchase would you be willing to pay for electronic products?

2. Find three items that you regularly use once and then throw away. Are there other reusable items that you could use in place of these disposable items? For each item, calculate and compare the cost of using the disposable option for a year versus the cost of using the reusable alternative. Write a brief report summarizing your findings.

3. Do you think that you could consume less by refusing to buy some of the things you regularly buy? If so, what are three of those things? Do you think that this is something you ought to do? Explain your reasoning.

4. A company called Changing World Technologies has built a pilot plant to test a process it has developed for converting a mixture of discarded computers, old tires, turkey bones and feathers, and other wastes into oil by mimicking and speeding up natural processes for converting biomass into oil. Explain why this recycling process, if it turns out to be technologically and economically feasible, could lead to increased waste production.

5. Would you oppose having **(a)** a hazardous waste landfill, **(b)** a waste treatment plant, **(c)** a deep-injection well, or **(d)** an incinerator in your community? For each of these facilities, explain your answer. If you oppose having such disposal facilities in your community, how do you believe the hazardous waste generated in your community should be managed?

6. How does your school dispose of its solid and hazardous wastes? Does it have a recycling program? How well does it work? Does your school encourage reuse? If so, how? Does it have a hazardous waste collection system? If so, describe it. List three ways in which you would improve your school's waste-reduction and management systems.

7. List three ways in which you could apply **Concept 21-6** to making your lifestyle more environmentally sustainable.

8. Congratulations! You are in charge of the world. List the three most important components of your strategy for dealing with **(a)** solid waste and **(b)** hazardous waste.

Doing Environmental Science

Collect the trash (excluding food waste) that you generate in a typical week. Measure its total weight and volume. Sort it into major categories such as paper, plastic, metal, and glass. Then weigh each category and calculate its percentage by weight of the total amount of trash that you have measured. What percentage by weight of this waste consists of materials that could be recycled? What percentage consists of materials for which you could have used a reusable substitute, such as a coffee mug instead of a disposable cup? What percentage by weight of the items could you have done without? Compare your answers to these questions with those of your classmates. Together with your classmates, combine all the results and do the same analysis for the entire class. Use these results to estimate the same values for the entire student population at your school. (If you wish, you could include food wastes in your research, but you would need to find a way to store it such that it won't become a nuisance or a health hazard.)

Global Environment Watch Exercise

Within the GREENR database, go to the *E-Waste* topic portal. Research and find statistics on how rapidly the world's production of e-waste (**Core Case Study**) is growing and how rapidly e-waste production is growing in the United States. Write a brief report on what the United States and one other country of your choice are doing to deal with this growing waste problem. Compare the two approaches in terms of how successful they are.

Ecological Footprint Analysis

Researchers estimate that the average daily municipal solid waste production per person in the United States is 3.2 kilograms (4.34 pounds). Use the data in the pie chart below to get an idea of a typical annual MSW ecological footprint for each American by calculating the total weight in kilograms (and pounds) for each category generated during 1 year (1 kilogram = 2.20 pounds). Use the table below to enter your answers.

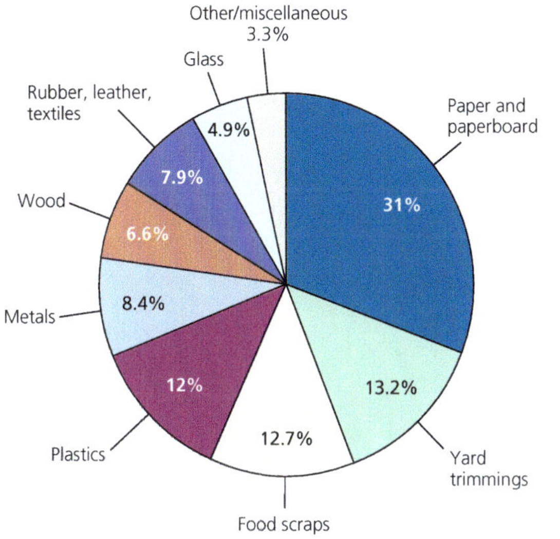

Composition of a typical sample of U.S. municipal solid waste, 2010.
(Compiled by the authors using data from U.S. Environmental Protection Agency.)

Waste Category	Annual MSW Footprint per Person
Paper and paperboard	
Yard trimmings	
Food scraps	
Plastics	
Metals	
Wood	
Rubber, leather, and textiles	
Glass	
Other/miscellaneous	

SUPPLEMENT 1 MEASUREMENT UNITS

Length

Metric

1 kilometer (km) = 1,000 meters (m)
1 meter (m) = 100 centimeters (cm)
1 meter (m) = 1,000 millimeters (mm)
1 centimeter (cm) = 0.01 meter (m)
1 millimeter (mm) = 0.001 meter (m)

English

1 foot (ft) = 12 inches (in)
1 yard (yd) = 3 feet (ft)
1 mile (mi) = 5,280 feet (ft)
1 nautical mile = 1.15 miles

Metric–English

1 kilometer (km) = 0.621 mile (mi)
1 meter (m) = 39.4 inches (in) or 3.28 feet (ft)
1 inch (in) = 2.54 centimeters (cm)
1 foot (ft) = 0.305 meter (m)
1 yard (yd) = 0.914 meter (m)
1 nautical mile = 1.85 kilometers (km)

Area

Metric

1 square kilometer (km^2) = 1,000,000 square meters (m^2)
1 square meter (m^2) = 1,000,000 square millimeters (mm^2)
1 square meter (m^2) = 10,000 square centimeters (cm^2)
1 hectare (ha) = 10,000 square meters (m^2)
1 hectare (ha) = 0.01 square kilometer (km^2)

English

1 square foot (ft^2) = 144 square inches (in^2)
1 square yard (yd^2) = 9 square feet (ft^2)
1 square mile (mi^2) = 27,880,000 square feet (ft^2)
1 acre (ac) = 43,560 square feet (ft^2)

Metric–English

1 hectare (ha) = 2.471 acres (ac)
1 square kilometer (km^2) = 0.386 square mile (mi^2)
1 square meter (m^2) = 1.196 square yards (yd^2)
1 square meter (m^2) = 10.76 square feet (ft^2)
1 square centimeter (cm^2) = 0.155 square inch (in^2)

Volume

Metric

1 cubic kilometer (km^3) = 1,000,000,000 cubic meters (m^3)
1 cubic meter (m^3) = 1,000,000 cubic centimeters (cm^3)
1 cubic meter (m^3) = 1,000 liters (L)
1 liter (L) = 1,000 milliliters (mL) = 1,000 cubic centimeters (cm^3)
1 cubic meter (m^3) = 1,000 liters (L)
1 milliliter (mL) = 0.001 liter (L)
1 milliliter (mL) = 1 cubic centimeter (cm^3)

English

1 gallon (gal) = 4 quarts (qt)
1 quart (qt) = 2 pints (pt)

Metric–English

1 liter (L) = 0.265 gallon (gal)
1 liter (L) = 1.06 quarts (qt)
1 liter (L) = 0.0353 cubic foot (ft^3)
1 cubic meter (m^3) = 35.3 cubic feet (ft^3)
1 cubic meter (m^3) = 1.30 cubic yards (yd^3)
1 cubic kilometer (km^3) = 0.24 cubic mile (mi^3)
1 barrel (bbl) = 159 liters (L)
1 barrel (bbl) = 42 U.S. gallons (gal)

Mass

Metric

1 kilogram (kg) = 1,000 grams (g)
1 gram (g) = 1,000 milligrams (mg)
1 gram (g) = 1,000,000 micrograms (μg)
1 milligram (mg) = 0.001 gram (g)
1 microgram (μg) = 0.000001 gram (g)
1 metric ton (mt) = 1,000 kilograms (kg)

English

1 ton (t) = 2,000 pounds (lb)
1 pound (lb) = 16 ounces (oz)

Metric–English

1 metric ton (mt) = 2,200 pounds (lb) = 1.1 tons (t)
1 kilogram (kg) = 2.20 pounds (lb)
1 pound (lb) = 454 grams (g)
1 gram (g) = 0.035 ounce (oz)

Energy and Power

Metric

1 kilojoule (kJ) = 1,000 joules (J)
1 kilocalorie (kcal) = 1,000 calories (cal)
1 calorie (cal) = 4.184 joules (J)

Metric–English

1 kilojoule (kJ) = 0.949 British thermal unit (Btu)
1 kilojoule (kJ) = 0.000278 kilowatt-hour (kW-h)
1 kilocalorie (kcal) = 3.97 British thermal units (Btu)
1 kilocalorie (kcal) = 0.00116 kilowatt-hour (kW-h)
1 kilowatt-hour (kW-h) = 860 kilocalories (kcal)
1 kilowatt-hour (kW-h) = 3,400 British thermal units (Btu)
1 quad (Q) = 1,050,000,000,000,000 kilojoules (kJ)
1 quad (Q) = 293,000,000,000 kilowatt-hours (kW-h)

Temperature Conversions

Fahrenheit (°F) to Celsius (°C): $°C = (°F - 32.0) \div 1.80$
Celsius (°C) to Fahrenheit (°F): $°F = (°C \times 1.80) + 32.0$

SUPPLEMENT 2 READING GRAPHS AND MAPS

Graphs and Maps Are Important Visual Tools

A **graph** is a tool for conveying information that we can summarize numerically by illustrating that information in a visual format. This information, called *data,* is collected in experiments, surveys, and other information-gathering activities. Graphing can be a powerful tool for summarizing and conveying complex information.

In this textbook, we use three major types of graphs: *line graphs, bar graphs,* and *pie graphs.* Here, you will explore each of these types of graphs and learn how to read them.

An important visual tool used to summarize data that vary over small or large areas is a **map**. We discuss some aspects of reading maps relating to environmental science at the end of this supplement.

Line Graphs

Line graphs usually represent data that fall in some sort of sequence such as a series of measurements over time or distance. In most such cases, units of time or distance lie on the horizontal *x-axis.* The possible measurements of some quantity or variable such as temperature that changes over time or distance usually lie on the vertical *y-axis.*

In Figure 1, the x-axis shows the years between 1950 and 2009, and the y-axis displays the possible values for the annual amounts of oil consumed worldwide during that time in millions of tons, ranging from 0 to 4,000 million (or 4 billion) tons. Usually, the y-axis appears on the left end of the x-axis, although y-axes can appear on the right end, in the middle, or on both ends of the x-axis.

The curving line on a line graph represents the measurements taken at certain time or distance intervals. In Figure 1, the curve represents changes in oil consumption between 1950 and 2009. To find the oil consumption for any year, find that year on the x-axis (a point called the *abscissa*) and run a vertical line from the axis to the curve. At the point where your line intersects the curve, run a horizontal line to the y-axis. The value at that point on the y-axis, called the *ordinate,* is the amount you are seeking. You can go through the same process in reverse to find a year in which oil consumption was at a certain point.

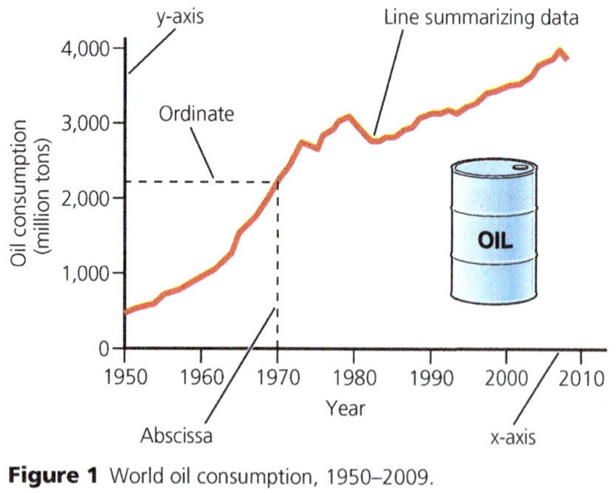

Figure 1 World oil consumption, 1950–2009.

(Compiled by the authors using data from U.S. Energy Information Administration, International Energy Agency, and United Nations.)

Questions

1. What was the total amount of oil consumed in the world in 1990? In about what year between 1950 and 2000 did oil consumption first start declining?
2. About how much oil was consumed in 2009? Roughly how many times more oil was consumed in 2009 than in 1970? How many times more oil was consumed in 2009 than in 1950?

Line graphs have several important uses. One of the most common applications is to compare two or more variables. Figure 2 compares two variables: monthly temperature and precipitation (rain and snowfall) during a typical year in a temperate deciduous forest. However, in this case the variables are measured on two different scales, so there are two y-axes. The y-axis on the left end of the graph shows a Centigrade temperature scale, and the y-axis on the right shows the range of precipitation measurements in millimeters. The x-axis displays the first letters of each of the 12 months' names.

Questions

1. In which month does most precipitation fall? Which is the driest month of the year? Which is the hottest month?

Temperate deciduous forest

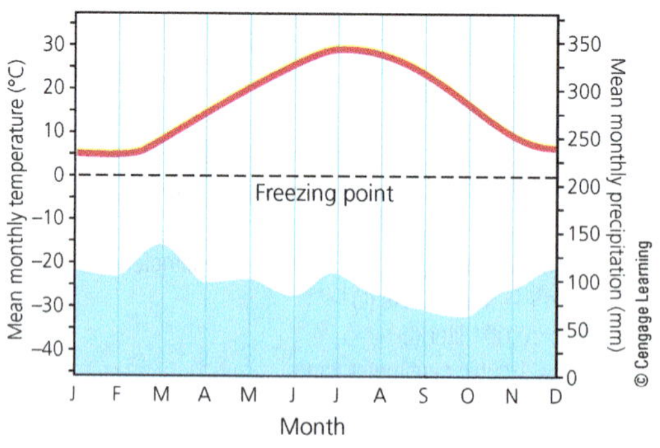

Figure 2 Climate graph showing typical variations in annual temperature (red) and precipitation (blue) in a temperate deciduous forest.

2. If the temperature curve were almost flat, running throughout the year at roughly its highest point of about 30°C, how do you think this forest would change from what it is now (see Figure 7-13, center, p. 156)? If the annual precipitation suddenly dropped and remained under 25 centimeters all year, what do you think would eventually happen to this forest?

It is also important to consider what aspect of a set of data is being displayed on a graph. The creator of a graph can take two different aspects of one data set and create two very different-looking graphs that would give two different interpretations of the same phenomenon. For example, when talking about any type of growth we must be careful to distinguish the question of whether something is growing from the question of how fast it is growing. While a quantity can keep growing continuously, its rate of growth can go up and down.

One of many important examples of growth used in this book is human population growth. For example, the graph in Figure 1-16 (p. 17) gives you the impression that human population growth has, for the most part, been continuous and uninterrupted. However, consider Figure 3, which plots the rate of growth of the human population since 1950. Note that all of the numbers on the y-axis, even the smallest ones, represent growth. The lower end of the scale represents slower growth and the higher end faster growth. Thus, while one graph tracks population growth in terms of numbers of people, the other tracks the rate of growth.

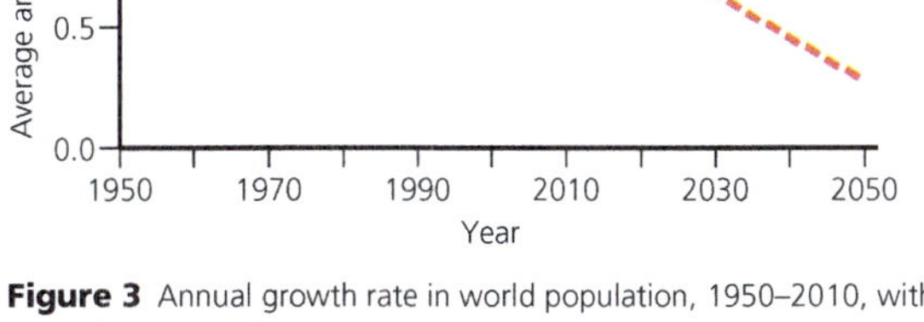

Figure 3 Annual growth rate in world population, 1950–2010, with projections to 2050.

(Compiled by the authors using data from UN Population Division, U.S. Census Bureau, and Population Reference Bureau.)

Questions

1. If this graph were presented to you as a picture of human population growth, what would be your first impression?
2. Do you think that reaching a growth rate of 0.5% would relieve those who are concerned about overpopulation? Why or why not?

Bar Graphs

The *bar graph* is used to compare measurements for one or more variables across categories. Unlike the line graph, a bar graph typically does not involve a sequence of measurements over time or distance. The measurements compared on a bar graph usually represent data collected at some point in time or during a well-defined period. For instance, we can compare the *net primary productivity (NPP)*, a measure of chemical energy produced by plants in an ecosystem, for different ecosystems, as represented in Figure 4.

In most bar graphs, the categories to be compared are laid out on the x-axis, and the range of measurements for the variable under consideration lies along the y-axis. In our example in Figure 4, the categories (ecosystems) are on the y-axis, and the variable range (NPP) lies on the x-axis. In either case, reading the graph is straightforward. Simply run a line perpendicular to the bar you are reading from the top of that bar (or the right or left end, if it lies horizontally) to the variable value axis. In Figure 4, you can see that the NPP for continental shelf, for example, is close to 1,600 kcal/m^2/yr.

Questions

1. What are the two terrestrial ecosystems that are closest in NPP value of all pairs of such ecosystems? About how many times greater is the NPP in a tropical rain forest than the NPP in a savanna?
2. What is the most productive of aquatic ecosystems shown here? What is the least productive?

An important application of the bar graph used in this book is the *age-structure diagram* (see Figure 6-11, p. 131), which describes a population by showing the numbers of males and females in certain age groups (see Chapter 6, pp. 131–132).

Pie Graphs

Like bar graphs, *pie graphs*, or *pie charts*, illustrate numerical values for two or more categories. But in addition to that, they can also show each category's proportion of the total of all measurements. The categories are usually ordered on the graph from largest to smallest, for ease of comparison, although this is not always the case. Also, as with bar graphs, pie graphs are generally snapshots of a

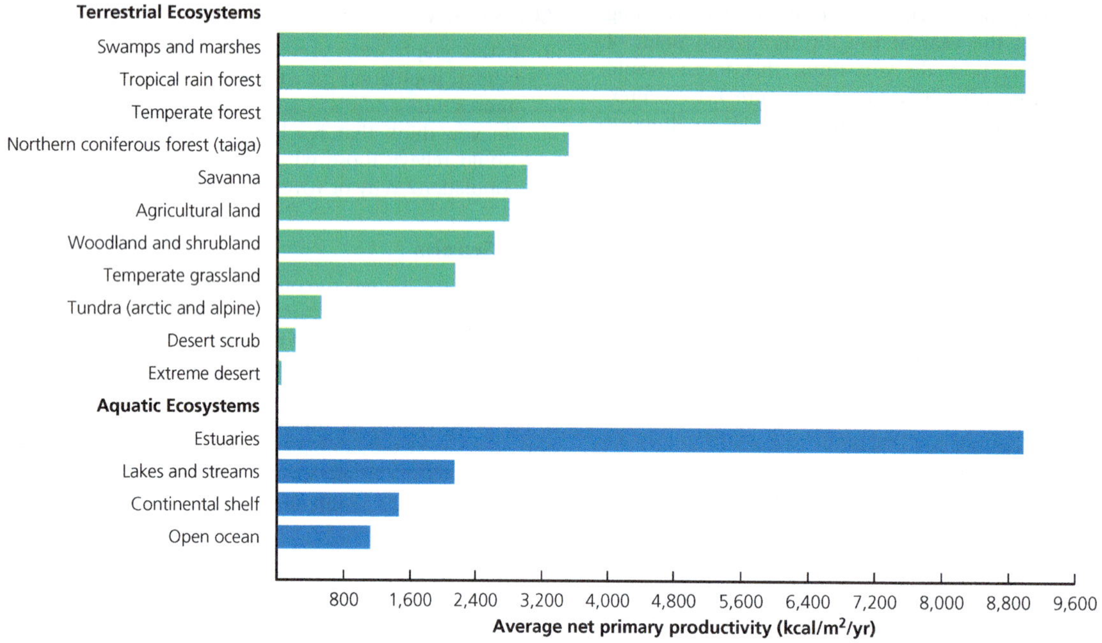

Figure 4 Estimated annual average net primary productivity (NPP) in major life zones and ecosystems, expressed as kilocalories of energy produced per square meter per year (kcal/m^2/yr).

(Compiled by the authors using data from R. H. Whittaker, *Communities and Ecosystems*, 2nd ed., New York: Macmillan, 1975.)

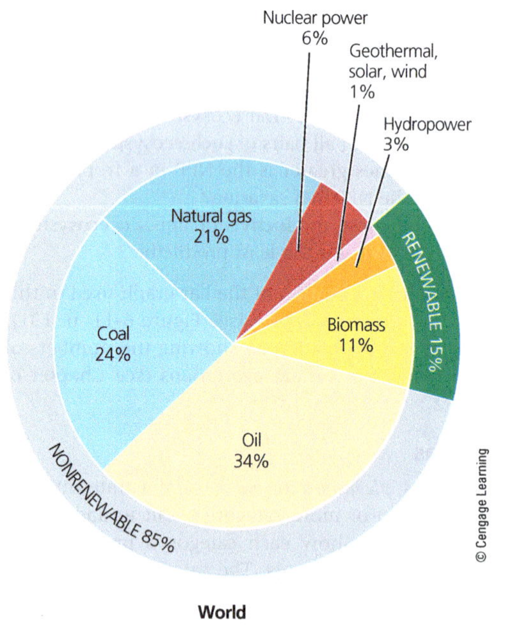

Figure 5 Pie graph showing world energy use by source in 2010.

data set at a point in time or during a defined time period. Unlike line graphs, one pie graph cannot show changes over time.

For example, Figure 5 shows how much each major energy source contributed to the world's total amount of energy used in 2010. This graph includes the numerical data used to construct it: the percentages of the total taken up by each part of the pie. But we can use pie graphs without including the numerical data and we can roughly estimate such percentages. The pie graph thereby provides a generalized picture of the composition of a data set.

Questions

1. Can you tell from this graph whether the use of renewable energy sources is growing or shrinking? Explain.
2. About how many times bigger is oil use than biomass use?

Reading Maps

We can use maps for considerably more than showing where places are relative to one another. For example, in environmental science, maps can be very helpful in comparing how people in different areas are affected by environmental problems such as air pollution and acid deposition (a form of air pollution). Figure 6 is a map of the United States showing the relative numbers of premature deaths due to air pollution in the various regions of the country.

Questions

1. Which part of the country generally has the lowest level of premature deaths due to air pollution?
2. Which part of the country has the highest level? What is the level in the area where you live or go to school?

In some of the data analysis exercises that appear at the ends of the chapters in this book, you will have opportunities to apply much of the information from this supplement.

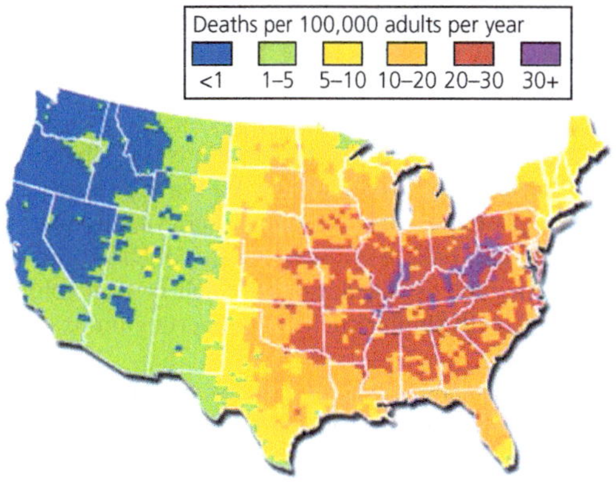

Figure 6 Map showing comparative numbers of premature deaths from air pollution in the United States.

(Compiled by the authors using data from U.S. Environmental Protection Agency.)

Chemists Use the Periodic Table to Classify Elements on the Basis of Their Chemical Properties

Matter consists of elements and compounds (see Chapter 2, pp. 35–38). Chemists have developed a way to classify the elements according to their chemical behavior in what is called the *Periodic Table of Elements* (Figure 1). Each horizontal row in the table is called a *period*. Each vertical column lists elements with similar chemical properties and is called a *group*.

The Periodic Table in Figure 1 shows how the elements can be classified as *metals, nonmetals,* and *metalloids*. Examples of metals are sodium (Na), calcium (Ca), aluminum (Al), iron (Fe), lead (Pb), silver (Ag), and mercury (Hg).

Atoms of metals tend to lose one or more of their electrons to form positively charged ions such as Na^+, Ca^{2+}, and Al^{3+}. For example, an atom of the metallic element sodium (Na, atomic number 11) with 11 positively charged protons and 11 negatively charged electrons can lose one of its electrons. It then becomes a sodium ion with a positive charge of 1 (Na^+) because it now has 11 positive charges (protons) but only 10 negative charges (electrons).

Examples of *nonmetals* are hydrogen (H), carbon (C), nitrogen (N), oxygen (O), phosphorus (P), sulfur (S),

Figure 1 *Periodic Table of Elements.* Elements in the same vertical column, called a *group*, have similar chemical properties.

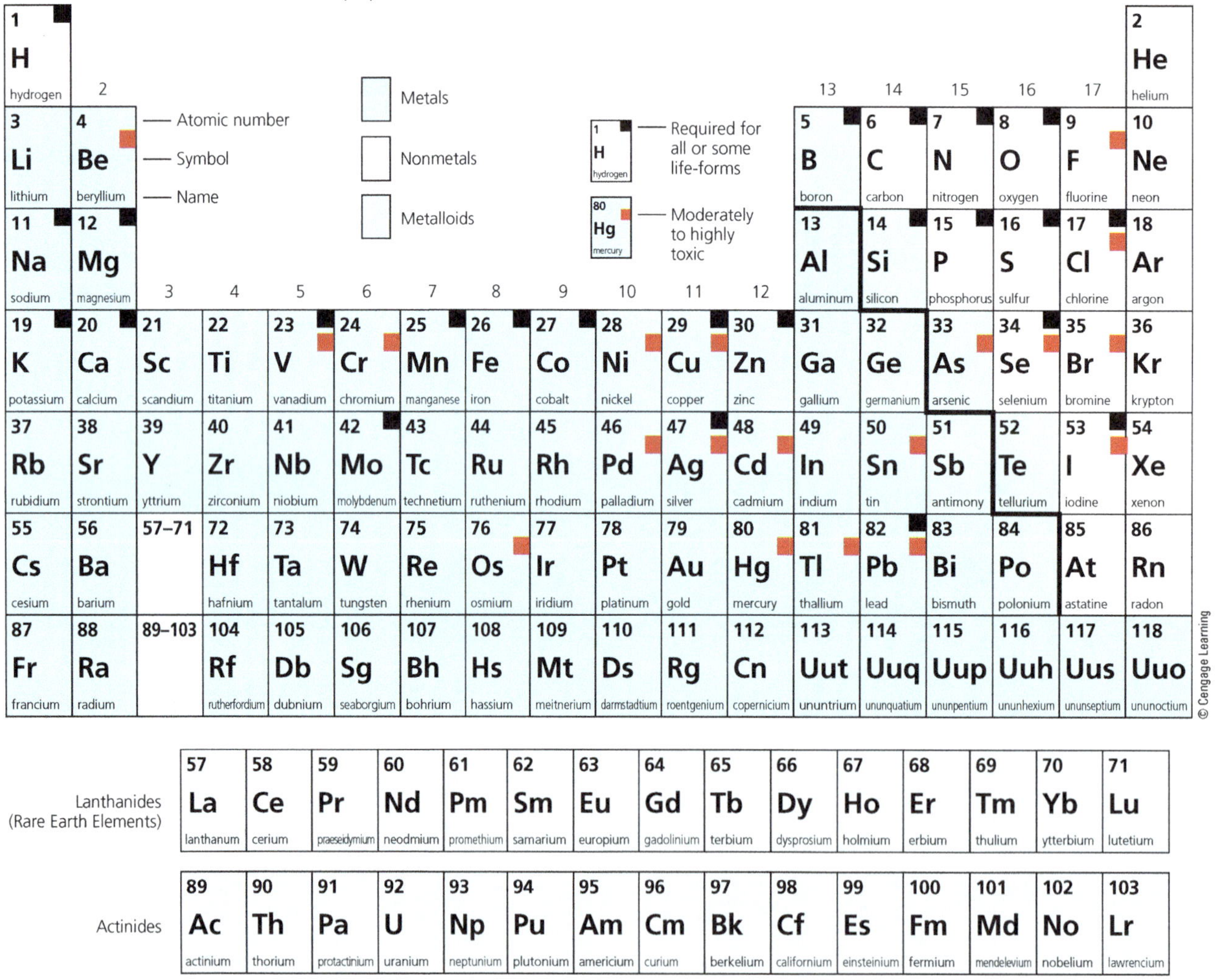

chlorine (Cl), and fluorine (F). Atoms of some nonmetals such as chlorine, oxygen, and sulfur tend to gain one or more electrons lost by metallic atoms to form negatively charged ions such as O^{2-}, S^{2-}, and Cl^-. For example, an atom of the nonmetallic element chlorine (Cl, atomic number 17) can gain an electron and become a chlorine ion. The ion has a negative charge of 1 (Cl^-) because it has 17 positively charged protons and 18 negatively charged electrons. Atoms of nonmetals can also combine with one another to form molecules in which they share one or more pairs of their electrons. Hydrogen, a nonmetal, is placed by itself above the center of the table because it does not fit very well into any of the groups.

The elements arranged in a diagonal staircase pattern between the metals and nonmetals have a mixture of metallic and nonmetallic properties and are called *metalloids*. Examples are germanium (Ge) and arsenic (As).

Figure 1 also identifies the elements required as *nutrients* (marked by small black squares) for all or some forms of life, and elements that are moderately or highly toxic (marked by small red squares) to all or most forms of life. Note that some elements such as copper (Cu) serve as nutrients, but can also be toxic at high enough doses. Six nonmetallic elements—carbon (C), oxygen (O), hydrogen (H), nitrogen (N), sulfur (S), and phosphorus (P)—make up about 99% of the atoms of all living things.

CONSIDER THIS. . .

THINKING ABOUT The Periodic Table

Use the Periodic Table to identify by name and symbol two elements that should have chemical properties similar to those of **(a)** Ca, **(b)** potassium, **(c)** S, **(d)** lead.

Ionic and Covalent Bonds Hold Compounds Together

Sodium chloride (NaCl) consists of a three-dimensional network of oppositely charged *ions* (Na^+ and Cl^-) held together by the forces of attraction between opposite charges (see Figure 2-6, p. 38). The strong forces of attraction between such oppositely charged ions are called *ionic bonds*. They are formed when an electron is transferred from a metallic atom such as sodium (Na) to a nonmetallic element such as chlorine (Cl). Because ionic compounds consist of ions formed from atoms of metallic elements (positive ions) and nonmetallic elements (negative ions), they can be described as *metal–nonmetal compounds*.

Sodium chloride and many other ionic compounds tend to dissolve in water and break apart into their individual ions (Figure 2).

NaCl sodium chloride (in water)	$\longrightarrow$	Na^+ sodium ion	+	Cl^- chloride ion

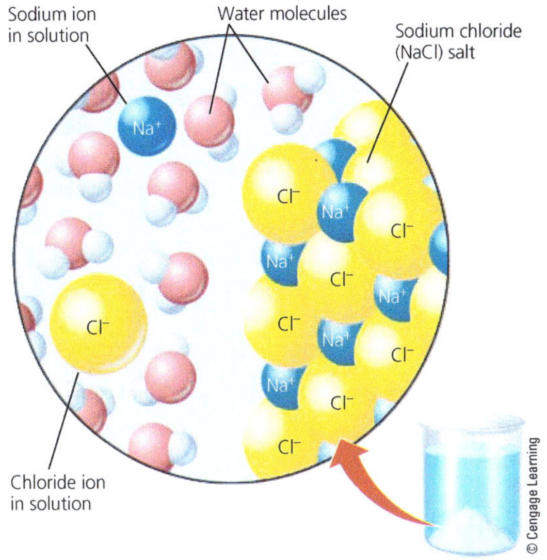

Figure 2 This model illustrates how a salt dissolves in water.

Water, a *covalent compound*, consists of molecules made up of uncharged atoms of hydrogen (H) and oxygen (O). Each water molecule consists of two hydrogen atoms chemically bonded to an oxygen atom, yielding H_2O molecules. The bonds between the atoms in such molecules are called *covalent bonds* and form when the atoms in the molecule share one or more pairs of their electrons. Because they are formed from atoms of nonmetallic elements (see Figure 1), covalent compounds can be described as *nonmetal–nonmetal compounds*. Figure 3 shows the chemical formulas and shapes of the molecules that are the building blocks for several common *covalent compounds*.

What Makes Solutions Acidic? Hydrogen Ions and pH

The *concentration*, or number of hydrogen ions (H^+) in a specified volume of a solution (typically a liter), is a measure of its acidity. Pure water (not tap water or rainwater) has an equal number of hydrogen (H^+) and hydroxide (OH^-) ions. It is called a **neutral solution**. An **acidic solution** has more hydrogen ions than hydroxide ions per liter. A **basic solution** has more hydroxide ions than hydrogen ions per liter.

Scientists use **pH** as a measure of the acidity of a solution based on its concentration of hydrogen ions (H^+). By definition, a neutral solution has a pH of 7, an acidic solution has a pH of less than 7, and a basic solution has a pH greater than 7.

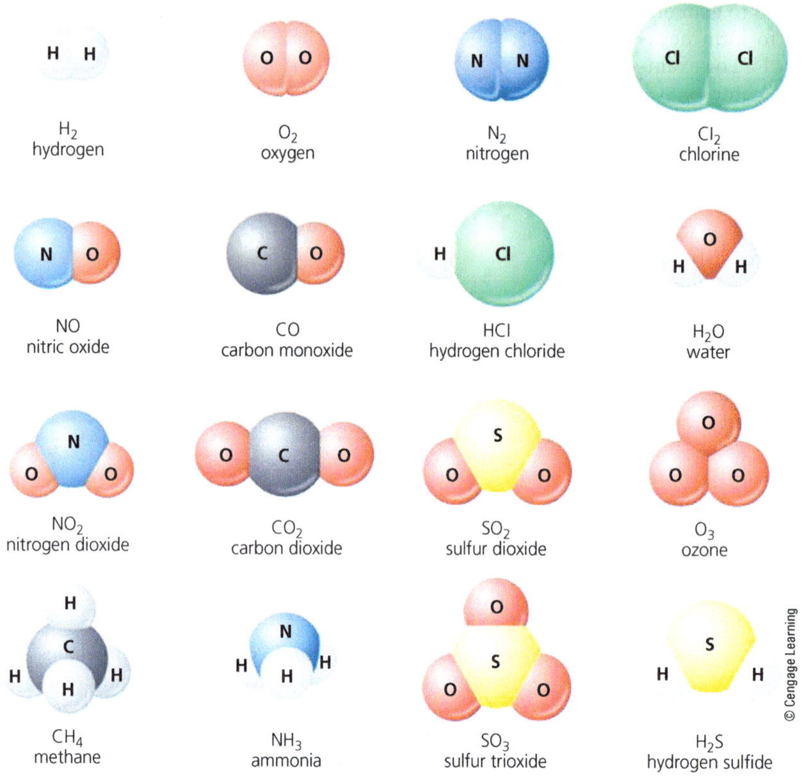

Figure 3 Chemical formulas and shapes for some *covalent compounds* formed when atoms of one or more nonmetallic elements combine with one another. The bonds between the atoms in such molecules are called *covalent bonds.*

Each single unit change in pH represents a tenfold increase or decrease in the concentration of hydrogen ions per liter. For example, an acidic solution with a pH of 3 is 10 times more acidic than a solution with a pH of 4. Figure 4 shows the approximate pH and hydrogen ion concentration per liter of solutions for various common substances.

CONSIDER THIS. . .

THINKING ABOUT pH

A solution has a pH of 2. How many times more acidic is this solution than one with a pH of 6?

The measurement of acidity is important in the study of environmental science, because environmental changes involving acidity can have serious environmental impacts. For example, when coal and oil are burned, they give off acidic compounds that can return to the earth as *acid deposition* (see Figure 18-12, p. 485, and Figure 18-13, p. 486), which has become a major environmental problem.

There Are Weak Forces of Attraction between Some Molecules

Ionic and covalent bonds form between the ions or atoms *within* a compound. There are also weaker forces of attraction *between* the molecules of covalent compounds (such as water) resulting from an unequal sharing of electrons by two atoms.

For example, an oxygen atom has a much greater attraction for electrons than does a hydrogen atom. Thus, the electrons shared between the oxygen atom and its two hydrogen atoms in a water molecule are pulled closer to the oxygen atom, but not actually transferred to the oxygen atom. As a result, the oxygen atom in a water molecule has a slightly negative partial charge and its two hydrogen atoms have a slightly positive partial charge.

The slightly positive hydrogen atoms in one water molecule are then attracted to the slightly negative oxygen atoms in another water molecule. These forces of attraction *between* water molecules are called *hydrogen bonds* (see Figure 3-A, p. 64). They account for many of water's unique properties (see Chapter 3, Science Focus 3.2, p. 64). Hydrogen bonds also form between other covalent molecules or between portions of such molecules containing hydrogen and nonmetallic atoms with a strong ability to attract electrons.

Four Types of Large Organic Compounds Are the Molecular Building Blocks of Life

Larger and more complex organic compounds, called *polymers,* consist of a number of basic structural or molecular units *(monomers)* linked by chemical bonds, somewhat like rail cars linked in a freight train. Four types of macromolecules—complex carbohydrates, proteins, nucleic acids, and lipids—are the molecular building blocks of life.

Complex carbohydrates consist of two or more monomers of *simple sugars* (such as glucose, Figure 5) linked together. One example is the starches that plants use to store energy and also to provide energy for animals that feed on plants. Another is cellulose, the earth's most abundant organic compound, which is found in the cell walls of bark, leaves, stems, and roots.

Proteins are large polymer molecules formed by linking together long chains of monomers called *amino acids* (Figure 6). Living organisms use about 20 different amino acid molecules to build a variety of proteins, which play different roles. Some help to store energy. Some are components of the *immune system* that protects the body against diseases and harmful substances by forming antibodies that make invading agents harmless. Others are *hormones* that are used as chemical messengers in the bloodstreams of animals to turn various bodily functions on or off. In animals, proteins are also components of hair, skin, muscle, and tendons. In addition, some proteins act as *enzymes* that catalyze or speed up certain chemical reactions.

Nucleic acids are large polymer molecules made by linking hundreds to thousands of four types of monomers called *nucleotides*. Two nucleic acids—DNA (**d**eoxyribo**n**ucleic **a**cid) and RNA (**r**ibo**n**ucleic **a**cid)—participate in the building of proteins and carry hereditary information used to pass traits from parent to offspring. Each nucleotide consists of a *phosphate group*, a *sugar molecule* containing five carbon atoms (deoxyribose in DNA molecules and ribose in RNA molecules), and one of four different *nucleotide bases* (represented by A, G, C, and T, the first letter in each of their names, or A, G, C, and U in RNA) (Figure 7). The four basic nucleotides used to make various forms of DNA molecules differ in the types of nucleotide bases they contain—adenine (A), guanine (G), cytosine (C), and thymine (T). (Uracil, labeled U, occurs instead of thymine in RNA.) In the cells of living organisms, these nucleotide units combine in different numbers and sequences to form *nucleic acids* such as various types of DNA and RNA (Figure 8, p. S17).

Hydrogen bonds formed between parts of the four nucleotides in DNA hold two DNA strands together like a spiral staircase, forming a double helix (see Figure 8). DNA molecules can unwind and replicate themselves.

The total weight of the DNA needed to reproduce all of the world's people is only about 50 milligrams—the weight of a small match. If the DNA coiled in your body were unwound, it would stretch about 960 million kilometers (600 million miles)—more than 6 times the distance between the sun and the earth.

The different molecules of DNA that make up the millions of species found on the earth are like a vast and diverse

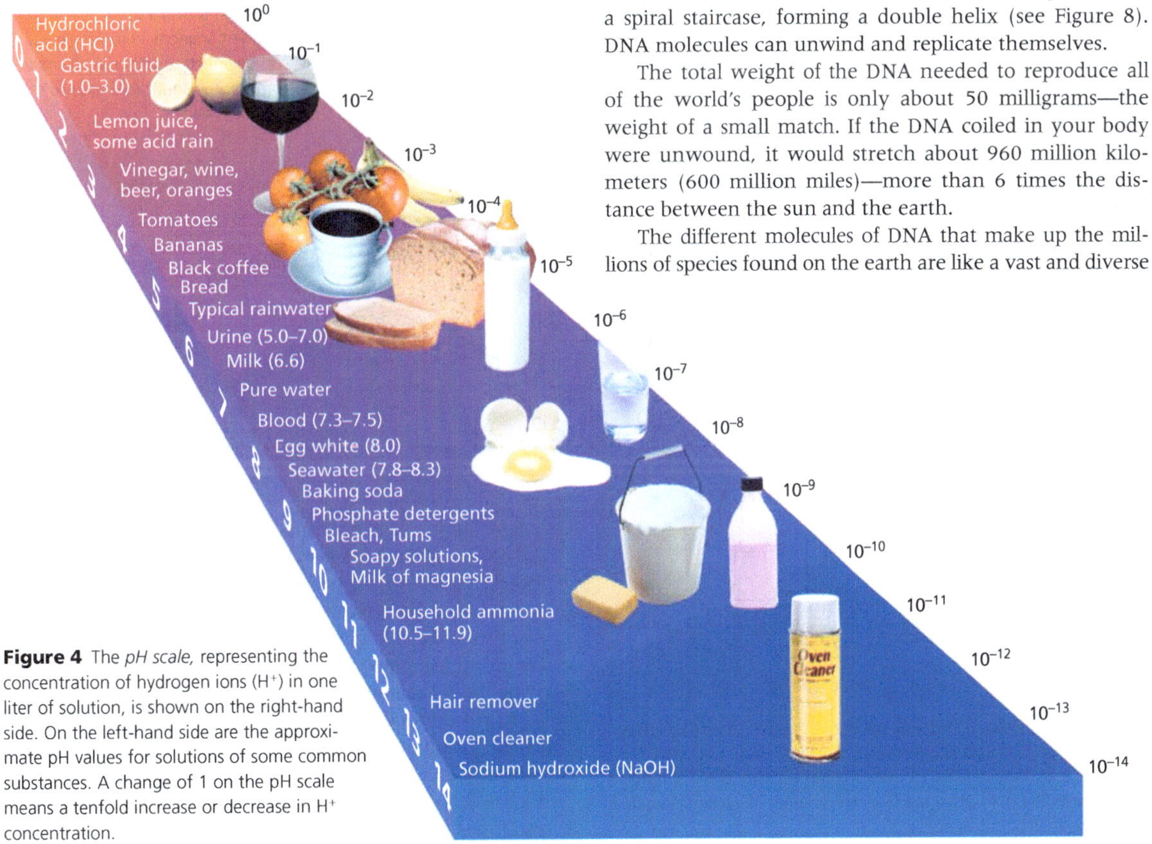

Figure 4 The *pH scale*, representing the concentration of hydrogen ions (H^+) in one liter of solution, is shown on the right-hand side. On the left-hand side are the approximate pH values for solutions of some common substances. A change of 1 on the pH scale means a tenfold increase or decrease in H^+ concentration.

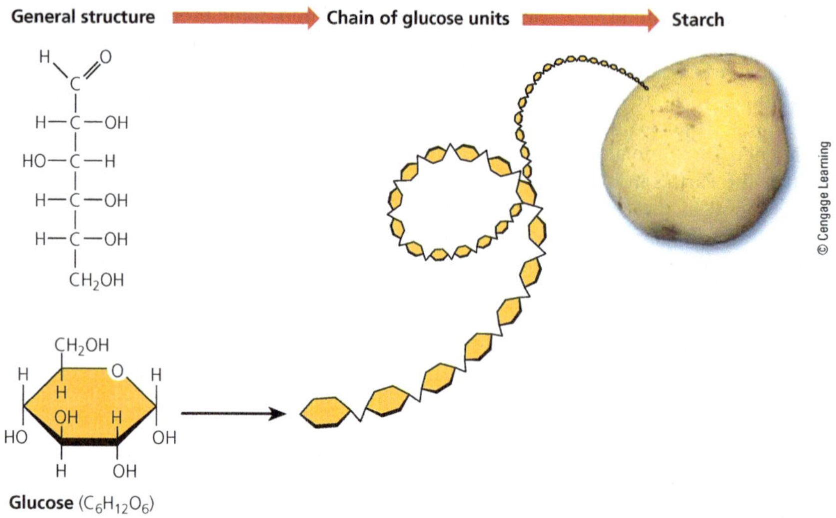

Figure 5 Straight-chain and ring structural formulas of glucose, a simple sugar that can be used to build long chains of complex carbohydrates such as starch and cellulose.

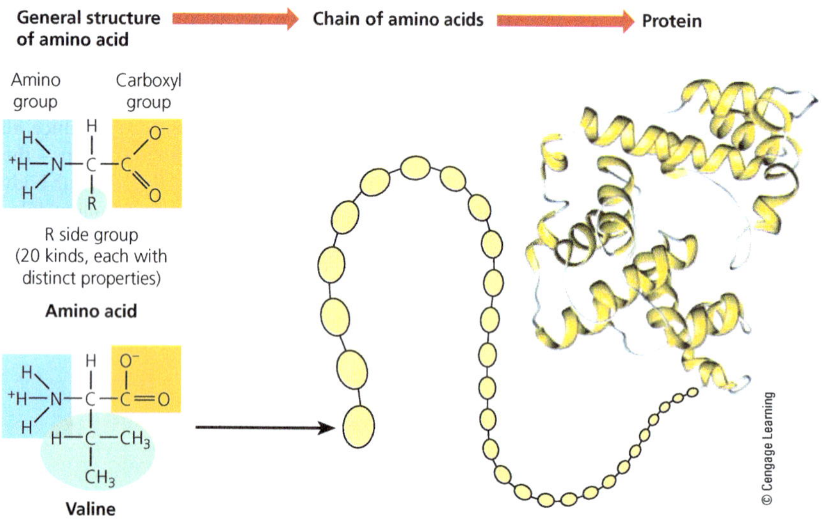

Figure 6 General structural formula of amino acids and a specific structural formula of one of the 20 different amino acid molecules that can be linked together in chains to form proteins that fold up into more complex shapes.

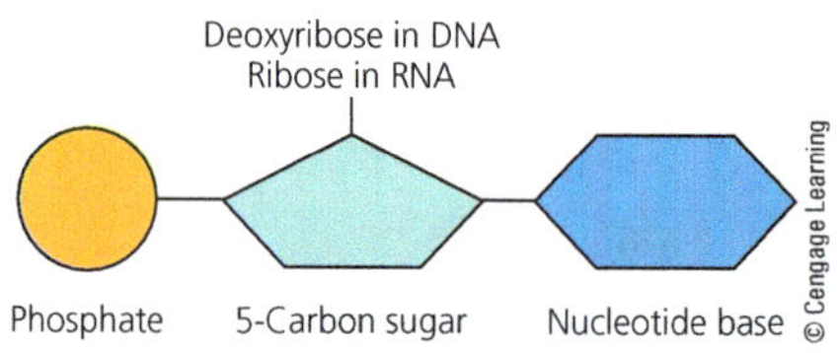

Figure 7 Generalized structures of the nucleotide molecules linked in various numbers and sequences to form large nucleic acid molecules such as various types of deoxyribonucleic acid (DNA) and ribonucleic acid (RNA). In DNA, the five-carbon sugar in each nucleotide is deoxyribose; in RNA it is ribose.

genetic library. Each species is a unique book in that library. The *genome* of a species is made up of the entire sequence of DNA "letters" or base pairs that combine to "spell out" the chromosomes in typical members of each species. In 2002, scientists were able to map out the genome for the human species by using powerful computers to help them in analyzing the 3.1 billion base sequences in human DNA.

Lipids, a fourth building block of life, are a chemically diverse group of large organic compounds that do not dissolve in water. Examples are *fats and oils* for storing energy (Figure 9), *waxes* for structure, and *steroids* for producing hormones.

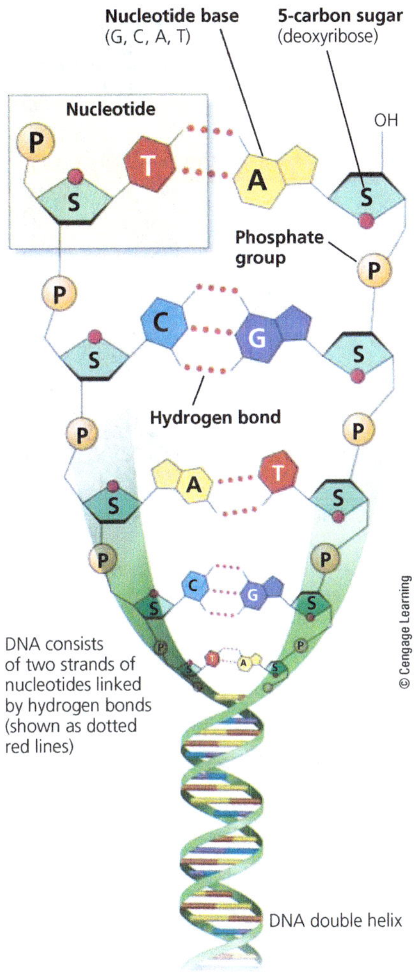

Figure 8 A portion of the double helix of a DNA molecule, which is composed of two spiral (helical) strands of nucleotides. Each nucleotide contains a unit of phosphate (P), deoxyribose (S), and one of four nucleotide bases: adenine (A), guanine (G), cytosine (C), and thymine (T). The two strands are held together by hydrogen bonds formed between various pairs of the nucleotide bases. Guanine (G) bonds with cytosine (C), and adenine (A) with thymine (T).

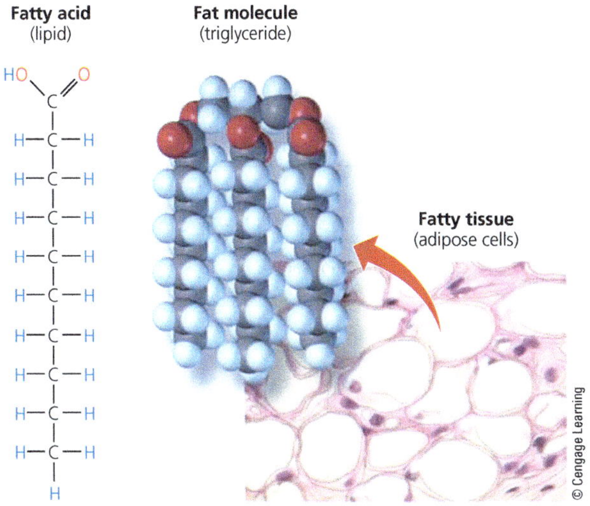

Figure 9 Structural formula of fatty acid, which is one form of lipid (left). Fatty acids are converted into more complex fat molecules (center) that are stored in adipose cells (right).

Figure 10 shows the relative sizes of simple and complex molecules, cells, and multi-celled organisms.

Certain Molecules Store and Release Energy in Cells

Chemical reactions occurring in plant cells during photosynthesis (see Chapter 3, p. 55) release energy that is absorbed by adenosine diphosphate (ADP) molecules and stored as chemical energy in adenosine triphosphate (ATP) molecules (Figure 11, left). When cellular processes require energy, ATP molecules release it to form ADP molecules (Figure 11, right).

Chemists Balance Chemical Equations to Keep Track of Atoms

Chemists use a shorthand system, or equation, to represent chemical reactions. These chemical equations are also used as an accounting system to verify that no atoms are created or destroyed in a chemical reaction as required by the law of conservation of matter (see Chapter 2, p.40). As a consequence, each side of a chemical equation must have the same number of atoms or ions of each element involved. Ensuring that this condition is met leads to what chemists call a *balanced chemical equation*. The equation for the burning of carbon ($C + O_2 \rightarrow CO_2$) is balanced because one atom of carbon and two atoms of oxygen are on both sides of the equation.

Consider the following chemical reaction. When electricity passes through water (H_2O), the latter can be broken down into hydrogen (H_2) and oxygen (O_2), as represented by the following equation:

H_2O	$\longrightarrow$	H_2	+	O_2
2 H atoms		2 H atoms		2 O atoms
1 O atom				

This equation is unbalanced because one atom of oxygen is on the left side of the equation but two oxygen atoms are on the right side. We cannot change the subscripts of any of the formulas to balance this equa-

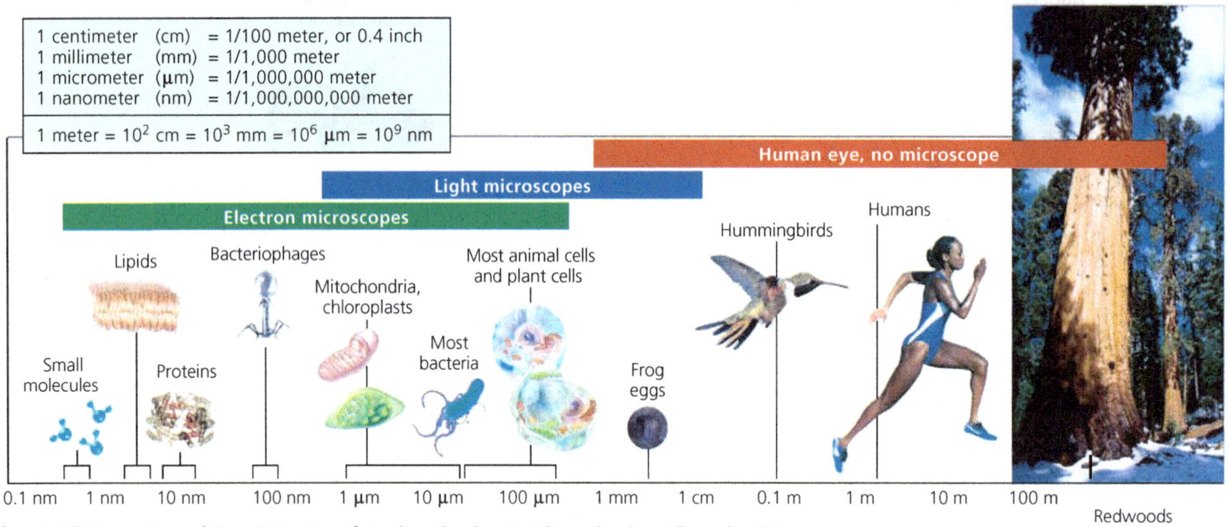

Figure 10 Comparison of the relative sizes of simple molecules, complex molecules, cells, and multicellular organisms. This scale is exponential, not linear. Each unit of measure is 10 times larger than the unit preceding it.

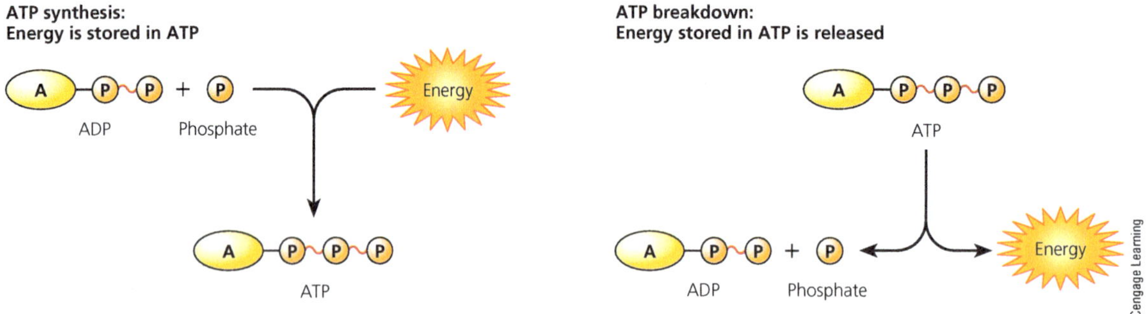

Figure 11 Models representing energy storage and release in cells.

tion because that would change the arrangements of the atoms, leading to different substances. Instead, we must use different numbers of the molecules involved to balance the equation. For example, we could use two water molecules:

$$2\,H_2O \longrightarrow H_2 + O_2$$

$2\,H_2O$	$\longrightarrow$	H_2	+	O_2
4 H atoms		2 H atoms		2 O atoms
2 O atoms				

This equation is still unbalanced. Although the numbers of oxygen atoms on both sides of the equation are now equal, the numbers of hydrogen atoms are not. We can correct this problem by having the reaction produce two hydrogen molecules:

$2\,H_2O$	$\longrightarrow$	$2\,H_2$	+	O_2
4 H atoms		4 H atoms		2 O atoms
2 O atoms				

Now the equation is balanced, and the law of conservation of matter has been observed. For every two molecules of water through which we pass electricity, two hydrogen molecules and one oxygen molecule are produced.

CONSIDER THIS. . .

THINKING ABOUT Chemical Equations

Try to balance the chemical equation for the reaction that combines nitrogen gas (N_2) and hydrogen gas (H_2) to form ammonia gas (NH_3).

SUPPLEMENT 5 WEATHER BASICS: EL NIÑO, TORNADOES, AND TROPICAL CYCLONES

Weather Is Affected by Moving Masses of Warm and Cold Air

Weather is the set of short-term atmospheric conditions—typically those occurring over hours or days—for a particular area. Examples of atmospheric conditions include temperature, pressure, moisture content, precipitation, sunshine, cloud cover, and wind direction and speed.

Meteorologists use equipment mounted on weather balloons, aircraft, ships, and satellites, as well as radar and stationary sensors, to obtain data on weather variables. They then feed these data into computer models to draw weather maps. Other computer models project the weather for a period of several days by calculating the probabilities that air masses, winds, and other factors will change in certain ways.

Much of the weather we experience results from interactions between the leading edges of moving masses of warm and cold air driven largely by uneven heating of the earth's surface by the sun (see Figure 7-3, p. 146). Weather changes as one air mass replaces or meets another. The most dramatic changes in weather occur along a **front**, the boundary between two air masses with different temperatures and densities.

A **warm front** is the boundary between an advancing warm air mass and the cooler one it is replacing (Figure 1, left). Because warm air is less dense (weighs less per unit of volume) than cool air, an advancing warm front rises up over a mass of cool air. As the warm front rises, its moisture begins condensing into droplets, forming layers of clouds at different altitudes. Gradually, the clouds thicken, descend to a lower altitude, and often release their moisture as rainfall. A moist warm front can bring days of cloudy skies and drizzle.

A **cold front** (Figure 1, right) is the leading edge of an advancing mass of cold air. Because cold air is denser than warm air, an advancing cold front stays close to the ground and wedges underneath less dense warmer air. An approaching cold front produces rapidly moving, towering clouds called *thunderheads,* with flat, anvil-like tops.

As a cold front passes through, it may cause high surface winds and thunderstorms. After it leaves the area, it usually results in cooler temperatures and a clear sky.

Near the top of the troposphere, hurricane-force winds circle the earth. These powerful winds, called *jet streams,* follow rising and falling paths that have a strong influence on weather patterns (Figure 2).

Weather Is Affected by Changes in Atmospheric Pressure

Changes in atmospheric pressure also affect weather. *Atmospheric pressure* results from molecules of gases (mostly nitrogen and oxygen) in the atmosphere zipping around at very high speeds and hitting and bouncing off everything they encounter.

Atmospheric pressure is greater near the earth's surface because the molecules in the atmosphere are squeezed together under the weight of the air above them. An air mass with high pressure, called a **high**, contains cool, dense air that descends slowly toward the earth's surface and becomes warmer. Because of this warming, condensation of moisture usually does not take place and clouds usually do not form. Fair weather with clear skies follows as long as this high-pressure air mass remains over the area.

In contrast, a low-pressure air mass, called a **low**, produces cloudy and sometimes stormy weather. Because of

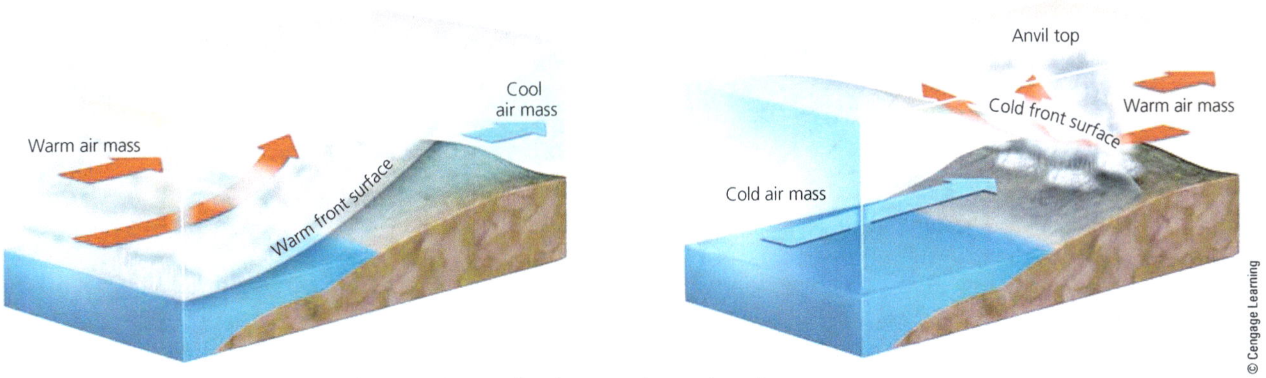

Figure 1 *Weather fronts:* A *warm front* (left) occurs when an advancing mass of warm air meets and rises up over a mass of denser cool air. A *cold front* (right) forms when a moving mass of cold air wedges beneath a mass of less dense warm air.

Figure 1 Countries of the world.

Map Analysis

1. Name three countries that border the Arctic Ocean.
2. Which countries surround **(a)** China, **(b)** Mexico, **(c)** Germany, and **(d)** Sudan?

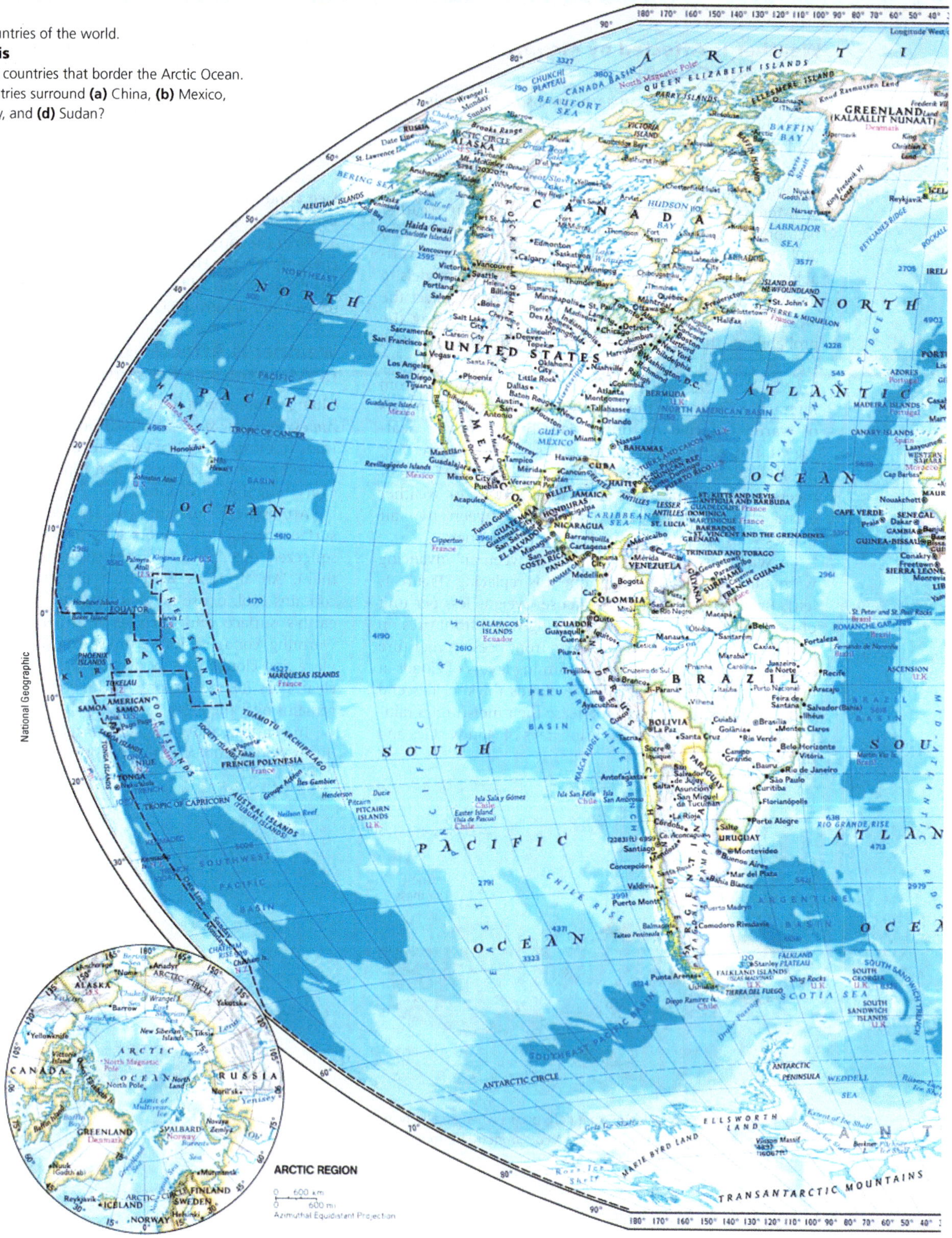

Longitude East of Greenwich
WINKEL TRIPEL PROJECTION, CENTRAL MERIDIAN 0°
KILOMETERS
STATUTE MILES
O C E A N
ARCTIC CIRCLE
R U S S I A
NORWEGIAN SEA
BARENTS SEA
KARA SEA
LAPTEV SEA
EAST SIBERIAN SEA
BERING SEA
SEA OF OKHOTSK
NEW SIBERIAN ISLANDS
FRANZ JOSEF LAND
SVALBARD
ICELAND
UNITED KINGDOM
IRELAND
FRANCE
SPAIN
PORTUGAL
ITALY
GERMANY
POLAND
UKRAINE
FINLAND
KAZAKHSTAN
MONGOLIA
CHINA
INDIA
JAPAN
NORTH KOREA
SOUTH KOREA
TURKEY
IRAN
SAUDI ARABIA
EGYPT
LIBYA
ALGERIA
SAHARA
MALI
NIGER
CHAD
SUDAN
ETHIOPIA
SOMALIA
KENYA
NIGERIA
DEMOCRATIC REPUBLIC OF THE CONGO
TANZANIA
ANGOLA
ZAMBIA
NAMIBIA
BOTSWANA
MOZAMBIQUE
MADAGASCAR
SOUTH AFRICA
PHILIPPINES
INDONESIA
MALAYSIA
PAPUA NEW GUINEA
AUSTRALIA
NEW ZEALAND
NORTH PACIFIC OCEAN
SOUTH PACIFIC OCEAN
TROPIC OF CANCER
EQUATOR
TROPIC OF CAPRICORN
I N D I A N O C E A N
ARABIAN SEA
SOUTH INDIAN BASIN
SOUTHWEST INDIAN RIDGE
ATLANTIC - INDIAN RIDGE
QUEEN MAUD LAND
WILKES LAND
TRANSANTARCTIC MOUNTAINS
VICTORIA LAND
ANTARCTIC CIRCLE
ANTARCTIC REGION
Azimuthal Equidistant Projection
• Research station
ANTARCTICA
WEST ANTARCTICA
EAST ANTARCTICA
South Pole

Figure 2 Composite satellite view of the earth showing its major terrestrial and aquatic features.

IC OCEAN
West of Greenwich
Longitude East of Greenwich
Iceland
Svalbard
Frans Josef Land
North Land
Novaya Zemlya
BARENTS SEA
Scandinavia
British Isles
North Sea
EUROPE
ALPS
URAL MOUNTAINS
SIBERIA
ASIA
ARCTIC CIRCLE
SEA OF OKHOTSK
Kamchatka Peninsula
Lake Baikal
Aral Sea
Caspian Sea
Black Sea
El'brus
MEDITERRANEAN SEA
GOBI
Plateau of Tibet
HIMALAYA
Mt. Everest
JAPAN
NORTH PACIFIC OCEAN
TROPIC OF CANCER
PHILIPPINE SEA
ARABIAN PENINSULA
Red Sea
INDIA
ARABIAN SEA
BAY OF BENGAL
SOUTH CHINA SEA
Philippine Islands
Challenger Deep
MICRONESIA
AFRICA
Gulf of Guinea
Lake Victoria
Congo Basin
Kilimanjaro
Lake Tanganyika
Lake Malawi
Madagascar
Kalahari Desert
Cape of Good Hope
INDONESIA
EQUATOR
New Guinea
MELANESIA
NINETYEAST RIDGE
INDIAN OCEAN
CORAL SEA
Fiji Islands
New Caledonia
TROPIC OF CAPRICORN
AUSTRALIA
Great Dividing Range
SOUTH PACIFIC OCEAN
Mt. Kosciuszko
Bass Strait
Tasmania
TASMAN SEA
North Island
South Island
NEW ZEALAND
SOUTHWEST INDIAN RIDGE
SOUTHEAST INDIAN RIDGE
ANTARCTIC CIRCLE
Wilkes Land
Winkel Tripel Projection, Central Meridian 0°
KILOMETERS
STATUTE MILES
Elevations in meters
ARCTICA

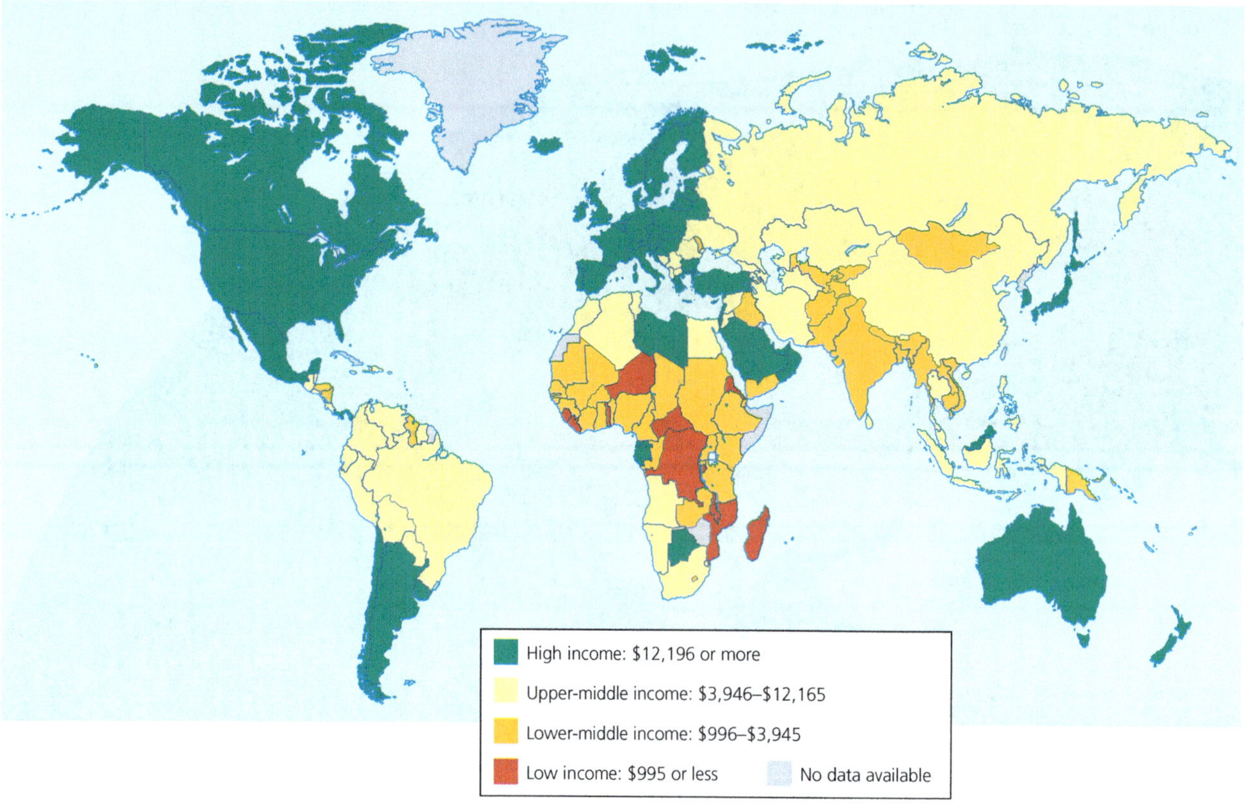

(Compiled by the authors using data from World Bank and International Monetary Fund.)

Figure 3 High-income, upper-middle-income, lower-middle-income, and low-income countries in terms of gross national income (GNI) purchasing power parity (PPP) per capita (U.S. dollars) in 2010.

Data and Map Analysis

1. In how many countries is the per capita average income $995 or less? Look at Figure 1 and find the names of three of these countries.
2. In how many instances does a lower-middle- or low-income country share a border with a high-income country? Look at Figure 1 and find the names of the countries that reflect three examples of this situation.

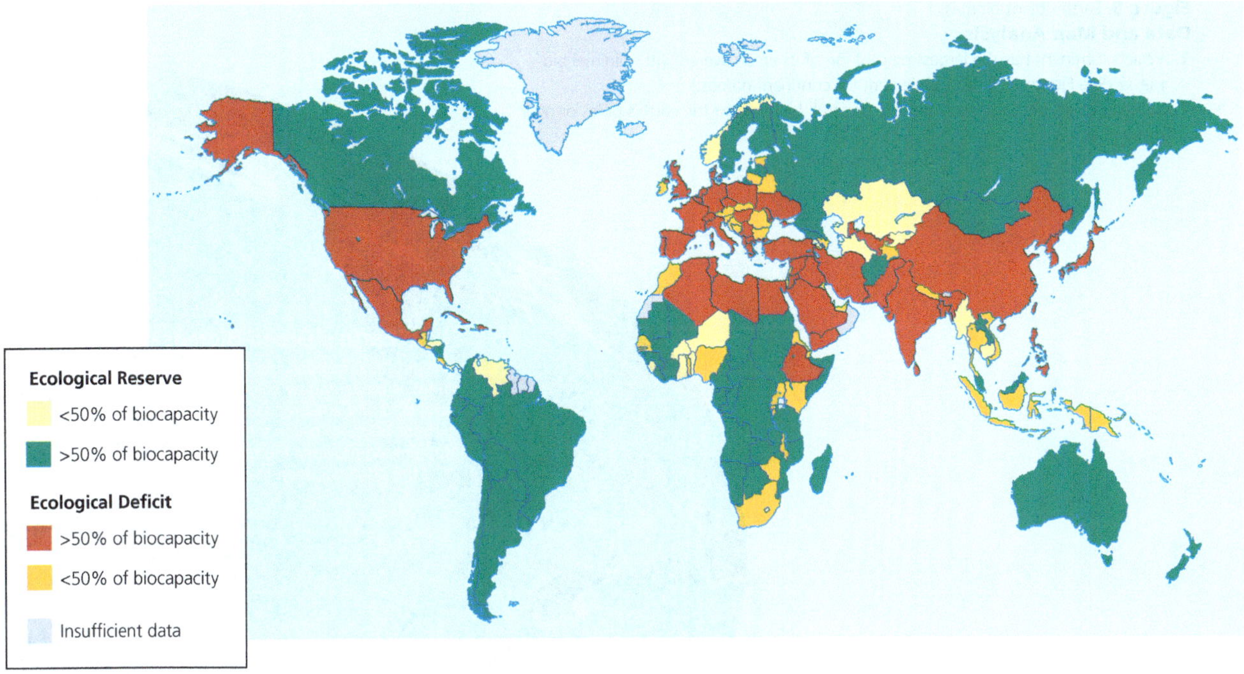

(Compiled by the authors using data from Global Footprint Network.)

Figure 4 *Ecological debtors and creditors:* The ecological footprints of some countries exceed their biocapacity, while other countries still have ecological reserves.

Data and Map Analysis

1. List five countries, including the three largest, in which the ecological deficit is greater than 50% of biocapacity. (See Figure 1 of this supplement for country names.)
2. On which two continents does land with ecological reserves of more than 50% of biocapacity occupy the largest percentage of total land area? (See Figure 2 of this supplement for continent names.)

Figure 5 Earth's plant biomass.

Data and Map Analysis

1. Which continent has the largest percentage of its area covered with minimal biomass? (See Figure 2 of this supplement for continent names.)
2. Is ocean chlorophyll concentration, in general, higher near the earth's poles or near its equator?

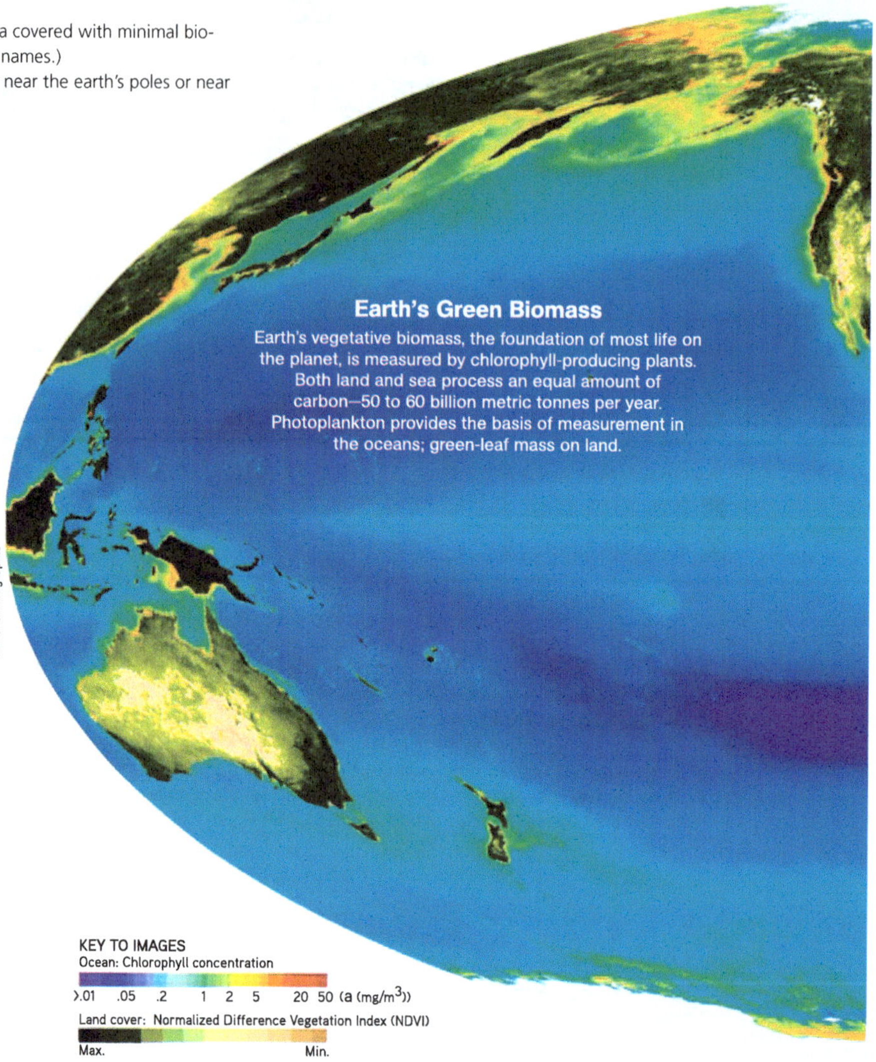

National Geographic

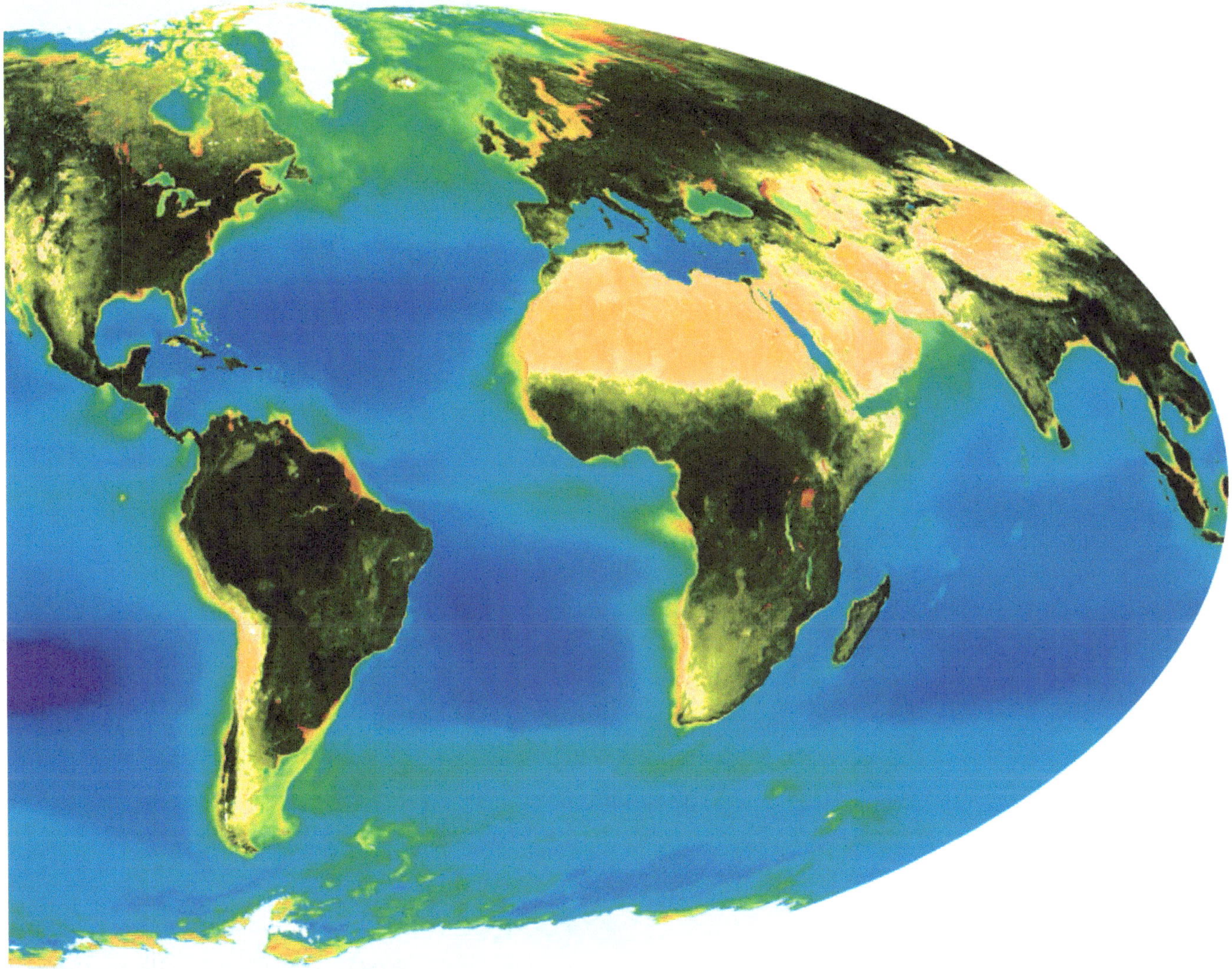

Figure 6 Land cover in North America.

Data and Map Analysis

1. Approximately what percentage of North America is covered by savanna?
2. Approximately what percentage of the continental United States is used for cropland?

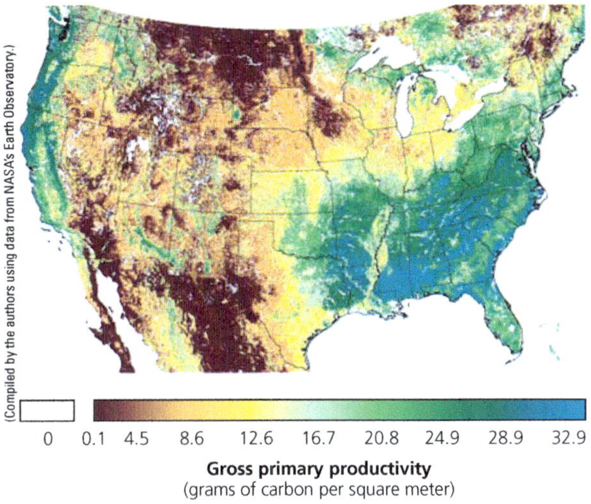

Figure 7 Natural capital: Gross primary productivity across the continental United States, based on remote satellite data. The differences roughly correlate with variations in moisture and soil types.

Data and Map Analysis

1. Comparing the five northwestern-most states with the five southeastern-most states, which of these regions has the greater variety in levels of gross primary productivity? Which of the regions has the highest levels overall?
2. Compare this map with that of Figure 6. Which biome in the United States is associated with the highest level of gross primary productivity?

Figure 8 Natural capital degradation: The human ecological footprint in North America. Colors represent the percentage of each area influenced by human activities.

(Compiled by the authors using data from Wildlife Conservation Society and the Center for International Earth Science Information Network at Columbia University.)

Data and Map Analysis

1. Which general area of the United States has the highest human footprint values?
2. What is the relative value of the human ecological footprint in the area where you live or go to school?

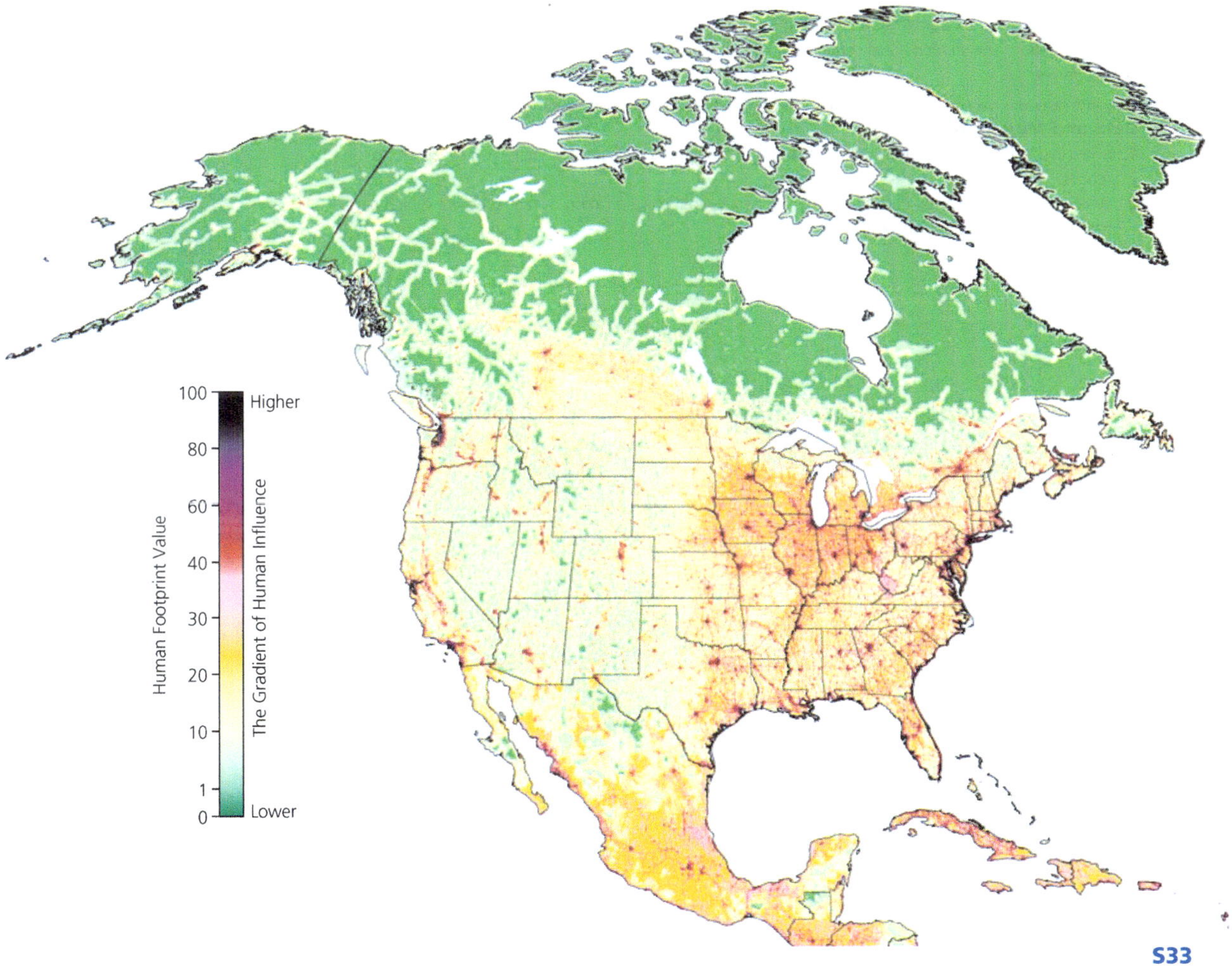

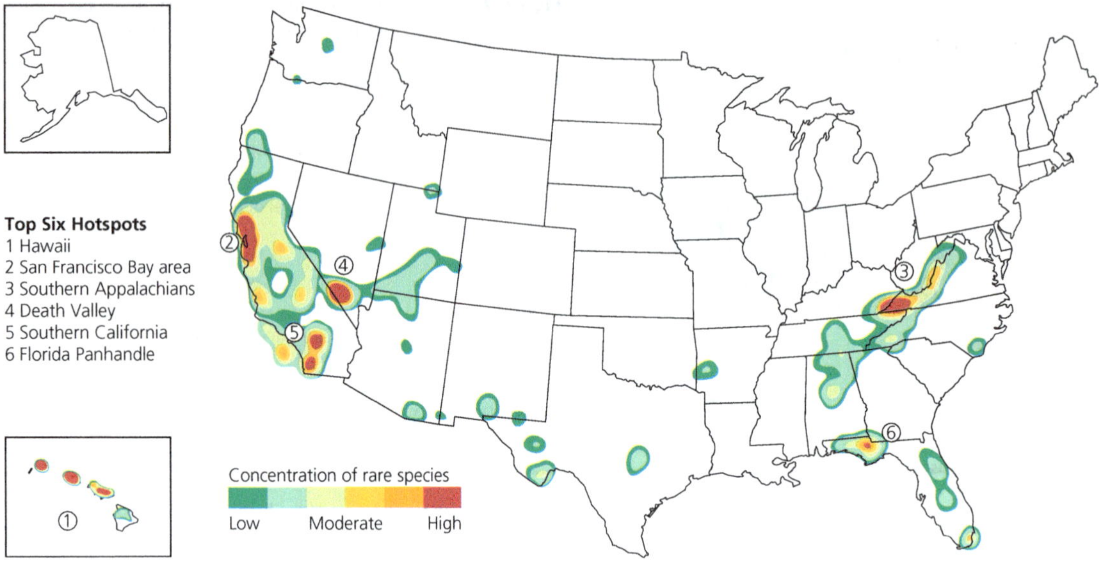

Figure 9 Endangered natural capital: Major biodiversity hotspots in the United States that need emergency protection. The shaded areas contain the largest concentrations of rare and potentially endangered species.

(Compiled by the authors using data from State Natural Heritage Programs, the Nature Conservancy, and Association for Biodiversity Information.)

Data and Map Analysis

1. If you live in the United States, which of the top six hotspots is closest to where you live or go to school?
2. Which general part of the country has the highest overall concentration of rare species? Which part has the second-highest concentration?

Figure 10 Land cover in South America.

Data and Map Analysis

1. Which type of land cover would you say occupies the largest area in South America?
2. Brazil was once almost completely covered by rain forest (evergreen broadleaf), much of which has been converted to savanna as it has been cleared by humans and has not grown back as forest. About what percentage of Brazil would you say is occupied by savanna?

Figure 11 Land cover in Europe.

Data and Map Analysis

1. What are the two most common types of land cover in Europe?
2. What percentage of European land would you estimate to be used for growing crops?

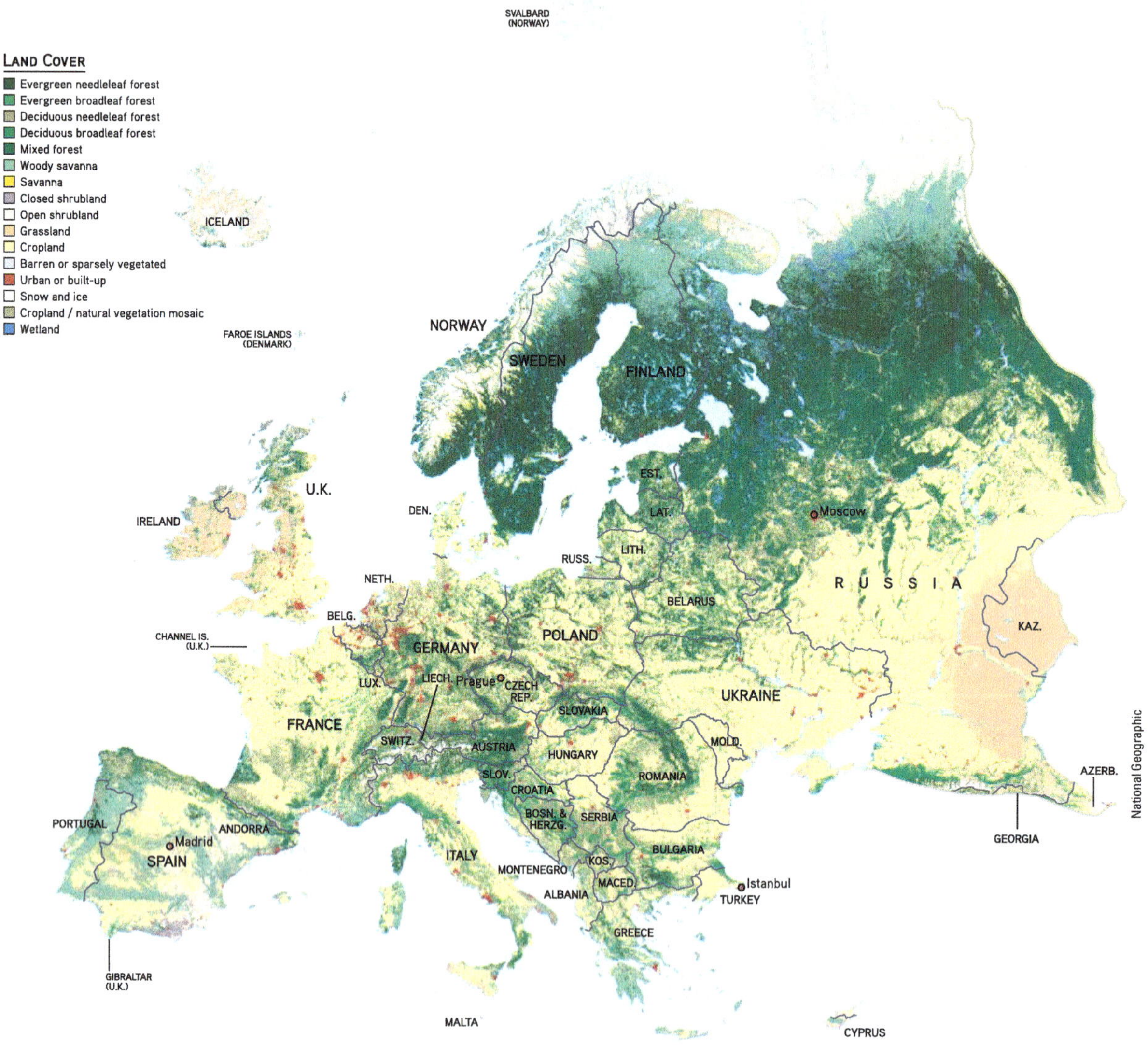

Figure 12 Land cover in Asia.

Data and Map Analysis

1. About what percentage of China's land area would you estimate to be classified as barren or sparsely vegetated?
2. About what percentage of India's land would you say is used for growing crops?

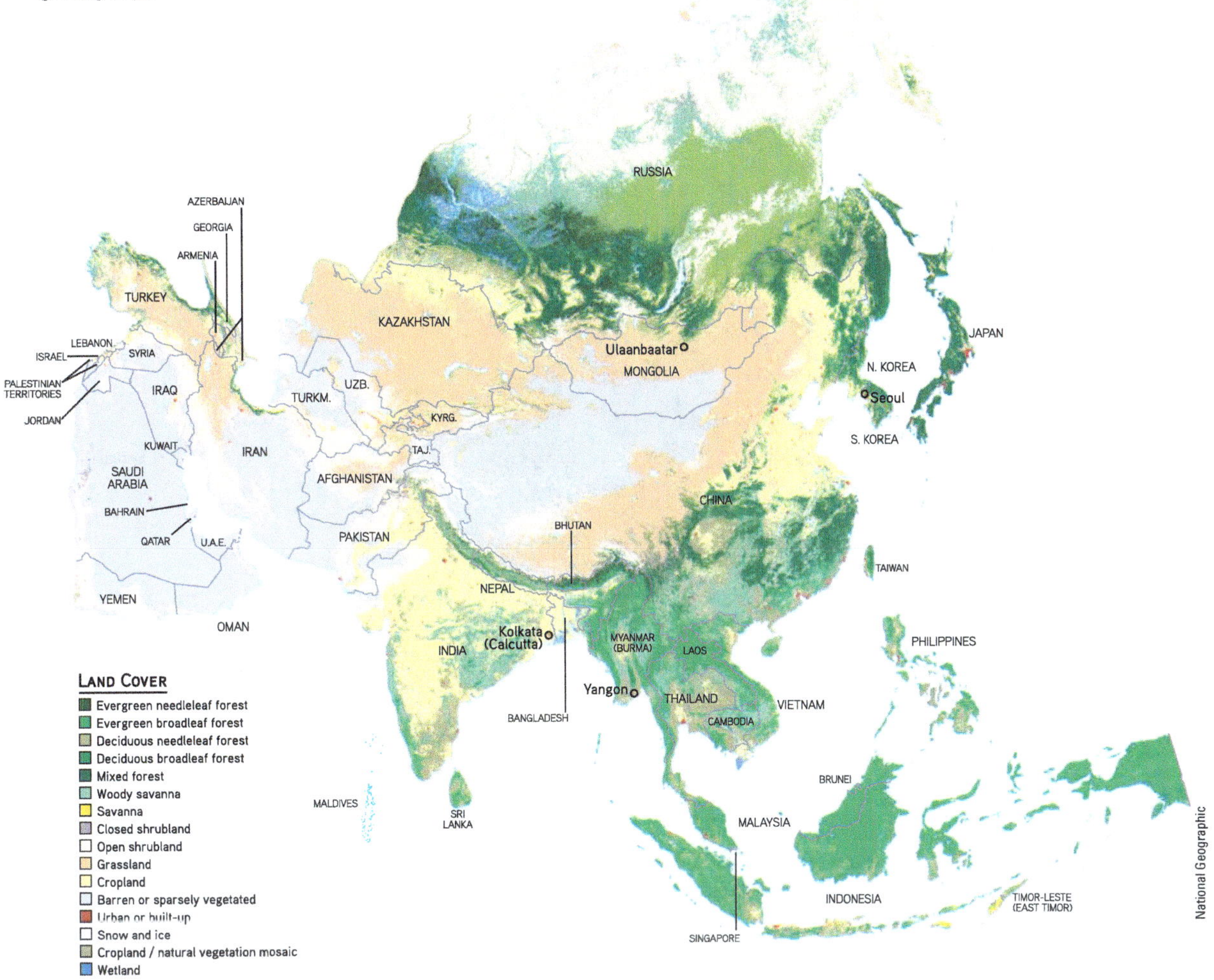

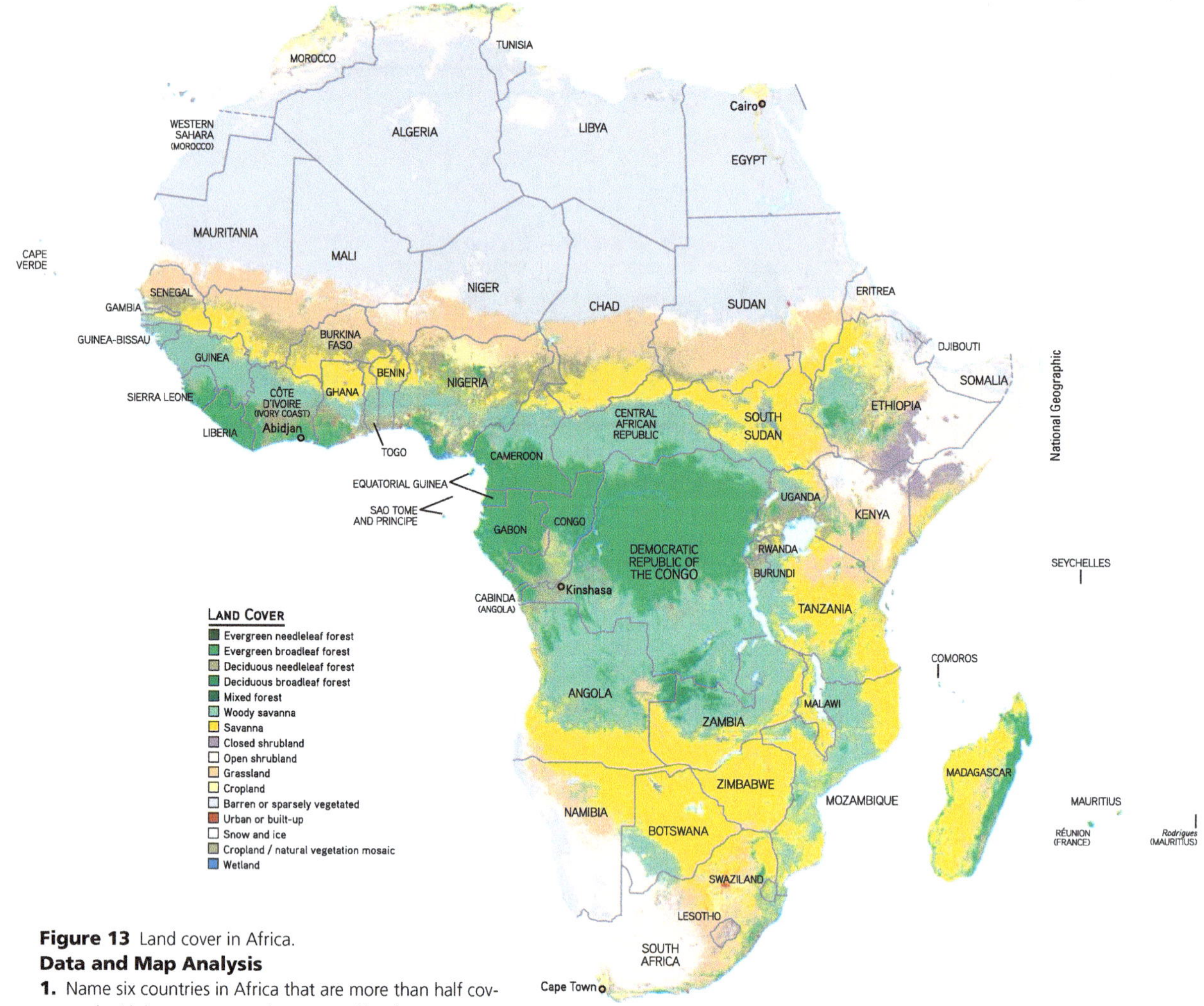

Figure 13 Land cover in Africa.

Data and Map Analysis

1. Name six countries in Africa that are more than half covered with barren or sparsely vegetated land.
2. Which African country has the largest area of forest land?

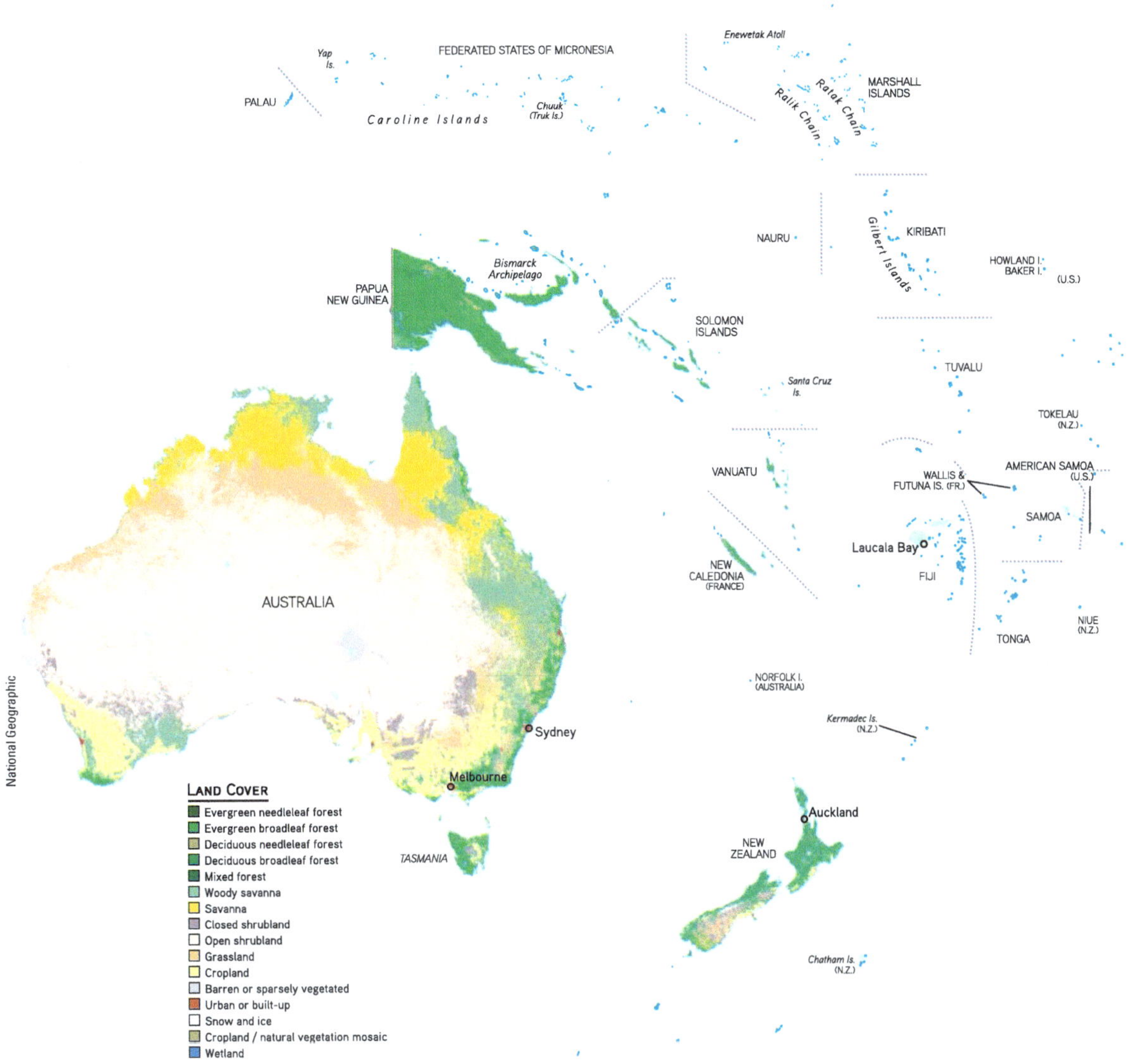

Figure 14 Land cover in Oceania and Australia.

Data and Map Analysis

1. What type of land cover occupies the most land in Australia?
2. What do all urban areas shown in red on this map have in common?

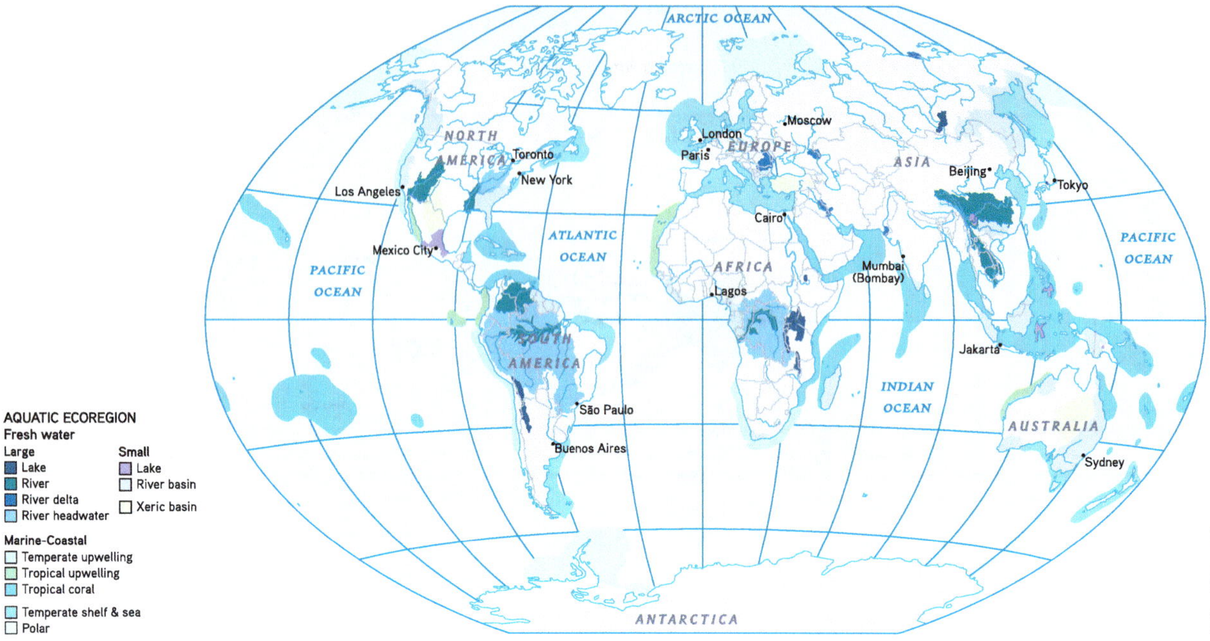

Figure 15 Major types of marine and freshwater habitats.

Data and Map Analysis

1. On which continent is the largest area of river habitat?
2. List the locations of three tropical upwelling areas.

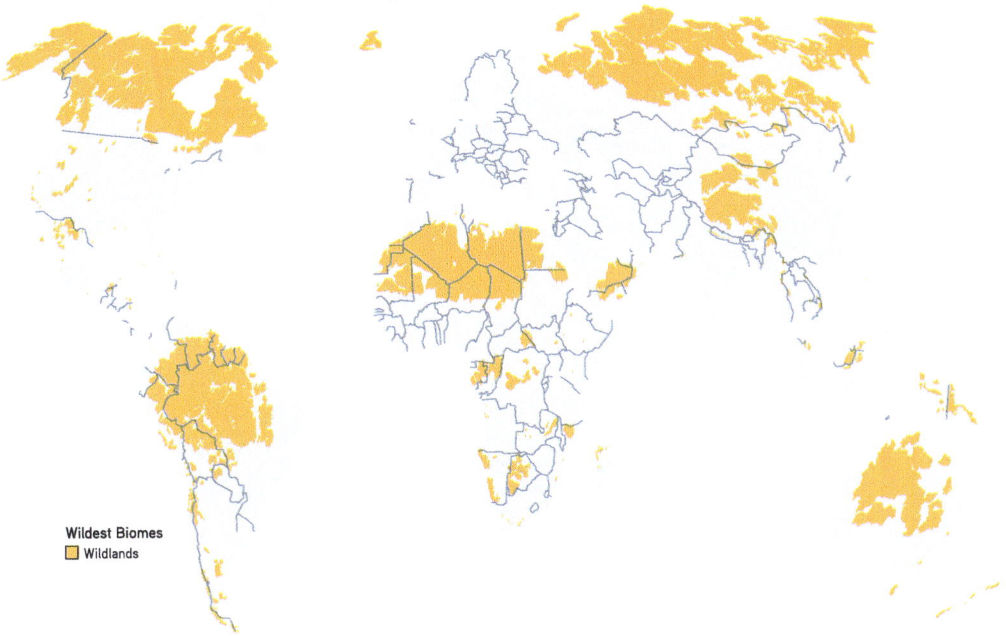

National Geographic

Figure 16 World's remaining wildlands.

Data and Map Analysis

1. Comparing this map with those in Figures 12–14, list the locations of three wild areas that are also classified as barren or sparsely vegetated. (See Figures 1 and 2 of this supplement for country and continent names.)
2. What continent appears to have the largest percentage of wildland?

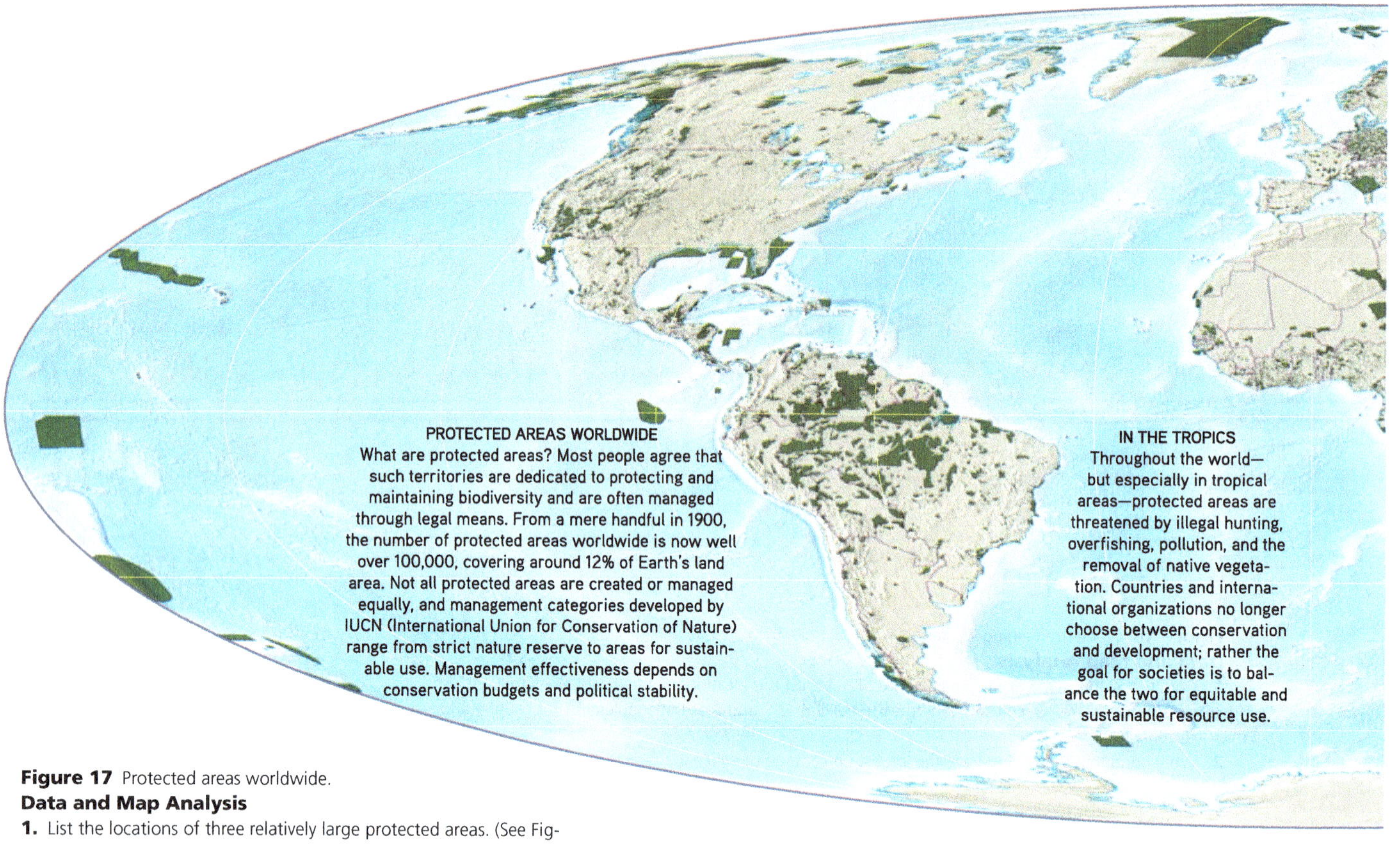

Figure 17 Protected areas worldwide.

Data and Map Analysis

1. List the locations of three relatively large protected areas. (See Figures 1 and 2 of this supplement for country and continent names.)
2. What continent appears to have the largest percentage of protected land? What continent appears to have the smallest percentage of protected land?

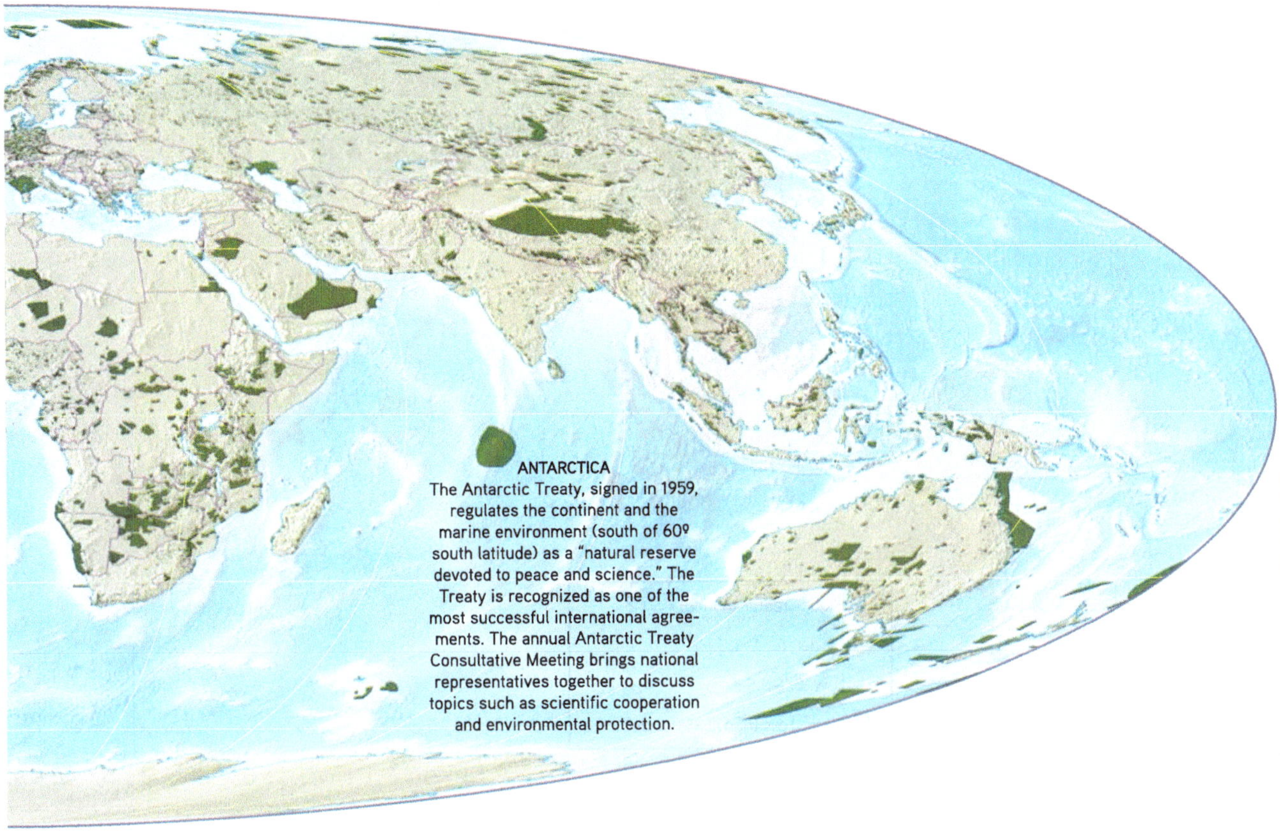

National Geographic

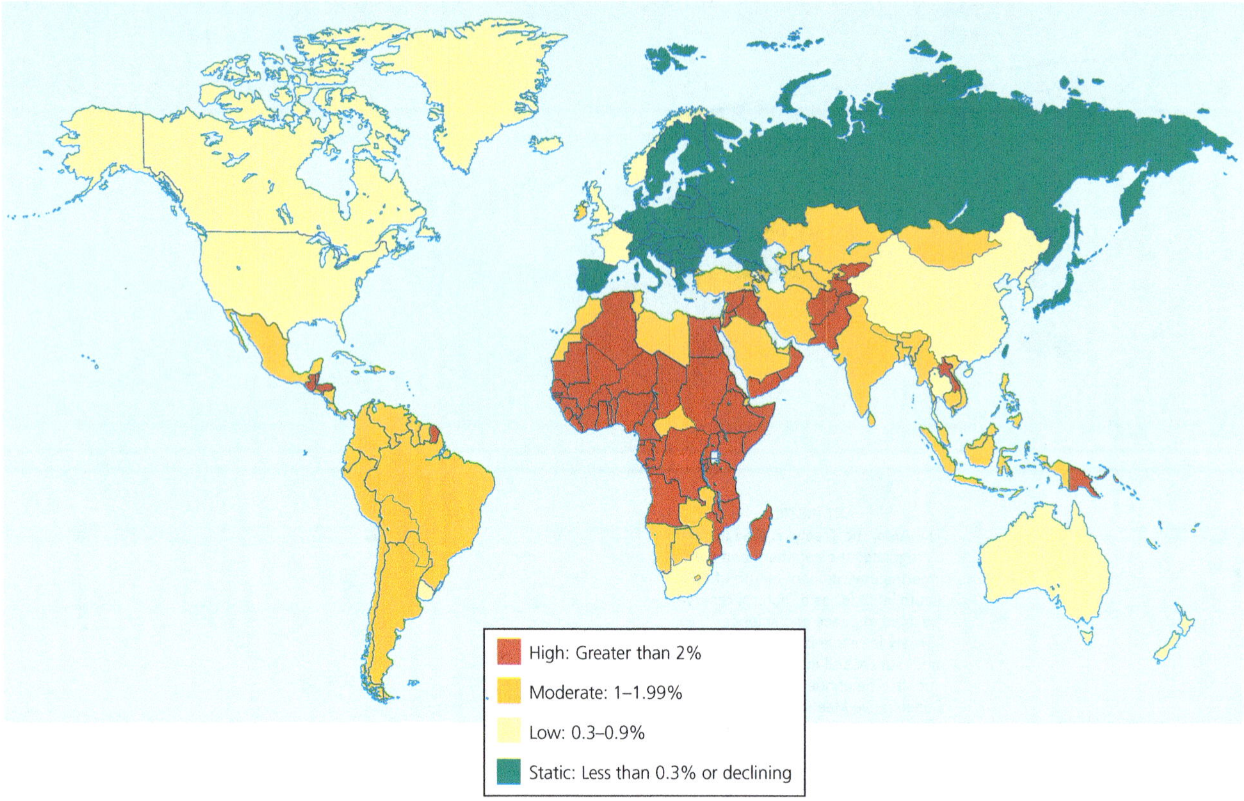

Figure 18 Comparative rates of population growth (%) throughout the world in 2012.

(Compiled by the authors using data from Population Reference Bureau and United Nations Population Division.)

Data and Map Analysis

1. Which continent has the greatest number of countries with high rates of population increase? Which continent has the greatest number of countries with static rates? (See Figure 2 of this supplement for continent names.)
2. For each category on this map, name the two countries that you think are largest in terms of total area (see Figure 1 for country names).

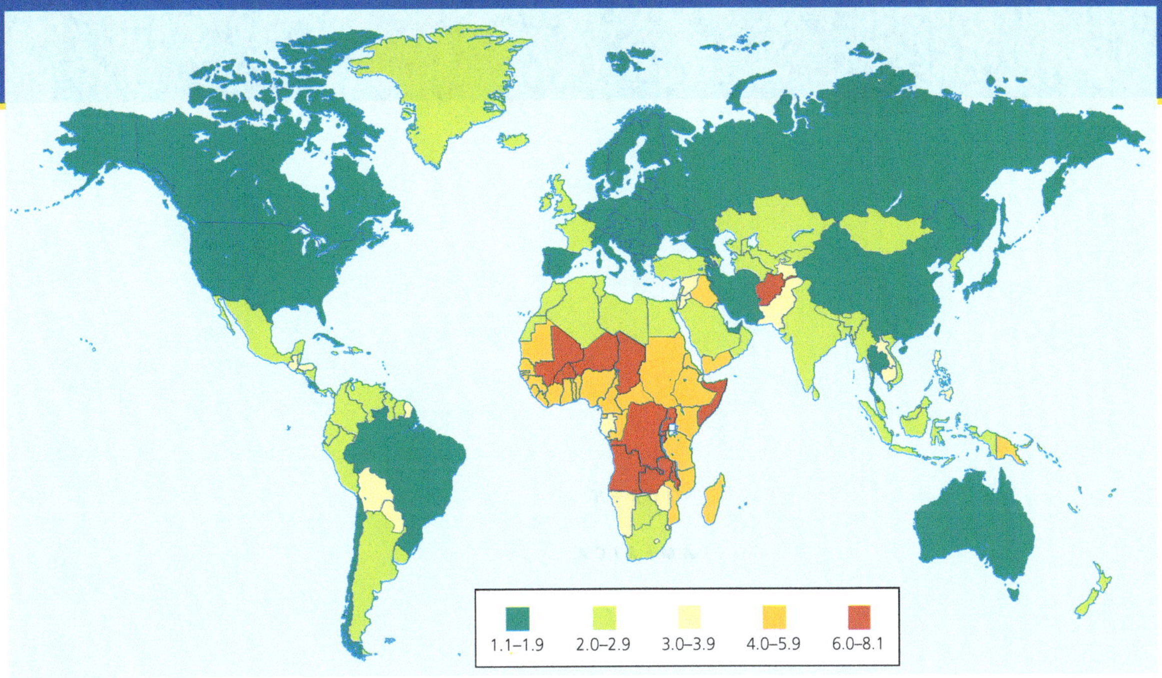

Figure 19 Comparison of total fertility rate (TFR), or average number of children born to the world's women throughout their lifetimes, around the world as measured in 2012.

(Compiled by the authors using data from Population Reference Bureau and United Nations Population Division.)

Data and Map Analysis

1. Which country in the highest TFR category borders two countries in the lowest TFR category? What are those two countries? (See Figure 1 of this supplement for country names.)
2. Do you see any geographic patterns on this map? Explain.

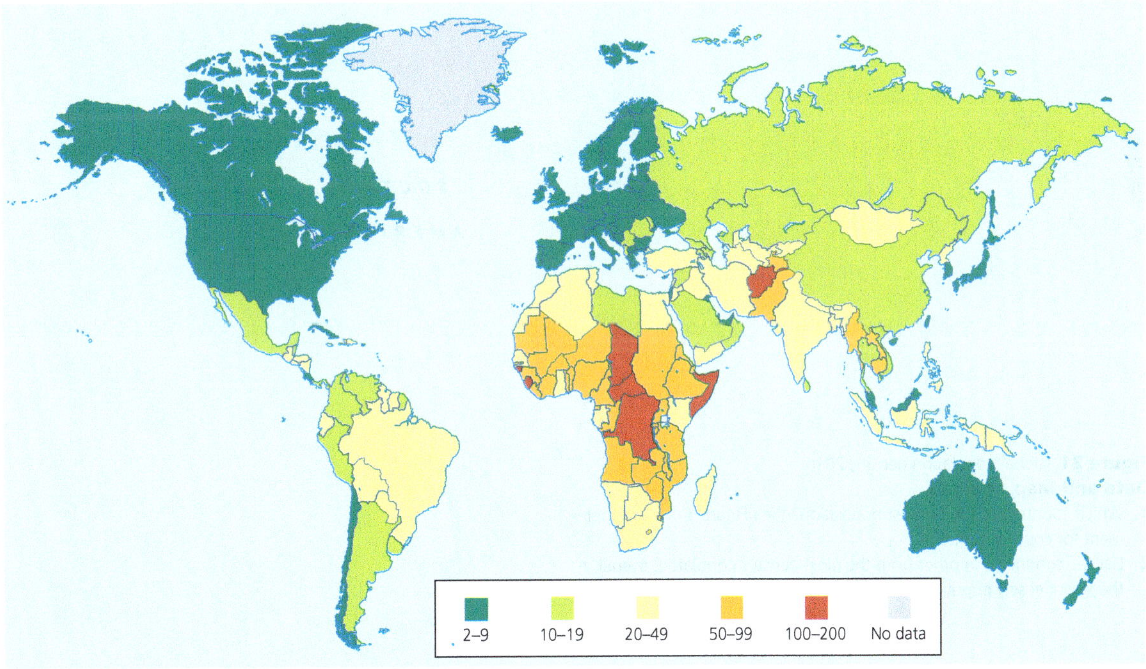

Figure 20 Comparison of global infant mortality rates around the world in 2012.

(Compiled by the authors using data from Population Reference Bureau and United Nations Population Division.)

Data and Map Analysis

1. Do you see a geographic pattern related to infant mortality rates as reflected on this map? Explain.
2. List any similarities that you see in geographic patterns between this map and the one in Figure 19.

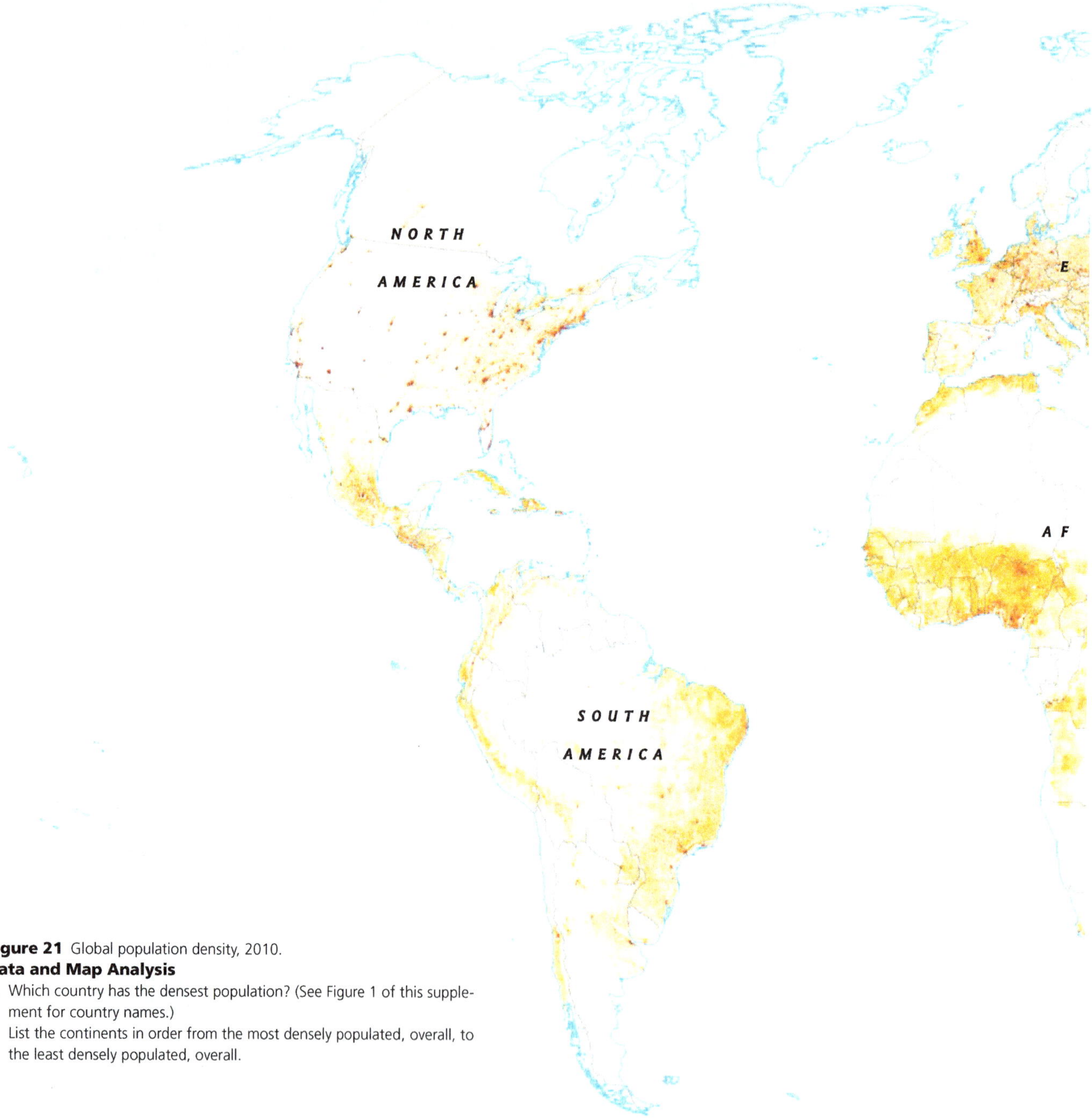

Figure 21 Global population density, 2010.

Data and Map Analysis

1. Which country has the densest population? (See Figure 1 of this supplement for country names.)
2. List the continents in order from the most densely populated, overall, to the least densely populated, overall.

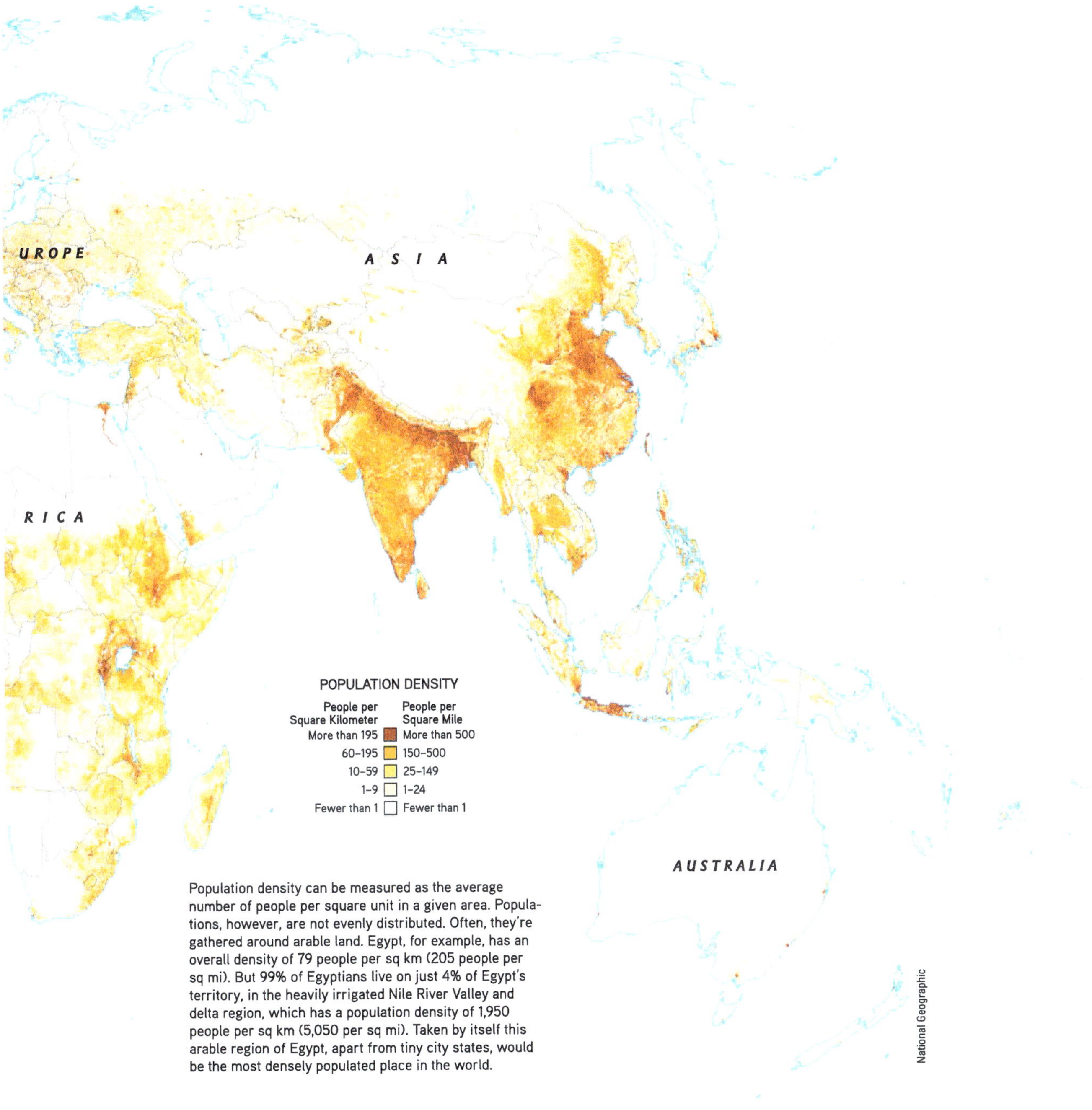

Population density can be measured as the average number of people per square unit in a given area. Populations, however, are not evenly distributed. Often, they're gathered around arable land. Egypt, for example, has an overall density of 79 people per sq km (205 people per sq mi). But 99% of Egyptians live on just 4% of Egypt's territory, in the heavily irrigated Nile River Valley and delta region, which has a population density of 1,950 people per sq km (5,050 per sq mi). Taken by itself this arable region of Egypt, apart from tiny city states, would be the most densely populated place in the world.

National Geographic

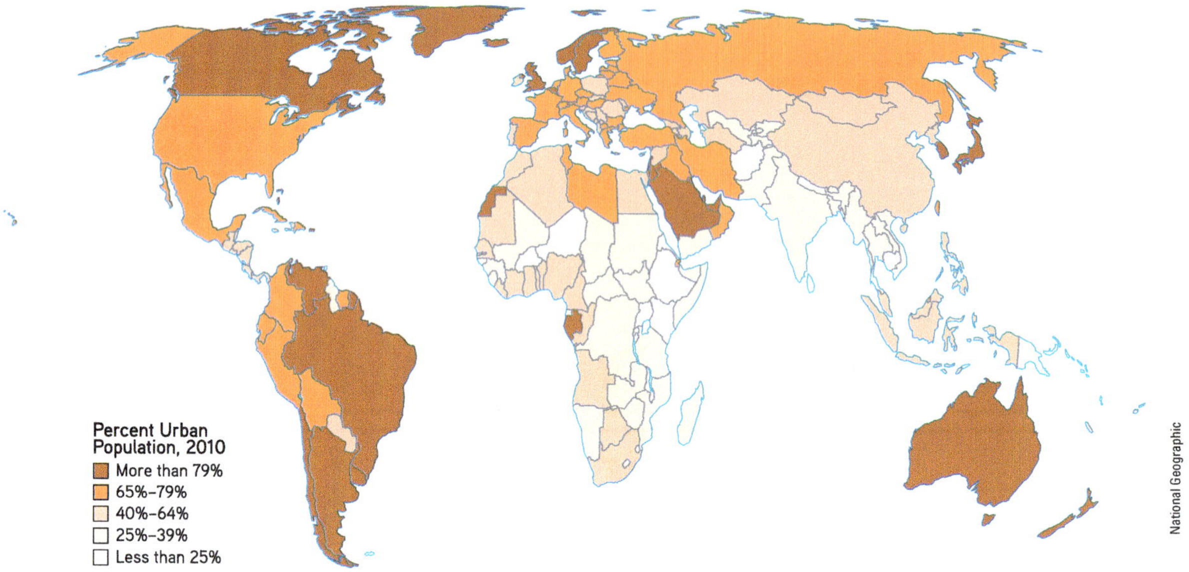

Figure 22 Urban population, 2010.

Data and Map Analysis

1. Name six countries whose populations are more than 79% urban. (See Figure 1 of this supplement for country names.)
2. List three pairs of bordering countries in which the population of one country is more than 79% urban and the population of the other is 39% or less urban.

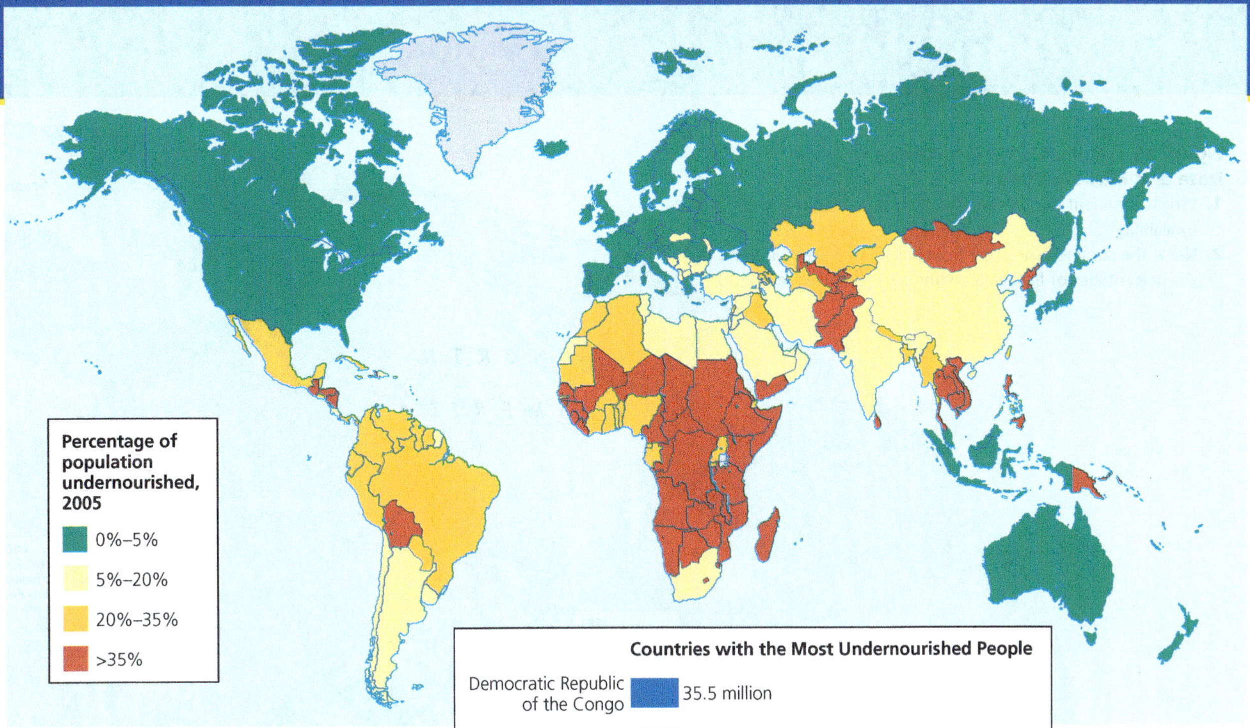

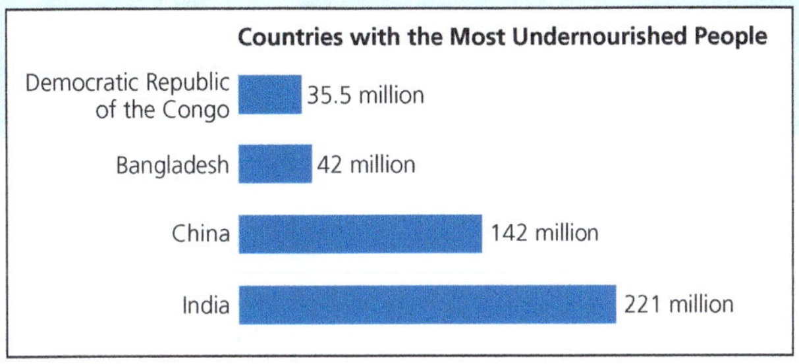

Figure 23 World hunger expressed as percentages of populations suffering from chronic hunger and malnutrition.

(Compiled by the authors using data from the UN Food and Agriculture Organization, World Health Organization, and U.S. Department of Agriculture.)

Data and Map Analysis

1. List the continents in order, starting with the one that has the highest percentage of undernourished people and ending with the one that has the lowest such percentage. (See Figure 2 of this supplement for continent names.)
2. On which continent is the largest block of countries that suffer the highest levels of undernourishment? List five of these countries.

Figure 24 U.S. farm area, 2007.

Data and Map Analysis

1. Name six states that are primarily more than 70% covered by farmland.
2. Name six states where there is a high variation in farmland cover, ranging from less than 10% to more than 70%.

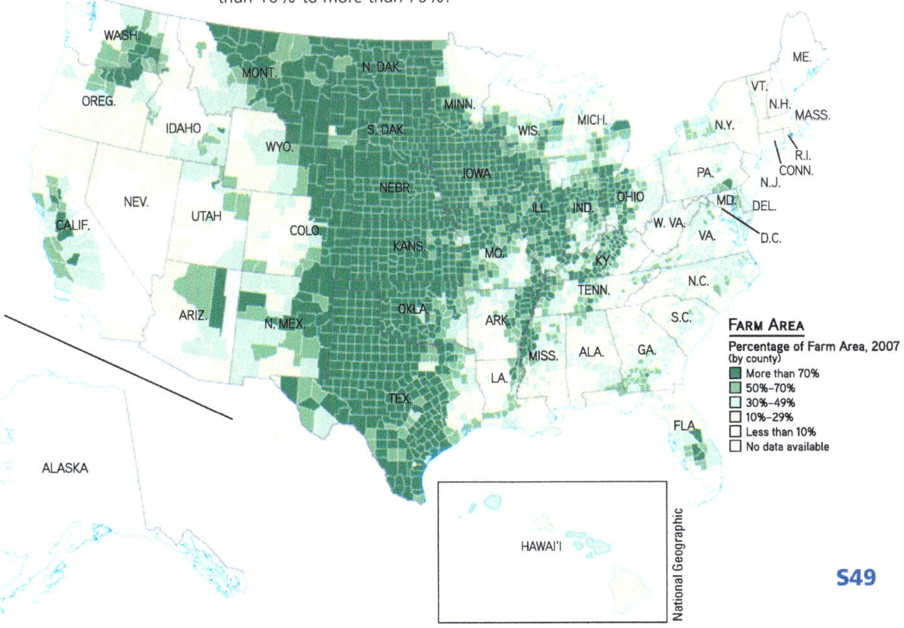

Figure 25 Global availability of freshwater.

Data and Map Analysis

1. Which continent has the largest area with the highest water availability?
2. Name the continents in order from highest to lowest, in terms of the percentages of freshwater withdrawn for industrial uses.

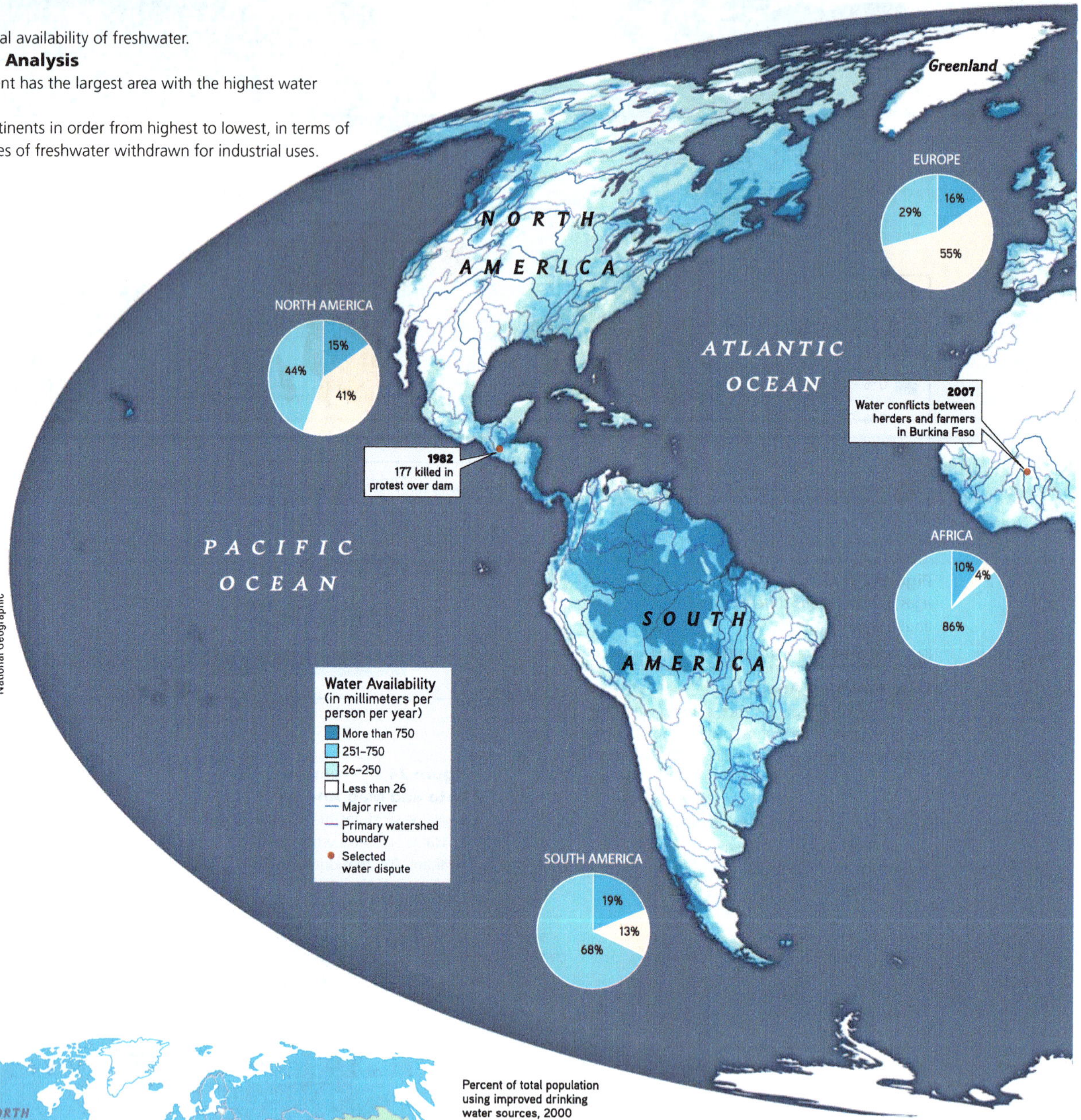

National Geographic

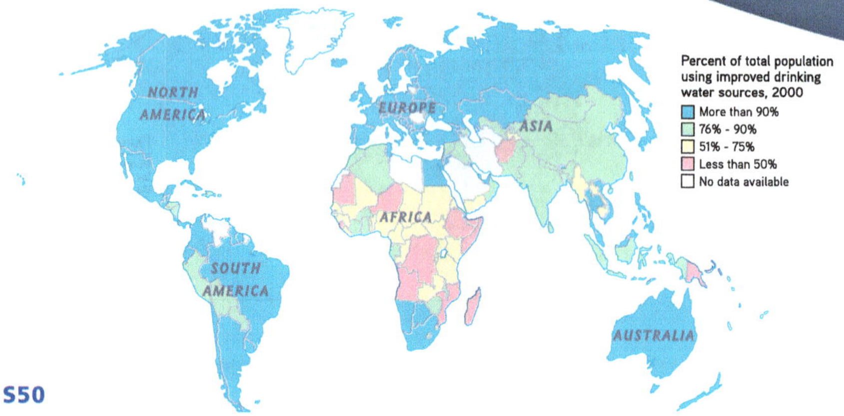

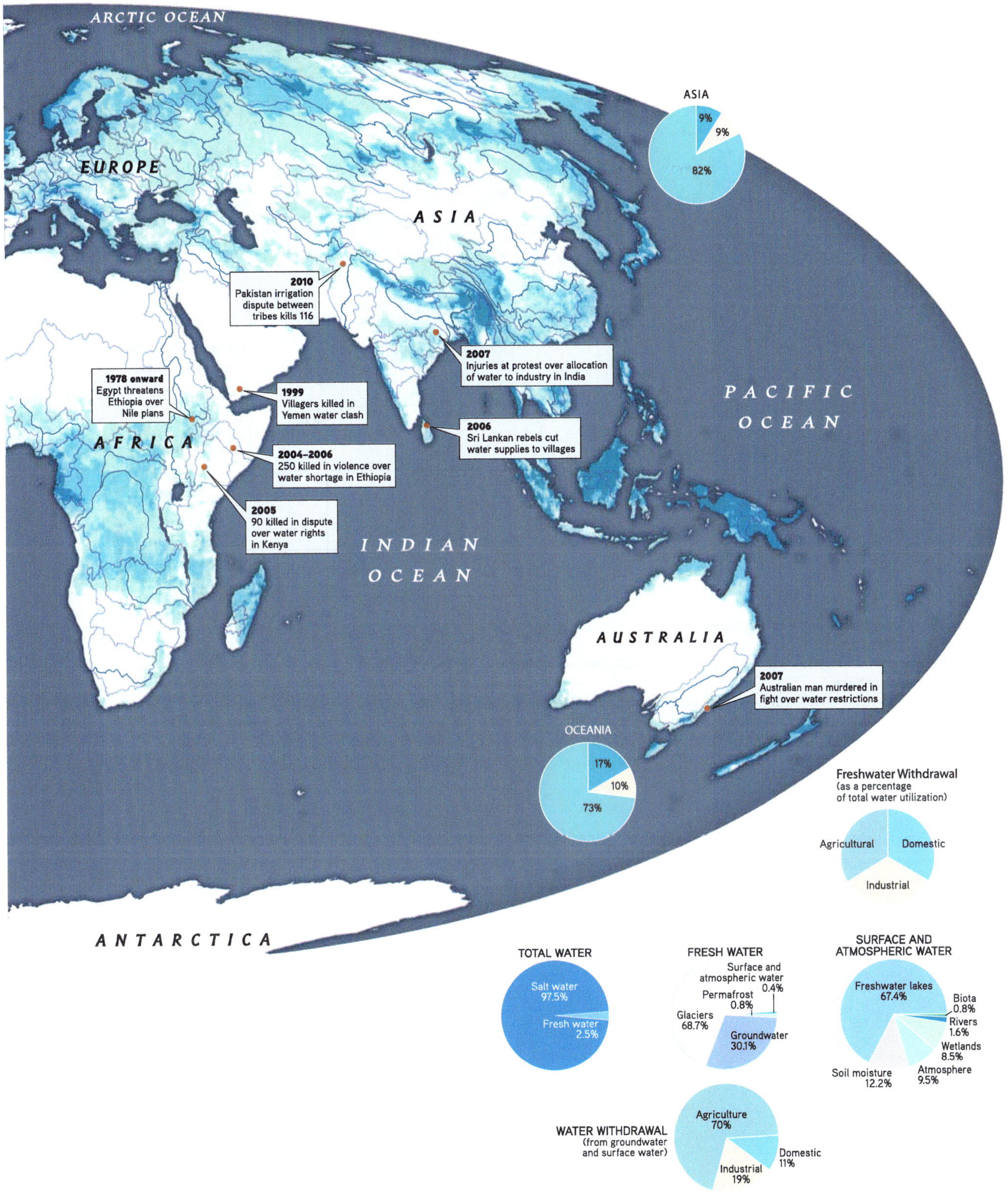
ARCTIC OCEAN
EUROPE
ASIA
AFRICA
PACIFIC OCEAN
INDIAN OCEAN
AUSTRALIA
ANTARCTICA
ASIA
9%
9%
82%
OCEANIA
17%
10%
73%
2010
Pakistan irrigation dispute between tribes kills 116
2007
Injuries at protest over allocation of water to industry in India
1978 onward
Egypt threatens Ethiopia over Nile plans
1999
Villagers killed in Yemen water clash
2006
Sri Lankan rebels cut water supplies to villages
2004–2006
250 killed in violence over water shortage in Ethiopia
2005
90 killed in dispute over water rights in Kenya
2007
Australian man murdered in fight over water restrictions
Freshwater Withdrawal
(as a percentage of total water utilization)
Agricultural
Domestic
Industrial
TOTAL WATER
Salt water 97.5%
Fresh water 2.5%
FRESH WATER
Surface and atmospheric water 0.4%
Permafrost 0.8%
Glaciers 68.7%
Groundwater 30.1%
SURFACE AND ATMOSPHERIC WATER
Freshwater lakes 67.4%
Biota 0.8%
Rivers 1.6%
Wetlands 8.5%
Atmosphere 9.5%
Soil moisture 12.2%
WATER WITHDRAWAL
(from groundwater and surface water)
Agriculture 70%
Domestic 11%
Industrial 19%

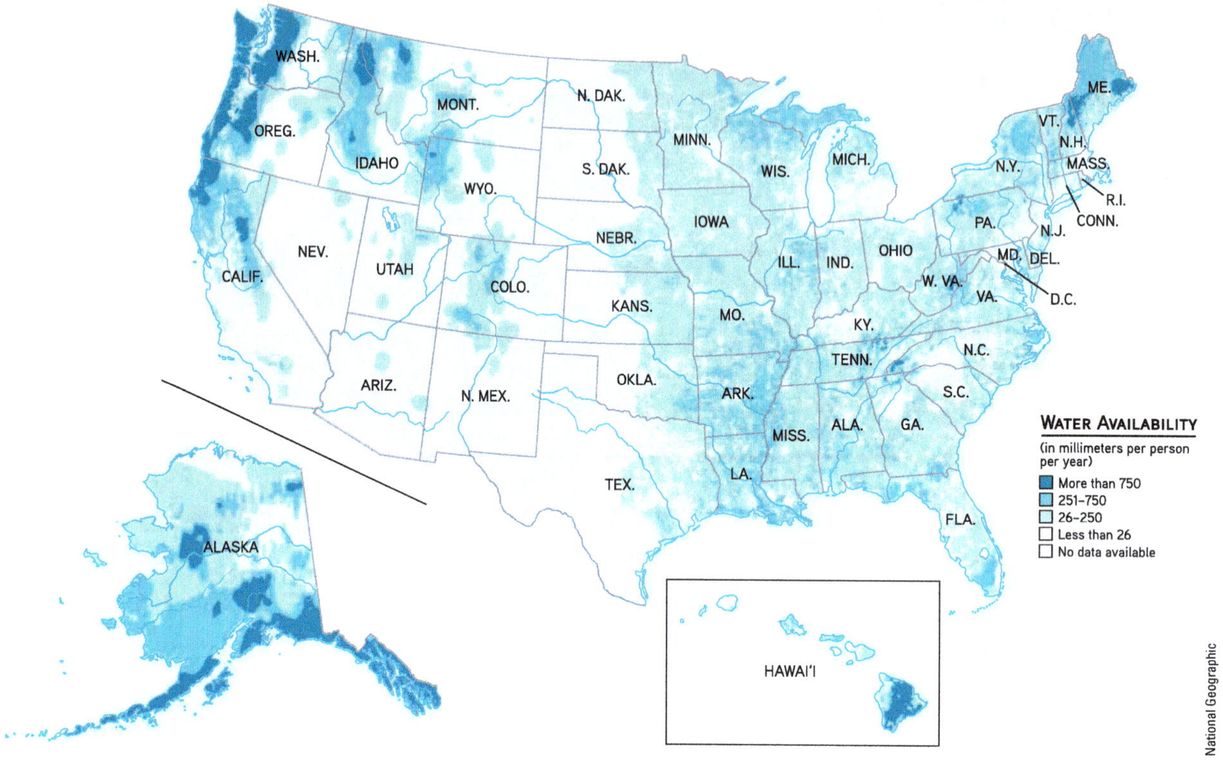

Figure 26 Availability of freshwater in the United States.

Data and Map Analysis

1. Name the three states with the largest areas of land where water availability is more than 750 milliliters per person per year.
2. Name six states where there are no areas of water availability of more than 250 milliliters per person per year.

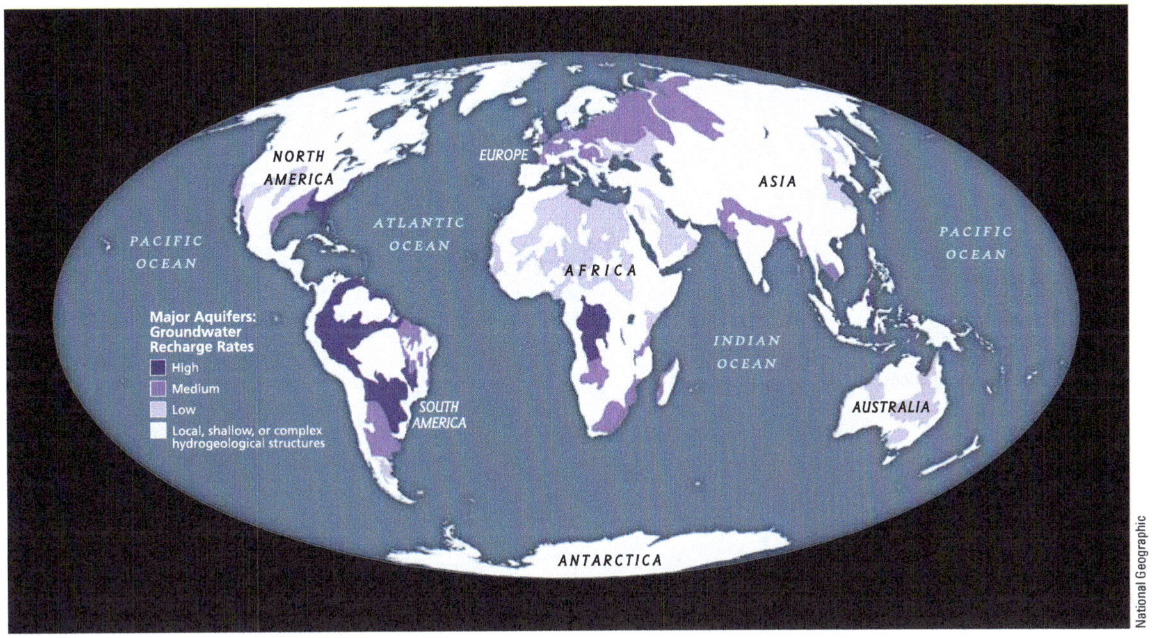

Figure 27 Recharge rates of the world's major aquifers.

Data and Map Analysis

1. Which continent has the largest area of aquifers with high recharge rates?

2. Which continents have no aquifers with high recharge rates?

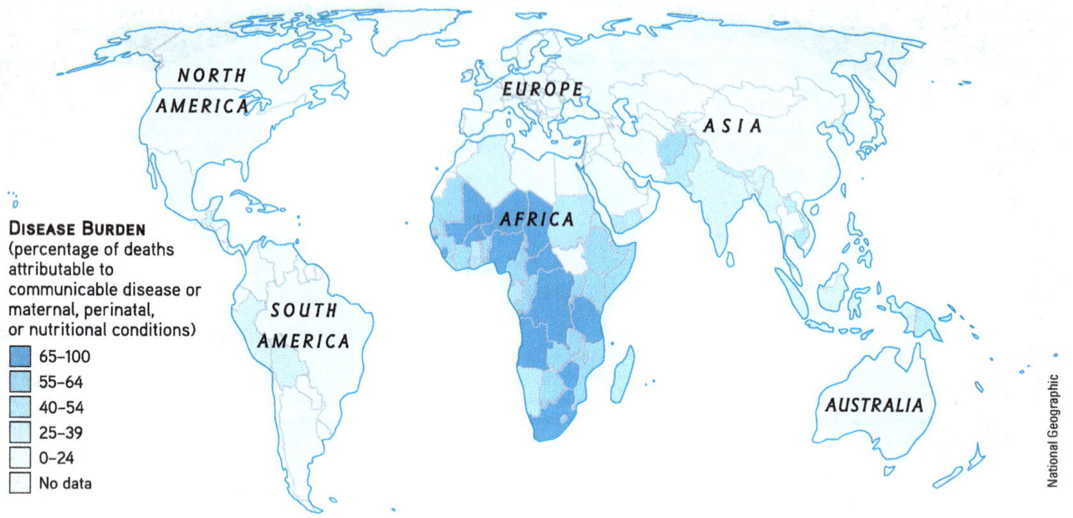

Figure 28 Global disease burden.

Data and Map Analysis

1. Name five countries that have the highest rates of communicable diseases. (See Figure 1 to find the names of countries.)
2. Name five countries that have the highest rates of noncommunicable diseases.

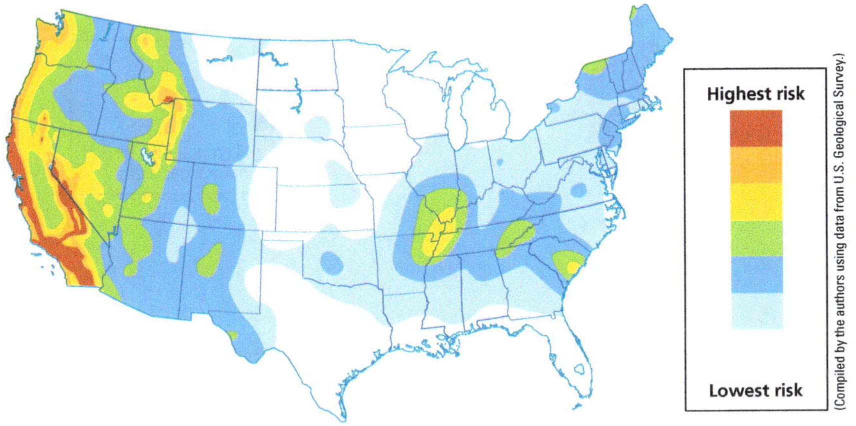

Figure 29 Earthquake (seismic) risk in various areas of the continental United States.

Data and Map Analysis

1. Speaking in general terms (northeast, southeast, central, west coast, etc.) which area has the highest earthquake risk and which area has the lowest risk?
2. For each of the categories of risk, list the number of states that fall into the category.

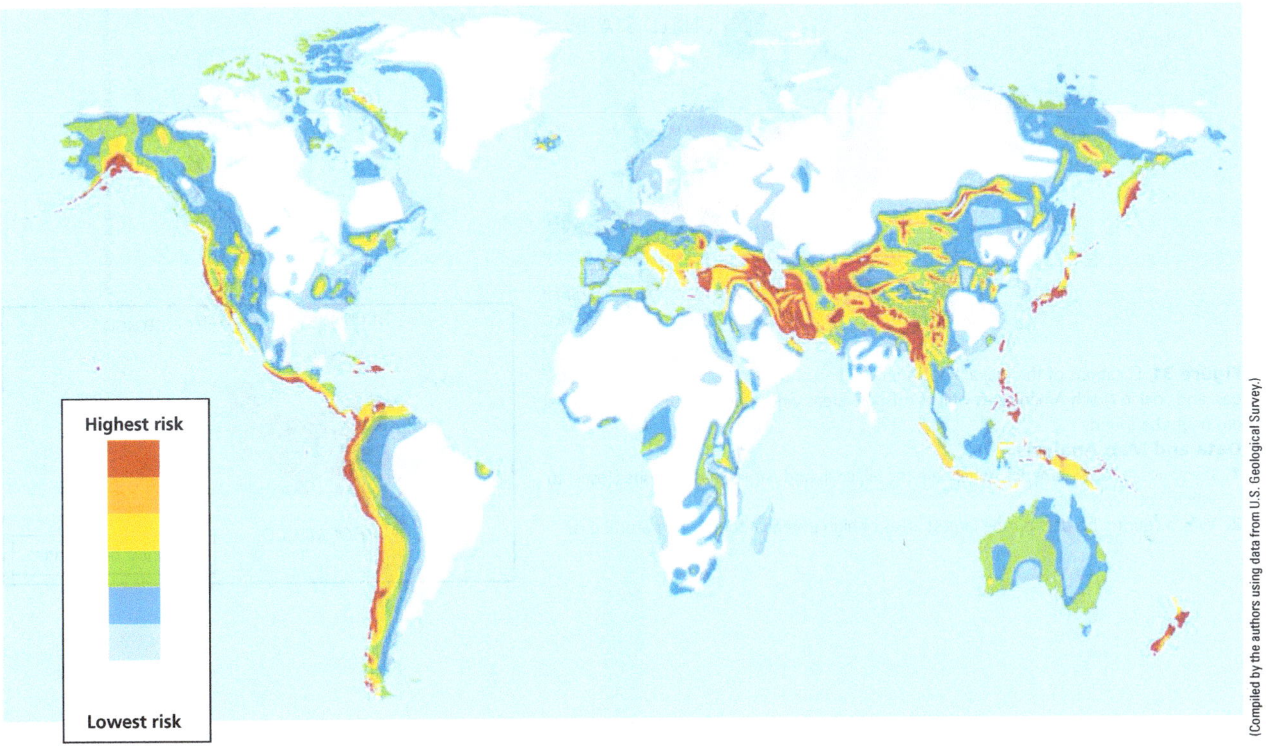

Figure 30 Earthquake (seismic) risk in the world.

Data and Map Analysis

1. How are these areas related to the boundaries of the earth's major tectonic plates as shown in Figure 14-18, p. 366?
2. Which continent has the longest coastal area subject to the highest possible risk? Which continent has the second longest such coastal area? (See Figure 2 of this supplement for continent names.)

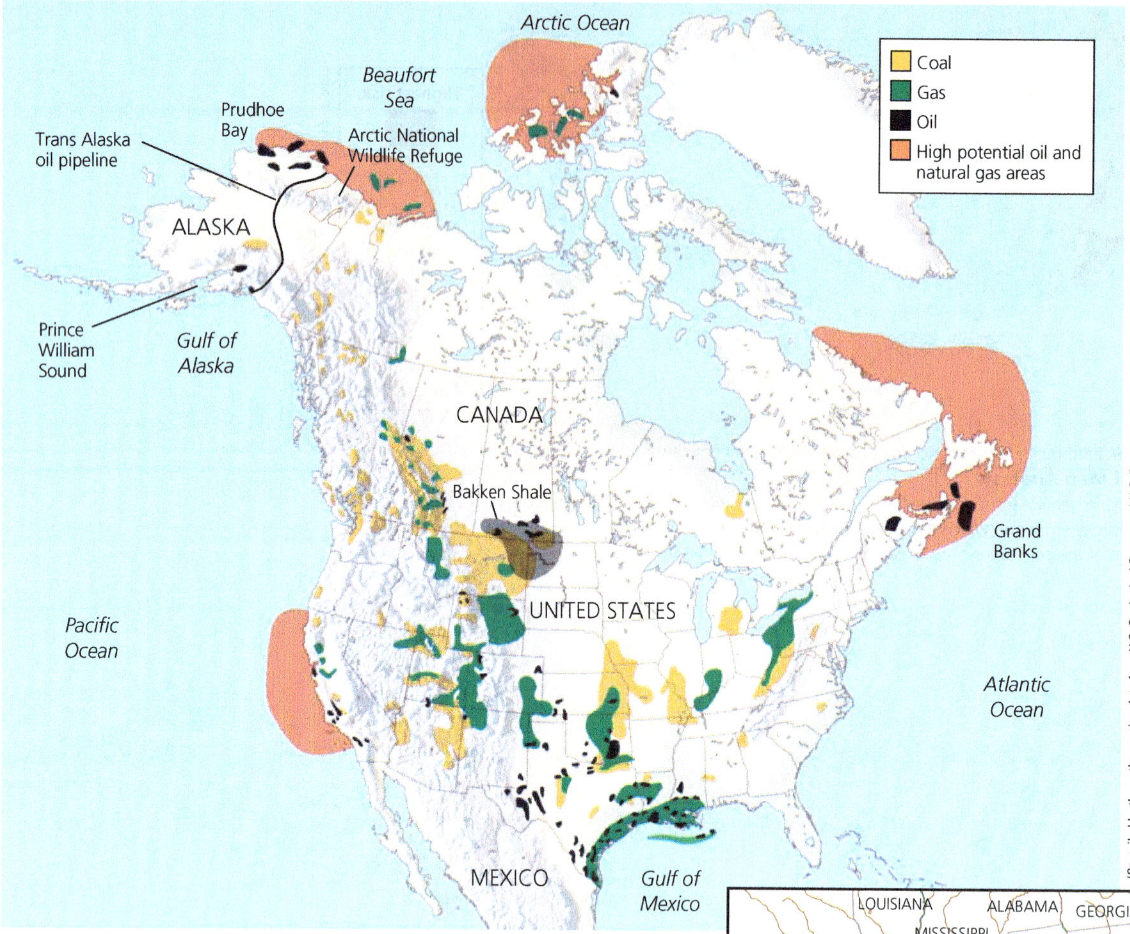

(Compiled by the authors using data from U.S. Geological Survey and U.S. Department of Energy.)

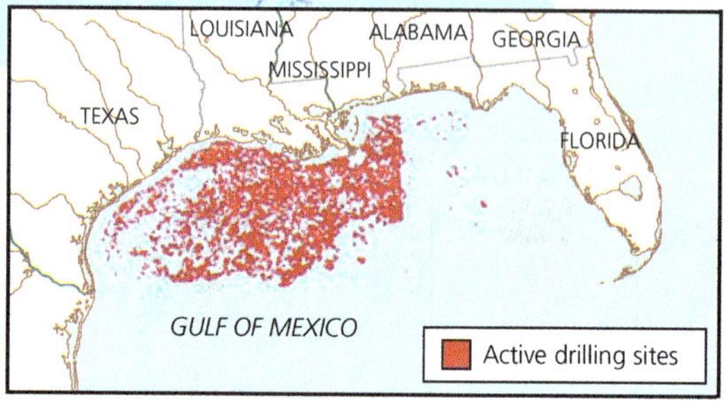

Figure 31 Locations of the major, known proven and unproven reserves of oil, natural gas, and coal in North America, as well as offshore areas where more crude oil and natural gas might be found.

Data and Map Analysis

1. If you live in North America, where are the oil, coal, and natural gas deposits closest to where you live?
2. Which country borders on the largest areas of high potential for oil and natural gas?

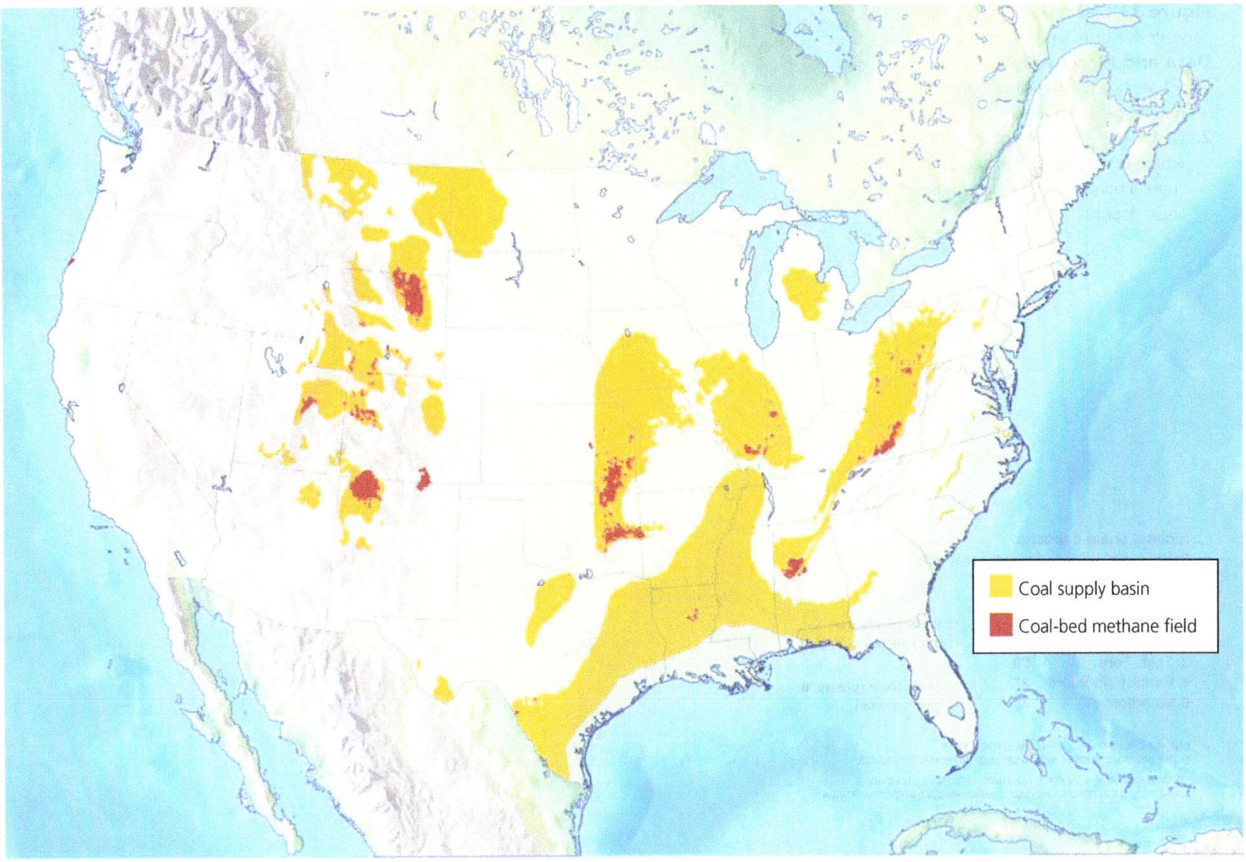

(Compiled by the authors using data from U.S. Department of Energy and U.S. Geological Survey.)

Figure 32 Major coal supply basins and coal-bed methane fields in the lower 48 states of the United States.

Data and Map Analysis

1. If you live in the United States, where are the coal-bed methane deposits closest to where you live?
2. Removing these deposits requires lots of water. Compare the locations of the major deposits of coal-bed methane with water-deficit areas shown in Figure 13-7, p. 322, and Figure 13-8, p. 323.

Figure 33 Major natural gas shale deposits in North America.

Data and Map Analysis

1. What state has the largest area of natural gas shale deposits?
2. Describe the locations of three areas where two states or two countries share a border over a natural gas shale deposit.

Richest shale deposits
Recoverable gas, in trillion cubic feet (tcf)
Total for continental U.S. **542***

1. Marcellus	141
2. Haynesville	66
3. Eagle Ford	50
4. Barnett-Woodford	27
5. Woodford	24

Shale play (active or expected drilling)
Shale basin (potential gas resource)

ONE TCF CAN HEAT 15 MILLION HOMES FOR A YEAR.
*TOTAL INCLUDES "PROVED RESERVES" AND "UNPROVED RESOURCES."
VIRGINIA W. MASON, NGM STAFF; ALEXANDER STEGMAIER, NG STAFF
SOURCES: U.S. EIA; FRACTRACKER; U.S. CENSUS BUREAU; STATISTICS CANADA

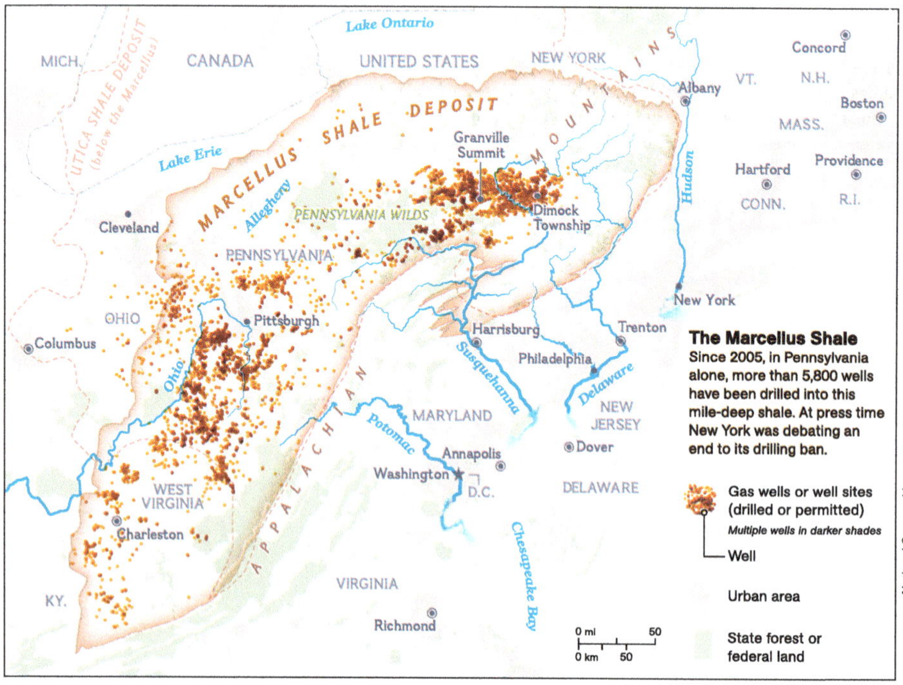

National Geographic

Figure 34 Locations of the 100 commercial nuclear power reactors in the United States.

(Compiled by the authors using data from U.S. Nuclear Regulatory Commission and U.S. Department of Energy.)

Data and Map Analysis

1. If you live in the United States, do you live or go to school within about 97 kilometers (60 miles) of a commercial nuclear power reactor?
2. Which state has the largest number of commercial nuclear power reactors?

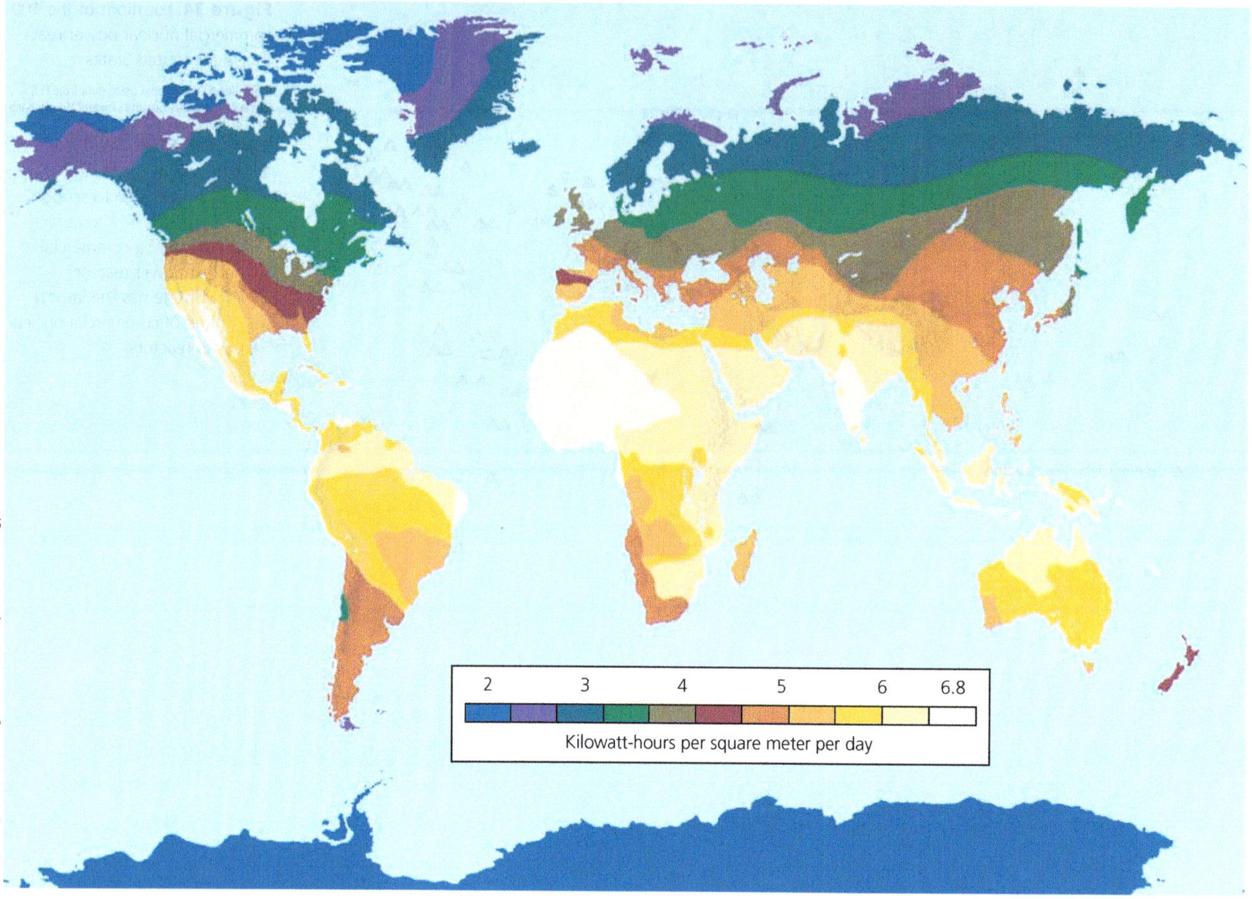

(Compiled by the authors using data from U.S. Department of Energy.)

Figure 35 Global availability of direct solar energy. Areas with more than 3.5 kilowatt-hours per square meter per day (see scale) are good candidates for passive and active solar heating systems and use of solar cells to produce electricity.

Data and Map Analysis

1. What is the potential for making greater use of solar energy to provide heat and produce electricity (with solar cells) where you live or go to school?
2. List the continents in order of overall availability of direct solar energy, from those with the highest to those with the lowest. (See Figure 2 of this supplement for continent names.)

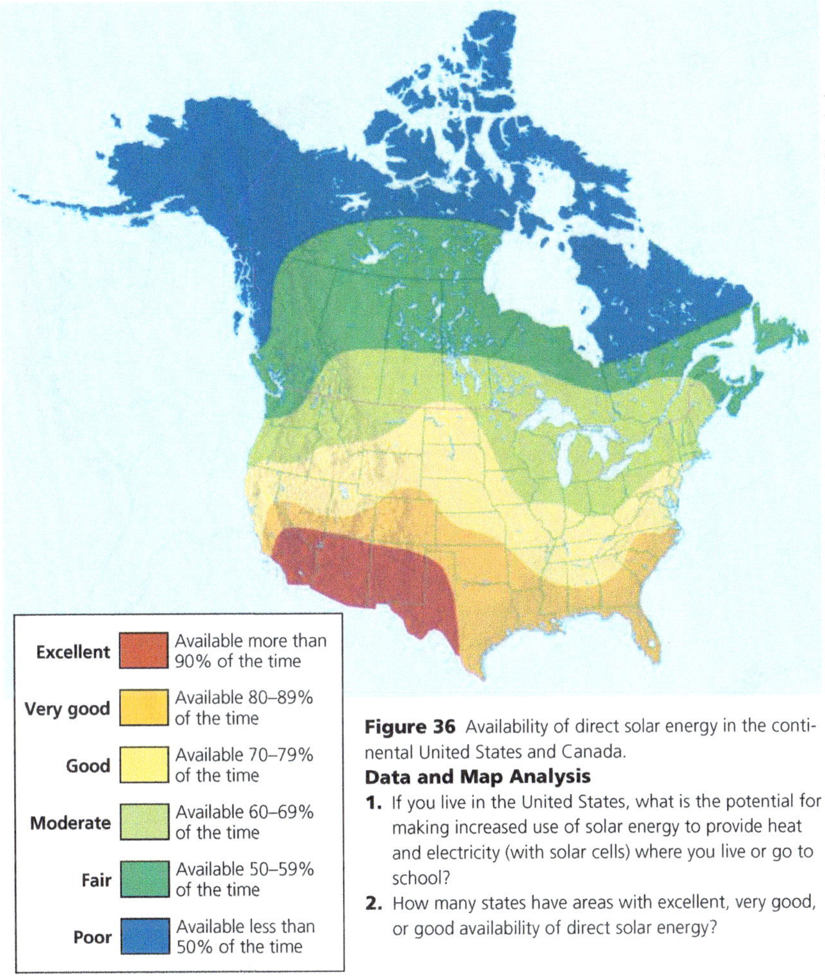

(Compiled by the authors using data from the U.S. Department of Energy and the National Wildlife Federation.)

Figure 36 Availability of direct solar energy in the continental United States and Canada.

Data and Map Analysis

1. If you live in the United States, what is the potential for making increased use of solar energy to provide heat and electricity (with solar cells) where you live or go to school?
2. How many states have areas with excellent, very good, or good availability of direct solar energy?

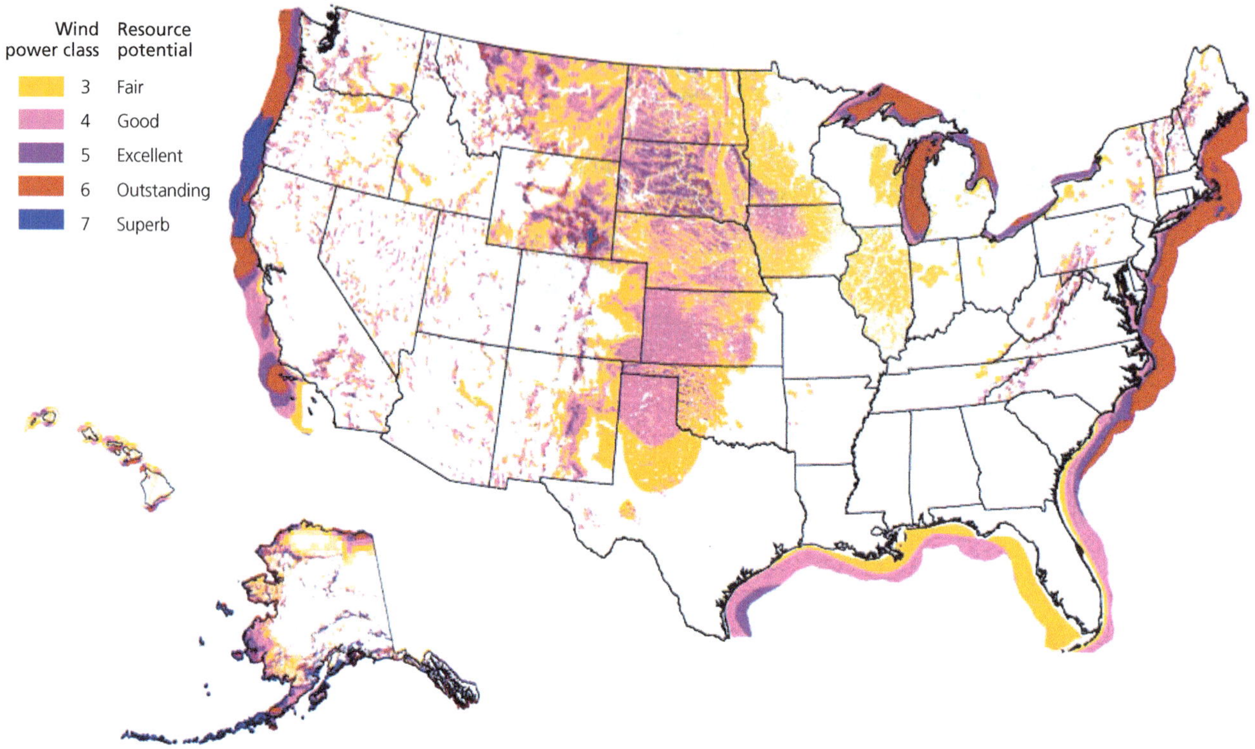

Figure 37 Potential supply of land- and ocean-based wind energy in the United States.

(Compiled by the authors using data from U.S. Geological Survey and U.S. Department of Energy.)

Data and Map Analysis

1. If you live in the United States, what is the general wind energy potential where you live or go to school?
2. How many states have areas with good or excellent potential for wind energy?

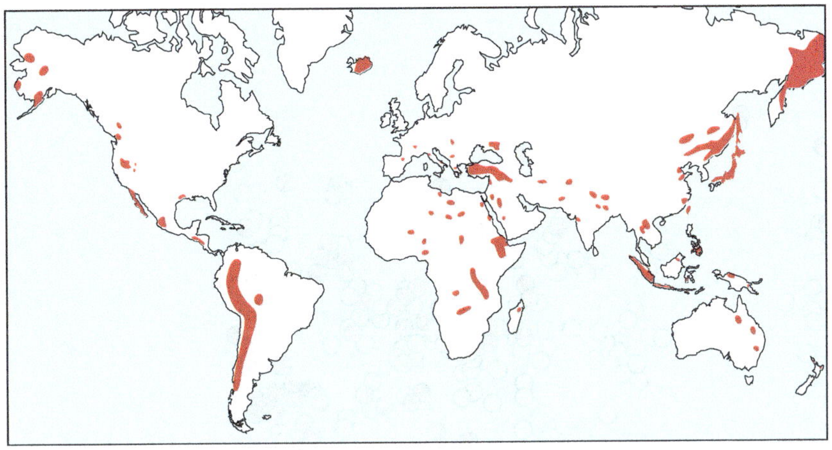

Figure 38 Known global reserves of moderate- to high-temperature geothermal energy.

(Compiled by the authors using data from Canadian Geothermal Resources Council, U.S. Geological Survey, and U.S. Department of Energy.)

Data and Map Analysis

1. Between North and South America, which continent appears to have the greater total potential for geothermal energy? (See Figure 2 of this supplement for continent names.)
2. Which country in Asia has the greatest known reserves of geothermal energy? (See Figure 1 of this supplement for country names.)

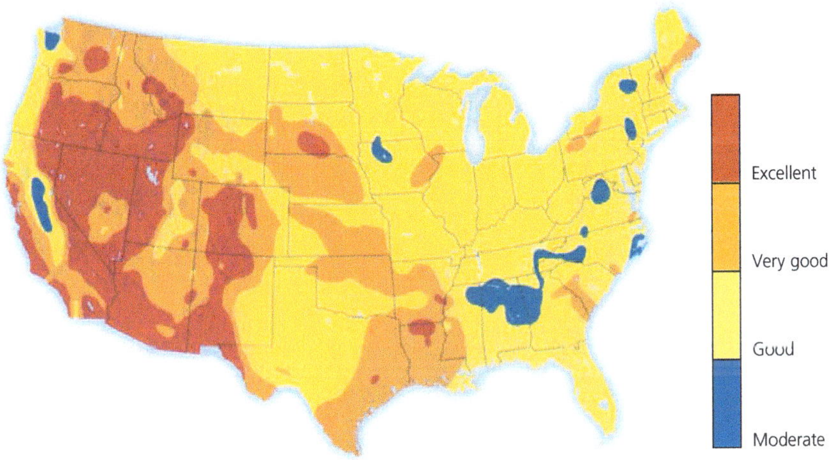

Figure 39 Potential geothermal energy resources in the continental United States.

(Compiled by the authors using data from U.S. Department of Energy and U.S. Geological Survey.)

Data and Map Analysis

1. If you live in the United States, what is the potential for using geothermal energy to provide heat or to produce electricity where you live or go to school?
2. How many states have areas with very good or excellent potential for using geothermal energy?

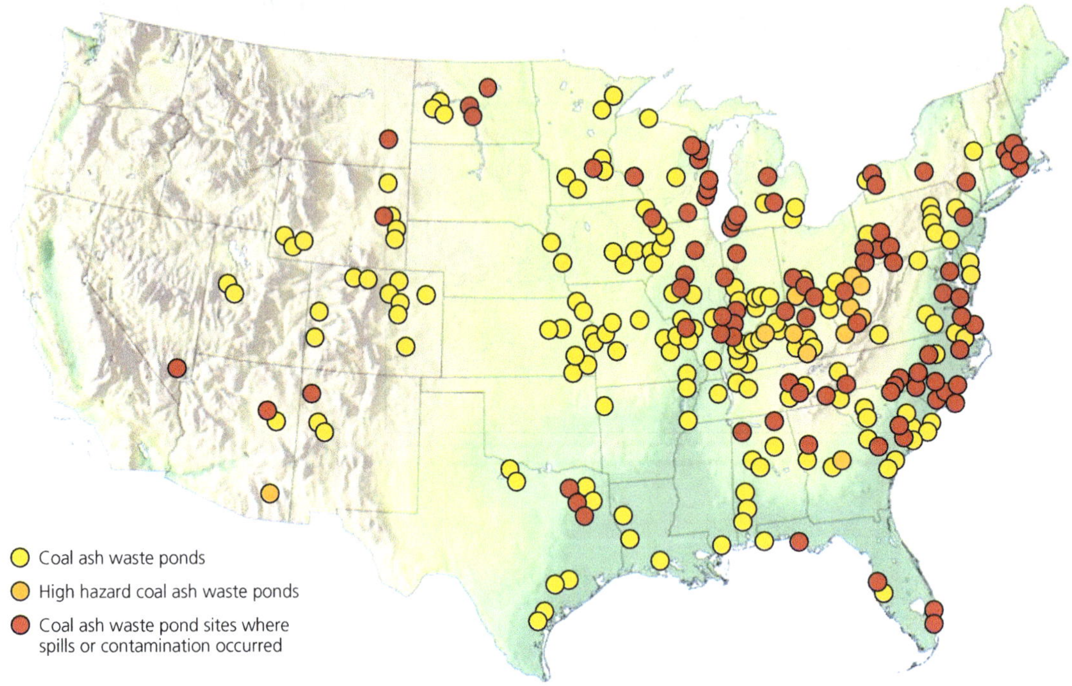

Figure 40 Coal ash waste pond sites in the United States. ***Question:*** Do you live or go to school near any of these ponds?

(Compiled by the authors using data from U.S. Environmental Protection Agency and the Sierra Club.)

Data and Map Analysis

1. In what part of the country (eastern third, central third, or western third) have most of the cases of contamination and spills occurred?
2. In what part of the country are most of the high-hazard coal ash ponds located?

SUPPLEMENT 7 ENVIRONMENTAL DATA AND DATA ANALYSIS

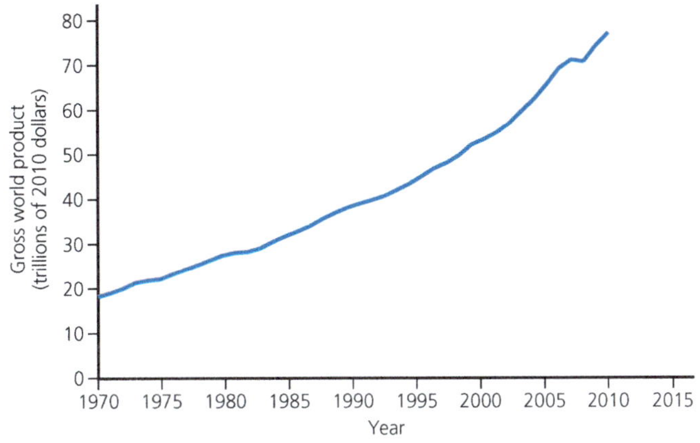

Figure 1 Gross world product (GWP), 1970–2011, in 2010 dollars.

(Compiled by the authors using data from the International Monetary Fund, World Bank, World Economic Fund, UN Population Division, and Earth Policy Institute, using purchasing power parity terms.)

Data and Graph Analysis

1. Roughly how many times bigger than the gross world product of 1985 was the gross world product of 2011?
2. If the current trend continues, about what do you think the gross world product will be in 2015, in trillions of dollars?

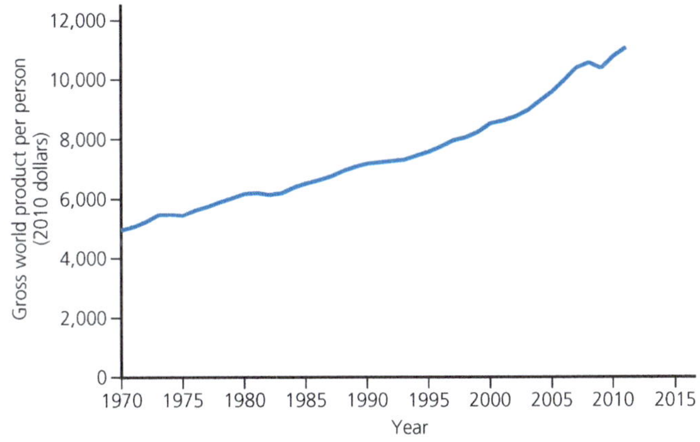

Figure 2 Gross world product (GWP) per person, 1970–2011, in 2010 dollars, using purchasing power parity terms.

(Compiled by the authors using data from the International Monetary Fund, World Bank, World Economic Fund, UN Population Division, and Earth Policy Institute.)

Data and Graph Analysis

1. Roughly how many years did it take for the GWP per person value in 1970 to double?
2. If the current trend continues, about what do you think the GWP per person will be in 2015?

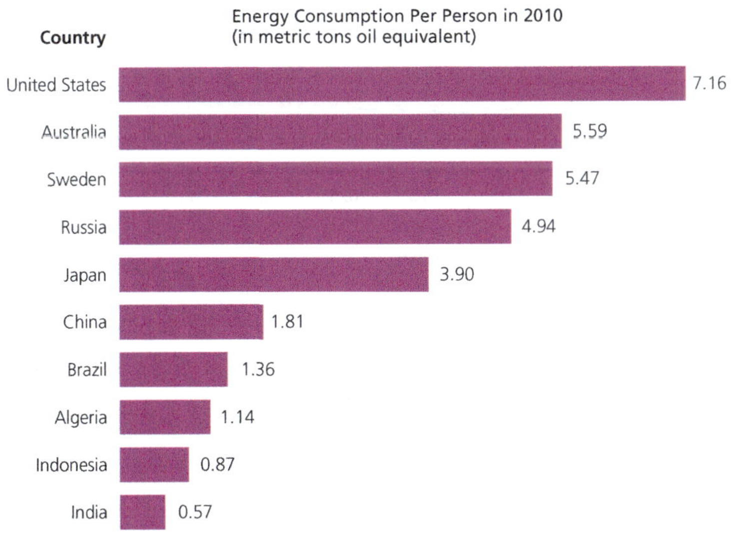

Figure 3 Energy consumption per person in 2010 for selected countries.

(Compiled by the authors using data from the World Bank.)

Data and Graph Analysis

1. On average, how many times more energy does an American consume per year than does a person in **(a)** China, **(b)** India, **(c)** Japan, and **(d)** Brazil?
2. On average, how many times more energy does a Chinese citizen consume per year than does a person in **(a)** India and **(b)** Algeria?

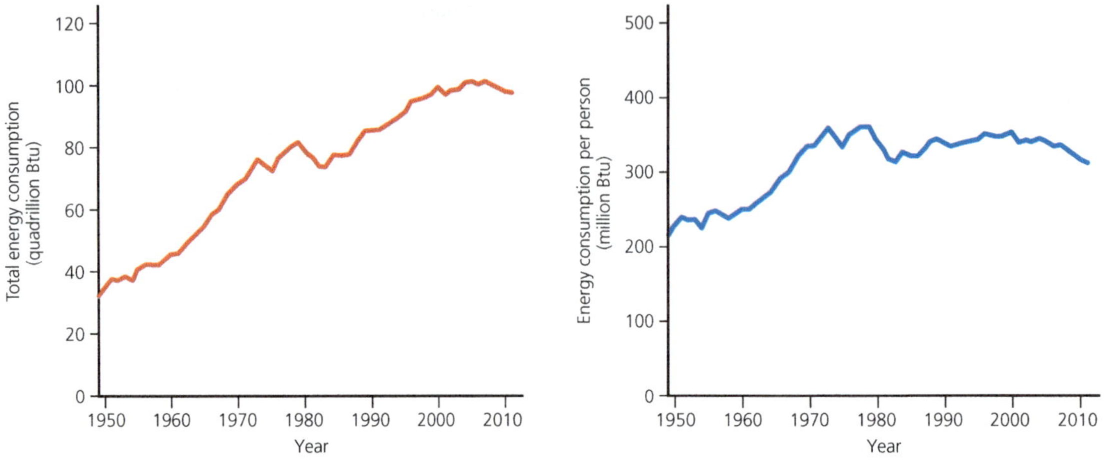

Figure 4 Total (left) and per capita (right) energy consumption in the United States, 1950–2011.

(Compiled by the authors using data from U.S. Energy Information Administration and the U.S. Census Bureau.)

Data and Graph Analysis

1. In what year or years did total U.S. energy consumption reach 80 quadrillion Btus?
2. In what year did energy consumption per person reach its highest level shown on this graph, and about what was that level of consumption?

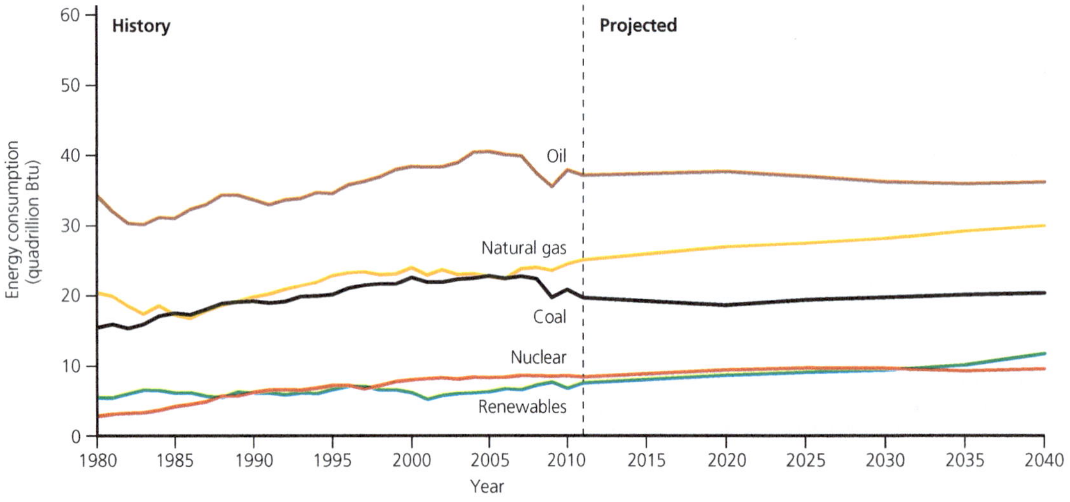

Figure 5 Energy consumption by fuel in the United States, 1980–2011, with projections to 2040.

(Compiled by the authors using data from U.S. Energy Information Administration Primary Energy Consumption Estimates by Source, 1949–2011, and Annual Energy Outlook 2013.)

Data and Graph Analysis

1. Usage of which energy source grew the most between 1990 and 2011? How much did it grow (in quadrillion Btus)?
2. For which two energy sources is usage expected to grow the fastest between 2011 and 2040?

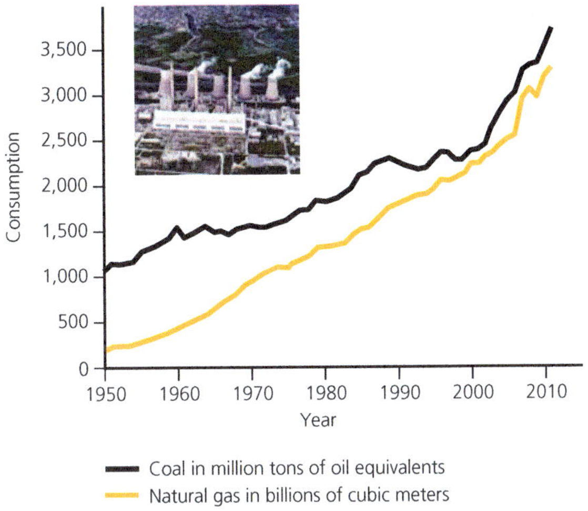

Figure 6 World coal and natural gas consumption for the period 1950–2011.

(Compiled by the authors using data from British Petroleum and International Energy Agency.)

Photo: airphoto.gr/Shutterstock.com

Data and Graph Analysis

1. Which energy source has grown more steadily—coal or natural gas? In what years has coal use grown most sharply?
2. In what year did coal use reach a level twice as high as it was in 1960?

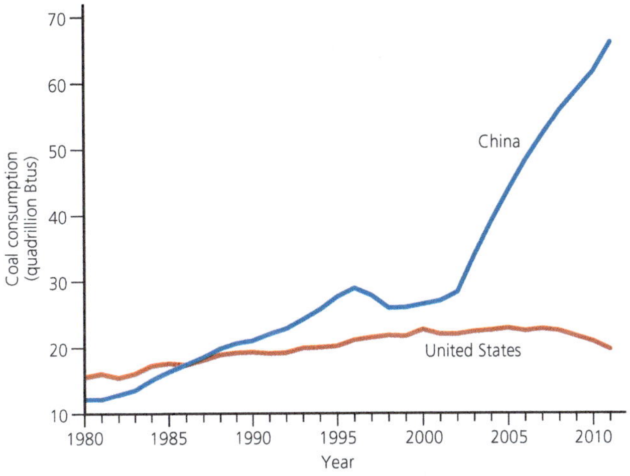

Figure 7 Coal consumption in China and the United States, 1980–2011.

(Compiled by the authors using data from Earth Policy Institute and British Petroleum.)

Data and Graph Analysis

1. By what percentage did coal consumption increase in China between 1980 and 2011?
2. By what percentage did coal consumption increase in the United States between 1980 and 2011?

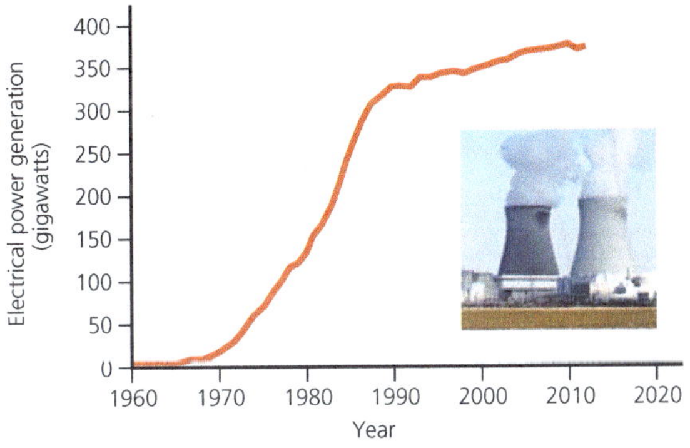

Figure 8 Global electrical generating capacity of nuclear power plants, 1960–2012.

(Compiled by the authors using data from International Energy Agency, Worldwatch Institute, and Earth Policy Institute.)

Photo: SpaceKris/Shutterstock.com

Data and Graph Analysis

1. After 1980, how long did it take to double that year's generating capacity?
2. Considering the decades of the 1970s, 1980s, 1990s, and 2000s, which decade saw the sharpest growth in generating capacity? During which decade did this growth level off?

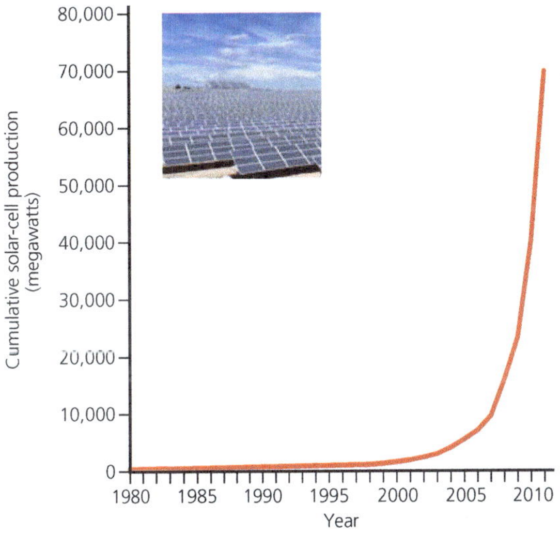

Figure 9 Global cumulative solar (photovoltaic) installations, 1980–2011.

(Compiled by the authors using data from U.S. Energy Information Administration, International Energy Agency, Worldwatch Institute, and Earth Policy Institute.)

Photo: pedrosala/Shutterstock.com

Data and Graph Analysis

1. About how many times more photovoltaic capacity was there in 2011, compared to the capacity in 2000?
2. How long did it take the world to go **(a)** from 0 to 5,000 megawatts in its cumulative installation of photovoltaic capacity and **(b)** from 5,000 to 40,000 megawatts?

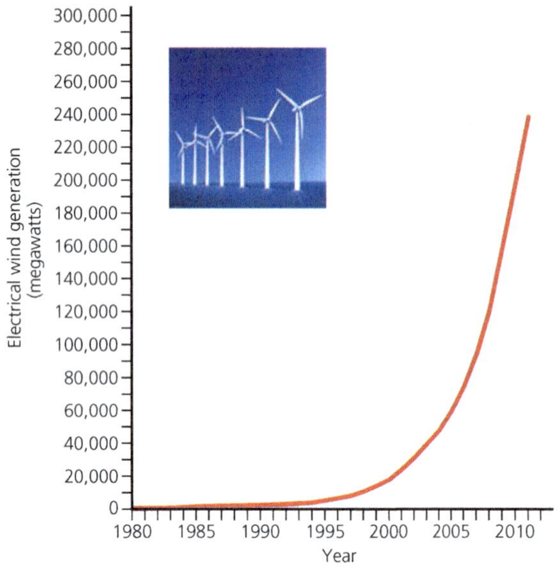

Figure 10 Global installed capacity for generation of electricity by wind energy, 1980–2011.

(Compiled by the authors using data from Global Wind Energy Council, European Wind Energy Association, American Wind Energy Association, Worldwatch Institute, World Wind Energy Association, and Earth Policy Institute.)

Photo: TebNad/Shutterstock.com

Data and Graph Analysis

1. How long did it take for the world to go from 0 to 100,000 megawatts of installed capacity for wind-generated electricity?
2. In 2011, the world's installed capacity for generating electricity by wind power was about how many times more than it was in 1995?

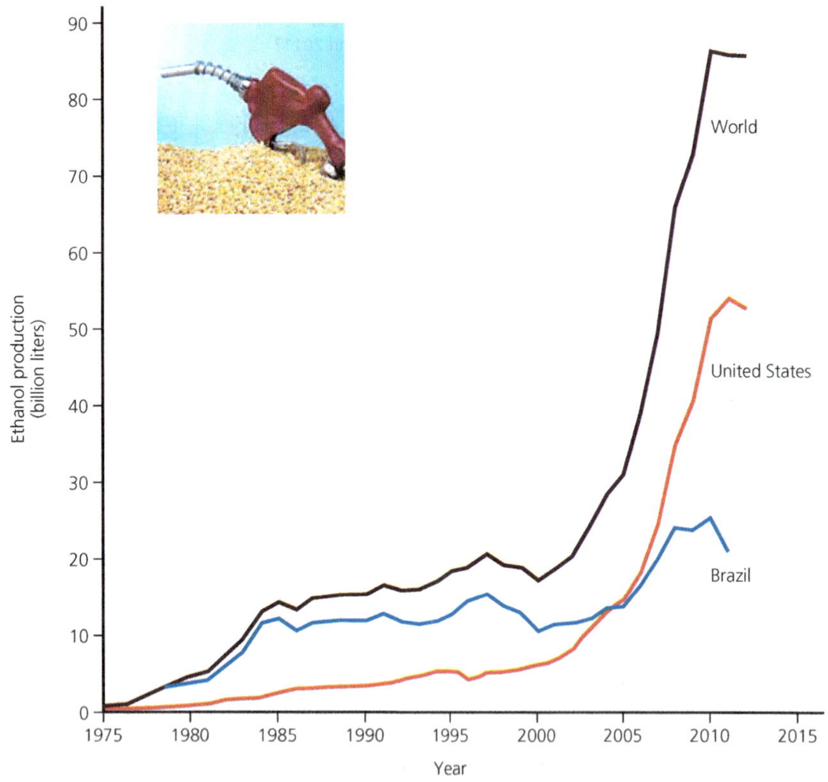

Figure 11 Production of ethanol motor fuel in the world, in Brazil, and in the United States, 1975–2012.

(Compiled by the authors using data from USDA, U.S. Energy Information Administration, Worldwatch Institute, and Earth Policy Institute.)

Photo: Jim Barber/Shutterstock.com

Data and Graph Analysis

1. By roughly what percentage did global ethanol production increase between 2000 and 2010?
2. In what years did ethanol production peak in **(a)** the United States, **(b)** Brazil, and **(c)** the world?

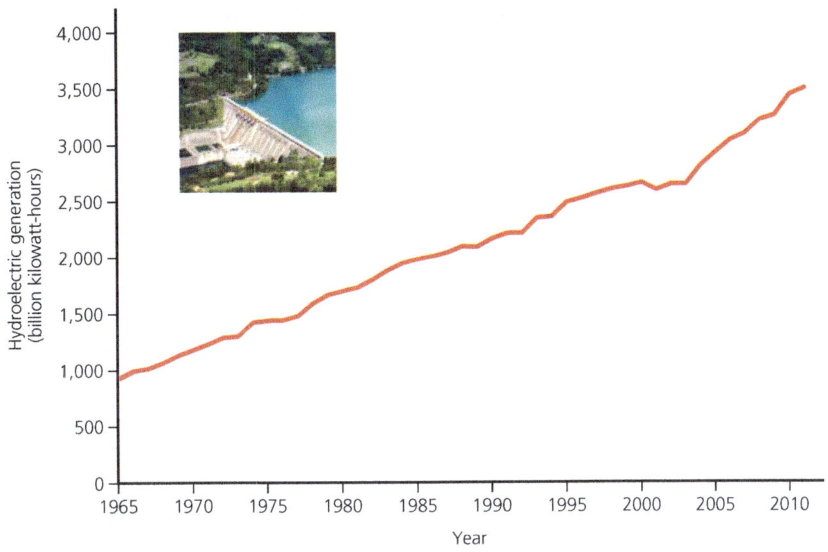

Figure 12 World hydroelectric generation, 1965–2011.

(Compiled by the authors using data from International Energy Agency, British Petroleum, Worldwatch Institute, and Earth Policy Institute.)

Photo: Zeljko Radokjo/Shutterstock.com

Data and Graph Analysis

1. After 1965, how many years did it take for the world to double its hydroelectric capacity?
2. Between what years was growth in hydroelectric capacity the sharpest?

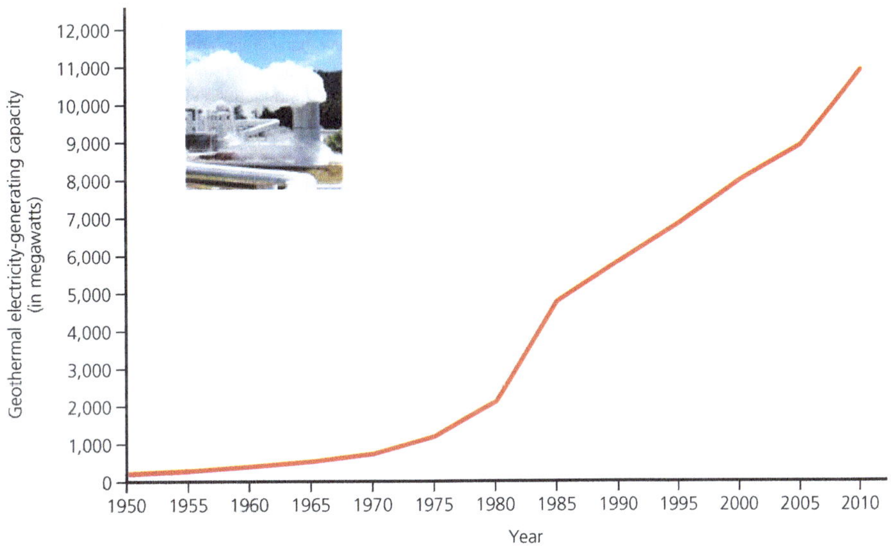

Figure 13 Global cumulative installed geothermal electricity–generating capacity, 1950–2010.

(Compiled by the authors using data from International Energy Agency, Worldwatch Institute, Earth Policy Institute, and Ruggero Bertani.)

Photo: N.Minton/Shutterstock.com

Data and Graph Analysis

1. About how many times more geothermal electricity–generating capacity was available in 2010 as there was in 1965?
2. Between what years was growth in this power source the sharpest?

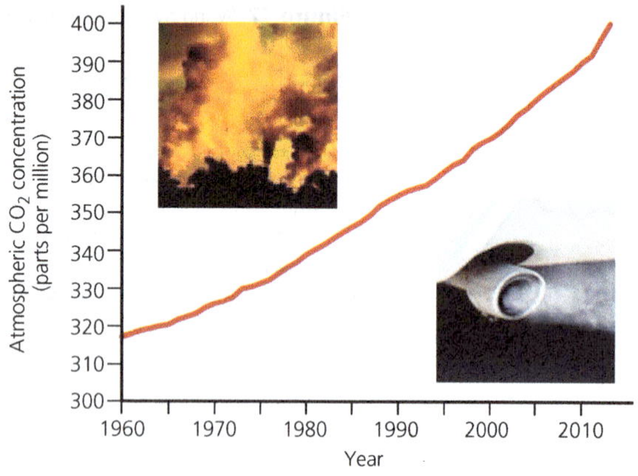

Figure 14 Atmospheric concentration of carbon dioxide (CO_2) measured at a major atmospheric research center in Mauna Loa, Hawaii, 1960–2013.

(Compiled by the authors using data from Scripps Institute of Oceanography, U.S. Energy Information Agency, and Earth Policy Institute.)

Top photo: Peter Weber/Shutterstock.com

Bottom photo: INSAGO/Shutterstock.com

Data and Graph Analysis

1. By how much did atmospheric CO_2 concentrations grow between 1960 and 2011 (in parts per million)?
2. Assuming that atmospheric CO_2 concentrations continue growing as is reflected on this graph, estimate the year in which such concentrations will reach 450 parts per million.

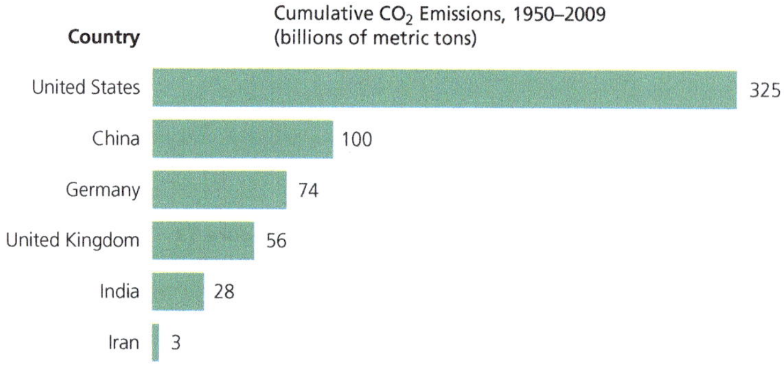

Figure 15 Cumulative carbon dioxide (CO_2) emissions from use of fossil fuels for selected countries, 1950–2009.

(Compiled by the authors using data from International Panel on Climate Change, World Resources Institute, Earth Policy Institute, and British Petroleum.)

Data and Graph Analysis

1. How many times higher are the cumulative CO_2 emissions of the United States than those of **(a)** China and **(b)** India?
2. How many times higher are the cumulative CO_2 emissions of China than those of **(a)** the United Kingdom and **(b)** India?

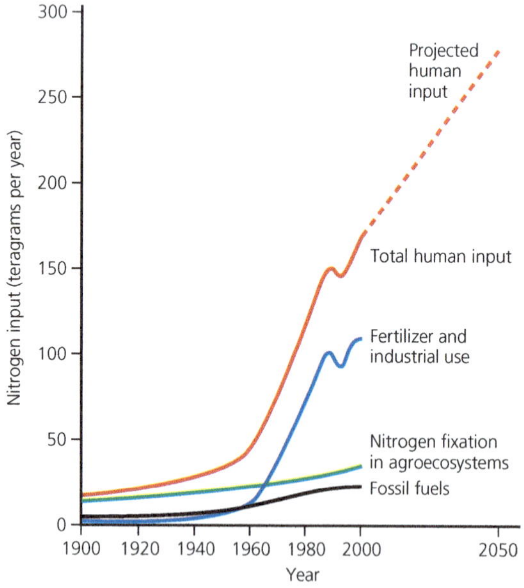

Figure 16 Global trends in the annual inputs of nitrogen into the environment from human activities, with projections to 2050.

(Compiled by the authors using data from Millennium Ecosystem Assessment and International Fertilizer Industry Association.)

Data and Graph Analysis

1. About how many times higher than the total human input of nitrogen to the environment in 1900 was the human input in 2000?
2. If nitrogen inputs continue as projected, about how many times higher will the total inputs be in 2050 than they were in 1980?

GLOSSARY

abiotic Nonliving. Compare *biotic*.

acid See *acid solution*.

acid deposition The falling of acids and acid-forming compounds from the atmosphere to the earth's surface. Acid deposition is commonly known as *acid rain*, a term that refers to the wet deposition of droplets of acids and acid-forming compounds.

acidity Chemical characteristic that helps determine how a substance dissolved in water (a solution) will interact with and affect its environment; based on the comparative amounts of hydrogen ions (H^+) and hydroxide ions (OH^-) contained in a particular volume of the solution. See *pH*.

acid rain See *acid deposition*.

acid solution Any water solution that has more hydrogen ions (H^+) than hydroxide ions (OH^-); any water solution with a pH less than 7. Compare *basic solution, neutral solution*.

active solar heating system System that uses solar collectors to capture energy from the sun and store it as heat for space heating and water heating. Liquid or air pumped through the collectors transfers the captured heat to a storage system such as an insulated water tank or rock bed. Pumps or fans then distribute the stored heat or hot water throughout a dwelling as needed. Compare *passive solar heating system*.

adaptation Any genetically controlled structural, physiological, or behavioral characteristic that helps an organism survive and reproduce under a given set of environmental conditions. It usually results from a beneficial mutation. See *biological evolution, differential reproduction, mutation, natural selection*.

adaptive trait See *adaptation*.

aerobic respiration Complex process that occurs in the cells of most living organisms, in which nutrient organic molecules such as glucose ($C_6H_{12}O_6$) combine with oxygen (O_2) to produce carbon dioxide (CO_2), water (H_2O), and energy. Compare *photosynthesis*.

affluence wealth that results in high levels of consumption and unnecessary waste of resources, based mostly on the assumption that buying more and more material goods will bring fulfillment and happiness.

age structure Percentage of the population (or number of people of each sex) at each age level in a population.

agrobiodiversity The genetic variety of plant and animal species used on farms to produce food. Compare *biodiversity*.

agroforestry Planting trees and crops together.

air pollution One or more chemicals in high enough concentrations in the air to harm humans, other animals, vegetation, or materials. Excess heat is also considered a form of air pollution. Such chemicals or physical conditions are called air pollutants. See *primary pollutant, secondary pollutant*.

albedo Ability of a surface to reflect light.

alien species See *nonnative species*.

alley cropping Planting of crops in strips with rows of trees or shrubs on each side.

alpha particle Positively charged matter, consisting of two neutrons and two protons, which is emitted as radioactivity from the nuclei of some radioisotopes. See also *beta particle, gamma rays*.

altitude Height above sea level. Compare *latitude*.

anaerobic respiration Form of cellular respiration in which some decomposers get the energy they need through the breakdown of glucose (or other nutrients) in the absence of oxygen. Compare *aerobic respiration*.

animal manure Dung and urine of animals used as a form of organic fertilizer. Compare *green manure*.

annual plant Plant that grows, sets seed, and dies in one growing season. Compare *perennial*.

Anthropocene A new era in which humans have become major agents of change in the functioning of the earth's life-support systems as their ecological footprints have spread over the earth. See *ecological footprint*. Compare *Holopocene*.

anthropocentric Human-centered.

aquaculture Growing and harvesting of fish and shellfish for human use in freshwater ponds, irrigation ditches, and lakes, or in cages or fenced-in areas of coastal lagoons and estuaries or in the open ocean. See *fish farming, fish ranching*.

aquatic Pertaining to water. Compare *terrestrial*.

aquatic life zone Marine and freshwater portions of the biosphere. Examples include freshwater life zones (such as lakes and streams) and ocean or marine life zones (such as estuaries, coastlines, coral reefs, and the open ocean).

aquifer Porous, water-saturated layers of sand, gravel, or bedrock that can yield an economically significant amount of water.

arable land Land that can be cultivated to grow crops.

area strip mining Type of surface mining used where the terrain is flat. An earthmover strips away the overburden, and a power shovel digs a cut to remove the mineral deposit. The trench is then filled with overburden, and a new cut is made parallel to the previous one. The process is repeated over the entire site. Compare *mountaintop removal, open-pit mining, subsurface mining*.

arid Dry. A desert or other area with an arid climate has little precipitation.

artificial selection Process by which humans select one or more desirable genetic traits in the population of a plant or animal species and then use *selective breeding* to produce populations containing many individuals with the desired traits. Compare *genetic engineering, natural selection*.

asthenosphere Zone within the earth's mantle made up of hot, partly melted rock that flows and can be deformed like soft plastic.

atmosphere Whole mass of air surrounding the earth. See *stratosphere, troposphere*. Compare *biosphere, geosphere, hydrosphere*.

atmospheric pressure Force or mass per unit area of air, caused by the bombardment of a surface by the molecules in air.

atom Minute unit made of subatomic particles that is the basic building block of all chemical elements and thus all matter; the smallest unit of an element that can exist and still have the unique characteristics of that element. Compare *ion, molecule*.

atomic number Number of protons in the nucleus of an atom. Compare *mass number*.

atomic theory Idea that all elements are made up of atoms; the most widely accepted scientific theory in chemistry.

autotroph See *producer*.

background extinction rate Normal extinction of various species as a result of changes in local environmental conditions. Compare *mass extinction*.

bacteria Prokaryotic, one-celled organisms. Some transmit diseases. Most act as decomposers and get the nutrients they need by breaking down complex organic compounds in the tissues of living or dead organisms into simpler inorganic nutrient compounds.

basic solution Water solution with more hydroxide ions (OH^-) than hydrogen ions (H^+); water solution with a pH greater than 7. Compare *acid solution, neutral solution*.

benthos Bottom-dwelling organisms. Compare *decomposer, nekton, plankton*.

beta particle Swiftly moving electron emitted by the nucleus of a radioactive isotope. See also *alpha particle, gamma ray*.

bioaccumulation An increase in the concentration of a chemical in specific organs or tissues at a level higher than would normally be expected. Compare *biomagnification*.

biocentric Life-centered. Compare *anthropocentric*.

biodegradable Capable of being broken down by decomposers.

biodegradable pollutant Material that can be broken down into simpler substances (elements and compounds) by bacteria or other decomposers. Paper and most organic wastes such as animal manure are biodegradable but can take decades to biodegrade in modern landfills. Compare *nondegradable pollutant*.

biodiversity Variety of different species (*species diversity*), genetic variability among individuals within each species (*genetic diversity*), variety of ecosystems (*ecological diversity*), and functions such as energy flow and matter cycling needed for the survival of species and biological communities (*functional diversity*).

biodiversity hotspot An area especially rich in plant species that are found nowhere else and are in great danger of extinction. Such areas suffer serious ecological disruption, mostly because of rapid human population growth and the resulting pressure on natural resources.

biofuel Gas (such as methane) or liquid fuel (such as ethyl alcohol or biodiesel) made from plant material (biomass).

biogeochemical cycle Natural processes that recycle nutrients in various chemical forms from the nonliving environment to living organisms and then back to the nonliving environment. Examples include the carbon, oxygen, nitrogen, phosphorus, sulfur, and hydrologic cycles.

biological amplification See *biomagnification*.

biological community See *community*.

biological diversity See *biodiversity*.

biological evolution Change in the genetic makeup of a population of a species in successive generations. If continued long enough, it can lead to the formation of a new species. Note that populations, not individuals, evolve. See also *adaptation, differential reproduction, natural selection, theory of evolution*.

biological extinction Complete disappearance of a species from the earth. It happens when a species cannot adapt and successfully reproduce under new environmental conditions or when a species evolves into one or more new species. Compare *speciation*. See also *endangered species, mass extinction, threatened species*.

biological pest control Control of pest populations by natural predators, parasites, or disease-causing bacteria and viruses (pathogens).

biomagnification Increase in concentration of DDT, PCBs, and other slowly degradable, fat-soluble chemicals in organisms at successively higher trophic levels of a food chain or web. Compare *bioaccumulation*.

biomass Organic matter produced by plants and other photosynthetic producers; total dry weight of all living organisms that can be supported at each trophic level in a food chain or web; dry weight of all organic matter in plants and animals in an ecosystem; plant materials and animal wastes used as fuel.

biome Terrestrial regions inhabited by certain types of life, especially vegetation. Examples include various types of deserts, grasslands, and forests.

biomimicry Process of observing certain changes in nature, studying how natural systems have responded to such changing conditions over many millions of years, and applying what is learned to dealing with some environmental challenge.

biosphere Zone of the earth where life is found. It consists of parts of the atmosphere (the troposphere), hydrosphere (mostly surface water and groundwater), and lithosphere (mostly soil and surface rocks and sediments on the bottoms of oceans and other bodies of water) where life is found. Compare *atmosphere, geosphere, hydrosphere*.

biotic Living organisms. Compare *abiotic*.

biotic pollution The effect of invasive species that can reduce or wipe out populations of many native species and trigger ecological disruptions.

biotic potential Maximum rate at which the population of a given species can increase when there are no limits on its rate of growth. See *environmental resistance*.

birth rate See *crude birth rate*.

bitumen Gooey, black, high-sulfur, heavy oil extracted from tar sand and then upgraded to synthetic fuel oil. See *oil sand*.

broadleaf deciduous plants Plants such as oak and maple trees that survive drought and cold by shedding their leaves and becoming dormant. Compare *broadleaf evergreen plants, coniferous evergreen plants*.

broadleaf evergreen plants Plants that keep most of their broad leaves year-round. An example is the trees found in the canopies of tropical rain forests. Compare *broadleaf deciduous plants, coniferous evergreen plants*.

buffer Substance that can react with hydrogen ions in a solution and thus hold the acidity or pH of a solution fairly constant. See *pH*.

calorie Unit of energy; amount of energy needed to raise the temperature of 1 gram of water by 1 C° (unit on Celsius temperature scale). See also *kilocalorie*.

cancer Group of more than 120 different diseases, one for each type of cell in the human body. Each type of cancer produces a tumor in which cells multiply uncontrollably and invade surrounding tissue.

carbon capture and storage (CCS) Process of removing carbon dioxide gas from coal-burning power and industrial plants and storing it somewhere (usually underground or under the seabed) so that it is not released into the atmosphere, essentially forever.

carbon cycle Cyclic movement of carbon in different chemical forms from the environment to organisms and then back to the environment.

carcinogen Chemicals, ionizing radiation, and viruses that cause or promote the development of cancer. See *cancer*. Compare *mutagen, teratogen*.

carnivore Animal that feeds on other animals. Compare *herbivore, omnivore*.

carrying capacity (*K*) Maximum population of a particular species that a given habitat can support over a given period. Compare *cultural carrying capacity*.

CCS See *carbon capture and storage*.

cell Smallest living unit of an organism. Each cell is encased in an outer membrane or wall and contains genetic material (DNA) and other parts to perform its life function. Organisms such as bacteria consist of only one cell, but most organisms contain many cells.

cell theory The idea that all living things are composed of cells; the most widely accepted scientific theory in biology.

CFCs See *chlorofluorocarbons*.

chain reaction Multiple nuclear fissions, taking place within a certain mass of a fissionable isotope, which release an enormous amount of energy in a short time.

chemical One of the millions of different elements and compounds found naturally and synthesized by humans. See *compound, element*.

chemical change Interaction between chemicals in which the chemical composition of the elements or compounds involved changes. Compare *nuclear change, physical change*.

chemical cycling The continual cycling of chemicals necessary for life through natural processes such as the water cycle and feeding interactions; processes that evolved due to the fact that the earth gets essentially no new inputs of these chemicals.

chemical formula Shorthand way to show the number of atoms (or ions) in the basic structural unit of a compound. Examples include H_2O, NaCl, and $C_6H_{12}O_6$.

chemical reaction See *chemical change*.

chemosynthesis Process in which certain organisms (mostly specialized bacteria) extract inorganic compounds from their environment and convert them into organic nutrient compounds without the presence of sunlight. Compare *photosynthesis*.

chlorinated hydrocarbon Organic compound made up of atoms of carbon, hydrogen, and chlorine. Examples include DDT and PCBs.

chlorofluorocarbons (CFCs) Organic compounds made up of atoms of carbon, chlorine, and fluorine. An example is Freon-12 (CCl_2F_2), which is used as a refrigerant in refrigerators and air conditioners and in making plastics such as Styrofoam. Gaseous CFCs can deplete the ozone layer when they slowly rise into the stratosphere and their chlorine atoms react with ozone molecules. Their use is being phased out.

CHP (Combined heat and power) See *cogeneration*.

chromosome A grouping of genes and associated proteins in plant and animal cells that carry certain types of genetic information. See *genes*.

chronic malnutrition Faulty nutrition, caused by a diet that does not supply an individual with enough protein, essential fats, vitamins, minerals, and other nutrients needed for good health. Compare *overnutrition, chronic undernutrition*.

chronic undernutrition Condition suffered by people who cannot grow or buy enough food to meet their basic energy needs. Most chronically undernourished children live in developing countries and are likely to suffer from mental retardation and stunted growth and to die from infectious diseases. Compare *chronic malnutrition, overnutrition*.

civil suit Court case brought to settle disputes or damages between one party and another.

clear-cutting Method of timber harvesting in which all trees in a forested area are removed in a single cutting. Compare *selective cutting, strip cutting*.

climate Physical properties of the troposphere of an area based on analysis of its weather records over a long period (at least 30 years). The two main factors determining an area's climate are its average *temperature*, with its seasonal variations, and the average amount and distribution of *precipitation*. Compare *weather*.

climate tipping point Point at which an environmental problem reaches a threshold level where scientists fear it could cause irreversible climate disruption.

climax community See *mature community*.

closed-loop recycling See *primary recycling*.

coal Solid, combustible mixture of organic compounds with 30–98% carbon by weight, mixed with various amounts of water and small amounts of sulfur and nitrogen compounds. It forms in several stages as the remains of plants are subjected to heat and pressure over millions of years.

coal gasification Conversion of solid coal to synthetic natural gas (SNG).

coal liquefaction Conversion of solid coal to a liquid hydrocarbon fuel such as synthetic gasoline or methanol.

coastal wetland Land along a coastline, extending inland from an estuary that is covered with saltwater all or part of the year. Examples include marshes, bays, lagoons, tidal flats, and mangrove swamps. Compare *inland wetland*.

coastal zone Warm, nutrient-rich, shallow part of the ocean that extends from the high-tide mark on land to the edge of a shelflike extension of continental land masses known as the continental shelf. Compare *open sea*.

coevolution Evolution in which two or more species interact and exert selective pressures on each other that can lead each species to undergo adaptations. See *evolution, natural selection*.

cogeneration Production of two useful forms of energy, such as high-temperature heat or steam and electricity, from the same fuel source.

cold front Leading edge of an advancing mass of cold air. Compare *warm front*.

colony collapse disorder (CCD) Loss through death or disappearance of all or most of the European honeybees in a particular colony due to unknown causes; a phenomenon that has resulted in large losses of European honeybees in the United States and in parts of Europe.

combined heat and power (CHP) production See *cogeneration*.

commensalism An interaction between organisms of different species in which one type of organism benefits and the other type is neither helped nor harmed to any great degree. Compare *mutualism*.

commercial extinction Depletion of the population of a wild species used as a resource to a level at which it is no longer profitable to harvest the species.

commercial forest See *tree plantation*.

commercial inorganic fertilizer Commercially prepared mixture of inorganic plant nutrients such as nitrates, phosphates, and potassium applied to the soil to restore fertility and increase crop yields. Compare *organic fertilizer*.

common law A body of unwritten rules and principles derived from court decisions along with commonly accepted practices, or norms, within a society. Compare *statutory law*.

common-property resource Resource that is owned jointly by a large group of individuals. One example is the roughly one-third of the land in the United States that is owned jointly by all U.S. citizens and held and managed for them by the government. Another example is an area of land that belongs to a whole village and that can be used by anyone for grazing cows or sheep. Compare *open-access renewable resource*. See *tragedy of the commons*.

community Populations of all species living and interacting in an area at a particular time.

competition Two or more individual organisms of a single species (*intraspecific competition*) or two or more individuals of different species (*interspecific competition*) attempting to use the same scarce resources in the same ecosystem.

compost Partially decomposed organic plant and animal matter used as a soil conditioner or fertilizer.

composting See *compost*.

compound Combination of atoms, or oppositely charged ions, of two or more elements held together by attractive forces called chemical bonds. Examples are NaCl, CO_2, and $C_6H_{12}O_6$. Compare *element*.

concentration Amount of a chemical in a particular volume or weight of air, water, soil, or other medium.

coniferous evergreen plants Cone-bearing plants (such as spruces, pines, and firs) that keep some of their narrow, pointed leaves (needles) all year. Compare *broad-leaf deciduous plants, broad-leaf evergreen plants*.

coniferous trees Cone-bearing trees, mostly evergreens, that have needle-shaped or scalelike leaves. They produce wood known commercially as softwood. Compare *deciduous plants*.

conservation Sensible and careful use of natural resources by humans. People with this view are called *conservationists*.

conservation biology Multidisciplinary science created to deal with the crisis of maintaining the genes, species, communities, and ecosystems that make up earth's biological diversity. Its goals are to investigate human impacts on biodiversity and to develop practical approaches to preserving biodiversity.

conservationist Person concerned with using natural areas and wildlife in ways that sustain them for current and future generations of humans and other forms of life.

conservation-tillage farming Crop cultivation in which the soil is disturbed little (minimum-tillage farming) or not at all (no-till farming) in an effort to reduce soil erosion, lower labor costs, and save energy. Compare *conventional-tillage farming*.

consumer Organism that cannot synthesize the organic nutrients it needs and gets its organic nutrients by feeding on the tissues of producers or of other consumers; generally divided into *primary consumers* (herbivores), *secondary consumers* (carnivores), *tertiary* (*higher-level*) *consumers, omnivores*, and *detritivores* (decomposers and detritus feeders). In economics, one who uses economic goods. Compare *producer*.

contour farming Plowing and planting across the changing slope of land, rather than in straight lines, to help retain water and reduce soil erosion.

contour strip mining Form of surface mining used on hilly or mountainous terrain. A power shovel cuts a series of terraces into the side of a hill. An earthmover removes the overburden, and a power shovel extracts the coal. The overburden from each new terrace is dumped onto the one below. Compare *area strip mining, mountaintop removal, open-pit mining, subsurface mining*.

controlled burning Deliberately set, carefully controlled surface fires that reduce flammable litter and decrease the chances of damaging *crown fires. See ground fire, surface fire*.

conventional-tillage farming Crop cultivation method in which a planting surface is made by plowing land, breaking up the exposed soil, and then smoothing the surface. Compare *conservation-tillage farming*.

convergent plate boundary Area where the earth's lithospheric plates are pushed together. See subduction zone. *Compare divergent plate boundary, transform fault*.

coral reef Formation produced by massive colonies containing billions of tiny coral animals, called polyps, that secrete a stony substance (calcium carbonate) around themselves for protection. When the corals die, their empty outer skeletons form layers and cause the reef to grow. Coral reefs are found in the coastal zones of warm tropical and subtropical oceans.

core Inner zone of the earth. It consists of a solid inner core and a liquid outer core. Compare *crust, mantle*.

corrective feedback loop See *negative feedback loop*.

cost–benefit analysis A comparison of estimated costs and benefits of actions such as implementing a pollution control regulation, building a dam on a river, or preserving an area of forest.

crop rotation Planting a field, or an area of a field, with different crops from year to year to reduce soil nutrient depletion. A plant such as corn, tobacco, or cotton, which removes large amounts of nitrogen from the soil, is planted one year. The next year a legume such as soybeans, which adds nitrogen to the soil, is planted.

crown fire Extremely hot forest fire that burns ground vegetation and treetops. Compare *controlled burning, ground fire, surface fire*.

crude birth rate Annual number of live births per 1,000 people in the population of a geographic area at the midpoint of a given year. Compare *crude death rate*.

crude death rate Annual number of deaths per 1,000 people in the population of a geographic area at the midpoint of a given year. Compare *crude birth rate*.

crude oil Gooey liquid consisting mostly of hydrocarbon compounds and small amounts of compounds containing oxygen, sulfur, and nitrogen. Extracted from underground accumulations, it is sent to oil refineries, where it is converted to heating oil, diesel fuel, gasoline, tar, and other materials.

crust Solid outer zone of the earth. It consists of oceanic crust and continental crust. Compare *core, mantle*.

cultural carrying capacity The limit on population growth that would allow most people in an area or the world to live in reasonable comfort and freedom without impairing the ability of the planet to sustain future generations. Compare *carrying capacity*.

cultural eutrophication Overnourishment of aquatic ecosystems with plant nutrients (mostly nitrates and phosphates) because of human activities such as agriculture, urbanization, and discharges from industrial plants and sewage treatment plants. See *eutrophication*.

culture Whole of a society's knowledge, beliefs, technology, and practices.

dam A structure built across a river to control the river's flow or to create a reservoir. See *reservoir*.

data Factual information collected by scientists.

DDT Dichlorodiphenyltrichloroethane, a chlorinated hydrocarbon that has been widely used as an insecticide but is now banned in some countries.

death rate See *crude death rate*.

debt-for-nature swap Agreement in which a certain amount of foreign debt is canceled in exchange for local currency investments that will improve natural resource management or protect certain areas in the debtor country from environmentally harmful development.

deciduous plants Trees, such as oaks and maples, and other plants that survive during dry or cold seasons by shedding their leaves. Compare *coniferous trees, succulent plants*.

decomposer Organism that digests parts of dead organisms, and cast-off fragments and wastes of living organisms by breaking down the complex organic molecules in those materials into simpler inorganic compounds and then absorbing the soluble nutrients. Producers return most of these chemicals to the soil and water for reuse. Decomposers consist of various bacteria and fungi. Compare *consumer, detritivore, producer*.

defendant The party in a court case being charged with creating a harm. See *plaintiff* and *civil suit*.

deforestation Removal of trees from a forested area.

degree of urbanization Percentage of the population in the world, or in a country, living in urban areas. Compare *urban growth*.

democracy Government by the people through their elected officials and appointed representatives. In a *constitutional democracy*, a constitution provides the basis of government authority and puts restraints on government power through free elections and freely expressed public opinion.

demographic transition Hypothesis that countries, as they become industrialized, have declines in death rates followed by declines in birth rates.

density Mass per unit volume.

depletion time The time it takes to use a certain fraction (usually 80%) of the known or estimated supply of a nonrenewable resource at an assumed rate of use. Finding and extracting the remaining 20% usually costs more than it is worth.

desalination Purification of saltwater or brackish (slightly salty) water by removal of dissolved salts.

desert Biome in which evaporation exceeds precipitation and the average amount of precipitation is less than 25 centimeters (10 inches) per year. Such areas have little vegetation or have widely spaced, mostly low vegetation. Compare *forest, grassland*.

desertification Conversion of rangeland, rain-fed cropland, or irrigated cropland to desertlike land, with a drop in agricultural productivity of 10% or more. It usually is caused by a combination of overgrazing, soil erosion, prolonged drought, and climate change.

detritivore Consumer organism that feeds on detritus, parts of dead organisms, and cast-off fragments and wastes of living organisms. Examples include earthworms, termites, and crabs. Compare *decomposer*.

detritus Parts of dead organisms and cast-off fragments and wastes of living organisms.

detritus feeder See *detritivore*.

deuterium (D; hydrogen-2) Isotope of the element hydrogen, with a nucleus containing one proton and one neutron and a mass number of 2.

developed country See *more-developed country*.

developing country See *less-developed country*.

dieback Sharp reduction in the population of a species when its numbers exceed the carrying capacity of its habitat. See *carrying capacity*.

differential reproduction Phenomenon in which individuals with adaptive genetic traits produce more living offspring than do individuals without such traits. See *natural selection*.

dioxins Family of 75 chlorinated hydrocarbon compounds formed as unwanted by-products in chemical reactions involving chlorine and hydrocarbons, usually at high temperatures.

discount rate An estimate of a resource's future economic value compared to its present value; based on the idea that having something today may be worth more than it will be in the future.

dissolved oxygen (DO) content Amount of oxygen gas (O_2) dissolved in a given volume of water at a particular temperature and pressure, often expressed as a concentration in parts of oxygen per million parts of water.

disturbance An event that disrupts an ecosystem or community. Examples of *natural disturbances* include fires, hurricanes, tornadoes, droughts, and floods. Examples of *human-caused disturbances* include deforestation, overgrazing, and plowing.

divergent plate boundary Area where the earth's lithospheric plates move apart in opposite directions. Compare *convergent plate boundary, transform fault*.

DNA (deoxyribonucleic acid) Large molecules in the cells of living organisms that carry genetic information.

domesticated species Wild species tamed or genetically altered by crossbreeding for use by humans for food (cattle, sheep, and food crops), as pets (dogs and cats), or for enjoyment (animals in zoos and plants in botanical gardens). Compare *wild species*.

dose Amount of a potentially harmful substance an individual ingests, inhales, or absorbs through the skin. Compare *response. See dose-response curve, median lethal dose*.

dose-response curve Plot of data showing the effects of various doses of a toxic agent on a group of test organisms. See *dose, median lethal dose, response*.

doubling time Time it takes (usually in years) for the quantity of something growing exponentially to double. It can be calculated by dividing the annual percentage growth rate into 70.

drainage basin See *watershed*.

drift-net fishing Catching fish in huge nets that drift in the water.

drought Condition in which an area does not get enough water because of lower-than-normal precipitation or higher-than-normal temperatures that increase evaporation.

earthquake Shaking of the ground resulting from the fracturing and displacement of subsurface rock, which produces a fault, or from subsequent movement along the fault.

ecological diversity The variety of forests, deserts, grasslands, oceans, streams, lakes, and other biological communities interacting with one another and with their nonliving environment. See *biodiversity*. Compare *functional diversity, genetic diversity, species diversity*.

ecological footprint Amount of biologically productive land and water needed to supply a population with the renewable resources it uses and to absorb or dispose of the wastes from such resource use. It is a measure of the average environmental impact of populations in different countries and areas. See *per capita ecological footprint*.

ecological niche Total way of life or role of a species in an ecosystem. It includes all physical, chemical, and biological conditions that a species needs to live and reproduce in an ecosystem. See *fundamental niche, realized niche*.

ecological restoration Deliberate alteration of a degraded habitat or ecosystem to restore as much of its ecological structure and function as possible.

ecological succession Process in which communities of plant and animal species in a particular area are replaced over time by a series of different and often more complex communities. See *primary ecological succession, secondary ecological succession*.

ecological tipping point Point at which an environmental problem reaches a threshold level, which causes an often irreversible shift in the behavior of a natural system.

ecologist Biological scientist who studies relationships between living organisms and their environment.

ecology Biological science that studies the relationships between living organisms and their environment; study of the structure and functions of nature.

economic depletion Exhaustion of 80% of the estimated supply of a nonrenewable resource. Finding, extracting, and processing the remaining 20% usually costs more than it is worth. May also apply to the depletion of a renewable resource, such as a fish or tree species.

economic development Improvement of human living standards by economic growth. Compare *economic growth, environmentally sustainable economic development*.

economic growth Increase in the capacity to provide people with goods and services; an increase in gross domestic product (GDP). Compare *economic development, environmentally sustainable economic development*. See *gross domestic product*.

economic resources Natural resources, capital goods, and labor used in an economy to produce material goods and services. See *natural resources*.

economics Social science that deals with the production, distribution, and consumption of goods and services to satisfy people's needs and wants.

economic system Method that a group of people uses to choose which goods and services to produce, how to produce them, how much to produce, and how to distribute them to people.

economy System of production, distribution, and consumption of economic goods.

ecosphere See *biosphere*.

ecosystem One or more communities of different species interacting with one another and with the chemical and physical factors making up their nonliving environment.

ecosystem services Natural services or natural capital that support life on the earth and are essential to the quality of human life and the functioning of the world's economies. Examples are the chemical cycles, natural pest control, and natural purification of air and water. See *natural resources*.

electromagnetic radiation Forms of kinetic energy traveling as electromagnetic waves. Examples include radio waves, TV waves, microwaves, infrared radiation, visible light, ultraviolet radiation, X-rays, and gamma rays. Compare *ionizing radiation, nonionizing radiation*.

electron (e) Tiny particle moving around outside the nucleus of an atom. Each electron has one unit of negative charge and almost no mass. Compare *neutron, proton*.

element Chemical, such as hydrogen (H), iron (Fe), sodium (Na), carbon (C), nitrogen (N), or oxygen (O), whose distinctly different atoms serve as the basic building blocks of all matter. Two or more elements combine to form the compounds that make up most of the world's matter. Compare *compound*.

elevation Distance above sea level.

emigration Movement of people out of a specific geographic area. Compare *immigration, migration*.

endangered species Wild species with so few individual survivors that the species could soon become extinct in all or most of its natural range. Compare *threatened species*.

endemic species Species that is found in only one area. Such species are especially vulnerable to extinction.

energy Capacity to do work by performing mechanical, physical, chemical, or electrical tasks or to cause a heat transfer between two objects at different temperatures.

energy conservation Reducing or eliminating the unnecessary waste of energy.

energy efficiency Percentage of the total energy input that does useful work and is not converted into low-quality, generally useless heat in an energy conversion system or process. See *energy quality, net energy*. Compare *material efficiency*.

energy productivity See *energy efficiency*.

energy quality Ability of a form of energy to do useful work. High-temperature heat and the chemical energy in fossil fuels and nuclear fuels are concentrated high-quality energy. Low-quality energy such as low-temperature heat is dispersed or diluted and cannot do much useful work. See *high-quality energy, low-quality energy*.

environment All external conditions, factors, matter, and energy, living and nonliving, that affect any living organism or other specified system.

environmental degradation Depletion or destruction of a potentially renewable resource such as soil, grassland, forest, or wildlife that is used faster than it is naturally replenished. If such use continues, the resource becomes nonrenewable (on a human time scale) or nonexistent (extinct). See also *sustainable yield.*

environmental ethics Human beliefs about what is right or wrong with how we treat the environment.

environmentalism Social movement dedicated to protecting the earth's life-support systems for us and other species.

environmentalist Person who is concerned about the impacts of human activities on the environment.

environmental justice Fair treatment and meaningful involvement of all people, regardless of race, color, sex, national origin, or income, with respect to the development, implementation, and enforcement of environmental laws, regulations, and policies.

environmental law A body of laws and treaties that broadly define what is acceptable environmental behavior for individuals, groups, businesses, and nations.

environmentally sustainable economic development Development that meets the basic needs of the current generations of humans and other species without preventing future generations of humans and other species from meeting their basic needs. It is the economic component of an *environmentally sustainable society.* Compare *economic development, economic growth.*

environmentally sustainable society Society that meets the current and future needs of its people for basic resources in a just and equitable manner without compromising the ability of future generations of humans and other species from meeting their basic needs.

environmental movement Citizens organized to demand that political leaders enact laws and develop policies to curtail pollution, clean up polluted environments, and protect unspoiled areas from environmental degradation.

environmental policy Laws, rules, and regulations related to an environmental problem that are developed, implemented, and enforced by a particular government body or agency.

environmental resistance All of the limiting factors that act together to limit the growth of a population. See *biotic potential, limiting factor.*

environmental revolution Cultural change that includes halting population growth and altering lifestyles, political and economic systems, and the way we treat the environment with the goal of living more sustainably. It requires working with the rest of nature by learning more about how nature sustains itself.

environmental science Interdisciplinary study that uses information and ideas from the physical sciences (such as biology, chemistry, and geology) with those from the social sciences and humanities (such as economics, politics, and ethics) to learn how nature works, how we interact with the environment, and how we can to help deal with environmental problems.

environmental scientist Scientist who uses information from the physical sciences and social sciences to understand how the earth works, learn how humans interact with the earth, and develop solutions to environmental problems. See *environmental science.*

environmental wisdom worldview Worldview holding that humans are part of and totally dependent on nature and that nature exists for all species, not just for us. Our success depends on learning how the earth sustains itself and integrating such environmental wisdom into the ways we think and act. Compare *frontier worldview, planetary management worldview, stewardship worldview.*

environmental worldview Set of assumptions and beliefs about how people think the world works, what they think their role in the world should be, and what they believe is right and wrong environmental behavior (environmental ethics). See *environmental wisdom worldview, frontier worldview, planetary management worldview, stewardship worldview.*

EPA U.S. Environmental Protection Agency; responsible for managing federal efforts to control air and water pollution, radiation and pesticide hazards, environmental research, hazardous waste, and solid waste disposal.

epidemiology Study of the patterns of disease or other harmful effects from exposure to toxins and diseases caused by pathogens within defined groups of people to find out why some people get sick and some do not.

epiphyte Plant that uses its roots to attach itself to branches high in trees, especially in tropical forests.

erosion Process or group of processes by which loose or consolidated earth materials, especially topsoil, are dissolved, loosened, or worn away and removed from one place and deposited in another. See *weathering.*

estuary Partially enclosed coastal area at the mouth of a river where its freshwater, carrying fertile silt and runoff from the land, mixes with salty seawater.

eukaryotic cell Cell that is surrounded by a membrane and has a distinct nucleus. Compare *prokaryotic cell.*

euphotic zone Upper layer of a body of water through which sunlight can penetrate and support photosynthesis.

eutrophication Physical, chemical, and biological changes that take place after a lake, estuary, or slow-flowing stream receives inputs of plant nutrients—mostly nitrates and phosphates—from natural erosion and runoff from the surrounding land basin. See *cultural eutrophication.*

eutrophic lake Lake with a large or excessive supply of plant nutrients, mostly nitrates and phosphates. Compare *mesotrophic lake, oligotrophic lake.*

evaporation Conversion of a liquid into a gas.

evergreen plants Plants that keep some of their leaves or needles throughout the year. Examples include cone-bearing trees (conifers) such as firs, spruces, pines, redwoods, and sequoias. Compare *deciduous plants, succulent plants.*

evolution See *biological evolution.*

exhaustible resource See *nonrenewable resource.*

exotic species See *nonnative species.*

experiment Procedure a scientist uses to study some phenomenon under known conditions. Scientists conduct some experiments in the laboratory and others in nature. The resulting scientific data or facts must be verified or confirmed by repeated observations and measurements, ideally by several different investigators.

exponential growth Growth in which some quantity, such as population size or economic output, increases at a constant rate per unit of time. An example is the growth sequence 2, 4, 8, 16, 32, 64, and so on, which increases by 100% at each interval. When the increase in quantity over time is plotted, this type of growth yields a curve shaped like the letter J. Compare *linear growth.*

external benefit Beneficial social effect of producing and using an economic good that is not included in the market price of the good. Compare *external cost, full cost.*

external cost Harmful environmental, economic, or social effect of producing and using an economic good that is not included in the market price of the good. Compare *external benefit, full cost, internal cost.*

extinction See *biological extinction.*

extinction rate Percentage or number of species that go extinct within a certain period of time such as a year.

family planning Providing information, clinical services, and contraceptives to help people choose the number and spacing of children they want to have.

famine Widespread malnutrition and starvation in a particular area because of a shortage of food, usually caused by drought, war, flood, earthquake, or other catastrophic events that disrupt food production and distribution.

feedback Any process that increases (positive feedback) or decreases (negative feedback) a change to a system.

feedback loop Occurs when an output of matter, energy, or information is fed back into the system as an input and leads to changes in that system. See *positive feedback loop* and *negative feedback loop.*

feedlot Confined outdoor or indoor space used to raise hundreds to thousands of domesticated livestock.

fermentation See *anaerobic respiration.*

fertility rate Number of children born to an average woman in a population during her lifetime. Compare *replacement-level fertility.*

fertilizer Substance that adds inorganic or organic plant nutrients to soil and improves its ability to grow crops, trees, or other vegetation. See *commercial inorganic fertilizer, organic fertilizer.*

first law of thermodynamics Whenever energy is converted from one form to another in a physical or chemical change, no energy is created or destroyed, but energy can be changed from one form to another; you cannot get more energy out of something than you put in; in terms of energy quantity, you cannot get something for nothing. This law does not apply to nuclear changes, in which large amounts of energy can be produced from small amounts of matter. See *second law of thermodynamics.*

fishery Concentration of particular aquatic species suitable for commercial harvesting in a given ocean area or inland body of water.

fish farming See *aquaculture.*

fishprint Area of ocean needed to sustain the consumption of an average person, a nation, or the world. Compare *ecological footprint.*

fissionable isotope Isotope that can split apart when hit by a neutron at the right speed and thus undergo nuclear fission. Examples include uranium-235 and plutonium-239.

floodplain Flat valley floor next to a stream channel. For legal purposes, the term often applies to any low area that has the potential for flooding, including certain coastal areas.

flows See *throughputs.*

food chain Series of organisms in which each eats or decomposes the preceding one. Compare *food web.*

food insecurity Condition under which people live with chronic hunger and malnutrition that threatens their ability to lead healthy and productive lives. Compare *food security.*

food security Condition under which every person in a given area has daily access to enough nutritious food to have an active and healthy life. Compare *food insecurity.*

food web Complex network of many interconnected food chains and feeding relationships. Compare *food chain.*

forest Biome with enough average annual precipitation to support the growth of tree species and smaller forms of vegetation. Compare *desert, grassland.*

fossil fuel Products of partial or complete decomposition of plants and animals; occurs as crude oil, coal, natural gas, or heavy oils as a result of exposure to heat and pressure in the earth's crust over millions of years. See *coal, crude oil, natural gas.*

fossils Skeletons, bones, shells, body parts, leaves, seeds, or impressions of such items that provide recognizable evidence of organisms that lived long ago.

foundation species Species that plays a major role in shaping a community by creating and enhancing a habitat that benefits other species. Compare *indicator species, keystone species, native species, nonnative species.*

fracking Freeing oil or natural gas that is tightly held in rock deposits by using perforated drilling well tubes with explosive charges to create fissures in rock and then using high pressure pumps to shoot a mixture of water, sand, and chemicals into the well to hold the rock fractures open and release the oil or natural gas, which flows back to the surface along with a mixture of water, sand, fracking chemicals, and other chemicals (some of them hazardous) that are released from the rock. See *horizontal drilling.*

free-access resource See *open-access renewable resource.*

Freons See *chlorofluorocarbons.*

freshwater Water that contains very low levels of dissolved salts.

freshwater life zones Aquatic systems where water with a dissolved salt concentration of less than 1% by volume accumulates on or flows through the surfaces of terrestrial biomes. Examples include *standing* (lentic) bodies of freshwater such as lakes, ponds, and inland wetlands and *flowing* (lotic) systems such as streams and rivers. Compare *biome.*

front The boundary between two air masses with different temperatures and densities. See *cold front, warm front.*

frontier science See *tentative science.*

frontier worldview View held by European colonists settling North America in the 1600s that the continent had vast resources and was a wilderness to be conquered by settlers clearing and planting land.

full-cost pricing Finding ways to include the harmful environmental and health costs of producing and using goods in their market prices. See *external cost, internal cost.*

functional diversity Biological and chemical processes or functions such as energy flow and matter cycling needed for the survival of species and biological communities. See *biodiversity, ecological diversity, genetic diversity, species diversity.*

fungicide Chemical that kills fungi.

game species Type of wild animal that people hunt or fish as a food source or for sport or recreation.

gamma ray Form of ionizing electromagnetic radiation with a high energy content emitted by some radioisotopes. It readily penetrates body tissues. See *alpha particle, beta particle.*

GDP See *gross domestic product.*

gene mutation See *mutation.*

gene pool Sum total of all genes found in the individuals of the population of a particular species.

generalist species Species with a broad ecological niche. They can live in many different places, eat a variety of foods, and tolerate a wide range of environmental conditions. Examples include flies, cockroaches, mice, rats, and humans. Compare *specialist species.*

genes Coded units of information about specific traits that are passed from parents to offspring during reproduction. They consist of segments of DNA molecules found in chromosomes.

gene splicing See *genetic engineering.*

genetic adaptation Changes in the genetic makeup of organisms of a species that allow the species to reproduce and gain a competitive advantage under changed environmental conditions. See *differential reproduction, evolution, mutation, natural selection.*

genetically modified organism (GMO) Organism whose genetic makeup has been altered by genetic engineering.

genetic diversity Variability in the genetic makeup among individuals within a single species. See *biodiversity.* Compare *ecological diversity, functional diversity, species diversity.*

genetic engineering Insertion of an alien gene into an organism to give it a beneficial genetic trait. Compare *artificial selection, natural selection.*

genuine progress indicator (GPI) GDP plus the estimated value of beneficial transactions that meet basic needs, but in which no money changes hands, minus the estimated harmful environmental, health, and social costs of all transactions. Compare *gross domestic product.*

geographic isolation Separation of populations of a species into different areas for long periods of time.

geology Study of the earth's dynamic history. Geologists study and analyze rocks and the features and processes of the earth's interior and surface.

geosphere Earth's intensely hot core, thick mantle composed mostly of rock, and thin outer crust that contains most of the earth's rock, soil, and sediment. Compare *atmosphere, biosphere, hydrosphere.*

geothermal energy Heat transferred from the earth's underground concentrations of dry steam (steam with no water droplets), wet steam (a mixture of steam and water droplets), or hot water trapped in fractured or porous rock.

global climate change Broad term referring to long-term changes in any aspects of the earth's climate, especially temperature and precipitation. Compare *weather.*

global warming Warming of the earth's lower atmosphere (troposphere) because of increases in the concentrations of one or more greenhouse gases. It can result in climate change that can last for decades to thousands of years. See *greenhouse effect, greenhouse gases, natural greenhouse effect.*

GMO See *genetically modified organism.*

GPI See *genuine progress indicator.*

GPP See *gross primary productivity.*

grassland Biome found in regions where there is enough annual average precipitation to support the growth of grass and small plants but not enough to support large stands of trees. Compare *desert, forest.*

greenhouse effect Natural effect that releases heat in the atmosphere near the earth's surface. Water vapor, carbon dioxide, ozone, and other gases in the lower atmosphere (troposphere) absorb some of the infrared radiation (heat) radiated by the earth's surface. Their molecules vibrate and transform the absorbed energy into longer-wavelength infrared radiation in the troposphere. If the atmospheric concentrations of these greenhouse gases increase and other natural processes do not remove them, the average temperature of the lower atmosphere will increase. Compare *global warming.*

greenhouse gases Gases in the earth's lower atmosphere (troposphere) that cause the greenhouse effect. Examples include carbon dioxide, chlorofluorocarbons, ozone, methane, water vapor, and nitrous oxide.

green manure Freshly cut or still-growing green vegetation that is plowed into the soil to increase the organic matter and humus available to support crop growth. Compare *animal manure.*

green revolution Popular term for the introduction of scientifically bred or selected varieties of grain (rice, wheat, maize) that, with adequate inputs of fertilizer and water, can greatly increase crop yields.

greenwashing Deceptive practice that some businesses use to spin environmentally harmful products as green, clean, or environmentally beneficial.

gross domestic product (GDP) Annual market value of all goods and services produced by all firms and organizations, foreign and domestic, operating within a country. See *per capita GDP.* Compare genuine progress indicator (GPI).

gross primary productivity (GPP) Rate at which an ecosystem's producers capture and store a given amount of chemical energy as biomass in a given length of time. Compare *net primary productivity.*

ground fire Fire that burns decayed leaves or peat deep below the ground's surface. Compare *crown fire, surface fire.*

groundwater Water that sinks into the soil and is stored in slowly flowing and slowly renewed underground reservoirs called *aquifers;* underground water in the zone of saturation, below the water table. Compare *runoff, surface water.*

habitat Place or type of place where an organism or population of organisms lives. Compare *ecological niche.*

habitat fragmentation Breakup of a habitat into smaller pieces, usually as a result of human activities.

hazard Something that can cause injury, disease, economic loss, or environmental damage. See also *risk.*

hazardous chemical Chemical that can cause harm because it is flammable or explosive, can irritate or damage the skin or lungs (such as strong acidic or alkaline substances), or can cause allergic reactions of the immune system (allergens). See also *toxic chemical.*

hazardous waste Any solid, liquid, or containerized gas that can catch fire easily, is corrosive to skin tissue or metals, is unstable and can explode or release toxic fumes, or has harmful concentrations of one or more toxic materials that can leach out. These substances are usually by-products of manufacturing processes. See also *toxic waste.*

heat Total kinetic energy of all randomly moving atoms, ions, or molecules within a given substance, excluding the overall motion of the whole object. Heat always flows spontaneously from a warmer sample of matter to a colder sample of matter. This is one way to state the *second law of thermodynamics.* Compare *temperature.*

herbicide Chemical that kills a plant or inhibits its growth.

herbivore Plant-eating organism. Examples include deer, sheep, grasshoppers, and zooplankton. Compare *carnivore, omnivore.*

heterotroph See *consumer.*

high Air mass with a high pressure. Compare *low.*

high-grade ore Ore containing a large amount of a desired mineral. Compare *low-grade ore.*

high-input agriculture See *industrialized agriculture.*

high-quality energy Energy that is concentrated and has great ability to perform useful work. Examples include high-temperature heat and the energy in electricity, coal, oil, gasoline, sunlight, and nuclei of uranium-235. Compare *low-quality energy.*

high-quality matter Matter that is concentrated and contains a high concentration of a useful resource. Compare *low-quality matter.*

high-throughput economy Economic system in most advanced industrialized countries, in which ever-increasing economic growth is sustained by maximizing the rate at which matter and energy resources are used, with little emphasis on pollution prevention, recycling, reuse, reduction of unnecessary waste, and other forms of resource conservation. Compare *low-throughput economy, matter-recycling economy.*

high-waste economy See *high-throughput economy.*

HIPPCO Acronym used by conservation biologists for the six most important secondary causes of premature extinction: Habitat destruction, degradation, and fragmentation; Invasive (nonnative) species; Population growth (too many people consuming too many resources); Pollution; Climate change; and Overexploitation.

Holocene A geological period of relatively stable climate and other environmental conditions following the last glacial period. It began about 12,000 years ago. Compare *Anthropocene.*

horizontal drilling Method for extracting oil or natural gas from underground deposits by first drilling down and then using a flexible drilling bore to drill horizontally to gain greater access to oil and gas deposits. See *fracking.*

host Plant or animal on which a parasite feeds.

human capital People's physical and mental talents that provide labor, innovation, culture, and organization. Compare *manufactured capital, natural capital.*

human resources See *human capital.*

humus Slightly soluble residue of undigested or partially decomposed organic material in topsoil. This material helps retain water and water-soluble nutrients, which can be taken up by plant roots.

hunger See *chronic undernutrition.*

hunter–gatherers People who get their food by gathering edible wild plants and other materials and by hunting wild animals and catching fish.

hydraulic fracturing See *fracking.*

hydrocarbon Organic compound made of hydrogen and carbon atoms. The simplest hydrocarbon is methane (CH_4), the major component of natural gas.

hydroelectric power plant Structure in which the energy of falling or flowing water spins a turbine generator to produce electricity.

hydrologic cycle Biogeochemical cycle that collects, purifies, and distributes the earth's fixed supply of water from the environment to living organisms and then back to the environment.

hydroponics Form of agriculture in which farmers grow plants by exposing their roots to a nutrient-rich water solution instead of soil.

hydropower Electrical energy produced by falling or flowing water. See *hydroelectric power plant.*

hydrosphere Earth's *liquid water* (oceans, lakes, other bodies of surface water, and underground water), *frozen water* (polar ice caps, floating ice caps, and ice in soil, known as permafrost), and *water vapor* in the atmosphere. See also *hydrologic cycle.* Compare *atmosphere, biosphere, geosphere.*

igneous rock Rock formed when molten rock material (magma) wells up from the earth's interior, cools, and solidifies into rock masses. Compare *metamorphic rock, sedimentary rock.* See *rock cycle.*

immature community Community at an early stage of ecological succession. It usually has a low number of species and ecological niches and cannot capture and use energy and cycle critical nutrients as efficiently as more complex, mature communities. Compare *mature community.*

immigrant species See *nonnative species.*

immigration Migration of people into a country or area to take up permanent residence. Compare *emigration.*

indicator species Species whose decline serves as early warnings that a community or ecosystem is being degraded. Compare *foundation species, keystone species, native species, nonnative species.*

industrialized agriculture Production of large quantities of crops and livestock for domestic and foreign sale; involves use of large inputs of energy from fossil fuels (especially oil and natural gas), water, fertilizer, and pesticides. Compare *subsistence farming.*

industrial smog Type of air pollution consisting mostly of a mixture of sulfur dioxide, suspended droplets of sulfuric acid formed from some of the sulfur dioxide, and suspended solid particles. Compare *photochemical smog.*

industrial solid waste Solid waste produced by mines, factories, refineries, food growers, and businesses that supply people with goods and services. Compare *municipal solid waste*

inertia See *persistence.*

inexhaustible resource See *perpetual resource.* Compare *nonrenewable resource, renewable resource.*

infant mortality rate Number of babies out of every 1,000 born each year who die before their first birthday.

infectious disease Disease caused when a pathogen such as a bacterium, virus, or parasite invades the body and multiplies in its cells and tissues. Examples are flu, HIV, malaria, tuberculosis, and measles. See *transmissible disease.* Compare *nontransmissible disease.*

infiltration Downward movement of water through soil.

inherent value See *intrinsic value.*

inland wetland Land away from the coast, such as a swamp, marsh, or bog, that is covered all or part of the time with freshwater. Compare *coastal wetland.*

inorganic compounds All compounds not classified as organic compounds. See *organic compounds.*

inorganic fertilizer See *commercial inorganic fertilizer.*

input Matter, energy, or information entering a system. Compare *output, throughput.*

input pollution control See *pollution prevention*.

insecticide Chemical that kills insects.

instrumental value Value of an organism, species, ecosystem, or the earth's biodiversity based on its usefulness to humans. Compare *intrinsic value*.

integrated pest management (IPM) Combined use of biological, chemical, and cultivation methods in proper sequence and timing to keep the size of a pest population below the level that causes economically unacceptable loss of a crop or livestock animal.

integrated waste management Variety of strategies for both waste reduction and waste management designed to deal with the solid wastes we produce.

intercropping Growing two or more different crops at the same time on a plot. For example, a carbohydrate-rich grain that depletes soil nitrogen and a protein-rich legume that adds nitrogen to the soil may be intercropped. Compare *monoculture, polyculture*.

internal cost Direct cost paid by the producer and the buyer of an economic good. Compare *external benefit, external cost, full cost*.

interspecific competition Attempts by members of two or more species to use the same limited resources in an ecosystem. See *competition, intraspecific competition*.

intertidal zone The area of shoreline between low and high tides.

intraspecific competition Attempts by two or more organisms of a single species to use the same limited resources in an ecosystem. See *competition, interspecific competition*.

intrinsic rate of increase (*r*) Rate at which a population could grow if it had unlimited resources. Compare *environmental resistance*.

intrinsic value Value of an organism, species, ecosystem, or the earth's biodiversity based on its existence, regardless of whether it has any usefulness to humans. Compare *instrumental value*.

invasive species See *nonnative species*.

inversion See *temperature inversion*.

invertebrates Animals that have no backbones. Compare *vertebrates*.

ion Atom or group of atoms with one or more positive (+) or negative (–) electrical charges. Examples are Na^+ and Cl^-. Compare *atom, molecule*.

ionizing radiation Fast-moving alpha or beta particles or high-energy radiation (gamma rays) emitted by radioisotopes. They have enough energy to dislodge one or more electrons from the atoms they hit, thereby forming charged ions in tissue that can react with and damage living tissue. Compare *nonionizing radiation*.

IPM See *integrated pest management*.

irrigation Mix of methods used to supply water to crops by artificial means.

isotopes Two or more forms of a chemical element that have the same number of protons but different mass numbers because they have different numbers of neutrons in their nuclei.

J-shaped curve Curve with a shape similar to that of the letter J; can represent prolonged exponential growth. See *exponential growth*.

junk science See *unreliable science*.

kerogen Solid, waxy mixture of hydrocarbons found in oil shale rock. Heating the rock to high temperatures causes the kerogen to vaporize. The vapor is condensed, purified, and then sent to a refinery to produce gasoline, heating oil, and other products. See also *oil shale, shale oil*.

keystone species Species that play roles affecting many other organisms in an ecosystem. Compare *foundation species, indicator species, native species, nonnative species*.

kilocalorie (kcal) Unit of energy equal to 1,000 calories. See *calorie*.

kilowatt (kW) Unit of electrical power equal to 1,000 watts. See *watt*.

kinetic energy Energy that matter has because of its mass and speed, or velocity. Compare *potential energy*.

lake Large natural body of standing freshwater formed when water from precipitation, land runoff, or groundwater flow fills a depression in the earth created by glaciation, earth movement, volcanic activity, or a giant meteorite. See *eutrophic lake, mesotrophic lake, oligotrophic lake*.

land degradation Decrease in the ability of land to support crops, livestock, or wild species in the future as a result of natural or human-induced processes.

landfill See *sanitary landfill*.

land-use planning Planning to determine the best present and future uses of each parcel of land.

latitude Distance from the equator. Compare *altitude*.

law of conservation of energy See *first law of thermodynamics*.

law of conservation of matter In any physical or chemical change, matter is neither created nor destroyed but merely changed from one form to another; in physical and chemical changes, existing atoms are rearranged into different spatial patterns (physical changes) or different combinations (chemical changes).

law of nature See *scientific law*.

law of tolerance Existence, abundance, and distribution of a species in an ecosystem are determined by whether the levels of one or more physical or chemical factors fall within the range tolerated by the species. See *threshold effect*.

LD50 See *median lethal dose*.

LDC See *less-developed country*. Compare more-developed country.

leaching Process in which various chemicals in upper layers of soil are dissolved and carried to lower layers and, in some cases, to groundwater.

less-developed country Country that has low to moderate industrialization and low to moderate per capita GDP. Most are located in Africa, Asia, and Latin America. Compare *more-developed country*.

life-cycle cost Initial cost plus lifetime operating costs of an economic good. Compare *full cost*.

life expectancy Average number of years a newborn infant can be expected to live.

limiting factor Single factor that limits the growth, abundance, or distribution of the population of a species in an ecosystem. See *limiting factor principle*.

limiting factor principle Too much or too little of any abiotic factor can limit or prevent growth of a population of a species in an ecosystem, even if all other factors are at or near the optimal range of tolerance for the species.

linear growth Growth in which a quantity increases by some fixed amount during each unit of time. An example is growth that increases by 2 units in the sequence 2, 4, 6, 8, 10, and so on. Compare *exponential growth*.

liquefied natural gas (LNG) Natural gas converted to liquid form by cooling it to a very low temperature.

liquefied petroleum gas (LPG) Mixture of liquefied propane (C_3H_8) and butane (C_4H_{10}) gas removed from natural gas and used as a fuel.

lithosphere Outer shell of the earth, composed of the crust and the rigid, outermost part of the mantle outside the asthenosphere; material found in the earth's plates. See *crust, geosphere, mantle*.

LNG See *liquefied natural gas*.

lobbying Process in which individuals or groups use public pressure, personal contacts, and political action to persuade legislators to vote or act in their favor.

logistic growth Pattern in which exponential population growth occurs when the population is small, and population growth decreases steadily with time as the population approaches the carrying capacity. See *S-shaped curve*.

low Air mass with a low pressure. Compare *high*.

low-grade ore Ore containing a small amount of a desired mineral. Compare *high-grade ore*.

low-input agriculture See *sustainable agriculture*.

low-quality energy Energy that is dispersed and has little ability to do useful work. An example is low-temperature heat. Compare *high-quality energy*.

low-quality matter Matter that is dilute or dispersed or contains a low concentration of a useful resource. Compare *high-quality matter*.

low-throughput economy Economy based on working with nature by recycling and reusing discarded matter; preventing pollution; conserving matter and energy resources by reducing unnecessary waste and use; and building things that are easy to recycle, reuse, and repair. Compare *high-throughput economy, matter-recycling economy*.

low-waste economy See *low-throughput economy*.

LPG See *liquefied petroleum gas*.

magma Molten rock below the earth's surface.

malnutrition See *chronic malnutrition*.

mangrove swamps Swamps found on the coastlines in warm tropical climates. They are dominated by mangrove trees, any of about 55 species of trees and shrubs that can live partly submerged in the salty environment of coastal swamps.

mantle Zone of the earth's interior between its core and its crust. Compare *core, crust*. See *geosphere, lithosphere*.

manufactured capital See *manufactured resources*.

manufactured inorganic fertilizer See *commercial inorganic fertilizer*.

manufactured resources Manufactured items made from natural resources and used to produce and distribute economic goods and services bought by consumers. They include tools, machinery, equipment, factory buildings, and transportation and distribution facilities. Compare *human resources, natural resources*.

manure See *animal manure, green manure*.

marine life zone See *saltwater life zone*.

mass Amount of material in an object.

mass extinction Catastrophic, widespread, often global event in which major groups of species are wiped out over a short time compared with normal (background) extinctions. Compare *background extinction*.

mass number Sum of the number of neutrons (n) and the number of protons (p) in the nucleus of an atom. It gives the approximate mass of that atom. Compare *atomic number*.

mass transit Buses, trains, trolleys, and other forms of transportation that carry large numbers of people.

material efficiency Total amount of material needed to produce each unit of goods or services. Also called *resource productivity*. Compare *energy efficiency*.

matter Anything that has mass (the amount of material in an object) and takes up space. On the earth, where gravity is present, we weigh an object to determine its mass.

matter quality Measure of how useful a matter resource is, based on its availability and concentration. See *high-quality matter, low-quality matter*.

matter-recycling-and-reuse economy Economy that emphasizes recycling the maximum amount of all resources that can be recycled and reused. The goal is to allow economic growth to continue without depleting matter resources and without producing excessive pollution and environmental degradation. Compare *high-throughput economy, low-throughput economy*.

mature community Fairly stable, self-sustaining community in an advanced stage of ecological succession; usually has a diverse array of species and ecological niches; captures and uses energy and cycles critical chemicals more efficiently than simpler, immature communities. Compare *immature community*.

maximum sustainable yield See *sustainable yield*.

MDC See *more-developed country*.

median lethal dose (LD50) Amount of a toxic material per unit of body weight of test animals that kills half the test population in a certain time.

megacity City with 10 million or more people.

meltdown Melting of the highly radioactive core of a nuclear reactor.

mesotrophic lake Lake with a moderate supply of plant nutrients. Compare *eutrophic lake, oligotrophic lake*.

metabolism Ability of a living cell or organism to capture and transform matter and energy from its environment to supply its needs for survival, growth, and reproduction.

metamorphic rock Rock produced when a preexisting rock is subjected to high temperatures (which may cause it to melt partially), high pressures, chemically active fluids, or a combination of these agents. Compare *igneous rock, sedimentary rock*. See *rock cycle*.

metastasis Spread of malignant (cancerous) cells from a tumor to other parts of the body.

metropolitan area See *urban area*.

microorganisms Organisms such as bacteria that are so small that it takes a microscope to see them.

micropower systems Systems of small-scale decentralized units that generate 1–10,000 kilowatts of electricity. Examples include microturbines, fuel cells, wind turbines, and household solar-cell panels and solar-cell roofs.

migration Movement of people into and out of specific geographic areas. Compare *emigration* and *immigration*.

mineral Any naturally occurring inorganic substance found in the earth's crust as a crystalline solid. See *mineral resource*.

mineral resource Concentration of naturally occurring solid, liquid, or gaseous material in or on the earth's crust in a form and amount such that extracting and converting it into useful materials or items is currently or potentially profitable. Mineral resources are classified as *metallic* (such as iron and tin ores) or *nonmetallic* (such as fossil fuels, sand, and salt).

minimum-tillage farming See *conservation-tillage farming.*

mixture Combination of one or more elements and compounds.

model Approximate representation or simulation of a system being studied.

molecule Combination of two or more atoms of the same chemical element (such as O_2) or different chemical elements (such as H_2O) held together by chemical bonds. Compare *atom, ion.*

monoculture Cultivation of a single crop, usually on a large area of land. Compare *polyculture.*

more-developed country Country that is highly industrialized and has a high per capita GDP. Compare *less-developed country.*

mountaintop removal mining Type of surface mining that uses explosives, massive power shovels, and large machines called draglines to remove the top of a mountain and expose seams of coal underneath a mountain. Compare *area strip mining, contour strip mining.*

MSW See *municipal solid waste.*

multiple use Use of an ecosystem such as a forest for a variety of purposes such as timber harvesting, wildlife habitat, watershed protection, and recreation. Compare *sustainable yield.*

municipal solid waste (MSW) Solid materials discarded by homes and businesses in or near urban areas. See *solid waste.* Compare *industrial solid waste.*

mutagen Chemical or form of radiation that causes inheritable changes (mutations) in the DNA molecules in genes. See *carcinogen, mutation, teratogen.*

mutation Random change in DNA molecules making up genes that can alter anatomy, physiology, or behavior in offspring. See *mutagen.*

mutualism Type of species interaction in which both participating species generally benefit. Compare *commensalism.*

nanotechnology The use of science and engineering to manipulate and create materials out of atoms and molecules at the ultra-small scale of less than 100 nanometers. A nanometer is one-millionth of a meter.

native species Species that normally live and thrive in a particular ecosystem. Compare *foundation species, indicator species, keystone species, nonnative species.*

natural capital Natural resources and natural services that keep us and other species alive and support our economies. See *natural resources, natural services.*

natural capital degradation See *environmental degradation.*

natural gas Underground deposits of gases consisting of 50–90% by weight methane gas (CH_4) and small amounts of heavier gaseous hydrocarbon compounds such as propane (C_3H_8) and butane (C_4H_{10}).

natural greenhouse effect See *greenhouse effect.*

natural income Renewable resources such as plants, animals, and soil provided by natural capital.

natural law See *scientific law.*

natural radioactive decay Nuclear change in which unstable nuclei of atoms spontaneously shoot out particles (usually alpha or beta particles) or energy (gamma rays) at a fixed rate.

natural rate of extinction See *background extinction.*

natural recharge Natural replenishment of an aquifer by precipitation, which percolates downward through soil and rock. See *recharge area.*

natural resources Materials such as air, water, and soil and energy in nature that are essential or useful to humans. See *natural capital.*

natural selection Process by which a particular beneficial gene (or set of genes) is reproduced in succeeding generations more than other genes. The result of natural selection is a population that contains a greater proportion of organisms better adapted to certain environmental conditions. See *adaptation, biological evolution, differential reproduction, mutation.*

natural services Processes of nature, such as purification of air and water and pest control, which support life and human economies. See *natural capital.*

negative feedback loop Feedback loop that causes a system to change in the opposite direction from which is it moving. Compare *positive feedback loop.*

nekton Strongly swimming organisms found in aquatic systems. Compare *benthos, plankton.*

net energy yield Total amount of useful energy available from an energy resource or energy system over its lifetime, minus the amount of energy *used* (the first energy law), *automatically wasted* (the second energy law), and *unnecessarily wasted* in finding, processing, concentrating, and transporting it to users.

net primary productivity (NPP) Rate at which all the plants in an ecosystem produce net useful chemical energy; equal to the difference between the rate at which the plants in an ecosystem produce useful chemical energy (gross primary productivity) and the rate at which they use some of that energy through cellular respiration. Compare *gross primary productivity.*

neurotoxin Chemical that can harm the human *nervous system* (brain, spinal cord, peripheral nerves).

neutral solution Water solution containing an equal number of hydrogen ions (H^+) and hydroxide ions (OH^-); water solution with a pH of 7. Compare *acid solution, basic solution.*

neutron (n) Elementary particle in the nuclei of all atoms (except hydrogen-1). It has a relative mass of 1 and no electric charge. Compare *electron, proton.*

niche See *ecological niche.*

nitric oxide (NO) Colorless gas that forms when nitrogen and oxygen gas in air react at the high-combustion temperatures in automobile engines and coal-burning plants. Lightning and certain bacteria in soil and water also produce NO as part of the *nitrogen cycle.*

nitrogen cycle Cyclic movement of nitrogen in different chemical forms from the environment to organisms and then back to the environment.

nitrogen dioxide (NO_2) Reddish-brown gas formed when nitrogen oxide reacts with oxygen in the air.

nitrogen fixation Conversion of atmospheric nitrogen gas, by lightning, bacteria, and cyanobacteria, into forms useful to plants; it is part of the nitrogen cycle.

nitrogen oxides (NO_x) See *nitric oxide* and *nitrogen dioxide.*

noise pollution Any unwanted, disturbing, or harmful sound that impairs or interferes with hearing, causes stress, hampers concentration and work efficiency, or causes accidents.

nondegradable pollutant Material that is not broken down by natural processes. Examples include the toxic elements lead and mercury. Compare *biodegradable pollutant.*

nonionizing radiation Forms of radiant energy such as radio waves, microwaves, infrared light, and ordinary light that do not have enough energy to cause ionization of atoms in living tissue. Compare *ionizing radiation.*

nonnative species Species that migrate into an ecosystem or are deliberately or accidentally introduced into an ecosystem by humans. Compare *native species.*

nonpoint sources Broad and diffuse areas, rather than points, from which pollutants enter bodies of surface water or air. Examples include runoff of chemicals and sediments from cropland, livestock feedlots, logged forests, urban streets, parking lots, lawns, and golf courses. Compare *point source.*

nonrenewable energy Energy from resources that can be depleted and are not replenished by natural processes within a human time scale. Examples are energy produced by the burning of oil, coal, and natural gas, and nuclear energy released when the nuclei of heavy elements such as uranium are split apart (nuclear fission) or when the nuclei of light atoms such as hydrogen are forced together (nuclear fusion). Compare *renewable energy.*

nonrenewable resource Resource that exists in a fixed amount (stock) in the earth's crust and has the potential for renewal by geological, physical, and chemical processes taking place over hundreds of millions to billions of years. Examples include copper, aluminum, coal, and oil. We classify these resources as exhaustible because we are extracting and using them at a much faster rate than they are formed. Compare *renewable resource.*

nontransmissible disease Disease that is not caused by living organisms and does not spread from one person to another. Examples include most cancers, diabetes, cardiovascular disease, and malnutrition. Compare *transmissible disease.*

no-till farming See *conservation-tillage farming.*

NPP See *net primary productivity.*

nuclear change Process in which nuclei of certain isotopes spontaneously change, or are forced to change, into one or more different isotopes. The three principal types of nuclear change are natural radioactivity, nuclear fission, and nuclear fusion. Compare *chemical change, physical change.*

nuclear energy Energy released when atomic nuclei undergo a nuclear reaction such as the spontaneous emission of radioactivity, nuclear fission, or nuclear fusion.

nuclear fission Nuclear change in which the nuclei of certain isotopes with large mass numbers (such as uranium-235 and plutonium-239) are split apart into lighter nuclei when struck by a neutron. This process releases more neutrons and a large amount of energy. Compare *nuclear fusion.*

nuclear fuel cycle Includes the mining of uranium, processing and enriching the uranium to make nuclear fuel, using it in the reactor, safely storing the resulting highly radioactive wastes for thousands of years until their radioactivity falls to safe levels, and retiring the highly radioactive nuclear plant by taking it apart and storing its high- and moderate-level radioactive material safely for thousands of years.

nuclear fusion Nuclear change in which two nuclei of isotopes of elements with a low mass number (such as hydrogen-2 and hydrogen-3) are forced together at extremely high temperatures until they fuse to form a heavier nucleus (such as helium-4). This process releases a large amount of energy. Compare *nuclear fission.*

nucleus Extremely tiny center of an atom, making up most of the atom's mass. It contains one or more positively charged protons and one or more neutrons with no electrical charge (except for a hydrogen-1 atom, which has one proton and no neutrons in its nucleus).

nutrient Any chemical an organism must take in to live, grow, or reproduce.

nutrient cycle See *biogeochemical cycle.*

nutrient cycling The circulation of chemicals necessary for life, from the environment (mostly from soil and water) through organisms and back to the environment.

ocean acidification Increasing levels of acid in world's oceans due to their absorption of much of the CO_2 emitted into the atmosphere by human activities, especially the burning of carbon-containing fossil fuels. The CO_2 reacts with ocean water to form a weak acid and decreases the levels of carbonate ions (CO_3^{2-}) needed to form coral and the shells and skeletons of organisms such as crabs, oysters, and some phytoplankton.

ocean currents Mass movements of surface water produced by prevailing winds blowing over the oceans.

oil See *crude oil.*

oil reserves See *proven oil reserves.*

oil sand See *tar sand.*

oil shale Fine-grained rock containing various amounts of kerogen, a solid, waxy mixture of hydrocarbon compounds. Heating the rock to high temperatures converts the kerogen into a vapor that can be condensed to form a slow-flowing heavy oil called shale oil. See *kerogen, shale oil.*

old-growth forest Virgin and old, second-growth forests containing trees that are often hundreds—sometimes thousands—of years old. Examples include forests of Douglas fir, western hemlock, giant sequoia, and coastal redwoods in the western United States. Compare *second-growth forest, tree plantation.*

oligotrophic lake Lake with a low supply of plant nutrients. Compare *eutrophic lake, mesotrophic lake.*

omnivore Animal that can use both plants and other animals as food sources. Examples include pigs, rats, cockroaches, and humans. Compare *carnivore, herbivore.*

open-access renewable resource Renewable resource owned by no one and available for use by anyone at little or no charge. Examples include clean air, underground water supplies, the open ocean and its fish, and the ozone layer. Compare *common-property resource.*

open dump Fields or holes in the ground where garbage is deposited and sometimes covered with soil. They are rare in developed countries, but are widely used in many developing countries, especially to handle wastes from megacities. Compare *sanitary landfill.*

open-pit mining Removing minerals such as gravel, sand, and metal ores by digging them out of the earth's surface and leaving an open pit behind. Compare *area strip mining, contour strip mining, mountaintop removal, subsurface mining.*

open sea Part of an ocean that lies beyond the continental shelf. Compare *coastal zone*.

ore Part of a metal-yielding material that can be economically extracted from a mineral; typically containing two parts: the ore mineral, which contains the desired metal, and waste mineral material (gangue). See *high-grade ore, low-grade ore*.

organic agriculture Growing crops with limited or no use of synthetic pesticides and synthetic fertilizers; genetically modified crops, raising livestock without use of synthetic growth regulators and feed additives; and using organic fertilizer (manure, legumes, compost) and natural pest controls (bugs that eat harmful bugs, plants that repel bugs and environmental controls such as crop rotation). See *sustainable agriculture*.

organic compounds Compounds containing carbon atoms combined with each other and with atoms of one or more other elements such as hydrogen, oxygen, nitrogen, sulfur, phosphorus, chlorine, and fluorine. All other compounds are called *inorganic compounds*.

organic farming See *organic agriculture* and *sustainable agriculture*.

organic fertilizer Organic material such as animal manure, green manure, and compost applied to cropland as a source of plant nutrients. Compare *commercial inorganic fertilizer*.

organism Any form of life.

output Matter, energy, or information leaving a system. Compare *input, throughput*.

output pollution control See *pollution cleanup*.

overburden Layer of soil and rock overlying a mineral deposit. Surface mining removes this layer.

overfishing Harvesting so many fish of a species, especially immature individuals, that not enough breeding stock is left to replenish the species and it becomes unprofitable to harvest them.

overgrazing Destruction of vegetation when too many grazing animals feed too long on a specific area of pasture or rangeland and exceed the carrying capacity of a rangeland or pasture area.

overnutrition Diet so high in calories, saturated (animal) fats, salt, sugar, and processed foods, and so low in vegetables and fruits that the consumer runs a high risk of developing diabetes, hypertension, heart disease, and other health hazards. Compare *malnutrition, undernutrition*.

oxygen-demanding wastes Organic materials that are usually biodegraded by aerobic (oxygen-consuming) bacteria if there is enough dissolved oxygen in the water.

ozone (O_3) Colorless and highly reactive gas and a major component of photochemical smog. Also found in the ozone layer in the stratosphere. See *photochemical smog*.

ozone depletion Decrease in concentration of ozone (O_3) in the stratosphere. See *ozone layer*.

ozone layer Layer of gaseous ozone (O_3) in the stratosphere that protects life on earth by filtering out most harmful ultraviolet radiation from the sun.

PANs Peroxyacyl nitrates; group of chemicals found in photochemical smog.

parasite Consumer organism that lives on or in, and feeds on, a living plant or animal, known as the host, over an extended period. The parasite draws nourishment from and gradually weakens its host; it may or may not kill the host. See *parasitism*.

parasitism Interaction between species in which one organism, called the parasite, preys on another organism, called the host, by living on or in the host. See *host, parasite*.

particulates Also known as suspended particulate matter (SPM); variety of solid particles and liquid droplets small and light enough to remain suspended in the air for long periods. About 62% of the SPM in outdoor air comes from natural sources such as dust, wild fires, and sea salt. The remaining 38% comes from human sources such as coal-burning electric power and industrial plants, motor vehicles, plowed fields, road construction, unpaved roads, and tobacco smoke.

parts per billion (ppb) Number of parts of a chemical found in 1 billion parts of a particular gas, liquid, or solid.

parts per million (ppm) Number of parts of a chemical found in 1 million parts of a particular gas, liquid, or solid.

parts per trillion (ppt) Number of parts of a chemical found in 1 trillion parts of a particular gas, liquid, or solid.

passive solar heating system System that, without the use of mechanical devices, captures sunlight directly within a structure and converts it into low-temperature heat for space heating or for heating water for domestic use. Compare *active solar heating system*.

pasture Managed grassland or enclosed meadow that usually is planted with domesticated grasses or other forage to be grazed by livestock. Compare *feedlot*.

pathogen Living organism that can cause disease in another organism. Examples include bacteria, viruses, and parasites.

PCBs See *polychlorinated biphenyls*.

peak production Point in time when the pressure in an oil well drops and its rate of conventional crude oil production starts declining, usually a decade or so; for a group of wells or for a nation, the point at which all wells on average have passed peak production.

peer review Process of scientists reporting details of the methods and models they used, the results of their experiments, and the reasoning behind their hypotheses for other scientists working in the same field (their peers) to examine and criticize.

per capita ecological footprint Amount of biologically productive land and water needed to supply each person or population with the renewable resources they use and to absorb or dispose of the wastes from such resource use. It measures the average environmental impact of individuals or populations in different countries and areas. Compare *ecological footprint*.

per capita GDP Annual gross domestic product (GDP) of a country divided by its total population at midyear. It gives the average slice of the economic pie per person. Used to be called per capita gross national product (GNP). See *gross domestic product*. Compare *genuine progress indicator (GPI)*.

per capita GDP PPP (Purchasing Power Parity) Measure of the amount of goods and services that a country's average citizen could buy in the United States.

percolation Passage of a liquid through the spaces of a porous material such as soil.

perennial Plant that can live for more than 2 years. Compare *annual*.

permafrost Perennially frozen layer of the soil that forms when the water there freezes. It is found in arctic tundra.

perpetual resource Essentially inexhaustible resource on a human time scale because it is renewed continuously. Solar energy is an example. Compare *nonrenewable resource, renewable resource*.

persistence Ability of a living system such as a grassland or forest to survive moderate disturbances. Compare *resilience*.

perverse subsidy Subsidy that leads to environmental damage or damage to human health or well-being. See *subsidy*.

pest Unwanted organism that directly or indirectly interferes with human activities.

pesticide Any chemical designed to kill or inhibit the growth of an organism that people consider undesirable. See *fungicide, herbicide, insecticide*.

petrochemicals Chemicals obtained by refining (distilling) crude oil. They are used as raw materials in manufacturing most industrial chemicals, fertilizers, pesticides, plastics, synthetic fibers, paints, medicines, and many other products.

petroleum See *crude oil*.

pH Numeric value that indicates the relative acidity or alkalinity of a substance on a scale of 0 to 14, with the neutral point at 7. Acid solutions have pH values lower than 7; basic or alkaline solutions have pH values greater than 7.

phosphorus cycle Cyclic movement of phosphorus in different chemical forms from the environment to organisms and then back to the environment.

photochemical smog Complex mixture of air pollutants produced in the lower atmosphere by the reaction of hydrocarbons and nitrogen oxides under the influence of sunlight. Especially harmful components include ozone, peroxyacyl nitrates (PANs), and various aldehydes. Compare *industrial smog*.

photosynthesis Complex process that takes place in cells of green plants. Radiant energy from the sun is used to combine carbon dioxide (CO_2) and water (H_2O) to produce oxygen (O_2), carbohydrates (such as glucose, $C_6H_{12}O_6$), and other nutrient molecules. Compare *aerobic respiration, chemosynthesis*.

photovoltaic (PV) cell Device that converts radiant (solar) energy directly into electrical energy. Also called a solar cell.

physical change Process that alters one or more physical properties of an element or a compound without changing its chemical composition. Examples include changing the size and shape of a sample of matter (crushing ice and cutting aluminum foil) and changing a sample of matter from one physical state to another (boiling and freezing water). Compare *chemical change, nuclear change*.

phytoplankton Small, drifting plants, mostly algae and bacteria, found in aquatic ecosystems. Compare *plankton, zooplankton*.

pioneer community First integrated set of plants, animals, and decomposers found in an area undergoing primary ecological succession. See *immature community, mature community*.

pioneer species First hardy species—often microbes, mosses, and lichens—that begin colonizing a site as the first stage of ecological succession. See *ecological succession, pioneer community*.

plaintiff Party in a court case bringing charges or seeking to collect damages for injuries to health or for economic loss; may also seek an injunction, by which the party being charged would be required to stop whatever action is causing harm. See *defendant* and *civil suit*.

planetary management worldview Worldview holding that humans are separate from nature, that nature exists mainly to meet our needs and increasing wants, and that we can use our ingenuity and technology to manage the earth's life-support systems, mostly for our benefit. It assumes that economic growth is unlimited. Compare *environmental wisdom worldview, stewardship worldview*.

plankton Small plant organisms (phytoplankton) and animal organisms (zooplankton) that float in aquatic ecosystems.

plantation agriculture Growing specialized crops such as bananas, coffee, and cacao in tropical developing countries, primarily for sale to developed countries.

plates See *tectonic plates*.

plate tectonics Theory of geophysical processes that explains the movements of lithospheric plates and the processes that occur at their boundaries. See *lithosphere, tectonic plates*.

point source Single identifiable source that discharges pollutants into the environment. Examples include the smokestack of a power plant or an industrial plant, drainpipe of a meatpacking plant, chimney of a house, or exhaust pipe of an automobile. Compare *nonpoint source*.

poison Chemical that adversely affects the health of a living human or animal by causing injury, illness, or death.

policies programs, and the laws and regulations through which they are enacted, that a government enforces and funds.

politics Process through which individuals and groups try to influence or control government policies and actions that affect the local, state, national, and international communities.

pollutant Particular chemical or form of energy that can adversely affect the health, survival, or activities of humans or other living organisms. See *pollution*.

pollution Undesirable change in the physical, chemical, or biological characteristics of air, water, soil, or food that can adversely affect the health, survival, or activities of humans or other living organisms.

pollution cleanup Device or process that removes or reduces the level of a pollutant after it has been produced or has entered the environment. Examples include automobile emission control devices and sewage treatment plants. Compare *pollution prevention*.

pollution prevention Device, process, or strategy used to prevent a potential pollutant from forming or entering the environment or to sharply reduce the amount entering the environment. Compare *pollution cleanup*.

polychlorinated biphenyls (PCBs) Group of 209 toxic, oily, synthetic chlorinated hydrocarbon compounds that can be biologically amplified in food chains and webs.

polyculture Complex form of intercropping in which a large number of different plants maturing at different times are planted together. See also *intercropping*. Compare *monoculture*.

population Group of individual organisms of the same species living in a particular area.

population change Increase or decrease in the size of a population. It is equal to (Births + Immigration) − (Deaths + Emigration).

population crash Dieback of a population that has used up its supply of resources, exceeding the carrying capacity of its environment. See *carrying capacity*.

population density Number of organisms in a particular population found in a specified area or volume.

population dispersion General pattern in which the members of a population are arranged throughout its habitat.

population distribution Variation of population density over a particular geographic area or volume. For example, a country has a high population density in its urban areas and a much lower population density in its rural areas.

population dynamics Major abiotic and biotic factors that tend to increase or decrease the population size and affect the age and sex composition of a species.

population size Number of individuals making up a population's gene pool.

positive feedback loop Feedback loop that causes a system to change further in the same direction. Compare *negative feedback loop*.

potential energy Energy stored in an object because of its position or the position of its parts. Compare *kinetic energy*.

poverty Inability of people to meet their basic needs for food, clothing, and shelter.

ppb See *parts per billion*.

ppm See *parts per million*.

ppt See *parts per trillion*.

prairie See *grassland*.

precautionary principle When there is significant scientific uncertainty about potentially serious harm from chemicals or technologies, decision makers should act to prevent harm to humans and the environment. See *pollution prevention*.

precipitation Water in the form of rain, sleet, hail, and snow that falls from the atmosphere onto land and bodies of water.

predation Interaction in which an organism of one species (the predator) captures and feeds on some or all parts of an organism of another species (the prey).

predator Organism that captures and feeds on some or all parts of an organism of another species (the prey).

predator–prey relationship Relationship that has evolved between two organisms, in which one organism has become the prey for the other, the latter called the predator. See *predator, prey*.

prey Organism that is killed by an organism of another species (the predator) and serves as its source of food.

primary consumer Organism that feeds on some or all parts of plants (herbivore) or on other producers. Compare *detritivore, omnivore, secondary consumer*.

primary ecological succession Ecological succession in a area without soil or bottom sediments. See *ecological succession*. Compare *secondary ecological succession*.

primary forest See *old-growth forest*.

primary pollutant Chemical that has been added directly to the air by natural events or human activities and occurs in a harmful concentration. Compare *secondary pollutant*.

primary productivity See *gross primary productivity, net primary productivity*.

primary recycling Process in which materials are recycled into new products of the same type—turning used aluminum cans into new aluminum cans, for example.

primary sewage treatment Mechanical sewage treatment in which large solids are filtered out by screens and suspended solids settle out as sludge in a sedimentation tank. Compare *secondary sewage treatment*.

principles of sustainability See *scientific principles of sustainability, social science principles of sustainability*.

probability Mathematical statement about how likely it is that something will happen.

producer Organism that uses solar energy (green plants) or chemical energy (some bacteria) to manufacture the organic compounds it needs as nutrients from simple inorganic compounds obtained from its environment. Compare *consumer, decomposer*.

prokaryotic cell Cell containing no distinct nucleus or organelles. Compare *eukaryotic cell*.

proton (p) Positively charged particle in the nuclei of all atoms. Each proton has a relative mass of 1 and a single positive charge. Compare *electron, neutron*.

proven oil reserves Identified deposits from which conventional crude oil can be extracted profitably at current prices with current technology.

PV cell See *photovoltaic cell*.

pyramid of energy flow Diagram representing the flow of energy through each trophic level in a food chain or food web. With each energy transfer, only a small part (typically 10%) of the usable energy entering one trophic level is transferred to the organisms at the next trophic level.

radiation Fast-moving particles (particulate radiation) or waves of energy (electromagnetic radiation). See *alpha particle, beta particle, gamma ray*.

radioactive decay Change of a radioisotope to a different isotope by the emission of radioactivity.

radioactive isotope See *radioisotope*.

radioactive waste Waste products of nuclear power plants, research, medicine, weapon production, or other processes involving nuclear reactions. See *radioactivity*.

radioactivity Nuclear change in which unstable nuclei of atoms spontaneously shoot out "chunks" of mass, energy, or both at a fixed rate. The three principal types of radioactivity are gamma rays and fast-moving alpha particles and beta particles.

radioisotope Isotope of an atom that spontaneously emits one or more types of radioactivity (alpha particles, beta particles, gamma rays).

rain shadow effect Low precipitation on the leeward side of a mountain when prevailing winds flow up and over a high mountain or range of high mountains, creating semiarid and arid conditions on the leeward side of a high mountain range.

rangeland Land that supplies forage or vegetation (grasses, grasslike plants, and shrubs) for grazing and browsing animals and is not intensively managed. Compare *feedlot, pasture*.

range of tolerance Range of chemical and physical conditions that must be maintained for populations of a particular species to stay alive and grow, develop, and function normally. See *law of tolerance*.

recharge area Any area of land allowing water to percolate down through it and into an aquifer. See *aquifer, natural recharge*.

reconciliation ecology Science of inventing, establishing, and maintaining habitats to conserve species diversity in places where people live, work, or play.

recycle To collect and reprocess a resource so that it can be made into new products; one of the four R's of resource use. An example is collecting aluminum cans, melting them down, and using the aluminum to make new cans or other aluminum products. See *primary recycling, secondary recycling*. Compare *reduce and reuse*.

reduce To consume less of a good or service in order to reduce one's environmental impact and to save money. Compare *recycle, refuse, reuse*.

refining Complex process in which crude oil is heated and vaporized in giant columns and separated, by use of varying boiling points, into various products such as gasoline, heating oil, and asphalt. See *petrochemicals*.

reforestation Renewal of trees and other types of vegetation on land where trees have been removed; can be done naturally by seeds from nearby trees or artificially by planting seeds or seedlings.

refuse To refrain from buying or using a good or service in order to reduce one's ecological impact and to save money. Compare *recycle, reduce, reuse*.

reliable runoff Surface runoff of water that generally can be counted on as a stable source of water from year to year. See *runoff*.

reliable science Concepts and ideas that are widely accepted by experts in a particular field of the natural or social sciences. Compare *tentative science, unreliable science*.

renewable energy Energy that comes from resources that are replenished by natural processes continually or in a relatively short time. Examples are solar energy (sunlight), wind, moving water, heat from the earth's interior (geothermal energy), firewood from trees, tides, and waves. Compare *nonrenewable energy*.

renewable resource Resource that can be replenished rapidly (hours to several decades) through natural processes as long as it is not used up faster than it is replaced. Examples include trees in forests, grasses in grasslands, wild animals, fresh surface water in lakes and streams, most groundwater, fresh air, and fertile soil. If such a resource is used faster than it is replenished, it can be depleted and converted into a nonrenewable resource. Compare *nonrenewable resource* and *perpetual resource*. See also *environmental degradation*.

replacement-level fertility rate Average number of children a couple must bear to replace themselves. The average for a country or the world usually is slightly higher than two children per couple (2.1 in the United States and 2.5 in some developing countries) mostly because some children die before reaching their reproductive years. See also *total fertility rate*.

reproduction Production of offspring by one or more parents.

reproductive isolation Long-term geographic separation of members of a particular sexually reproducing species.

reproductive potential See *biotic potential*.

reserves Resources that have been identified and from which a usable mineral can be extracted profitably at present prices with current mining or extraction technology.

reservoir Artificial lake created when a stream is dammed. See *dam*.

resilience Ability of a living system such as a forest or pond to be restored through secondary ecological succession after a severe disturbance. See *secondary ecological succession*. Compare *persistence*.

resource Anything obtained from the environment to meet human needs and wants. It can also be applied to other species.

resource partitioning Process of dividing up resources in an ecosystem so that species with similar needs (overlapping ecological niches) use the same scarce resources at different times, in different ways, or in different places. See *ecological niche*.

resource productivity See *material efficiency*.

respiration See *aerobic respiration*.

response Amount of health damage caused by exposure to a certain dose of a harmful substance or form of radiation. See *dose, dose-response curve, median lethal dose*.

restoration ecology Research and scientific study devoted to restoring, repairing, and reconstructing damaged ecosystems.

reuse To use a product over and over again in the same form. An example is collecting, washing, and refilling glass beverage bottles. One of the 4 Rs. Compare *recycle, reduce,* and *refuse*.

riparian zone A thin strip or patch of vegetation that surrounds a stream. These zones are very important habitats and resources for wildlife.

risk Probability that something undesirable will result from deliberate or accidental exposure to a hazard. See *risk analysis, risk assessment, risk management*.

risk analysis Identifying hazards, evaluating the nature and severity of risks associated with the hazards (*risk assessment*), ranking risks (*comparative risk analysis*), using this and other information to determine options and make decisions about reducing or eliminating risks (*risk management*), and communicating information about risks to decision makers and the public (*risk communication*).

risk assessment Process of gathering data and making assumptions to estimate short- and long-term harmful effects on human health or the environment from exposure to hazards associated with the use of a particular product or technology.

risk communication Communicating information about risks to decision makers and the public. See *risk, risk analysis*.

risk management Use of risk assessment and other information to determine options and make decisions about reducing or eliminating risks. See *risk, risk analysis, risk communication*.

rock Any solid material that makes up a large, natural, continuous part of the earth's crust. See *mineral*.

rock cycle Largest and slowest of the earth's cycles, consisting of geologic, physical, and chemical processes that form and modify rocks and soil in the earth's crust over millions of years.

rule of 70 Doubling time (in years) = 70/(percentage growth rate). See *doubling time, exponential growth*.

runoff Freshwater from precipitation and melting ice that flows on the earth's surface into nearby streams, lakes, wetlands, and reservoirs. See *reliable runoff, surface runoff, surface water*. Compare *groundwater*.

salinity Amount of various salts dissolved in a given volume of water.

salinization Accumulation of salts in soil that can eventually make the soil unable to support plant growth.

saltwater intrusion Movement of saltwater or brackish (slightly salty) water into freshwater aquifers in coastal and inland areas as groundwater is withdrawn faster than it is recharged by precipitation.

saltwater life zones Aquatic life zones associated with oceans: oceans and their accompanying bays, estuaries, coastal wetlands, shorelines, coral reefs, and mangrove forests.

sanitary landfill Waste disposal site on land in which waste is spread in thin layers, compacted, and covered with a fresh layer of clay or plastic foam each day. Compare *open dump.*

scavenger Organism that feeds on dead organisms that were killed by other organisms or died naturally. Examples include vultures, flies, and crows. Compare *detritivore.*

science Attempts to discover order in nature and use that knowledge to make predictions about what is likely to happen in nature. See *reliable science, scientific data, scientific hypothesis, scientific law, scientific methods, scientific model, scientific theory, tentative science, unreliable science.*

scientific data Facts obtained by making observations and measurements. Compare *scientific hypothesis, scientific law, scientific methods, scientific model, scientific theory.*

scientific hypothesis An educated guess that attempts to explain a scientific law or certain scientific observations. Compare *scientific data, scientific law, scientific methods, scientific model, scientific theory.*

scientific law Description of what scientists find happening in nature repeatedly in the same way, without known exception. See *first law of thermodynamics, law of conservation of matter, second law of thermodynamics.* Compare *scientific data, scientific hypothesis, scientific methods, scientific model, scientific theory.*

scientific methods The ways scientists gather data and formulate and test scientific hypotheses, models, theories, and laws. See *scientific data, scientific hypothesis, scientific law, scientific model, scientific theory.*

scientific model A simulation of complex processes and systems. Many are mathematical models that are run and tested using computers.

scientific principles of sustainability To live more sustainably we need to rely on solar energy, preserve biodiversity, and recycle the chemicals that we use. These three principles of sustainability are scientific lessons from nature based on observing how life on the earth has survived and thrived for 3.5 billion years. See *biodiversity, chemical cycling, solar energy.* Compare *social science principles of sustainability.*

scientific theory A well-tested and widely accepted scientific hypothesis. Compare *scientific data, scientific hypothesis, scientific law, scientific methods, scientific model.*

secondary consumer Organism that feeds only on primary consumers. Compare *detritivore, omnivore, primary consumer.*

secondary ecological succession Ecological succession in an area in which natural vegetation has been removed or destroyed but the soil or bottom sediment has not been destroyed. See *ecological succession.* Compare *primary ecological succession.*

secondary pollutant Harmful chemical formed in the atmosphere when a primary air pollutant reacts with normal air components or other air pollutants. Compare *primary pollutant.*

secondary recycling A process in which waste materials are converted into different products; for example, used tires can be shredded and turned into rubberized road surfacing. Compare *primary recycling.*

secondary sewage treatment Second step in most waste treatment systems in which aerobic bacteria decompose as much as 90% of degradable, oxygen-demanding organic wastes in wastewater. It usually involves bringing sewage and bacteria together in trickling filters or in the activated sludge process. Compare *primary sewage treatment.*

second-growth forest Stands of trees resulting from secondary ecological succession. Compare *old-growth forest, tree farm.*

second law of thermodynamics Whenever energy is converted from one form to another in a physical or chemical change, we end up with lower-quality or less usable energy than we started with. In any conversion of heat energy to useful work, some of the initial energy input is always degraded to lower-quality, more dispersed, less useful energy—usually low-temperature heat that flows into the environment; you cannot break even in terms of energy quality. See *first law of thermodynamics.*

sedimentary rock Rock that forms from the accumulated products of erosion and in some cases from the compacted shells, skeletons, and other remains of dead organisms. Compare *igneous rock, metamorphic rock.* See *rock cycle.*

selective cutting Cutting of intermediate-aged, mature, or diseased trees in an uneven-aged forest stand, either singly or in small groups. This encourages the growth of younger trees and maintains an uneven-aged stand. Compare *clear-cutting, strip cutting.*

septic tank Underground tank for treating wastewater from a home in rural and suburban areas. Bacteria in the tank decompose organic wastes, and the sludge settles to the bottom of the tank. The effluent flows out of the tank into the ground through a field of drainpipes.

sexual reproduction Reproduction in organisms that produce offspring by combining sex cells or *gametes* (such as ovum and sperm) from both parents. It produces offspring that have combinations of traits from their parents. Compare *asexual reproduction.*

shale oil Slow-flowing, dark brown, heavy oil obtained when kerogen in oil shale is vaporized at high temperatures and then condensed. Shale oil can be refined to yield gasoline, heating oil, and other petroleum products. See *kerogen, oil shale.*

shelterbelt See *windbreak.*

slash-and-burn agriculture Cutting down trees and other vegetation in a patch of forest, leaving the cut vegetation on the ground to dry, and then burning it. The ashes that are left add nutrients to the nutrient-poor soils found in most tropical forest areas. Crops are planted between tree stumps. Plots must be abandoned after a few years (typically 2–5 years) because of loss of soil fertility or invasion of vegetation from the surrounding forest.

sludge Gooey mixture of toxic chemicals, infectious agents, and settled solids removed from wastewater at a sewage treatment plant.

smart growth Form of urban planning that recognizes that urban growth will occur but uses zoning laws and other tools to prevent sprawl, direct growth to certain areas, protect ecologically sensitive and important lands and waterways, and develop urban areas that are more environmentally sustainable and more enjoyable places to live.

smelting Process in which a desired metal is separated from the other elements in an ore mineral.

smog Originally a combination of smoke and fog but now used to describe other mixtures of pollutants in the atmosphere. See *industrial smog, photochemical smog.*

SNG See *synthetic natural gas.*

social capital Result of getting people with different views and values to talk and listen to one another, find common ground based on understanding and trust, and work together to solve environmental and other problems.

social science principles of sustainability To live more sustainably we **(1)** need to include the harmful health and environmental costs of producing the goods and services in their market prices *(full-cost pricing),* **(2)** learn to work together to focus on solutions to environmental problems that will benefit the largest number of people and the environment now and in the future *(win-win solutions),* and **(3)** accept our responsibility to future generations to leave the planet's life-support systems in at least as good a shape as what we now enjoy *(responsibility to future generations).*

soil Complex mixture of inorganic minerals (clay, silt, pebbles, and sand), decaying organic matter, water, air, and living organisms.

soil conservation Methods used to reduce soil erosion, prevent depletion of soil nutrients, and restore nutrients previously lost by erosion, leaching, and excessive crop harvesting.

soil erosion Movement of soil components, especially topsoil, from one place to another, usually by wind, flowing water, or both. This natural process can be greatly accelerated by human activities that remove vegetation from soil. Compare *soil conservation.*

soil horizons Horizontal zones, or layers, that make up a particular mature soil. Each horizon has a distinct texture and composition that vary with different types of soils. See *soil profile.*

soil profile Cross-sectional view of the horizons in a soil. See *soil horizon.*

soil salinization Gradual accumulation of salts in upper soil layers that can stunt crop growth, lower crop yields, and can eventually kill plants and ruin the land.

solar capital Solar energy that warms the planet and supports photosynthesis, the process that plants use to provide food for themselves and for us and other animals. This direct input of solar energy also produces indirect forms of renewable solar energy such as wind and flowing water. Compare *natural capital.*

solar cell See *photovoltaic cell.*

solar collector Device for collecting radiant energy from the sun and converting it into heat. See *active solar heating system, passive solar heating system.*

solar energy Direct radiant energy from the sun and a number of indirect forms of energy produced by the direct input of such radiant energy. Principal indirect forms of solar energy include wind, falling and flowing water (hydropower), and biomass (solar energy converted into chemical energy stored in the chemical bonds of organic compounds in trees and other plants)—none of which would exist without direct solar energy.

solar thermal system System that uses any of various methods to collect and concentrate solar energy in order to boil water and produce steam for generating electricity. Compare *solar cell.*

solid waste Any unwanted or discarded material that is not a liquid or a gas. See *industrial solid waste, municipal solid waste.*

sound science See *reliable science.*

spaceship-earth worldview View of the earth as a spaceship: a machine that we can understand, control, and change at will by using advanced technology. See *planetary management worldview.* Compare *environmental wisdom worldview, stewardship worldview.*

specialist species Species with a narrow ecological niche. They may be able to live in only one type of habitat, tolerate only a narrow range of climatic and other environmental conditions, or use only one type or a few types of food. Compare *generalist species.*

speciation Formation of two species from one species because of divergent natural selection in response to changes in environmental conditions; usually takes thousands of years. Compare *extinction.*

species Group of similar organisms, and for sexually reproducing organisms, they are a set of individuals that can mate and produce fertile offspring. Every organism is a member of a certain species.

species diversity Number of different species (species richness) combined with the relative abundance of individuals within each of those species (species evenness) in a given area. See *biodiversity, species evenness, species richness.* Compare *ecological diversity, genetic diversity.*

species equilibrium model See *theory of island biogeography.*

species evenness Degree to which comparative numbers of individuals of each of the species present in a community are similar. See *species diversity.* Compare *species richness.*

species richness Variety of species, measured by the number of different species contained in a community. See *species diversity.* Compare *species evenness.*

spoils Unwanted rock and other waste materials produced when a material is removed from the earth's surface or subsurface by mining, dredging, quarrying, or excavation.

S-shaped curve Leveling off of an exponential, J-shaped curve when a rapidly growing population reaches or exceeds the carrying capacity of its environment and ceases to grow.

statistics Mathematical tools used to collect, organize, and interpret numerical data.

statutory laws Laws developed and passed by legislative bodies such as federal and state governments. Compare *common law.*

stewardship worldview Worldview holding that we can manage the earth for our benefit but that we have an ethical responsibility to be caring and responsible managers, or *stewards,* of the earth. It calls for encouraging environmentally beneficial forms of economic growth and discouraging environmentally harmful forms. Compare *worldview, environmental wisdom worldview, planetary management worldview.*

stratosphere Second layer of the atmosphere, extending about 17–48 kilometers (11–30 miles) above the earth's surface. It contains small amounts of gaseous ozone (O_3), which filters out about 95% of the incoming harmful ultraviolet radiation emitted by the sun. Compare *troposphere.*

stream Flowing body of surface water. Examples are creeks and rivers.

strip-cropping Planting regular crops and close-growing plants, such as hay or nitrogen-fixing legumes, in alternating rows or bands to help reduce depletion of soil nutrients.

strip cutting Variation of clear-cutting in which a strip of trees is clear-cut along the contour of the land, with the corridor being narrow enough to allow natural regeneration within a few years. After regeneration, another strip is cut above the first, and so on. Compare *clear-cutting, selective cutting.*

strip mining Form of surface mining in which bulldozers, power shovels, or stripping wheels remove large chunks of the earth's surface in strips. See *area strip mining, contour strip mining, surface mining.* Compare *subsurface mining.*

subatomic particles Extremely small particles—electrons, protons, and neutrons—that make up the internal structure of atoms.

subduction zone Area in which the oceanic lithosphere is carried downward (subducted) under an island arc or continent at a convergent plate boundary. A trench ordinarily forms at the boundary between the two converging plates. See *convergent plate boundary.*

subsidence Slow or rapid sinking of part of the earth's crust that is not slope-related.

subsidy Payment intended to help a business grow and thrive; typically provided by a government in the form of a grant or tax break, in order to give the company an advantage in the marketplace over its competitors. See *perverse subsidy.*

subsistence farming See *traditional subsistence agriculture.*

subsurface mining Extraction of a metal ore or fuel resource such as coal from a deep underground deposit. Compare *surface mining.*

succession See *ecological succession, primary ecological succession, secondary ecological succession.*

succulent plants Plants, such as desert cacti, that survive in dry climates by having no leaves, thus reducing the loss of scarce water through *transpiration.* They store water and use sunlight to produce the food they need in the thick, fleshy tissue of their green stems and branches. Compare *deciduous plants, evergreen plants.*

sulfur cycle Cyclic movement of sulfur in various chemical forms from the environment to organisms and then back to the environment.

sulfur dioxide (SO_2) Colorless gas with an irritating odor. About one-third of the SO_2 in the atmosphere comes from natural sources as part of the sulfur cycle. The other two-thirds comes from human sources, mostly combustion of sulfur-containing coal in electric power and industrial plants and from oil refining and smelting of sulfide ores.

superinsulated house House that is heavily insulated and extremely airtight. Typically, active or passive solar collectors are used to heat water, and an air-to-air heat exchanger prevents buildup of excessive moisture and indoor air pollutants.

surface fire Forest fire that burns only undergrowth and leaf litter on the forest floor. Compare *crown fire, ground fire.* See *controlled burning.*

surface mining Removing soil, subsoil, and other strata and then extracting a mineral deposit found fairly close to the earth's surface. See *area strip mining, contour strip mining, mountaintop removal, open-pit mining.* Compare *subsurface mining.*

surface runoff Water flowing off the land into bodies of surface water. See *reliable runoff.*

surface water Precipitation that does not infiltrate the ground or return to the atmosphere by evaporation or transpiration. See *runoff.* Compare *groundwater.*

survivorship curve Graph showing the number of survivors in different age groups for a particular species.

suspended particulate matter See *particulates.*

sustainability Ability of earth's various systems, including human cultural systems and economies, to survive and adapt to changing environmental conditions indefinitely.

sustainability revolution Major cultural change in which people learn how to reduce their ecological footprints and live more sustainably, largely by copying nature and using the six principles of sustainability to guide their lifestyles and economies. See *principles of sustainability.*

sustainable agriculture Method of growing crops and raising livestock based on organic fertilizers, soil conservation, water conservation, biological pest control, and minimal use of nonrenewable fossil-fuel energy.

sustainable development See *environmentally sustainable economic development.*

sustainable living Taking no more potentially renewable resources from the natural world than can be replenished naturally and not overloading the capacity of the environment to cleanse and renew itself by natural processes.

sustainable society Society that manages its economy and population size without doing irreparable environmental harm by overloading the planet's ability to absorb environmental insults, replenish its resources, and sustain human and other forms of life over a specified period, indefinitely. During this period, the society satisfies the needs of its people without depleting natural resources and thereby jeopardizing the prospects of current and future generations of humans and other species.

sustainable yield (sustained yield) Highest rate at which a potentially renewable resource can be used indefinitely without reducing its available supply. See also *environmental degradation.*

synergistic interaction Interaction of two or more factors or processes so that the combined effect is greater than the sum of their separate effects.

synergy See *synergistic interaction.*

synfuels Synthetic gaseous and liquid fuels produced from solid coal or sources other than natural gas or crude oil.

synthetic natural gas (SNG) Gaseous fuel containing mostly methane produced from solid coal.

system Set of components that function and interact in some regular and theoretically predictable manner.

tailings Rock and other waste materials removed as impurities when waste mineral material is separated from the metal in an ore.

tar sand Deposit of a mixture of clay, sand, water, and varying amounts of a tarlike heavy oil known as bitumen. Bitumen can be extracted from tar sand by heating. It is then purified and upgraded to synthetic crude oil. See *bitumen.*

tectonic plates Various-sized areas of the earth's lithosphere that move slowly around with the mantle's flowing asthenosphere. Most earthquakes and volcanoes occur around the boundaries of these plates. See *lithosphere, plate tectonics.*

temperature Measure of the average speed of motion of the atoms, ions, or molecules in a substance or combination of substances at a given moment. Compare *heat.*

temperature inversion Layer of dense, cool air trapped under a layer of less dense, warm air. It prevents upward-flowing air currents from developing. In a prolonged inversion, air pollution in the trapped layer may build up to harmful levels.

tentative science Preliminary scientific data, hypotheses, and models that have not been widely tested and accepted. Compare *reliable science, unreliable science.*

teratogen Chemical, ionizing agent, or virus that causes birth defects. Compare *carcinogen, mutagen.*

terracing Planting crops on a long, steep slope that has been converted into a series of broad, nearly level terraces with short vertical drops from one to another that run along the contour of the land to retain water and reduce soil erosion.

terrestrial Pertaining to land. Compare *aquatic.*

tertiary (higher-level) consumers Animals that feed on animal-eating animals. They feed at high trophic levels in food chains and webs. Examples include hawks, lions, bass, and sharks. Compare *detritivore, primary consumer, secondary consumer.*

theory of evolution Widely accepted scientific idea that all life-forms developed from earlier life-forms. It is the way most biologists explain how life has changed over the past 3.6–3.8 billion years and why it is so diverse today.

theory of island biogeography Widely accepted scientific theory holding that the number of different species (species richness) found on an island is determined by the interactions of two factors: the rate at which new species immigrate to the island and the rate at which species become *extinct,* or cease to exist, on the island. See *species richness.*

thermal energy The energy generated and measured by heat. See *heat.*

thermal inversion See *temperature inversion.*

threatened species Wild species that is still abundant in its natural range but is likely to become endangered because of a decline in numbers. Compare *endangered species.*

threshold effect Harmful or fatal effect of a small change in environmental conditions that exceeds the limit of tolerance of an organism or population of a species. See *law of tolerance.*

throughput Rate of flow of matter, energy, or information through a system. Compare *input, output.*

time delay In a complex system, the period of time between the input of a feedback stimulus and the system's response to it. See *tipping point.*

tipping point Threshold level at which an environmental problem causes a fundamental and irreversible shift in the behavior of a system. See *climate tipping point, ecological tipping point.*

tolerance limits Minimum and maximum limits for physical conditions (such as temperature) and concentrations of chemical substances beyond which no members of a particular species can survive. See *law of tolerance.*

topsoil The uppermost layer of soil as a soil's A-horizon layer. It contains the organic and inorganic nutrients that plants need for their growth and development.

total fertility rate (TFR) Estimate of the average number of children who will be born alive to a woman during her lifetime if she passes through all her childbearing years (ages 15–44) conforming to age-specific fertility rates of a given year. More simply, it is an estimate of the average number of children that women in a given population will have during their childbearing years.

toxic chemical See *poison, carcinogen, hazardous chemical, mutagen, teratogen.*

toxicity Measure of the harmfulness of a substance.

toxicology Study of the adverse effects of chemicals on health.

toxic waste Form of hazardous waste that causes death or serious injury (such as burns, respiratory diseases, cancers, or genetic mutations). See *hazardous waste.*

toxin See *poison.*

traditional intensive agriculture Production of enough food for a farm family's survival and a surplus that can be sold. This type of agriculture uses higher inputs of labor, fertilizer, and water than traditional subsistence agriculture. See *traditional subsistence agriculture.* Compare *industrialized agriculture.*

traditional subsistence agriculture Production of enough crops or livestock for a farm family's survival. Compare *industrialized agriculture, traditional intensive agriculture.*

tragedy of the commons Depletion or degradation of a potentially renewable resource to which people have free and unmanaged access. An example is the depletion of commercially desirable fish species in the open ocean beyond areas controlled by coastal countries. See *common-property resource, open-access renewable resource.*

trait Characteristic passed on from parents to offspring during reproduction in an animal or plant.

transform fault Area where the earth's lithospheric plates move in opposite but parallel directions along a fracture (fault) in the lithosphere. Compare *convergent plate boundary, divergent plate boundary.*

transgenic organisms See *genetically modified organisms.*

transmissible disease Disease that is caused by living organisms (such as bacteria, viruses, and parasitic worms) and can spread from one person to another by air, water, food, or body fluids (or in some cases by insects or other organisms). Compare *nontransmissible disease.*

transpiration Process in which water is absorbed by the root systems of plants, moves up through the plants, passes through pores (stomata) in their leaves or other parts, and evaporates into the atmosphere as water vapor.

tree farm See *tree plantation.*

tree plantation Site planted with one or only a few tree species in an even-aged stand. When the stand matures it is usually harvested by clear-cutting and then replanted. These farms normally raise rapidly growing tree species for fuelwood, timber, or pulpwood. Compare *old-growth forest, second-growth forest.*

trophic level All organisms that are the same number of energy transfers away from the original source of energy (for example, sunlight) that enters an ecosystem. For example, all producers belong to the first trophic level and all herbivores belong to the second trophic level in a food chain or a food web.

troposphere Innermost layer of the atmosphere. It contains about 75% of the mass of earth's air and extends about 17 kilometers (11 miles) above sea level. Compare *stratosphere.*

true cost See *full cost.*

tsunami Series of large waves generated when part of the ocean floor suddenly rises or drops.

turbidity Cloudiness in a volume of water; a measure of water clarity in lakes, streams, and other bodies of water.

undernutrition See *chronic undernutrition.*

unreliable science Scientific results or hypotheses presented as reliable science without having undergone the rigors of the peer review process. Compare *reliable science, tentative science.*

upwelling Movement of nutrient-rich bottom water to the ocean's surface. It can occur far from shore but usually takes place along certain steep coastal areas where the warm surface layer of ocean water is pushed away from shore and replaced by cold, nutrient-rich bottom water.

urban area Geographic area containing a community with a population of 2,500 or more. The number of people used in this definition may vary, with some countries setting the minimum number of people at 10,000–50,000.

urban growth Rate of growth of an urban population. Compare *degree of urbanization.*

urbanization Creation or growth of urban areas, or cities, and their surrounding developed land. See *degree of urbanization, urban area.*

urban sprawl Growth of low-density development on the edges of cities and towns. See *smart growth.*

virtual water Water that is not directly consumed but is used to produce food and other products.

virus Microorganism that can transmit an infectious disease by invading a cell and taking over its genetic machinery to copy itself and then spread throughout the body. Compare *bacteria.*

volatile organic compounds (VOCs) Organic compounds that exist as gases in the atmosphere and act as pollutants, some of which are hazardous.

volcano Vent or fissure in the earth's surface through which magma, liquid lava, and gases are released into the environment.

warm front Boundary between an advancing warm air mass and the cooler one it is replacing. Because warm air is less dense than cool air, an advancing warm front rises over a mass of cool air. Compare *cold front.*

waste management Managing wastes to reduce their environmental harm without seriously trying to reduce the amount of waste produced. See *integrated waste management.* Compare *waste reduction.*

waste reduction Reducing the amount of waste produced; wastes that are produced are viewed as potential resources that can be reused, recycled, or composted. See *integrated waste management.* Compare *waste management.*

water cycle See *hydrologic cycle.*

water footprint A rough measure of the volume of water that we use directly and indirectly to keep a person or group alive and to support their lifestyles.

waterlogging Saturation of soil with irrigation water or excessive precipitation so that the water table rises close to the surface.

water pollution Any physical or chemical change in surface water or groundwater that can harm living organisms or make water unfit for certain uses.

watershed Land area that delivers water, sediment, and dissolved substances via small streams to a major stream (river).

water table Upper surface of the zone of saturation, in which all available pores in the soil and rock in the earth's crust are filled with water. See *zone of aeration, zone of saturation.*

watt Unit of power, or rate at which electrical work is done. See *kilowatt.*

weather Short-term changes in the temperature, barometric pressure, humidity, precipitation, sunshine, cloud cover, wind direction and speed, and other conditions in the troposphere at a given place and time. Compare *climate.*

weathering Physical and chemical processes in which solid rock exposed at earth's surface is changed to separate solid particles and dissolved material, which can then be moved to another place as sediment. See *erosion.*

wetland Land that is covered all or part of the time with saltwater or freshwater, excluding streams, lakes, and the open ocean. See *coastal wetland, inland wetland.*

wilderness Area where the earth and its ecosystems have not been seriously disturbed by humans and where humans are only temporary visitors.

wildlife All free, undomesticated species. Sometimes the term is used to describe animals only.

wildlife resources Wildlife species that have actual or potential economic value to people.

wild species Species found in the natural environment. Compare *domesticated species.*

windbreak Row of trees or hedges planted to partially block wind flow and reduce soil erosion on cultivated land.

wind farm Cluster of wind turbines in a windy area on land or at sea, built to capture wind energy and convert it into electrical energy.

worldview How people think the world works and what they think their role in the world should be. See *environmental wisdom worldview, planetary management worldview, stewardship worldview.*

zone of aeration Zone in soil that is not saturated with water and that lies above the water table. See *water table, zone of saturation.*

zone of saturation Zone where all available pores in soil and rock in the earth's crust are filled by water. See *water table, zone of aeration.*

zoning Designating parcels of land for particular types of use.

zooplankton Animal plankton; small floating herbivores that feed on plant plankton (phytoplankton). Compare *phytoplankton.*